# Titles of related interest

ISBN: 978-1-118-51628-7

The perfect companion for study and revision for pre-registration children's nursing students from the publishers of the market-leading *at a Glance* series, this book explores assessment and screening, working with families, the newborn infant, the developing child, child health policy, nursing the sick child and young person, and chronic and life-limiting conditions.

The Great Ormond Street Hospital
**Manual of Children's Nursing Practices**

Edited by Susan Macqueen, Elizabeth Anne Bruce, Faith Gibson

ISBN: 978-1-4051-0932-1

'This manual builds on the knowledge, skills and expertise of one of the UK's finest children's hospitals and explains comprehensive, evidence-based care clearly.' (*Nursing Children and Young People*, 1 October 2012)

ISBN: 978-1-118-48136-3

*Nursing Practice* is the essential, textbook to support you throughout your entire nursing degree, from your first year onwards. It explores all the clinical and professional issues that you need to know in one complete volume, covering all fields of nursing: Adult, Child, Mental Health, Learning Disabilities and also Maternity care, in both acute and community settings.

ISBN: 978-0-470-67054-5

'A well-written resource that covers a range of pertinent topics in pediatric pain...This clinical guide is a welcome addition to the pediatric pain literature and would serve well as a textbook for a course in pediatric pain.' (Pediatric Pain Letter)

This title is also available as an e-book. For more details, please see
www.wiley.com/buy/9781118625057
or scan this QR code:

Fundamentals of

# Children's Anatomy and Physiology

In memory of our dear colleague Debbie Davies.

# Fundamentals of Children's Anatomy and Physiology

## A Textbook for Nursing and Healthcare Students

EDITED BY

**IAN PEATE**

Visiting Professor of Nursing, Head of the School of Health Studies, Gibraltar and Editor in Chief, *British Journal of Nursing*

AND

**ELIZABETH GORMLEY-FLEMING**

Associate Dean of School, Academic Quality Assurance, Children's Nursing, School of Health and Social Work, University of Hertfordshire, Hatfield, UK

WILEY Blackwell

*Registered Office*
John Wiley & Sons Ltd, The Atrium, Southern Gate, Chichester, West Sussex, PO19 8SQ, UK

*Editorial Offices*
350 Main Street, Malden, MA 02148-5020, USA
9600 Garsington Road, Oxford, OX4 2DQ, UK
The Atrium, Southern Gate, Chichester, West Sussex, PO19 8SQ, UK

For details of our global editorial offices, for customer services, and for information about how to apply for permission to reuse the copyright material in this book please see our website at www.wiley.com/wiley-blackwell.



*Library of Congress Cataloging-in-Publication Data*

Fundamentals of children's anatomy and physiology: a textbook for nursing and healthcare students / edited by Ian Peate and Elizabeth Gormley-Fleming.
p. ; cm.
Includes bibliographical references and index.
ISBN 978-1-118-62505-7 (pbk.)
I. Peate, Ian, editor. II. Gormley-Fleming, Elizabeth, editor.
[DNLM: 1. Anatomy. 2. Child. 3. Physiological Phenomena. QS 4]
QP34.5
612–dc23

2014038945

A catalogue record for this book is available from the British Library.

Cover image: iStock/© Yuri_Arcurs

Set in 10/12pt MyriadPro by Toppan Best-set Premedia Limited, Hong Kong
Printed and bound in Singapore by Markono Print Media Pte Ltd

1 2015

# Contents

# About the series

Wiley's *Fundamentals* series are a wide-ranging selection of textbooks written to support pre-registration nursing and other healthcare students throughout their course. Packed full of useful features such as learning objectives, activities to test knowledge and understanding, and clinical scenarios, the titles are also highly illustrated and fully supported by interactive MCQs, and each one includes access to a **Wiley E-Text: powered by VitalSource** – an interactive digital version of the book including downloadable text and images and highlighting and note-taking facilities. Accessible on your laptop, mobile phone or tablet device, the *Fundamentals* series is *the* most flexible, supportive textbook series available for nursing and healthcare students today.

# Contributors

**Elizabeth Akers RN B. Nursing, BSc (Hons), Diploma in Tropical Nursing PG Cert (Teaching and Learning)**
Practice Educator, Bear Ward, Cardiac Unit, Great Ormond Street Hospital NHS Foundation Trust, London

Born in central New South Wales, Elizabeth grew up in the great Australian outdoors. Elizabeth began her general nurse training at the University of Sydney in 1993 and completed a post-graduate year at Sydney Children's Hospital in 1996. On an adventurous whim Elizabeth accepted a post at Great Ormond Street Hospital in London and has since moved to the Bedfordshire countryside to raise her family. Elizabeth still works at Great Ormond Street Hospital in the cardiac unit, where she is one of a team of practice educators. Elizabeth's special interests are in simulation, general paediatrics and congenital heart disease.

**Ann L. Bevan RN, RSCN, PhD**
Senior Lecturer, School of Health and Social Care, Bournemouth University, Bournemouth, UK

Ann began her nursing career at Poole General Hospital before then moving on to various other positions and further education. She has worked in various places around the UK and in Hong Kong, before moving to Canada where she worked in clinical practice for 5 years before moving into education. Ann returned to her current post in 2010. Her key interests are child health, childhood nutrition and obesity, health promotion and action research. Ann's publications are growing, and she is a Board member and Treasurer of Phi Mu Honour Society for Nurses in England.

**Mary Brady RN, RSCN, CHSM, BSc, PGCHE, MSc**
Senior Lecturer, Faculty of Health, Social Care and Education, Kingston University, London, UK

After completing RN training, Mary undertook RSCN training. Afterwards, she worked on a general paediatric ward for infants, then moved to Birmingham Children's Hospital to work on a neonatal and infant surgical ward as a staff nurse and then sister. She completed the paediatric intensive care course at the Hospital for Sick Children (Great Ormond Street) and worked on the cardiac ICU. In 1991 she moved to a neonatal unit as a surgical sister. From 1995 to 2004 she was ward manager/senior sister on a general children's ward.

In 2004, Mary moved into nurse education, where she is currently Field lead for children's nursing. She teaches pre- and post-registration nurses, midwives and paramedics.

Mary is an external examiner at Huddersfield University.

**Petra Brown RGN DPSN BSc (Hons) MA**
Lecturer, School of Health and Social Care, Bournemouth University, Bournemouth, UK

Petra began her nursing a career in 1988 at Salisbury School of Nursing, becoming a staff nurse. She has worked in a variety of clinical areas, including recovery, intensive & coronary care and telephone triage. On completing her Critical Care degree, Petra started her career in nurse education as a Practice Educator for critical care, A&E and orthopaedics at Bournemouth and Christchurch NHS Hospital. On completing a Master's Degree in Health and Social Care Practice

Education, she was appointed as a lecturer at Bournemouth University on the overseas and pre-registration nursing courses. Her key areas of interest are nursing practice, overseas nursing, nurse education, practice development, respiratory and critical care.

**Mary L. Donnelly SRN, RSCN, DipEd, PG Cert Ed, BSc (Hons), MA**
Senior Lecturer in Children's Health, Alternate Professional Lead and Programme Field Tutor for Children's Nursing, School of Health and Social Work, University of Hertfordshire, Hatfield, UK

Mary began her nursing career in 1978 training to be a state registered nurse at Edgware General Hospital's School of Nursing. On completion of training she became a staff nurse in the Accident and Emergency department at Edgware General and later became a senior staff nurse on the children's ward at the same hospital. In 1986 she became an Industrial Nursing Officer, but returned to accident and emergency nursing in 1990. While working in the accident and emergency department at Barnet General Hospital, Mary studied for her second registration as a Registered Sick Children's Nurse, becoming an Accident and Emergency Sister and Paediatric Nurse Specialist for the same hospital. In 2001 she became a Nurse Facilitator for the North Central London Workforce Development Confederation and later went on to become the acting lead nurse for the cadet nursing scheme for the same NHS organization. Mary has worked as a lecturer in Child Health at the University of Hertfordshire since 2003 and is currently the Alternate Professional Leader and Programme Field Tutor for Children's Nursing at the same institution.

**Elizabeth Gormley-Fleming RGN, RSCN, BSc (Hons), MA (Keele), PG Dip HE (Herts)**
Associate Dean of School, Academic Quality Assurance, Children's Nursing, School of Health and Social Work, University of Hertfordshire, Hatfield, UK

Liz commenced her nursing career in Ireland, where she qualified as an RGN and RSCN. After working in Our Lady's Hospital for Sick Children in Dublin, she moved to London to work at Northwest London NHS Trust initially in adult nursing, but then moved to Children's Services where she had a variety of roles from staff nurse to lead nurse. Liz also worked in Northern Ireland during the 1990s, where she worked in childrens' orthopaedics and as a school nurse. Liz has worked in education since 2001, initially as a clinical facilitator before moving into full-time HE in 2003. Her areas of interest are care of the acutely ill child, healthcare law and ethics, the use of technology in higher education and practice-based learning.

**Debbie Martin MSc, BSc (Hons), EN(G), RSCN, PGDEd, DipON(Oxford)**
Senior Lecturer in Children's Nursing, School of Health and Social Work, University of Hertfordshire, Hatfield, UK

Debbie commenced her nursing career in 1979 at The Nuffield Orthopaedic Centre, Oxford; after completing a Diploma in Orthopaedic nursing she went on to Mount Vernon Hospital and became an Enrolled Nurse. She worked part time for a number of years, at Mount Vernon and St Albans Hospital whilst bringing up a family, then undertook further training to become a Registered Sick Children's Nurse. She worked as a Staff Nurse and a Junior Sister at Hemel Hempstead Hospital and then moved into a placement support and practice development role. She moved into nurse education at the University of Hertfordshire in 2006. Her key areas of interest are pre-registration nurse education, simulation in nurse education, adolescent health and care of siblings of children with long-term conditions.

**Alison Mosenthal RGN, RSCN, Dip N (London), Diploma Nursing Education, MSc**
Senior Lecturer in Children's Nursing, School of Health and Social Work, University of Hertfordshire, Hatfield, UK

Alison began her nursing career at St Thomas' Hospital London before undertaking her RSCN training at Great Ormond Street in 1979. After qualifying she worked in the respiratory intensive care unit and then moved into nurse education in the School of Nursing at Great Ormond Street in 1983.

After a career break raising her family, Alison returned to clinical nursing working as a clinical nurse specialist in paediatric Immunology nursing at St Georges' Healthcare NHS Trust in 1996. She remains in clinical practice there, but returned to teaching in higher education at the University of Hertfordshire in 2010. She currently works there part time as a senior lecturer on the children's nursing course.

**Joanne Outteridge RN (Child), ENB 415, BN (Hons), PG Dip Health Care Ethics, PG Dip HE, MSc**
Senior Lecturer, Department of Family and Community Studies, Anglia Ruskin University, Cambridge, UK

Jo began her nursing career as a children's nurse at the Evelina Children's Hospital, London, working in paediatric cardiology and then paediatric intensive care. She then moved to teaching children's pre-registration and respiratory nursing at City University, London, becoming a lecturer practitioner on the children's medical wards at the Royal London Hospital.

Jo now works at Anglia Ruskin University, teaching children's high-dependency and intensive care nursing, and leading an interprofessional MSc Children and Young People.

**Ian Peate EN(G) RGN DipN (Lond) RNT BEd (Hons) MA (Lond) LLM**
Visiting Professor of Nursing, Head of the School of Health Studies, Gibraltar and Editor in Chief, *British Journal of Nursing*

Ian began his nursing a career in 1981 at Central Middlesex Hospital, becoming an Enrolled Nurse working in an intensive care unit. He later undertook 3 years' student nurse training at Central Middlesex and Northwick Park Hospitals, becoming a Staff Nurse and then a Charge Nurse. He has worked in nurse education since 1989. His key areas of interest are nursing practice and theory, men's health, sexual health and HIV. Ian has published widely; he is Visiting Professor of Nursing, Editor in Chief of the *British Journal of Nursing* and Head of the School of Health Studies Gibraltar.

**Julia Petty RGN, RSCN, BSc (Hons), MSc, PGCE, MA**
Senior Lecturer, School of Health and Social Work, University of Hertfordshire, Hatfield, UK

Julia began her children's nursing career at Great Ormond Street Hospital before moving on to work in various other positions in children's and neonatal care. She worked in clinical practice for 8 years before moving into higher education. Julia worked as a Senior Lecturer at City University, London, for 12 years before commencing her current post in April 2013. Her key interests are neonatal health, outcome of early care and the impact in childhood, and most recently the development of online learning resources in neonatal/children's nursing care. Julia has a considerable publication portfolio and has developed an online, interactive learning resource in neonatal care. She is a resuscitation instructor for the UK Resuscitation Council in newborn life support, a member of the Neonatal Nurses Association Special Interest Education group as well as being on the editorial board as education editor for the *Journal of Neonatal Nursing*. Her recent research interest involves exploring learners' needs and satisfaction in interactive online learning and how user involvement can inform the development of resources for children's nurses.

**Sheila Roberts RGN RSCN RNT BA (Hons) MA**
Senior Lecturer, Children's Nursing, School of Health and Social Work, University of Hertfordshire, Hatfield, UK

Sheila began her nursing career in 1979 at the Queen Elizabeth School of Nursing in Birmingham, working primarily at Birmingham Children's Hospital. She moved to general paediatrics at Kidderminster, Ipswich and finally at Bedford Hospital, where she became ward sister. Sheila moved into nurse education in 2006. Her areas of interest include nursing practice and teaching nursing skills to students, along with an interest in the cardiac and respiratory systems and child development. More recently, Sheila has been involved in projects involving children as service users within the pre-registration nursing curriculum.

**Peter Vickers Cert Ed, Dip CD, SRN, RSCN, BA, PhD, FHEA**

After training as a schoolteacher, and undertaking many jobs both within and without teaching, Peter began his nursing career in 1980 at York District Hospital, followed by Great Ormond Street Children's Hospital, London, where he specialized in Immunology and Infectious Diseases, eventually becoming a Clinical Nurse Specialist in Paediatric Immunology. In 1999 he was awarded a PhD following research undertaken into children with Severe Combined Immune Deficiency (SCID) in the UK and Germany. He commenced working in nurse education in 2001. His key areas of interest are immunology, infectious diseases, genetics and research, and he has published widely – both as the author of a book on SCID, and co-author of a book on research, as well as contributing many chapters to academic textbooks. Although now retired, he is still active in writing, and in 2012 he was elected as President of INGID (the international organization for nurses working in primary immunodeficiencies).

**Lisa Whiting DHRes, MSc, BA (Hons) RGN, RSCN, RNT, LTCL, FHEA**
Professional Lead, Children's Nursing, School of Health and Social Work, University of Hertfordshire, Hatfield, UK

Lisa completed the 4-year RGN/RSCN programme at the Queen Elizabeth Medical Centre, Birmingham, in 1983. She gained a range of paediatric clinical experience, developing her career within the area of critical care nursing and becoming a ward sister on the children's cardiac intensive care unit at Guy's Hospital. Lisa has worked in education for a number of years, teaching students undertaking a range of children's nursing programmes. In addition, Lisa has studied academically, nurturing an interest in child health promotion; she has completed a Doctorate in Health Research that drew on an ethnographic approach to facilitate the emergence of assets that underpin children's wellbeing.

Lisa is now Professional Lead for Children's Nursing at the University of Hertfordshire.

# Preface

Caring for children (young people) and their families, sick or well, requires the nurse to have an understanding of a range of complex issues. It is important that you have insight into the anatomy and physiology of the child. The anatomical and physiological systems of children are different to those of the adult. In some instances there are marked differences and in others these are subtle.

Children are not little adults. Recognizing and responding to their anatomical and physiological differences is essential when caring for a child sick or well (Rzucidlo and Shirk, 2004). The body of a child is in a constant state of development and maturation and progressive growth, with dynamic biochemical processes.

The term child will be adopted throughout this text; this refers to children and young people aged between 0 and 16 years of age.

An individual is seen and treated as a whole, however; the human body is composed of organic and inorganic molecules that are organized at a variety of different structural levels. If the nurse is to ensure children and their families are to receive appropriate and timely care, they need to be educated in such a way that they can recognize illness, provide effective treatment and make appropriate referrals with the child at the centre of all they do.

In the nursing and healthcare literature (the psycho-social sciences) children have been well represented; however, there is a dearth of texts that focus on the anatomy and physiology of the child. If nurses are to be prepared to be effective children's nurses then they must demonstrate a sound knowledge of child-related anatomy and physiology with the intention of providing safe and effective nursing care.

The overall aim of this succinct and concise text is to provide you with a sound understanding of the fundamentals associated with the anatomy and physiology of children and the related biological sciences that will enable you to develop your practical caring skills and to enhance your knowledge in order to become a caring, kind and compassionate children's nurse. Having developed your knowledge you will be able to deliver increasingly complex care for children, sick or well, in a range of settings that safeguards and promotes the welfare of vulnerable children in an appropriate, coordinated, multidisciplinary, integrated and family-centred manner.

As children grow, so they develop physically and psychologically. As the child progressively grows and develops, their immature systemic organs and biochemical processes will influence disease processes as well as any therapeutic strategies.

*Fundamentals of Children's Anatomy and Physiology* provides you with the opportunity to apply the content to the care of children, young people and their families. As you begin to understand how children respond or adapt to pathophysiological changes and stresses you will be able to appreciate that children (regardless of age) have specific biological needs.

The integration and application of evidence-based theory to practice is a key component of effective and safe health care. This goal cannot be achieved without an understanding of the anatomical and pathophysiological aspects associated with child health.

This text provides you with structure and a comprehensive approach to anatomy and physiology. Expert nurses who have a passion and commitment to the child, young person and family have written the chapters with you, the student, at the fore. The text is designed to be used as a reference text in the practice placement setting, the classroom or at home. It is not intended to be read from cover to cover in one sitting.

# Anatomy and physiology

Living systems can be defined from a number of perspectives. At the very smallest level, the chemical level, atoms, molecules and the chemical bonds connecting atoms provide the structure upon which life is based. The smallest unit of life is the cell. Tissue is a group of cells that are similar and they perform a common function. Organs are groups of different types of tissues performing together to carry out a specific activity. A system is two or more organs working together to carry out a particular activity. Another system that possesses the characteristics of living things is an organism, this having the capacity to obtain and process energy, the ability to react to changes in the environment and to reproduce.

As anatomy is associated with the function of a living organism, it is almost always inseparable from physiology. Physiology is the science dealing with the study of the function of cells, tissues, organs and organisms; in essence, it is the study of life.

This text focuses on human anatomy and physiology. The definition used here to define anatomy is the study of the structure and function of the human body. This allows reference to function as well as structure. In all biological organisms, structure and function are closely interconnected. The human body operates through interrelated systems and, as such, by and large, a systems approach is used in this text.

# The Nursing and Midwifery Council

Nursing practice is constantly changing and evolving. The Nursing and Midwifery Council (NMC, 2010), in detailing the field standard for competence for children's nurses, states that the nurse must be able to care safely and effectively for children and young people in all settings to deliver care that meets essential and complex physical and mental health needs informed by a deep understanding of biological, psychological and social factors throughout infancy, childhood and adolescence. This text will help you to further develop and consolidate your knowledge and prepare you to undertake care delivery activities in primary, secondary and tertiary settings.

Theory associated with the biological sciences provides the scientific basis for children's nursing practice; acquiring a sound, up-to-date biological theory is essential for safe, effective professional practice in all healthcare settings. When you apply the knowledge associated with the biological sciences to clinical care you are demonstrating that you are providing safe and effective care, a hallmark of the professional children's nurse in changing contemporary society. Safe, high-quality and effective care for all is something that all health-care professionals should be striving to provide; it is not possible to do this effectively if you do not fully appreciate the whole being, the whole person.

You are undertaking your programme of study in order to acquire the competencies required to meet the criteria for registration with the NMC permitting you to practice as a registered nurse. The application of biological sciences theory encourages critical thinking in practice related to children's nursing as well as helping to provide a rationale for interventions undertaken and to structure care provision that can minimize or avoid complications and adverse consequences.

# Chapter content

The essence of the text will be its 'student friendliness' and easy navigation. Each chapter commences with learning outcomes, helping you to pre-plan learning and understand the rationale for the discrete yet intertwined chapters.

There are a number of features provided that aim to help you learn, retain and recall information. Each chapter contains:

- 'Learning outcomes' at the beginning of the chapter.
- Ten 'test your prior knowledge' questions at the beginning of each chapter.
- Boxed clinical applications: applying the anatomy and physiology to common child health conditions to provide a clinical focus.
- Review questions and chapter activities to help reinforce retention and learning.
- A glossary of terms.
- A list of conditions is provided prompting you to make notes about each listed condition.
- A colour-coded format and layout has been used to help enhance learning.
- Full-colour illustrations.

# Web-based materials

The text will be supplemented with web-based materials that you are able to access; for example:

- MCQs (long and short answer)
- 'Label the diagram' flashcards
- Glossary of terms used throughout the printed book

There are 19 chapters; the majority of them are concerned with body systems. The first chapter recognizes that a child's health is greatly influenced by social, political and environmental factors (influencing factors), and these complex interrelated dynamics must be acknowledged. As such, the opening chapter provides an overview of the child and society, enabling and reminding you that the care of the child must always be placed in context.

Other chapters at the beginning of the text consider issues such as homeostasis and homeostatic mechanisms and how this important biological concept is evident at microscopic and macroscopic biological levels. Emphasis will be placed on functional and dysfunctional homeostasis. The various scientific principles are addressed early on, leading to chapters related to the cell, genetics and tissues before taking the reader in to the body systems.

We have very much enjoyed writing this text, and we hope that you enjoy reading it and then applying the contents to the care you have the privilege of delivering to children, young people and their families. It has allowed us to share with you our understanding of the anatomy and physiology of the child and young people.

## References

NMC (2010) Standards for Pre registration Nursing Education. Nursing and Midwifery Council, London, http://standards.nmc-uk.org/PublishedDocuments/Standards%20for%20pre-registration%20nursing%20education%2016082010.pdf (accessed July 2014).

Rzucidlo, S.E., Shirk, B.J. (2004) Trauma nursing: pediatric patients. *RN*, **76** (6), 36–41.

# Acknowledgements

Ian would like to thank his partner Jussi Lahtinen for his encouragement and Mrs Frances Cohen for her constant inspiration and support.

Liz would like to thank her husband, Kieran, and Kate and Eilis, her daughters, for their love, support and patience.

Ian and Liz would like to thank all their colleagues for their help, support and guidance with this book.

# How to use your textbook

## Features contained within your textbook

Every chapter begins with 10 **test your prior knowledge questions**.

### Test your prior knowledge

- Can you name two key Acts of Parliament from the last 25 years that focus on protecting children?
- Is involving children in decision making a professional, ethical and legal obligation for health-care professionals?
- Where is it stipulated that *'The child shall have the right to freedom of expression'*?
- Which law states *'Everyone has the right to freedom of expression'*?
- What was the name of the review that was published in 2010 to identify how health inequalities could be addressed within the UK?
- What were the five key outcomes identified within Every Child Matters (Department for Education and Skills [DfES], 2004)?
- Which Act of Parliament reflects the five key outcomes from Every Child Matters (DfES, 2004)?
- Where was the UK ranked out of 29 'rich countries' by UNICEF in 2013 in relation to a child's wellbeing?
- What are the three key areas that child public health focuses on?
- Which Nursing and Midwifery Council (NMC) standards state that children's nurses must *'Support and promote the health, wellbeing, rights and dignity of people, groups, communities and populations'*?

**Learning outcome boxes** give a summary of the topics covered in a chapter.

### Learning outcomes

On completion of this chapter the reader will be able to:

- Define and discuss the concept of 'childhood'.
- Consider the child's 'voice' and the importance of involving children in decision-making processes.
- Explore the role of family, friends and the local community in relation to children's overall wellbeing.
- Define and discuss the concepts of health and wellbeing within a child-focussed context.
- Define child public health and consider associated key policies.
- Reflect on the potential health promoting role of the nurse.
- Consider childhood morbidity and mortality within a 21st century context.

**Clinical application boxes** give inside information on a topic.

### Clinical application

**Obesity**

Obesity is of increasing concern across the world for all age groups and is known to have serious health consequences. In 2010, the number of overweight children under the age of 5 years was over 42 million (WHO, 2013b). It is thought that this may be due to either a genetic predisposition to how fat is stored and synthesized or an imbalance between the amount of energy required and the amount of fat consumed.

Obesity may arise when the body becomes resistant to hormones and the ensuing sensory nerve actions that regulate the perception of hunger and the size of meals. For instance, within the hypothalamus, food intake will be reduced when brain nuclei are stimulated by the hormones insulin and leptin. Leptin is produced by adipose tissue, binds to receptors within the hypothalamus and provides feedback regarding energy stores. Mutation of the gene for leptin has been associated with obesity and Type 2 diabetes (Clancy and McVicar, 2009).

Your textbook is full of **illustrations and tables.**

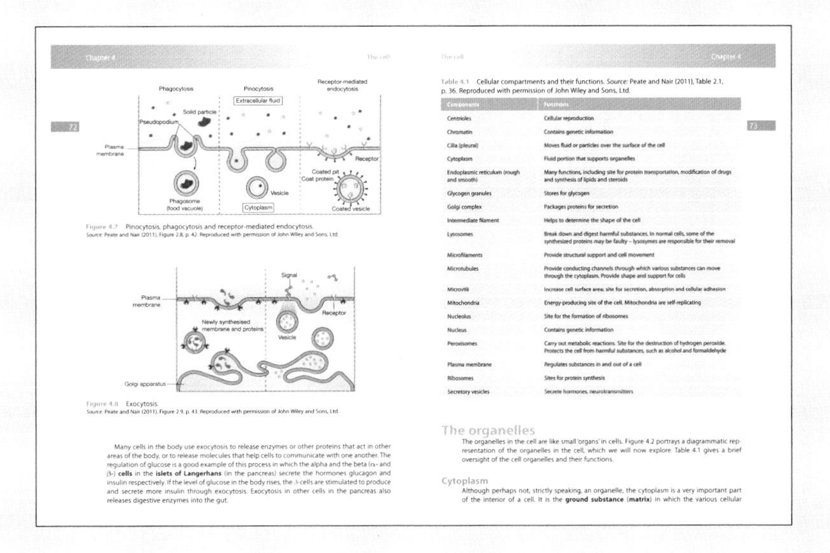
Chapter 4 The cell

72

Figure 4.7 Pinocytosis, phagocytosis and receptor-mediated endocytosis.
Source: Peate and Nair (2011), Figure 2.8, p. 42. Reproduced with permission of John Wiley and Sons, Ltd.

Figure 4.8 Exocytosis.
Source: Peate and Nair (2011), Figure 2.9, p. 43. Reproduced with permission of John Wiley and Sons, Ltd.

Many cells in the body use exocytosis to release enzymes or other proteins that act in other areas of the body, or to release molecules that help cells to communicate with one another. The regulation of glucose is a good example of this process in which the alpha and the beta (α- and β-) **cells** in the **islets of Langerhans** (in the pancreas) secrete the hormones glucagon and insulin respectively. If the level of glucose in the body rises, the β-cells are stimulated to produce and secrete more insulin through exocytosis. Exocytosis in other cells in the pancreas also releases digestive enzymes into the gut.

The cell Chapter 4

73

Table 4.1 Cellular compartments and their functions. *Source*: Peate and Nair (2011), Table 2.1, p. 36. Reproduced with permission of John Wiley and Sons, Ltd.

| Components | Functions |
|---|---|
| Centrioles | Cellular reproduction |
| Chromatin | Contains genetic information |
| Cilia (pleural) | Moves fluid or particles over the surface of the cell |
| Cytoplasm | Fluid portion that supports organelles |
| Endoplasmic reticulum (rough and smooth) | Many functions, including site for protein transportation, modification of drugs and synthesis of lipids and steroids |
| Glycogen granules | Stores for glycogen |
| Golgi complex | Packages proteins for secretion |
| Intermediate filament | Helps to determine the shape of the cell |
| Lysosomes | Break down and digest harmful substances. In normal cells, some of the synthesized proteins may be faulty – lysosymes are responsible for their removal |
| Microfilaments | Provide structural support and cell movement |
| Microtubules | Provide conducting channels through which various substances can move through the cytoplasm. Provide shape and support for cells |
| Microvilli | Increase cell surface area; site for secretion, absorption and cellular adhesion |
| Mitochondria | Energy-producing site of the cell. Mitochondria are self-replicating |
| Nucleolus | Site for the formation of ribosomes |
| Nucleus | Contains genetic information |
| Peroxisomes | Carry out metabolic reactions. Site for the destruction of hydrogen peroxide. Protects the cell from harmful substances, such as alcohol and formaldehyde |
| Plasma membrane | Regulates substances in and out of a cell |
| Ribosomes | Sites for protein synthesis |
| Secretory vesicles | Secrete hormones, neurotransmitters |

The organelles

The organelles in the cell are like small 'organs' in cells. Figure 4.2 portrays a diagrammatic representation of the organelles in the cell, which we will now explore. Table 4.1 gives a brief oversight of the cell organelles and their functions.

Cytoplasm

Although perhaps not, strictly speaking, an organelle, the cytoplasm is a very important part of the interior of a cell. It is the **ground substance** (**matrix**) in which the various cellular

End of chapter **activities** help you test yourself after each chapter.

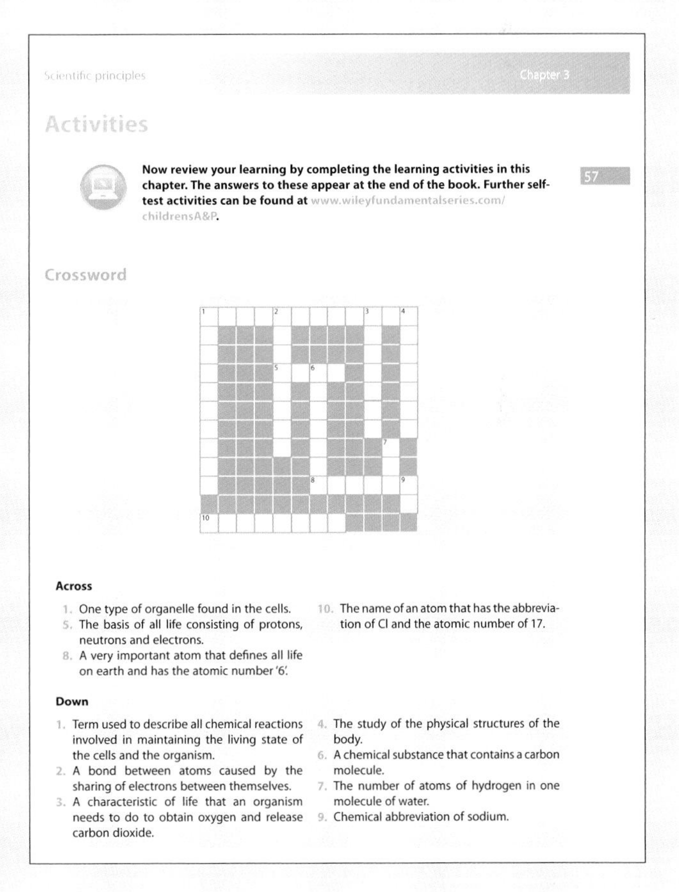
Scientific principles Chapter 3

Activities

57

**Now review your learning by completing the learning activities in this chapter. The answers to these appear at the end of the book. Further self-test activities can be found at www.wileyfundamentalseries.com/childrensA&P.**

Crossword

**Across**

1. One type of organelle found in the cells.
5. The basis of all life consisting of protons, neutrons and electrons.
8. A very important atom that defines all life on earth and has the atomic number '6'.
10. The name of an atom that has the abbreviation of Cl and the atomic number of 17.

**Down**

1. Term used to describe all chemical reactions involved in maintaining the living state of the cells and the organism.
2. A bond between atoms caused by the sharing of electrons between themselves.
3. A characteristic of life that an organism needs to do to obtain oxygen and release carbon dioxide.
4. The study of the physical structures of the body.
6. A chemical substance that contains a carbon molecule.
7. The number of atoms of hydrogen in one molecule of water.
9. Chemical abbreviation of sodium.

**The website icon** indicates that you can find accompanying resources on the book's companion website.

# The anytime, anywhere textbook

## Wiley E-Text

For the first time, your textbook comes with free access to a **Wiley E-Text: Powered by VitalSource** version – a digital, interactive version of this textbook which you own as soon as you download it. Your **Wiley E-Text** allows you to:

**Search:** Save time by finding terms and topics instantly in your book, your notes, even your whole library (once you've downloaded more textbooks)
**Note and Highlight:** Colour code, highlight and make digital notes right in the text so you can find them quickly and easily
**Organize:** Keep books, notes and class materials organized in folders inside the application
**Share:** Exchange notes and highlights with friends, classmates and study groups
**Upgrade:** Your textbook can be transferred when you need to change or upgrade computers
**Link:** Link directly from the page of your interactive textbook to all of the material contained on the companion website

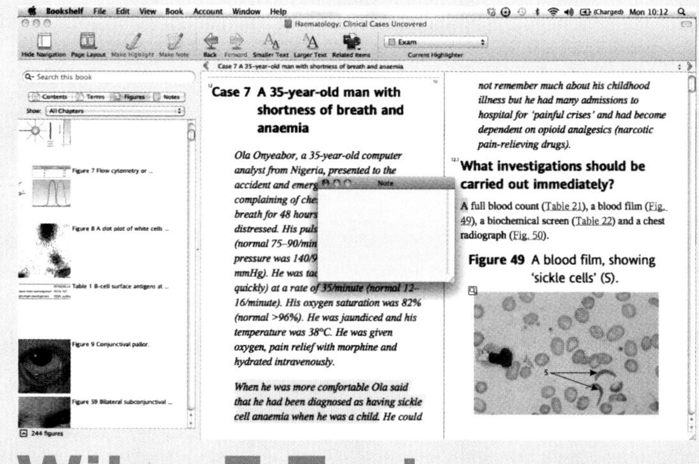


**Wiley E-Text**
Powered by VitalSource®

The **Wiley E-Text** version will also allow you to copy and paste any photograph or illustration into assignments, presentations and your own notes.

*To access your Wiley E-Text:*

- Find the redemption code on the inside front cover of this book and carefully scratch away the top coating of the label. Visit **www.vitalsource.com/software/bookshelf/downloads** to download the Bookshelf application to your computer, laptop, tablet or mobile device.
- If you have purchased this title as an e-book, access to your **Wiley E-Text** is available with proof of purchase within 90 days. Visit **http://support.wiley.com** to request a redemption code via the 'Live Chat' or 'Ask A Question' tabs.
- Open the Bookshelf application on your computer and register for an account.
- Follow the registration process and enter your redemption code to download your digital book.
- For full access instructions, visit **www.wileyfundamentalseries.com/childrenA&P**.

## The VitalSource Bookshelf can now be used to view your Wiley E-Text on iOS, Android and Kindle Fire!

- **For iOS:** Visit the app store to download the VitalSource Bookshelf: **http://bit.ly/17ib3XS**
- **For Android and Kindle Fire:** Visit the Google Play Market to download the VitalSource Bookshelf: **http://bit.ly/BSAAGP**

You can now sign in with the email address and password you used when you created your VitalSource Bookshelf Account.

Full E-Text support for mobile devices is available at: **http://support.vitalsource.com**

## CourseSmart

**CourseSmart** gives you instant access (via computer or mobile device) to this Wiley-Blackwell e-book and its extra electronic functionality, at 40% off the recommended retail print price. See all the benefits at **www.coursesmart.com/students.**

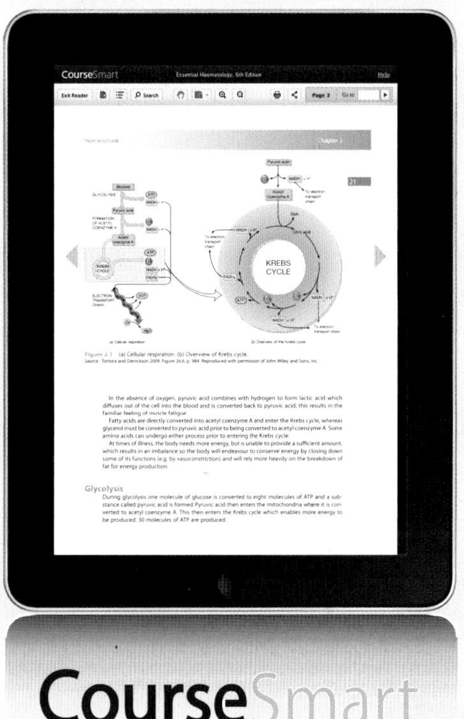

### *Instructors ... receive your own digital desk copies!*

**CourseSmart** also offers instructors an immediate, efficient, and environmentally-friendly way to review this textbook for your course.

For more information visit **www.coursesmart.com/instructors.**

With **CourseSmart**, you can create lecture notes quickly with copy and paste, and share pages and notes with your students. Access your **CourseSmart** digital textbook from your computer or mobile device instantly for evaluation, class preparation, and as a teaching tool in the classroom.

Simply sign in at **http://instructors.coursesmart.com/bookshelf** to download your Bookshelf and get started. To request your desk copy, hit 'Request Online Copy' on your search results or book product page.

We hope you enjoy using your new textbook. Good luck with your studies!

# About the companion website

Don't forget to visit the companion website for this book:

**www.wileyfundamentalseries.com/childrensA&P**

There you will find valuable material designed to enhance your learning, including:

- Interactive multiple choice questions
- 'Label the diagram' flashcards
- Searchable glossary

Scan this QR code to visit the companion website:

# Chapter 1

# The child in society: enhancing health and wellbeing

Lisa Whiting and Mary Donnelly

*School of Health and Social Work, University of Hertfordshire, Hatfield, UK*

## Aim

The aim of this chapter is to consider the concept of childhood and the enhancement of the overall health and wellbeing of children and young people.

## Learning outcomes

On completion of this chapter the reader will be able to:

- Define and discuss the concept of 'childhood'.
- Consider the child's 'voice' and the importance of involving children in decision-making processes.
- Explore the role of family, friends and the local community in relation to children's overall wellbeing.
- Define and discuss the concepts of health and wellbeing within a child-focussed context.
- Define child public health and consider associated key policies.
- Reflect on the potential health promoting role of the nurse.
- Consider childhood morbidity and mortality within a 21st century context.

*Fundamentals of Children's Anatomy and Physiology: A Textbook for Nursing and Healthcare Students*, First Edition. Edited by Ian Peate and Elizabeth Gormley-Fleming.

Companion website: www.wileyfundamentalseries.com/childrensA&P

## Test your prior knowledge

- Can you name two key Acts of Parliament from the last 25 years that focus on protecting children?
- Is involving children in decision making a professional, ethical and legal obligation for health-care professionals?
- Where is it stipulated that *'The child shall have the right to freedom of expression'*?
- Which law states *'Everyone has the right to freedom of expression'*?
- What was the name of the review that was published in 2010 to identify how health inequalities could be addressed within the UK?
- What were the five key outcomes identified within Every Child Matters (Department for Education and Skills [DfES], 2004)?
- Which Act of Parliament reflects the five key outcomes from Every Child Matters (DfES, 2004)?
- Where was the UK ranked out of 29 'rich countries' by UNICEF in 2013 in relation to a child's wellbeing?
- What are the three key areas that child public health focuses on?
- Which Nursing and Midwifery Council (NMC) standards state that children's nurses must *'Support and promote the health, wellbeing, rights and dignity of people, groups, communities and populations'*?

# Introduction

> *Health in childhood determines health throughout life and into the next generation. … Ill health or harmful lifestyle choices in childhood can lead to ill health throughout life, which creates health, financial and social burdens for countries today and tomorrow.*
>
> (World Health Organization [WHO], 2005: ix)

This quote confirms that the promotion and maintenance of children's health and wellbeing is of paramount importance, both now and for the future; this is something that has been widely recognized and has received considerable attention (e.g. the UK Department for Education and Skills [DfES], 2004; DfES and Department of Health [DH], 2004; Department for Children, Schools and Families [DCSF], 2007a; DCSF and DH, 2009; DH, 2010a–c; Department for Education [DfE], 2011).

Children are a fundamental and invaluable part of society. To promote their overall health, it is essential that key aspects of children's lives are appreciated, as well as some of the factors that have the potential to impact on their overall wellbeing. This chapter provides an introduction to the concept of childhood, reflecting on the child's 'voice' and the importance of involving children in any decisions that may affect them. The significance of the child's immediate environment, in terms of their family, friends and community, is considered; this is followed by a discussion focussing on children's health and wellbeing (including the public health agenda) and the potential health-promoting role of the nurse. The chapter concludes by briefly considering childhood morbidity and mortality, thus 'setting the scene' for the subsequent chapters.

The term 'child', used throughout the chapter, refers to children and young people aged between 0 and 18 years.

# The concept of childhood



This section introduces the concept of childhood with consideration being given to the meaning and evolvement of the term childhood over the years and the *'emergence of "children's voice"'* (Prout and Hallett, 2003: 1). In addition, the value and importance of involving children in decisions that impact on them is addressed.

A rudimentary dictionary definition of childhood is:

> *The condition of being a child; the period of life before puberty.*
>
> (*Collins Dictionary and Thesaurus*, 2000: 195)

However, it is also generally acknowledged that childhood spans four key phases – infancy and toddlerhood, early years, middle childhood and adolescence (Hutchison, 2011). Eminent psychologists, such as Erikson (1950), Piaget (1952) and Kohlberg (1984) have considered different aspects of children and young people's cognitive development.

Prout and James (1997: 8) offer more clarification and suggest that childhood is not simply about the organic maturation of children, but is a *'specific structural and cultural component of many societies.'* Importantly, Frønes (1993: 1) states that:

> *There is not one childhood, but many, formed at the intersection of different cultural, social and economic systems, natural and man-made physical environments. Different positions in society produce different childhoods, boys and girls experience different childhoods within the same family.*

This raises an important point: if children are referred to collectively within the term 'childhood', there is a danger that differences (e.g. in gender, age and ethnicity) will be lost (James and Prout, 1997). Frønes (1993) acknowledges the impact of society on the evolution of childhood, but also alludes to the individual experience and this perspective must surely be recognized.

There can be no doubt that the perception, understanding and recognition of childhood has changed considerably over the centuries. Authors (e.g. Ariès, 1962 and Cunningham, 2006) have considered the development of the meaning of the term childhood from the Middles Ages to more recent years, recognizing that it has been influenced by a number of factors; for example, the impact of Christianity in the 18th century meant that the child was often viewed as needing spiritual salvation from evil (Hendrick, 1997); in the Victorian era, as a result of the work of a range of reformists, there was a more overt drive to protect children (Cunningham, 2006). There has been a recurrent theme over the years of viewing children in terms of purity and innocence (Cunningham, 2006), something which Holt (1975: 23) suggests is not always the case for every child – he refers to the falsehood of the assumed *'happy, safe, protected, innocent childhood.'*

In more recent years, there has been a stronger focus on the protection of children with a variety of both legal and policy documents being published (e.g. the Children Act, 1989, 2004; DfES and DH, 2004; DfES, 2004; Royal College of Paediatrics and Child Health [RCPCH], 2010); in addition, there is now a wealth of literature that focuses on protecting children from a whole range of life events (e.g. environmental tobacco smoke [Botelho and Fiscella, 2005]; sun protection [Gritz *et al.*, 2005]; travel risks [Mathur and Kamat, 2005]; the internet [Pogue, 2005] and divorce [Vélez *et al.*, 2011]). It has been suggested that some aspects of protection could lead to reduced opportunities for children to socially interact, with the main conduits only existing within controlled settings such as schools and clubs (Smith, 2000). In support of this point, Palmer (2006) comments on the 'toxic' environment and the influence that this is having on childhood in the 21st century. Children's lives and the nature of their childhood (at both a

## Box 1.1 Key features of the paradigm of childhood (Prout and James, 1997: 8).

- Childhood is a framework for the contextualization of children's lives.
- Childhood cannot be separated from other variables in society, for example, gender and ethnicity.
- Children's social interactions should be studied and remain independent of the adult perspective.
- Children should be actively involved in decisions that may impact on their lives.
- Ethnography can be a valuable research approach for the study of childhood.
- A new paradigm of childhood necessitates the reconstruction of childhood.

societal and individual level) is different from that of previous generations; however, it could be argued that generational differences are not new and have existed for centuries. The most important issue is that we have a clear understanding of children's lives within the 21st century so that appropriate care can be provided by all healthcare professionals.

## The child's 'voice'

Whilst a range of literature has for many years demonstrated a strong interest in the lives of children across the age ranges, much of this has tended to focus on the adult perspective, rather than valuing the voice and contribution of the child (Prout and James, 1997). Prout and James (1997) have offered a new paradigm for childhood that has six key features (Box 1.1).

The work of James and Prout (1997) has been invaluable in raising the profile of children as participants who are capable of being involved in decisions that may impact on their lives. The need to involve children in a range of issues has grown in acceptance (Sinclair, 2004). It is now widely established that the views and experiences of children should be taken into account wherever possible, with a range of key documents advocating their involvement (e.g. the Children Act, 1989, 2004; The United Nations Convention on the Rights of the Child [UNCRC] [1989]; DfES and DH, 2004) (Table 1.1).

Whilst Prout and James (1997) comment on the importance of involving children when decisions may impact on them, it is also important to recognize that children themselves benefit from involvement by gaining a sense of achievement, increased self-esteem (Kirby, 2004; The National Youth Agency, 2007) and enhanced communication skills (Participation Works, 2007; Carnegie UK Trust, 2008).

# Fundamental aspects of children's lives

### The family

A range of prominent authors and organizations (e.g. Rutherford, 1998; United Nations, 1998; European Parliament, 2000) have acknowledged the potential impact of the family on children's growth, nurturing and development. In addition, research into the concept of attachment has suggested that children who feel secure are more likely to adhere to rules and boundaries set by parents (Thompson, 2006) and responsive parenting fosters responsive and cooperative

**Table 1.1** Key documents that advocate the involvement of children.

| | |
|---|---|
| **The United Nations Convention on the Rights of the Child (UNCRC) (1989)** | Article 12: Parties shall assure to the child who is capable of forming his or her own views the right to express those views freely in all matters affecting the child, the views of the child being given due weight in accordance with the age and maturity of the child |
| | Article 13: The child shall have the right to freedom of expression; this right shall include freedom to seek, receive and impart information and ideas of all kinds |
| | Article 42: Undertake to make the principles and provisions of the Convention widely known, by appropriate and active means, to adults and children alike |
| **Children Act (1989)** | Section 22(4): Before making any decision with respect to a child whom they are looking after, or proposing to look after, a local authority shall, so far as is reasonably practicable, ascertain the wishes and feelings of the child |
| **Children Act (2004)** | Section 17: To consult children and young people on Children and Young People's Plans |
| **National Service Framework for Children, Young People and Maternity Services: Core Standards (UK DfES and DH, 2004)** | Standard 3: Professionals communicating directly with children and young people, listening to them and attempting to see the world through their eyes |
| **Human Rights Act (1998)** | Article 10: Everyone has the right to freedom of expression |

children (Kochanska *et al.*, 2005). The acknowledgement of the family's contribution to children's overall wellbeing has been recognized within the literature and was one of the key findings from work by Parry *et al.*, (2010), Rees *et al.*, (2010) and Ipsos Mori and Nairn (2011).

However, it is important to recognize that the *'21st century family is not a static structure'* (Rigg and Pryor, 2007: 17). Statistics reveal that there were 117 558 divorces in England and Wales in 2011 (Office of National Statistics, 2012); although this represents a slight decrease on previous figures, it is clear that every year a great many children and young people are faced with a significant alteration in their family structure and their subsequent lifestyle, although they are the people who have been the least likely to initiate the changes (Rigg and Pryor, 2007).

It is important to understand children's perception of a family in order to appreciate the relationships and aspects of family life that are important to them. Very few studies have been conducted to determine children and young people's views of the family in the 21st century. However, Rigg and Pryor (2007) studied vignettes of, and interviews with, 111 New Zealand children (aged 9–13 years) from a range of family structures and backgrounds. They found that children frequently described affective factors – in other words, children perceived that family comprised of people who cared for and loved them (Rigg and Pryor, 2007); the children did not tend to make distinctions between couples being married or cohabitating.

Whilst the family has changed enormously in diversity and structure over the last century, it undoubtedly remains fundamental to children's lives and has many positive attributes; Whiting *et al.*, (2013) suggest that it is a complex concept that incorporates four key areas:

- **Family membership:** The family is an assumed presence that comprises of people who are always there and who are an integral part of children's lives; this undoubtedly brings stability. It is important to note that children's pets can also be regarded as family members.

- **Togetherness:** Children and young people undertake a range of activities with family members; these include weekly family treats, holidays, special occasions and days out; these experiences provide the opportunity for the family to be both physically and emotionally close to each other, fostering and enhancing their relationships.
- **Family influence:** This plays a substantive role in the development of children's lifestyle and interests, not just in the choice, but also in the provision of necessary resources such as food and accommodation. It is important to acknowledge that family influence can be both positive and negative.
- **Being busy:** In the 21st century, many families lead busy lives – parents frequently work and need to 'juggle' a home/work life balance; children and young people go to school/college and develop their own interests. Despite this, families learn to deal with their numerous commitments and frequently develop a team approach to manage the challenges associated with a busy lifestyle; this team approach can further reinforce relationships with members understanding and respecting each other's roles.

Appreciating the crucial role of the family in a child's life is fundamental to all child healthcare provision. Liaising and working in partnership with the people who the child perceives as their family is pivotal to the building of trusting, therapeutic, professional relationships – this in turn promotes high quality nursing care.

## Friendships

Friendships are an integral and crucial aspect of children's and young people's lives, with literature suggesting that friendships can enhance a child's overall wellbeing (Parry *et al.*, 2010; Rees *et al.*, 2010; Ipsos Mori and Nairn, 2011); friendships are also associated with other positive attributes such as enhanced social behaviour (Cillessin *et al.*, 2005).

Most children spend the majority of their lives within a relatively small community area – as a consequence, they become familiar with their local environment and this not only gives them confidence, but also contributes to the development and maintenance of friendships. Children tend to make friends readily and via a variety of mechanisms: school, local clubs (such as swimming lessons) and in the immediate vicinity of their homes; Troutman and Fletcher (2010) found that friendships were more likely to be maintained if they crossed different contexts (e.g. school, neighbourhood and extracurricular activities) as this provides children with the opportunity to interact within a variety of different circumstances. Children's friends are often viewed in a similar manner to a family member; it is therefore essential that professionals recognize the value placed on friendship and the potential contribution it can make to the enhancement of social and emotional wellbeing.

## The local community

The impact that the local community can have on children and their health and wellbeing has been acknowledged (Sellstrom and Bremberg, 2006; Counterpoint Research, 2008; Fattore *et al.*, 2009; Parry *et al.*, 2010). Eriksson *et al.*, (2010) conducted a Swedish qualitative study that aimed to identify how the local neighbourhood was perceived by children (11–12 years of age) who lived in rural areas and how social capital in their immediate community impacted on their wellbeing. The seven focus groups centred around four key categories (page 5) – '*community attachment*' (including the sense of belonging, the role of the school, community perceptions); '*community participation*' (including local clubs and activities); '*social networks*' (such as friends, family and neighbours). In summary, Eriksson *et al.*, (2010: 9) commented that:

*the children described the familiarity of the local communities as creating a trustworthy and secure atmosphere.*

Social capital is a term that has been used to describe the connections that people form in the communities in which they live; it enables people to either operate individually or together in order to enhance their situation and access resources. The familiarity that children can gain within their neighbourhood undoubtedly enables them to develop independence and responsibility within parental boundaries and reflects some of the underpinning ethos of social capital.

# Children's health and wellbeing

The concept of health as being more than the mere absence of disease has long been acknowledged. The focus within the 21st century has moved towards the inclusion of wellbeing and what it means to have a sense of purpose, to be able to build meaningful relationships with others and to fulfil one's potential (Ryff and Singer, 1998). Both the UK government and the WHO would concur with this as both of their definitions clearly identify that wellbeing is an aspect of health:

*We use a broad definition of health that encompasses both physical and mental health as well as wellbeing.*

(DH, 2010b: 6)

*A state of complete physical, mental and social well-being and not merely the absence of disease or infirmity.*

(WHO, 1948: 100)

Ryff *et al.*, (2004: 1383) suggest that:

*the core hypothesis of positive health, in fact, is that the experience of well-being contributes to the effective functioning of multiple biological systems.*

In other words, if people feel well and happy, this has the potential to positively impact on overall physical health – it is therefore crucial that healthcare professionals consider the psychological aspect of health as well as the physiological one. More recently, specific attention has been given to children's wellbeing. The previous UK Labour government published the document, 'Every Child Matters' (DfES, 2004), with the fundamental aim of improving children's wellbeing. This goal has since been reiterated:

*Enhance children and young people's wellbeing, particularly at key transition points in their lives.*

(DCSF, 2007a: 13)

Five key outcomes were identified (being healthy; staying safe; enjoying and achieving; making a positive contribution; economic wellbeing) together with 25 aims; these outcomes were reflected in the Children Act (2004) and the English Children's Commissioner was given responsibility for ascertaining the views of children in relation to these areas (see http://childrenscommissioner.gov.uk/search?search=wellbeing for more details).

Unfortunately, despite this work, UNICEF (2007) in *An Overview of Child Well-being in Rich Countries* revealed that the UK was ranked last of the 21 countries that were included; a more recent update from UNICEF (2013) indicated that the UK was ranked 16 out of 29 counties – whilst this demonstrates an improvement, there is still a long way to go until the UK reaches the higher rankings where children's wellbeing is concerned.

UNICEF UK commissioned Ipsos Mori and Nairn (2011) to undertake qualitative research in relation to children's wellbeing that aimed to understand the impact of inequality and materialism on children's lives. The research was underpinned by the work of Bronfenbrenner (1979) and compared the lives of children in the UK, Sweden and Spain; the children in all three countries highlighted the *'the importance of family time'*, *'friendship and companionship'* and *'being active and outdoors'* (pages 25–29); interestingly, the value of pets was also raised and the authors comment that they were frequently viewed as family members. The children perceived these three areas as being far more important than the materialistic resources that were also discussed within the study. UK parents were reported as struggling to spend as much time with children as their counterparts in Sweden and Spain (this was due to a range of reasons which included work commitments and financial resources); Ipsos Mori and Nairn (2011) comment that this has further ramifications since inequality within the UK is related to wealth, which in turn impacts on family life. In conclusion, the authors comment that:

> *The children in all three countries have the same needs and wants and concerns. Yet the response to these by each society is different. It seems that children are more likely to thrive where the social context makes it possible for them to have time with family and friends, to get out and about without having to spend money, to feel secure about who they are rather than what they own, and to be empowered to develop resilience to pressures to consume.*
> (page 73)

This is an important viewpoint to consider since the ultimate aim must surely be for children to maintain their health and wellbeing into adulthood. Although there has been little consideration of the potential long-term benefits of children's wellbeing, Richards and Huppert (2011) presented their findings from a British longitudinal study that analysed data relating to a stratified sample of 5362 people who had been born in England, Scotland and Wales within a particular week in March 1946. The aim was to ascertain whether positive wellbeing in childhood transferred into adulthood. Richards and Huppert (2011: 83) concluded that:

> *Children who were rated by teachers as being 'positive' at ages 13 or 15 years were significantly more likely than those who received no positive ratings to report satisfaction with their work in midlife, have regular contact with friends and family, and engage in regular social activities. Positive children were also much less likely to have a mental health problem throughout their lives.*

The work of Richards and Huppert (2011) affirms the potential impact and importance of wellbeing in childhood and the need, therefore, for current policy to focus on it.

# Child public health

Blair *et al.*, (2010: 2) define child public health as:

> *The art and science of promoting and protecting health and wellbeing and preventing disease in infants, children, and young people, through the skills and organized efforts of professionals, practitioners, their teams, wider organizations, and society as a whole.*

Child public health focusses on three key areas:

- **Prevention:** This includes vaccination programmes and education in relation to safe sex.
- **Promotion:** This encourages children and young people to live healthy lives by, for example, taking sufficient exercise, eating an appropriate diet and not smoking. It also promotes the child and young person's overall wellbeing, something that has received an increased worldwide commitment in recent years.

- **Protection:** The aim of this approach is to protect the child population from harm, including from factors existing in the environment such as air pollution.

In other words, child public health involves a multifaceted approach that includes a range of healthcare professionals, as well as those from education, social care and a variety of organizations. However, the role of policy makers, both at a local and national level, is crucial in terms of the promotion and implementation of public health policy.

## Public health policy relating to children and young people

Having clear policies relating to children's health and wellbeing is of paramount importance since it has been suggested that good health provision in the early years of life can benefit later outcomes (Ferri *et al.*, 2003; Muhajarine *et al.*, 2006). Despite the fact that children are entitled to health policy that is appropriate to their needs (Muhajarine *et al.*, 2006), clear child health-focussed policy aims and outcomes have only relatively recently received attention (Kurtz, 2003). When 'New Labour' came to power in the UK in 1997, there was an increased focus on the health of children (Kurtz, 2003) with a range of key papers being produced (e.g. DfES, 2004; DCSF, 2007a,b, 2008; DCSF and DH, 2009).

To facilitate the achievement of policy, the UK government identified a wide range of public service agreements (PSAs) – Glass (2001) commented that the PSAs were incomplete as they did not encapsulate everyone and therefore, it could be argued that they did not 'work'. Traditionally, target setting has been central to public health and goals have normally been identified in areas that can be measured. It could be argued that PSAs have underpinned and driven a deficit approach to health (in other words, waiting for a problem, such as obesity, to arise). Consideration is now being given to asset-based strategies that reflect a more positive approach to health (Whiting *et al.*, 2012).

Glass (2001) commented on the fact that 'what works' for children may be different from what may help families or society at large – he discussed the issue of child poverty and said that the raising of household income could primarily address parental poverty (rather than child poverty). This may have implications for the achievement of the PSA that stated that there will be an end to child poverty in 2020 (HM Treasury, 2002); in fact other issues may play a more direct role than household income (e.g. housing and the environment). It is interesting to note that in the UK in January 2011, the Health in Pregnancy Grant ceased (Directgov, 2011a) and in April 2011 restrictions were placed on eligibility for the Sure Start Maternity Grant (Directgov, 2011b) – these actions could have long-term implications for the health and wellbeing of children.

In the UK, the document, 'Fair Society, Healthy Lives. The Marmot Review' (Marmot, 2010) has the potential to have a significant impact on future UK government health policy. In November 2008, the Secretary of State for Health announced that the review was to be conducted to identify how health inequalities could be addressed and to highlight information that was fundamental to the development of future policies. The review identified six key areas for policy development – two having direct and immediate relevance to children:

- *Give every child the best start in life* (page 9): One of the objectives that this embraced was to *build the resilience and well-being of young children across the social gradient* (page 17).
- *Enable all children, young people and adults to maximise their capabilities and have control over their lives* (page 9).

It has clearly been identified that a high priority should be given to enhancing children's wellbeing and ensuring that children have the best start in life – as *'only then can the close links between early disadvantage and poor outcomes throughout life be broken'* (Marmot, 2010: 14).

The UK general election in May 2010 resulted in the establishment of a new coalition government. Since then, key documents with specific relevance to children have been published; one of the most important is 'Achieving Equity and Excellence for Children' (DH, 2010c). This states that:

> *In order to improve services for children and young people we need a system which works to achieve the outcomes that are important for their health and wellbeing.*
>
> (page 14)

It could be argued that this aim is not wholly attainable unless there is understanding of what is important to children themselves.

In the main, although not exclusively, it is the older age range of children who have been consulted in relation to policy making (Burfoot, 2003; Hill *et al.*, 2004); it is now acknowledged that they are able to offer valuable insight into their lives. In addition, there is wider recognition of the child's rights as a member of society (MacNaughton *et al.*, 2007) – the United Nations Committee on the Rights of the Child, in a General Comment statement (No 7) stated that children's views should be considered in *'the development of policies and services, including through research and consultation'* (Office of the High Commissioner for Human Rights [OHCHR], 2005: 7).

There is some evidence that children have successfully participated in policy development; for example, the HM Treasury and DCSF (2007) established 'Myplace' – this was an initiative that sought to develop new facilities for young people; it stated that all funding bids must identify how children and young people would participate in the project. In another instance, young people were involved in the development of the child version of the Department of Health Drug Strategy (DH, 2003).

However, Davey (2010) in a report, jointly commissioned by The Children's Rights Alliance for England (CRAE), The National Children's Bureau (NCB), The National Participation Forum (NPF), The Office of the Children's Commissioner and Participation Works, suggested that, although there has been a rise in recent years in children's participation in decision making, there is often a failure to involve children in key areas, including health. On a positive note, a Forum, on behalf of the UK DH (2012a), sought the views of children and young people (as well as professionals, parents and carers) in relation to the health outcomes that are most important to them. The Forum asked questions about four key areas of health (acutely ill children; mental health; children with disabilities and long-term conditions; public health): the overall aim was to report the findings to the UK government so that the future Children and Young People's Health Outcomes strategy could be appropriately informed (DH, 2013a).

In summary, there can be no doubt that health policies underpin and influence the lives that children live; therefore, the aim of policy must surely be to enable all children to optimize their potential. It is recognized that the current socioeconomic climate is challenging; therefore, it is more important than ever that the development of health policy is carefully considered to ensure that appropriate decisions are made for both the short and long term – taking children's perspectives into account is an essential aspect of this.

# Promoting child health: the role of the children's nurse

One of the key responsibilities of the nurse is to promote the health and wellbeing of children and young people. This is firmly embedded in England and Wales in the NMC Standards for

## Box 1.2 From the competencies for entry to the register: children's nursing (NMC, 2010).

Children's nurses must:

- *Support and promote the health, wellbeing, rights and dignity of people, groups, communities and populations* (NMC, 2010: 40).
- *Understand public health principles, priorities and practice in order to recognise and respond to the major causes and social determinants of health, illness and health inequalities. They must use a range of information and data to assess the needs of people, groups, communities and populations, and work to improve health, wellbeing and experiences of healthcare; secure equal access to health screening, health promotion and healthcare; and promote social inclusion* (NMC, 2010: 45).
- *Include health promotion, and illness and injury prevention, in their nursing practice. They must promote early intervention to address the links between early life adversity and adult ill health, and the risks to the current and future physical, mental, emotional and sexual health of children and young people* (NMC, 2010: 45).

Pre-registration Nursing Education (NMC, 2010) and therefore a requirement prior to registration (Box 1.2).

In the UK, the DH (2012b) launched an initiative, entitled Make Every Contact Count [MECC]. One of the aims of MECC is to encourage nurses to think about how they can improve public health. The DH (2012b) suggests that:

> *This is a programme with the potential to make a real difference. It's about giving every NHS employee the knowledge, skills and confidence they need to support patients in making healthier life choices. It's an approach that maximises every contact nurses make with the population and any nurse in any setting can apply it.*
>
> (http://cno.dh.gov.uk/2012/04/20/stepping-up-to-the-challenge-how-nurses-can-improve-public-health/)

The MECC approach has been embraced by a number of NHS Trusts across the UK with organizations such a Birmingham Children's Hospital suggesting that their aim will be to embed health promotion into routine practice, ensuring staff are competent and confident in delivering brief advice (National Health Service [NHS], 2012). It could be argued that MECC will be associated with a number of challenges, such as the time required to implement it; the impact that it could have on other roles and responsibilities; the confidence, sensitivity and knowledge required by health professionals. Despite these, it clearly identifies the expectations of nurses in terms of the promotion of the public's health – children's nurses are no exception to this and should therefore consider how this role can be integrated into their everyday work.

# Morbidity and mortality

Mortality refers to the number of deaths within a given population and geographical area, whereas morbidity measures the prevalence of disease or illness for a particular population within a certain area. The causes of childhood mortality and morbidity have changed considerably over the last century, with infectious diseases, for example, now causing fewer childhood deaths in the UK; in fact *'More children with serious illnesses and disabilities are surviving into adulthood and the infant mortality rate has fallen to less than a quarter of what it was at the beginning of the 1960s.'* (DH, 2013b: 2). Despite this, the government recognizes that:

> *It is a shocking fact that child mortality in Britain is the worst when compared to other similar European countries. There is unacceptable variation across the country in the quality of care for children – for example in the treatment of long-term conditions such as asthma and diabetes.*
>
> (Poulter, 2013; https://www.gov.uk/government/news/new-national-pledge-to-improve-children-s-health-and-reduce-child-deaths)

There is evidence that both pregnancy and the early childhood years are fundamental to the future health and wellbeing of children, with early interventions leading to a further reduction in childhood morbidity and mortality (DH, 2013a). The current UK government acknowledges that reducing childhood mortality is complex and needs a range of public health interventions (DH, 2013a). Valuable information, particularly in relation to childhood mortality is kept by the Office for National Statistics [ONS] However, this data is limited and provides little detail about the types of illnesses that children in the 21st century could die from; as a result, in 2011, the RCPCH was funded by the National Patient Safety Agency (NPSA) to conduct a review of mortality and morbidity in children and young people between the ages of 1 and 18 years. In addition, the UK government has invited a range of key organizations, such as the Royal College of Nursing, Public Health England and the RCPCH to sign a pledge to demonstrate their commitment to the improvement of health outcomes for children and young people so that they become amongst the best worldwide (DH, 2013b). The following chapters will provide information that will help you to achieve this goal.

# Conclusion

Children are key members of our society and are the future of our nation; it is therefore imperative that they are consulted about decisions that may impact on both themselves and their wider community. Recognizing and valuing the contribution that children and young people can make will serve to enrich the society in which we all live; however, at the same, children and young people remain a vulnerable group and it is imperative that healthcare professionals continue to strive to enhance children's health and wellbeing – both through care provision and the development of child health policy that is not only evidence based, but also informed, wherever possible, by children themselves. This approach is undoubtedly challenging and arguably time consuming, but it is not something that should be shied away from as it has the potential to facilitate the health of not just of our children and young people, but also the future adult population.

# Glossary

**Child:** Anyone below the age of 18 years.

**Family:** A group of people who may be (but not necessarily) affiliated by consanguinity and/or co-residence.

**Health:** An abstract concept that incorporates physical, emotional and social wellbeing. The concept means different things to different people.

**Health promotion:** The term used to embrace the extensive range of approaches used to enhance the health of people, communities and populations.

**Morbidity:** Refers to the incidence of ill health caused by disease in a particular population.

**Mortality:** Refers to the number of deaths during a particular timeframe and/or geographical area from a specific cause – can be referred to as *mortality rate.*

**Public health:** The promotion of health and wellbeing, and the prevention of disease, through the work of professionals, organisations and society as a whole.

**Wellbeing:** A subjective concept that describes a state of feeling good, doing well, being happy and being able to carry out meaningful and engaging activities. A person's feeling of wellbeing may vary across their lifespan as well as on a daily basis.

**Don't forget to visit the companion website for this book (www.wileyfundamentalseries.com/childrenA&P) where you can find self-assessment tests to check your progress, as well as lots of activities to practise your learning.**

## References

Ariès, P. (1962) *Centuries of Childhood*, Cape, London.

Blair, M., Stewart-Brown, S., Waterston, T., Crowther, R. (2010) *Child Public Health*, Open University Press, Oxford.

Botelho, R., Fiscella, K. (2005) Protect children from environmental tobacco smoke, but avoid stigmatization of parents: a commentary on Pyle *et al. Families, Systems and Health: The Journal of Collaborative Family Health Care*, **23** (1): 17–20.

Bronfenbrenner, U. (1979) *The Ecology of Human Development*, Harvard University Press, Cambridge, MA.

Burfoot, D. (2003) Children and young people's participation: Arguing for a better future. *Youth Studies Australia*, **22** (3): 44–51.

Carnegie UK Trust (2008) *Final Report of the Carnegie Young People Initiative. Empowering Young People*, Carnegie UK Trust, Fife.

Children Act (1989) The Stationery Office: London.

Children Act (2004) Available from: http://www.legislation.gov.uk/ukpga/2004/31/part/1 Accessed on 21st May 2013.

Cillessin, A.H.N., Jiang, X.L., West, T.V., Laszkowski, D.K. (2005) Predictors of dyadic friendship quality in adolescence. *International Journal of Behavioral Development*, **29** (2): 165–172.

*Collins Dictionary and Thesaurus* (2000) Harper Collins, Aylesbury.

Counterpoint Research (2008) *Childhood Wellbeing. Qualitative Research Study*, Department for Children, Schools and Families, London.

Cunningham, H. (2006) *The Invention of Childhood*, BBC Books, London.

Davey, C. (2010) *Children's Participation in Decision-Making. A Summary Report on Progress Made up to 2010*, Participation Works, London.

Department for Children, Schools and Families (2007a) *The Children's Plan. Building futures*, The Stationery Office, London.

Department for Children, Schools and Families (2007b) *Aiming High for Young People: A Ten Year Strategy for Positive Activities*, The Stationery Office, London.

Department for Children, Schools and Families (2008) *PE and Sport Strategy for Young People* [online]. Available from: https://www.education.gov.uk/publications/standard/Physicaleducation/Page1/DCSF-00131-2008 Accessed on 21st May 2013.

Department for Children, Schools and Families and Department of Health (2009) *Healthy Lives, Brighter Futures – The Strategy for Children and Young People's health*, Department of Health, London.

Department for Education (2011) *Healthy Schools*. Available from: http://www.education.gov.uk/schools/pupilsupport/pastoralcare/a0075278/healthy-schools Accessed on 21st May 2013.
Department for Education and Skills (2004) *Every Child Matters: Next Steps*, Department for Education and Skills, London.
Department for Education and Skills and Department of Health (2004) *National Services Framework for Children, Young People and Maternity Services: Core Standards*, The Stationery Office, London.
Department of Health (2003) *Listening, Hearing and Responding. Department of Health Action Plan 2003/04*, Department of Health, London.
Department of Health (2010a) *Healthy Lives, Healthy People. Our Strategy for Public Health in England*, The Stationery Office, London.
Department of Health (2010b) *Our Health and Wellbeing Today*, The Stationery Office, London.
Department of Health (2010c) *Achieving Equity and Excellence for Children*, The Stationery Office, London.
Department of Health (2012a) *Have Your Say on Children and Young People's Health Outcomes*. http://healthandcare.dh.gov.uk/children-say/ Accessed on 21st May 2013.
Department of Health (2012b) *Chief Nursing Officer Bulletin. Stepping Up to the Challenge: How Nurses can Improve Public Health*. Available at: http://cno.dh.gov.uk/2012/04/20/stepping-up-to-the-challenge-how-nurses-can-improve-public-health/ Accessed on 21st May 2013.
Department of Health (2013a) *Improving Children and Young People's Health Outcomes: A System Wide Approach*. Available from: https://www.gov.uk/government/publications/national-pledge-to-improve-children-s-health-and-reduce-child-deaths Accessed on 21st May 2013.
Department of Health (2013b) *Better Health Outcomes for Children and Young People. Our Pledge*. Available from: https://www.gov.uk/government/news/new-national-pledge-to-improve-children-s-health-and-reduce-child-deaths Accessed on 21st May 2013.
Directgov (2011a) *Health in Pregnancy Grant*. Available from: http://www.hmrc.gov.uk/hipg/ Accessed on 21st May 2013.
Directgov (2011b) *Sure Start Maternity Grant*. Available from: http://www.direct.gov.uk/en/MoneyTaxAndBenefits/BenefitsTaxCreditsAndOtherSupport/Expectingorbringingupchildren/DG_10018854 Accessed on 21st May 2013.
Erikson, E.H. (1950). *Childhood and Society*, Norton, New York.
Eriksson, U., Asplund, K., Sellström, W. (2010) Growing up in rural community – children's experiences of social capital from perspectives of wellbeing. *Rural and Remote Health*, **10** (1322): 1–12 Available from: http://www.rrh.org.au/articles/subviewnew.asp?ArticleID=1322 Accessed on 21st May 2013.
European Parliament, Council of the European Union and European Commission (2000) *Charter of Fundamental Rights of the European Union. Official Journal of European Communities* C364: Nice.
Fattore, T., Mason, J., Watson, E. (2009) When children are asked about their well-being: Towards a framework for guiding policy. *Child Indicators Research*, **2** (1): 57–77.
Ferri, E. Bynner, J., Wadsworth, M. (2003) *Changing Britain, Changing Lives. Three Generations at the Turn of the Century*, Institute of Education, University of London, London.
Frønes, I. (1993) Editorial: Changing childhood. *Childhood*, **1** (1): 1–2.
Glass, N. (2001) What works for children – the political issues. *Children and Society*, **15** (1): 14–20.
Gritz, E.R., Tripp, M.K., James, A.S., *et al.* (2005) An intervention for parents to promote preschool children's sun protection: effects of Sun Protection is Fun! *Preventive Medicine*, **41** (2): 357–366.
Hendrick, H. (1997) Constructions and reconstructions of British childhood: An interpretive survey, 1800 to the present day. In James, A., Prout, A. (eds) *Constructing and Reconstructing Childhood: Contemporary Issues in the Sociological Study of Childhood*, Routledge Falmer, London.
Hill, M., Davis, J., Prout, A., Tisdall, K. (2004) Moving the participation agenda forward. *Children and Society*, **18** (2): 77–96.
Holt, J. (1975) *Escape from Childhood*, Penguin, Harmondsworth.
HM Treasury (2002) *Spending Review: Public Service Agreements White Paper*, HM Treasury, London.
HM Treasury and DCSF (2007) *Aiming High for Young People: A Ten Year Strategy for Positive Activities*, Her Majesty's Treasury and Department for Children, Schools and Families, London.
Human Rights Act (1998) The Stationery Office: London.
Hutchison, E.D. (2011) (ed) *Dimensions of Human Behaviour. The Changing Lifecourse*, 4th edn. Sage Publications, Los Angeles, California.
Ipsos Mori and Nairn, A. (2011) *Children's Well-being in UK, Sweden and Spain: The Role of Inequality and Materialism*, Ipsos Mori, London.
James, A., Prout, A. (1997) (eds) *Constructing and Reconstructing Childhood: Contemporary Issues in the Sociological Study of Childhood*, Routledge Falmer, London.

Kirby, P. (2004) *A Guide to Actively Involving Young People in Research: For Researchers, Research Commissioners, and Managers*. INVOLVE, Hampshire.

Kochanska, G., Aksan, N., Carlson, J.J. (2005) Temperament, relationships, and young children's receptive cooperation with their parents. *Developmental Psychology*, **41** (4): 648–660.

Kohlberg, L. (1984) *Essays on Moral Development. Vol. 2, The Psychology of Moral Development: the Nature and Validity of Moral Stages*, Harper Row, New York.

Kurtz, A. (2003) Outcomes for children's health and well-being. *Children and Society*, **17** (3): 173–183.

MacNaughton, G., Hughes, P., Smith, K. (2007) Young children's rights and public policy: Practices and possibilities for citizenship in the early years. *Children and Society*, **21** (6): 458–469.

Marmot, M. (2010) *Fair Society, Healthy Lives. The Marmot review. Executive summary*, The Marmot Review, London.

Mathur, A., Kamat, D. (2005) Travel risks: how to help parents protect infants and young children. *Consultant*, **45** (8): 900–902.

Muhajarine, N., Vu, L., Labonte, R. (2006) Social contexts and children's health outcomes: Researching across boundaries. *Critical Public Health*, **16** (3): 205–218.

National Health Service (2012) *Making Every Contact Count. Examples from Practice*, NHS, London.

Nursing and Midwifery Council (2010) *Standards for Pre-registration Nursing*, NMC, London.

Office for National Statistics (2012) *Divorces in England and Wales – 2011*. Available from: http://www.ons.gov.uk/ons/rel/vsob1/divorces-in-england-and-wales/2011/stb-divorces-2011.html#tab-Key-findings Accessed on 21st May 2013.

Office of the High Commissioner for Human Rights [OHCHR] (2005) *General Comment No. 7 (2005: 01.11.05 Implementing Child Rights in Early Childhood)*. Available from: http://www.unhchr.ch/tbs/doc.nsf/(Symbol)/7cfcdace016acd61c1257214004713a4?Opendocument Accessed on 21st May 2013.

Palmer, S. (2006) *Toxic Childhood. How the Modern World is Damaging our Children and What we Can do About it*, Orion, London.

Participation Works (2007) *How to use creative methods for participation*, National Children's Bureau, London.

Parry, O., Warren, E., Madoc-Jones, I., *et al.* (2010) *Voices of Children and Young People in Wales Study: A Qualitative Study of Wellbeing Among Children and Young People under 25 Years of Age*, Welsh Assembly Government Social Research, Cardiff. Available from: http://wales.gov.uk/about/aboutresearch/social/latestresearch/cypwellbeingmonitor/?lang=en Accessed on 21st May 2013.

Piaget, J. (1952). *The Origins of Intelligence in Children*, International University Press, (Original work published 1936): New York.

Pogue, D. (2005) Parents report: surfing the web, spam, typos, protect your child. *Parents*, **80** (6): 148–151.

Poulter, D. (2013) *New National Pledge to Improve Children's Health and Reduce Child Deaths*. Available from: https://www.gov.uk/government/news/new-national-pledge-to-improve-children-s-health-and-reduce-child-deaths Accessed on 21st May 2013.

Prout, A., Hallett, C. (2003) Introduction. In: Hallett, C., Prout, A. (eds) *Hearing the Voices of Children. Social Policy for a New Century*, Routledge Falmer, London.

Prout, A., James, A. (1997) A new paradigm for the sociology of childhood? Provenance and problems. In James, A. and Prout, A. (eds) *Constructing and Reconstructing Childhood. Contemporary Issues in the Sociological Study of Childhood*, 2nd edn. Falmer Press, London.

Rees, G., Bradshaw, J., Goswami, H., Keung, A. (2010) *Understanding Children's Well-being: A National Survey of Young People's Well-being*, The Children's Society, London.

Richards, M., Huppert, F.A. (2011) Do positive children become positive adults? Evidence from a longitudinal birth cohort study. *Journal of Positive Psychology*, **6** (1): 75–87.

Rigg, A., Pryor, J. (2007) Children's perceptions of families: What do they really think? *Children and Society*, **21** (1): 17–30.

Royal College of Paediatrics and Child Health (2010) *Safeguarding Children and Young People: Roles and Competences for Health Care Staff. Intercollegiate Document*, Royal College of Paediatrics and Child Health, London.

Rutherford, D. (1998) Children's relationships. In: Taylor, J., Woods, M. (eds) *Early Childhood Studies. An Holistic Introduction*, Arnold, London.

Ryff, C.D., Singer, B. (1998) The contours of positive human health. *Psychological Inquiry*, **9** (1): 1–28.

Ryff, C.D., Singer, B.H., Love, G.D. (2004) Positive health: Connecting well-being with biology. *Philosophical Transactions of the Royal Society of Biological Sciences*, **359** (1449): 1383–1394.

Sellstrom, E., Bremberg, S. (2006) The significance of neighbourhood context to child and adolescent health and wellbeing: A systematic review of multilevel studies. *Scandinavian Journal of Public Health*, **34** (5): 544–554.

Sinclair, R. (2004) Participation in practice: Making it meaningful, effective and sustainable. *Children and Society*, **18** (2): 106–118.

Smith, R. (2000) Order and disorder: The contradictions of childhood. *Children and Society*, **14** : 3–10.

The National Youth Agency (2007) *Involving Children and Young People – An Introduction*, The National Youth Agency, Leicester.

Thompson, R.A. (2006) The development of the person: Social understanding, relationships, conscience, self. In: Damon, W., Lerner, R.M. (Serieseds) *Handbook of Child Psychology* (Volume ed: Eisenberg, N.) *Social, Emotional, and Personality Development*. Wiley, New York, pp. 348–365.

Troutman, D.R., Fletcher, A.C. (2010) Context and companionship in children's short-term versus long-term friendships. *Journal of Social and Personal Relationships*, **27** (8): 1060–1074.

UNICEF (2007) *Child Poverty in Perspective: An Overview of Child Well-being in Rich Countries, Innocenti Report Card 7*, UNICEF Innocenti Research Centre, Florence, Italy.

UNICEF (2013) *Child Well-being in Rich Countries. A Comparative Overview. Innocenti Report Card 11*, UNICEF Office of Research, Florence, Italy. Available from: http://www.unicef.org.uk/Latest/Publications/Report-Card-11-Child-well-being-in-rich-countries/ Accessed on 23rd May 2013.

United Nations (1998) *Fiftieth Anniversary of the Universal Declaration of Human Rights*, United Nations, Geneva.

United Nations Convention on the Rights of the Child (1989) Available from: http://www.unicef.org.uk/UNICEFs-Work/Our-mission/UN-Convention/ Accessed on 21st May 2013.

Vélez, C.E., Wolchik, S.A., Tein, J.Y., Sandler, I. (2011) Protecting children from the consequences of divorce: A longitudinal study of the effects of parenting on children's coping processes. *Child Development*, **82** (1): 244–257.

Whiting, L., Kendall, S., Wills, W. (2012) An asset-based approach: an alternative health promotion strategy? *Community Practitioner*, **85** (1): 25–28.

Whiting, L., Kendall, S., Wills, W. (2013) Rethinking children's public health: the development of an assets model. *Critical Public Health* Published on-line at http://dx.doi.org/10.1080/09581596.2013.777694 Accessed on 16th April 2013.

World Health Organization (1948) *Preamble to the Constitution of the World Health Organization as Adopted by the International Health Conference, New York, 19-22 June, 1946; Signed on 22 July 1946 by the Representatives of 61 States (Official Records of the World Health Organization, no. 2, p. 100) and Entered into Force on 7 April 1948*. Available from: http://www.who.int/suggestions/faq/en/index.html Accessed on 21st May 2013.

World Health Organization (2005) *The European Health Report 2005. Public Health Action for Healthier Children and Populations. Executive Summary*, WHO, Copenhagen.

# Chapter 2

# Homeostasis

Mary Brady

*Faculty of Health, Social Care and Education, Kingston University, London, UK*

## Aim

The aim of this chapter is to help you to develop insight and understanding of homeostatic mechanisms within the bodies of children and young people (0–18 years of age). By gaining further understanding and developing your insight you will be able to provide high quality safe and effective informed care.

## Learning outcomes

On completion of this chapter the reader will be able to:

- Describe homeostasis and how it is regulated.
- Describe what is meant by feedback mechanisms.
- Describe the homeostatic mechanisms.
- Describe how energy is produced.

## Test your prior knowledge

- What is homeostasis?
- How does the body receive feedback to ensure equilibrium is maintained within the body?
- What influence does the central nervous system have on homeostasis?
- What hormones are involved in homeostasis?
- How does the body maintain water balance?
- How does the body maintain its balance of salts?
- How does the body maintain its balance of oxygen and carbon dioxide?
- What is the stimulus for breathing in the newborn?
- What is the stimulus for breathing in the older child or young person?
- What are differences between the essential and non-essential amino acids?

*Fundamentals of Children's Anatomy and Physiology: A Textbook for Nursing and Healthcare Students*, First Edition. Edited by Ian Peate and Elizabeth Gormley-Fleming.

Companion website: www.wileyfundamentalseries.com/childrensA&P

# Introduction



Within the body, several mechanisms exist to ensure that the internal environment remains within a narrow set of parameters, regardless of the external environment. This process is called homeostasis and illness occurs when there is a disruption to this normal homeostatic control. It is often when homeostasis is disrupted that holistic nursing care is required. In order to do this well, the nurse must have a good understanding of the homeostatic mechanisms involved and how these can become impaired.

The child nurse needs to understand the physiological differences of the various age groups and how nursing interventions can help monitor, interpret and treat imbalances proactively to limit long-term damage. Since caring for the family and imparting knowledge is such an integral part of caring for the sick child, the child nurse needs to be able to help the family to understand often quite complex conditions.

This chapter will use the term 'child' to refer to those in the 0–18-year age range, except where specific information is provided for neonates and premature infants. It will also include details regarding:

- Homeostasis and how it is regulated.
- Feedback mechanisms.
- Homeostatic mechanisms.
- Energy production.
- Systematic approach to homeostasis.

# Regulation of homeostasis

Homeostasis is a dynamic process and to ensure that it is maintained, the body receives messages via a feedback system, where specialist sensor cells monitor for changes in the levels of oxygen, carbon dioxide, glucose, electrolytes, temperature, urea and water. Frequently the maintenance of homeostasis is controlled by more than one mechanism; for example, water balance within the body is controlled by a variety of receptors, hormones and reactions within organs. Any changes that result in levels that fall outside of the normal parameters are detected by a feedback mechanism.

## Feedback mechanisms

Most feedback systems are negative: a mechanism is switched off. Positive feedback systems ensure that an action continues and increases and therefore could be harmful, for example if body temperature exceeds 41 °C, the result is hyperthermia and heat stroke; vasodilation of the skin occurs with mental confusion and loss of muscular coordination. If the temperature continues to rise to 43 °C, death usually occurs. An example of a positive feedback system is seen in child birth as contractions need to continue.

Mulryan (2011) lists the following specialist sensor cells:

- Chemoreceptors which monitor chemical concentrations.
- Baroreceptors which monitor pressure.
- Osmoreceptors which monitor pressure or the amount of water present within the body.
- Thermoreceptors which monitor body temperature.

When an abnormality has been detected by a sensor, a message is sent, either by an electrical impulse from the central nervous system (CNS) or by the release of hormones that travel in the circulatory system to their destination, to restore normal limits.

# Homeostatic mechanisms



Each cell has a semipermeable membrane, made of lipids and proteins, which enables certain molecules and water to enter and leave the cell. Molecules can pass through this membrane by diffusion, facilitated diffusion and active transport. Other forces also assist with these mechanisms, such as hydrostatic, osmotic and colloid osmotic pressures. Finally, exocytosis, receptor-mediated endocytosis and pinocytosis are processes that also assist with the movement of fluids and particles into and out of the cells (Clancy and McVicar, 2009).

## Diffusion facilitated diffusion and active transport

Diffusion is the movement of small molecules through the semipermeable membrane from a high concentration to a low concentration, for example the movement of oxygen from the blood into a cell. Facilitated diffusion is the movement of large molecules such as glucose or amino acids either through specific gaps in the proteins of the membrane or by binding to a specific protein on one side of the cell membrane which then releases it on the other side (Clancy and McVicar, 2009). Active transport is the movement of substances such as sodium, potassium, magnesium and calcium from a low concentration to a high concentration; this process requires energy whereas diffusion and facilitated diffusion do not.

## Hydrostatic, osmotic and colloid osmotic pressures

Humans are relatively large mammals with a complex network of body systems interconnected by the circulatory system. The heart pumps blood around the body, causing a hydrostatic pressure that is higher on the arterial side of the circulation compared with the venous side; however, the osmotic pressure remains constant. This enables nutrients and water to be forced out into the interstitial fluid and cells on the arterial capillary side of the circulatory system and waste products to be taken up by cells on the venous side.

Water will move to a high solute concentration by osmosis unless prevented from doing so by applying pressure to the concentrated solution. The blood pressure exerts pressure within the blood vessels to force fluid out of them. The opposing force which draws fluid into the blood vessels is termed the colloid osmotic pressure of the fluid. This is expressed in the same units (millimetres of mercury or mmHg) as hydrostatic pressure and is maintained by plasma proteins which normally remain within the blood vessels due to their size in relation to the gap in the cell wall.

## Exocytosis, receptor-mediated endocytosis and pinocytosis

Exocytosis is a process where vesicles within the cell join with the cell membrane in order to release their contents out of the cell. Receptor-mediated endocytosis and pinocytosis are two different processes where the cell wall invaginates to bring substances into the cell. In receptor-mediated endocytosis, the receptors on the cell membrane bind with specific chemicals outside the cell, such as thyroxine. The cell then folds in on itself to form a vesicle called an endosome; the receptors are then released to return to the cell membrane. Pinocytosis is where the cell folds in on itself to create a small vesiclecalled a pinosome, containing fluid or large molecules such as proteins, which are digested by the cell.

Chapter 4 of this text discusses the cell in more detail.

## Osmoregulation

Osmosis is a process where water (a solvent) moves across a semipermeable membrane. Substances dissolved in the water (solutes) determine its movement from an area of low solute concentration to a higher concentration (Clancy and McVicar, 2009). Sodium is important in the osmotic balance within the body, since water will move across a semipermeable membrane from an area of low concentration of sodium to an area of high concentration.

A solution can only exert an osmotic pressure when it is separated from another solution via a membrane that is permeable to the solvent but not to the solute. Any solutes to which the membrane is permeable will also move with the osmotic flow of water. Their concentrations will not be changed by osmosis.

Fluid with a low osmotic pressure has a lower concentration of particles dissolved in it. Colloid osmotic pressure is an osmotic pressure that is exerted by proteins in the plasma; it acts in opposition to hydrostatic pressure. It is also known as oncotic pressure (Nair, 2011a). Approximately 70% of the total colloid osmotic pressure exerted by blood plasma on interstitial fluid is generated by the presence of high concentrations of albumin. The total colloid osmotic pressure of an average capillary is between 25 and 30 mmHg with albumin contributing approximately 22 mmHg of this. Normally blood proteins cannot escape through intact capillary endothelium; therefore, the colloid osmotic pressure of the capillaries tends to draw water into the blood vessels (Nair, 2011a). However, renal conditions and malnutrition can lead to loss of these proteins with subsequent reduced colloid osmotic pressure, causing fluid to move into the tissues and resulting in oedema.

# Energy production

Energy is vital for bodily function and is produced in the form of adenosine triphosphate (ATP), through cellular respiration, the Krebs cycle, glycolysis and the electron transport chain. Since ATP cannot be stored, the body relies on its synthesis from carbohydrates, fats or proteins as required. Indeed, without ATP cell death occurs with ensuing disruption to homeostasis.

## Cellular respiration

Initially food is digested by enzymes within the digestive tract into smaller components; then insulin, a hormone secreted by the beta cells within the pancreas, facilitates the passage of glucose into the cells. In the presence of oxygen, glycolysis occurs where glucose is aerobically converted to ATP; in the absence of oxygen, lactic acid is produced instead.

Energy is most efficiently produced from glucose (glycolysis; Figure 2.1a). If glucose is not available, energy can be generated by the oxidation of fatty acids within the liver (gluconeogenesis), but only if oxygen is present. Some cells, such as nerve cells, are unable to synthesize energy from fats, so they rely on ketones, also produced in the liver from fatty acids. However, ketones are acidic and alter the acid–base balance within the body with toxic effects on the cells.

## Krebs cycle

The Krebs cycle is a cycle of aerobic reactions that occur within the mitochondria where carbon dioxide and energy as ATP are released (Figure 2.1b). Two coenzymes called nicotinamide adenine dinucleotide ($NAD^+$) and flavine adenine dinucleotide (FAD) are involved in transferring the energy in the form of electrons to the electron transport chain and during the process are reduced to NADH and $FADH_2$.

In the mitochondria of the cell, pyruvic acid and hydrogen are released as well as carbon dioxide. The waste products are excreted by the body as water and carbon dioxide.

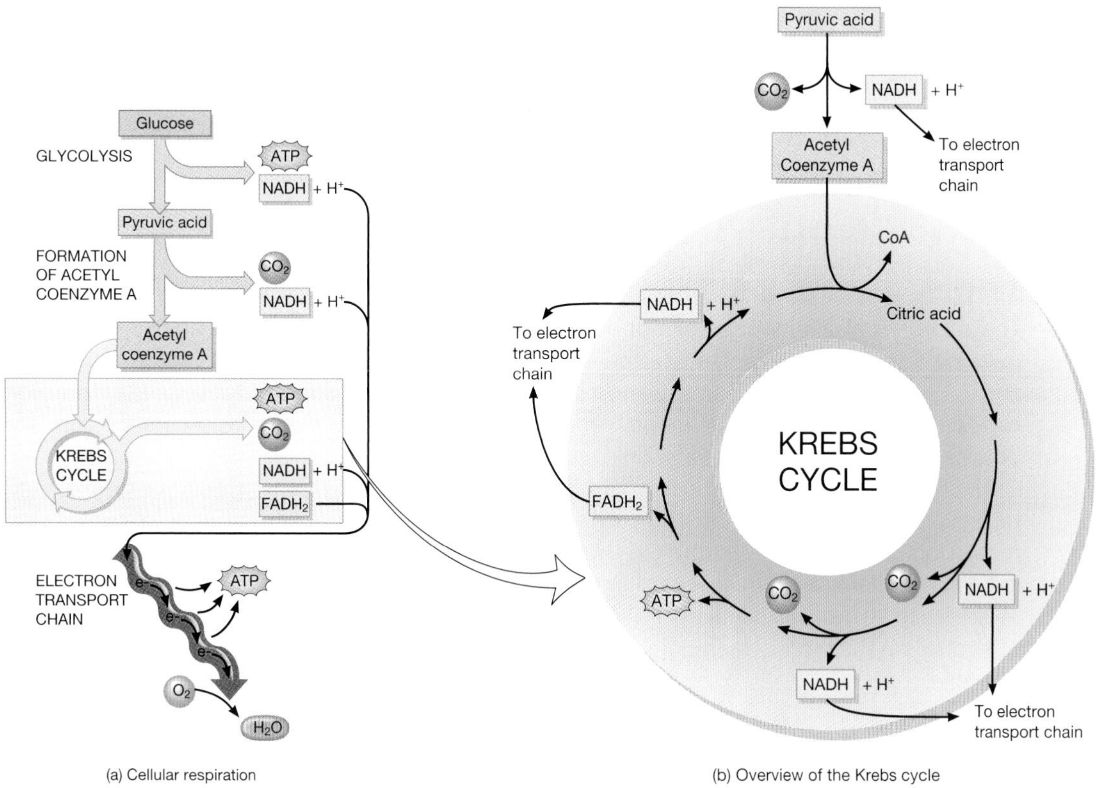

Figure 2.1 (a) Cellular respiration. (b) Overview of Krebs cycle.
*Source:* Tortora and Derrickson 2009, Figure 26.6, p. 984. Reproduced with permission of John Wiley and Sons, Inc.

In the absence of oxygen, pyruvic acid combines with hydrogen to form lactic acid which diffuses out of the cell into the blood and is converted back to pyruvic acid; this results in the familiar feeling of muscle fatigue.

Fatty acids are directly converted into acetyl coenzyme A and enter the Krebs cycle, whereas glycerol must be converted to pyruvic acid prior to being converted to acetyl coenzyme A. Some amino acids can undergo either process prior to entering the Krebs cycle.

At times of illness, the body needs more energy, but is unable to provide a sufficient amount, which results in an imbalance so the body will endeavour to conserve energy by closing down some of its functions (e.g. by vasoconstriction) and will rely more heavily on the breakdown of fat for energy production.

## Glycolysis

During glycolysis one molecule of glucose is converted to eight molecules of ATP and a substance called pyruvic acid is formed Pyruvic acid then enters the mitochondria where it is converted to acetyl coenzyme A. This then enters the Krebs cycle which enables more energy to be produced: 30 molecules of ATP are produced.

## Electron transport chain

In the membranes of the mitochondria electrons are released in the presence of oxygen and transferred through a series of chemical reactions involving cytochromes. This process is known as the electron transport chain.

# Systematic approach to homeostasis

## Respiratory system and acid–base balance

The acidity or alkalinity of a solution is measured in terms of the concentration of hydrogen ions, expressed as the pH. The plasma is slightly alkaline with a pH of 7.32–7.4 for a neonate and 7.35–7.45 for a child (Ramsay and Moules, 2008). The acid–base balance of the body can be assessed using a sample of blood. Venous blood can be used but more accurate results are obtained from arterial samples.

All chemical reactions that occur within the body produce acids that need to be excreted to prevent their accumulation and subsequent toxic effects on the body. Their excretion and regulation of acid–base is controlled by two organs – the lungs and kidneys – and a buffering system. A buffer is a chemical compound dissolved in a solution which can 'mop up' hydrogen ions released when an acid dissociates. Bicarbonate is the buffer that forms when carbonic acid dissociates in the blood. Haemoglobin also acts as a buffer when oxygen is displaced by hydrogen ions.

The child or young person breathes in air via the respiratory tract and gases are exchanged at the alveolar level via diffusion. The inspired air contains a higher concentration of oxygen (13.3 kPa), so this diffuses across the alveolar membrane into the blood cells where it oxidizes the haemoglobin molecule to produce oxyhaemoglobin. The oxygen is released when the oxyhaemoglobin reaches a cell with a low oxygen concentration. The reverse happens with carbon dioxide; it is in a high concentration within the cell and thus diffuses into the red blood cell and under the influence of carbonic anhydrase forms carbonic acid by combining with water. Carbonic acid is a weak acid and dissociates to produce hydrogen ions and bicarbonate. The hydrogen ions attach to deoxygenated haemoglobin. A small portion of oxygen and carbon dioxide is also transported within the plasma.

Carbon dioxide ($CO_2$) is a by-product of carbohydrate metabolism. As already mentioned, it exists combined with water as carbonic acid ($H_2CO_3$) and is mainly exhaled via the lungs. When difficulty in breathing occurs, such as when blood is unable to travel from the heart to the lungs as efficiently as expected due to an abnormality of cardiac outflow or the movement of oxygen to the circulatory system at alveolar level, carbonic acid accumulates causing the plasma to become acidic as it is a weak acid that dissociates to produce hydrogen ions ($H^+$) and bicarbonate ion ($HCO_3^-$).Both arms of the following equation are reversible:

$$H^+ + HCO_3^- \rightleftharpoons H_2CO_3 \rightleftharpoons H_2O + CO_2$$

The respiratory centre is located in the medulla oblongata and in the healthy child the chemoreceptors in the medulla oblongata are stimulated by a rising level of carbon dioxide or hydrogen ions in the cerebrospinal fluid. Peripheral chemoreceptors in the carotid sinus and the aortic arch also detect rising carbon dioxide (hypercapnia) and hydrogen ions and lowered oxygen levels (Watson and Fawcett, 2003). This initiates tachypnoea (faster, deeper breathing) to remove the excess carbon dioxide. When the level of circulating carbon dioxide or hydrogen ions is low,

Table 2.1 Normal blood gas values. *Source:* Adapted from Higgins (2007) and Moules and Ramsay (2008).

| | Neonate | | Child | | Adult | |
|---|---|---|---|---|---|---|
| pH | 7.31–7.47 | | 7.35–7.42 | | 7.35–7.45 | |
| $PCO_2$ | 3.8–6.5 kPa | 28.5–48.7 mmHg | 4.0–5.5 kPa | 30–41.3 mmHg | 4.7–6.0 kPa | 35.3–45 mmHg |
| $PO_2$ | 4.3–8.1 kPa | 32.3–60.8 mmHg | 11–14 kPa | 82.5–105 mmHg | 10.6–13.3 kPa | 79.5–99.8 mmHg |
| $HCO_3$ | 15–25 mmol/L | | 17–27 mmol/L | | 22–28 mmol/L | |



the respiratory centre is not stimulated and breathing is slower and shallower; carbon dioxide will build up and return the blood pH to normal values. In the newborn and those with chronic cardiac or respiratory problems, such as cystic fibrosis, the chemoreceptors are less sensitive to blood carbon dioxide levels and instead low levels of oxygen (hypoxia) act as the stimulus to breathe.

The majority of oxygen (98.5%) is carried around the body attached to haemoglobin (oxyhaemoglobin) in the red blood cell (erythrocyte) and each haemoglobin molecule has the capacity to bind four oxygen molecules (Waugh and Grant, 2010). When a haemoglobin molecule has bound four oxygen molecules it is described as saturated. The remaining 1.5% of oxygen circulates dissolved in plasma (Wheeldon, 2011).

Sampling of blood via an artery or capillary reveals the amount of gases present (Table 2.1). These are expressed as partial pressure of oxygen ($PO_2$) and carbon dioxide ($PCO_2$) in kiloPascals (kPa) and relate to the concentration of these gases in the blood sample; previously these were expressed as millimetres of mercury (mmHg) (1 kPa = 7.5 mmHg).

The amount of oxygen bound to haemoglobin can also been expressed as a percentage using a pulse oximeter and the readings plotted on an oxyhaemoglobin dissociation curve. However, the readings (and therefore the curve) can alter depending on the individual's pH level, carbon dioxide level and temperature (Figures 2.2a–c). It is worth remembering that fetal haemoglobin (which is gradually replaced in the first year of life) has a higher affinity for oxygen than adult haemoglobin. So, while in healthy children the oxygen saturation reading is usually between 95% and 99% (Peate and Nair, 2011), in premature neonates caution needs to be observed in deciding the acceptable oxygen saturation level (Neill and Knowles, 2004) as they are sensitive to hyperoxia which has been linked with the development of retinopathy of prematurity (Lloyd *et al.*, 2009); lower levels of 88–95% may be acceptable (Lissauer and Clayden, 2007). Pulse oximeters should always be used in conjunction with blood gas analysis. It is also worth noting that pulse oximetry readings can be affected by bright sunlight, phototherapy and nail polish.

Chapter 10 discusses the respiratory system in more detail.

## Gastrointestinal system

The body of the premature infant (1 kg) consists of up to 85% water, but this percentage reaches adult values by about 1 year of age (Lissauer and Clayden, 2007). The amount of protein and fat within the body builds up over the first year of life until it reaches adult ratios. During the first year of life there is a period of rapid growth where 30% of the energy intake is required for growth and by 1 year of age, two-thirds of the basal metabolic rate is due to brain activity and

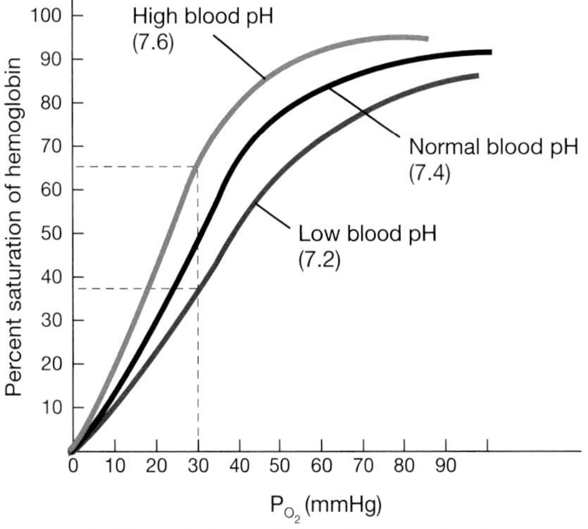

(a) Effect of pH on affinity of hemoglobin for oxygen

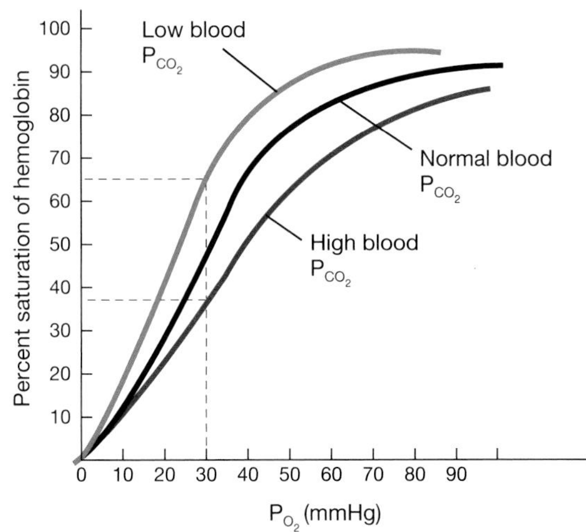

(b) Effect of $P_{CO_2}$ on affinity of hemoglobin for oxygen

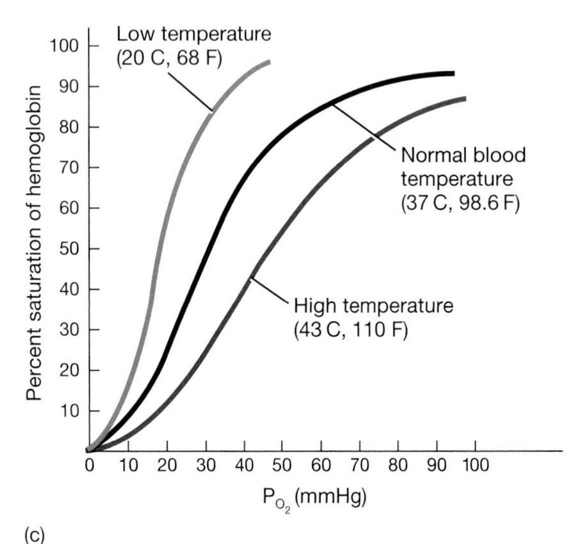

(c)

**Figure 2.2** Oxygen–haemoglobin dissociation curves showing the effect of (a) pH on affinity of haemoglobin for oxygen, (b) $PCO_2$ on affinity of haemoglobin for oxygen and (c) Temperature changes.
*Source:* Tortora and Derrickson 2009, Figures 23.20 and 23.21, pp. 902 and 903. Reproduced with permission of John Wiley and Sons, Inc.

development (Lissauer and Clayden, 2007). So, it follows that inadequate nourishment during this period will have a major impact on growth and development.

## *Amino acid metabolism*

Proteins are absorbed by the body following their digestion into amino acids, of which eight are essential in childhood and need to be obtained from the diet. The remaining amino acids can be synthesized by the liver by transamination. However, in the newborn this ability

Table 2.2 Essential and non-essential amino acids in children. *Source:* Adapted from Waugh and Grant (2010) and Cedar (2012).

| Essential | Synthesized by the body in inadequate quantities in childhood | Non-essential |
|---|---|---|
| Isoleucine | Arginine | Alanine |
| Leucine | Histidine | Asparagine |
| Lysine | | Aspartic acid |
| Methionine | | Cysteine |
| Phenylalanine | | Cystine |
| Threonine | | Glutamic acid |
| Tryptophan | | Glutamine |
| Valine | | Glycine |
| | | Hydroxyproline |
| | | Proline |
| | | Serine |
| | | Tyrosine |

to synthesize amino acids is limited and additional amino acids are also deemed essential: cysteine, histidine, and tyrosine (Table 2.2).

Amino acids are used in body growth and repair, and synthesized into enzymes, plasma proteins, antibodies and hormones or oxidized to produce energy (ATP) via the Krebs cycle and electron transport chain; any that are not required are converted into glucose by gluconeogenesis or into triglycerides by lipogenesis (Tortora and Derrickson 2009); the nitrogenous component being excreted by the kidneys.

Protein synthesis occurs within the ribosomes of most cells within the body under the control of deoxyribonucleic acid (DNA) and ribonucleic acid (RNA). It is stimulated by insulin-like growth factors, thyroid hormones, insulin, oestrogen and testosterone.

Proteins exist within the body as long chains of amino acids in different formations. Globular proteins form specific shapes alone or bind to other globular proteins to form enzymes. Enzymes are sensitive to body pH, temperature and osmolarity. At less than 37 °C, enzyme activity reduces; above 37 °C, activity initially increases but then falls dramatically as the shape of the protein changes (Watson and Fawcett, 2003).

Proteins released from worn out cells are changed by transamination into new amino acids and some of these will form other amino acids and new proteins. In addition, liver cells (hepatocytes) convert some to fatty acids, ketones and glucose.

Glycogenesis is the process where glucose is converted to glycogen for storage in the liver and skeletal muscles. However, when this store has been filled to capacity, hepatocytes transform glucose to glycerol and fatty acids to be used in the synthesis of triglycerides (lipogenesis) and these are stored anywhere in the body within adipose (fat) tissue. Glycogen is the only polysaccharide that is stored within the body. Insulin secreted by the pancreatic beta cells stimulates hepatocytes and skeletal muscle cells to synthesize glycogen. When energy is required, glycogen is broken down into glucose which is then oxidized during cellular respiration to produce ATP (glycogenolysis). This process is controlled by glucagon secreted from the alpha cells of the pancreas and epinephrine (adrenaline) secreted from the adrenal glands.

Skeletal muscle is unable to release energy from glycogen by this process; instead the glucose produced has to be catabolized via glycolysis and the Krebs cycle, as discussed earlier. Glycolysis, the Krebs cycle and the electron transport chain provide all the ATP required; however, the Krebs cycle and electron transport chain are aerobic activities.

Chapter 12 provides an in-depth discussion of the digestive system and nutrition.

## Clinical application

### Breast feeding

The World Health Organization (2013a) promotes breast feeding as the preferred method of infant feeding since it provides the correct infant nutrition and has many other health benefits for the mother and child. It also significantly improves global child survival rates, especially in developing countries, by reducing the possibility of gastrointestinal infections. It is worth remembering that in developing areas of the world clean drinking water to reconstitute formula feeds is not always accessible. It is expected that child nurses have the knowledge and skills to facilitate mothers in the breast feeding of their infants.

## Endocrine system

### *Control of glucose*

Glucose concentration is controlled by the two hormones insulin and glucagon, produced by the pancreas. In health it is maintained within a narrow range of 4–6 mmol/L (Clancy and McVicar, 2009). Fasting blood levels of glucose vary according to age as follows: newborn 2.2–3.3 mmol/L (at day 1), child: 3.3–5.5 mmol/L and adult: 3.9–5.8 mmol/L (Clancy and McVicar, 2009).

Insulin is produced by the beta cells in the pancreas and is secreted when blood glucose levels rise. Insulin enables glucose and amino acids to enter cells and increases cellular respiration and thereby the need for glucose. It also increases the conversion of glucose into fat for storage in adipose tissue and into glycogen for storage in the liver and muscle cells by glycogenesis.

Glucagon is produced by the alpha cells in the pancreas. When blood glucose levels drop, glucagon is secreted and acts on the enzyme phosphorylase in the liver, which initiates the breakdown of glycogen into glucose (glucogenolysis). Glucagon also promotes the breakdown of amino acids and glycerol to form glucose-6-phosphate (gluconeogenesis) (Kent, 2000).

The hormones epinephrine (adrenaline) and cortisol have similar actions to glucagon in that they can also trigger the release of glucose from glucagon. This is done in response to a requirement for energy during physical activity (Clancy and McVicar, 2009).

Thus, the body maintains blood glucose levels within a narrow parameter and has the ability to adjust to sudden demands for energy, such as during strenuous physical activity. In the event of the body failing to produce adequate amounts of insulin or none at all (such as with Type 1 diabetes), fats and proteins will be converted into glucose; however this results in the production of ketones and ketoacidosis (Clancy and McVicar, 2009). The rising level of glucose results in hyperglycaemia and as the renal glucose threshold is exceeded; the excess glucose is excreted in the urine as glycosuria. The hyperosmolar effect of glycosuria causes water loss via the urine, leading to increased thirst. Additionally, the breakdown of fats and protein for glucose release leads to weight loss. Thus, the symptoms of thirst, frequent micturition and weight loss become apparent as diabetes develops.

Diabetic ketoacidosis (DKA) is a potentially life-threatening event resulting from an infection or non-compliance with the insulin regimen. It manifests as metabolic acidosis due to the presence of ketone acids, dehydration due to glycosuria, and electrolyte imbalance due to the excretion of urine containing potassium. The rapid loss of fluid can also lead to cerebral oedema as sodium levels rise and fluid enters the brain cells (Lissauer and Clayden, 2007).

## Clinical application

### Obesity

Obesity is of increasing concern across the world for all age groups and is known to have serious health consequences. In 2010, the number of overweight children under the age of 5 years was over 42 million (WHO, 2013b). It is thought that this may be due to either a genetic predisposition to how fat is stored and synthesized or an imbalance between the amount of energy required and the amount of fat consumed.

Obesity may arise when the body becomes resistant to hormones and the ensuing sensory nerve actions that regulate the perception of hunger and the size of meals. For instance, within the hypothalamus, food intake will be reduced when brain nuclei are stimulated by the hormones insulin and leptin. Leptin is produced by adipose tissue, binds to receptors within the hypothalamus and provides feedback regarding energy stores. Mutation of the gene for leptin has been associated with obesity and Type 2 diabetes (Clancy and McVicar, 2009).

### *Control of appetite*

The amount that is eaten is controlled by hormones secreted by the gastrointestinal tract, such as cholecystokinin produced by the ileum and pancreas. Cholecystokinin decreases appetite and the hormone grehlin (produced by the stomach wall) increases appetite.

### *Control of fluid*

During vomiting, gastric acid is lost. Continued vomiting in young children can lead to alkalosis. Mucus, which is alkaline, is passed along with diarrhoea. Fluid is lost with both diarrhoea and vomiting and dehydration can ensue. In addition, electrolytes (sodium and chloride) and glucose are also lost from the body. To compensate, the body reduces the amount of fluid lost in urine, through the action of angiotensin, aldosterone and antidiuretic hormone (ADH).

Osmoreceptors within the hypothalamus in the brain detect the rising osmolarity of the blood. The individual will also feel thirsty and drink more. However, a baby or young child is unable to address this need for fluids and is reliant on a carer (usually their mother or father) to observe and respond to their need for fluids to replenish their thirst. The osmoreceptors within the hypothalamus also stimulate the release of antidiuretic hormone (ADH) which acts on the kidney to retain water.

It is important that the nurse can recognize a range of signs of dehydration in children of all ages, such as loss of skin turgor, sunken fontanelles in young infants, sunken eyes, reduced urinary output or dry nappies, dry mucous membranes, lethargy and irritability.

## Clinical application

### Preoperative fasting

Occasionally small infants require surgery and it is important that all the staff involved in the child's care are aware that the rate of gastric emptying varies according to the age of the child and the type of feed consumed. Feeds with a high fat content take longer to pass through the stomach and preterm infants take longer to pass food from their stomach than term infants (Neill and Knowles, 2004). Breast milk is easier to digest than formula feed and will therefore pass through the stomach faster. Current guidance acknowledges this and advises that breast-fed babies may be fed up to 4 hours prior to surgery, whereas formula-fed babies should be fed 6 hours prior to surgery (RCN 2005). This obviously has implications for providing babies and young children with intravenous fluids.

Further endocrine-related issues are discussed in Chapter 11.

## Excretion homeostasis

Excretion of waste and toxins occurs via the lungs (in exhalation), the skin (as sweat), in faeces and in the urine. There is a fine balance between consumption and excretion, and during trauma (blood loss) and illness, this may be disturbed, resulting in 'shock'. At other times, the homeostatic balance depends on the amount of physical activity undertaken and environmental ambient temperatures.

In childhood the most common type of shock is due to hypovolaemia (Elliott *et al.*, 2010), often due to severe haemorrhage, dehydration as a result of diarrhoea and vomiting, burn injuries and diabetic ketoacidosis.

## Renal system

The renal system consists of two kidneys, two ureters, a bladder and a urethra. Each kidney is made up of about 1 million nephrons which filter the urine. The amount of urine passed by the child varies according to their age and their fluid and dietary input. The minimum passed should be 1 mL/kg/h (Summers and Teasdale, 2009).

Although the neonatal and infant kidneys have the same number of nephrons as the adult kidney, they are immature and do not reach adult functionality until 2 years of age (Tucker Blackburn, 2013). This is due initially to a reduced blood supply and reduced glomerular filtration rate (GFR) of 30 mL/min/1.73 $m^2$. Over the first year of life, epithelial changes increase the GFR to 100 mL/min/1.73 $m^2$ by 9 months and to the adult value of 125 mL/min/1.73 $m^2$ by 1 year. In addition, the rapid growth of the kidney and subsequent lengthening of the nephrons increases the surface area, enabling better reabsorption in the loop of Henle and tubules. The response to the hormones atrial natriuretic peptide, aldosterone and ADH enables the kidney to concentrate urine (Figure 2.3).

There is a more detailed discussion concerning the renal system in Chapter 13.

### *Atrial natriuretic peptide*

Atrial natriuretic peptide (ANP) is a hormone that is secreted by heart muscle cells called myocytes in response to an increased fluid volume and subsequent increase in blood pressure. This stretches the baroreceptors in the carotid sinus and aortic arch, sending messages to the medulla in the brain stem regarding this increase (Figure 2.3). ANP causes the kidneys to excrete more sodium (natriuresis) and in so doing water is removed, which helps to reduce the blood

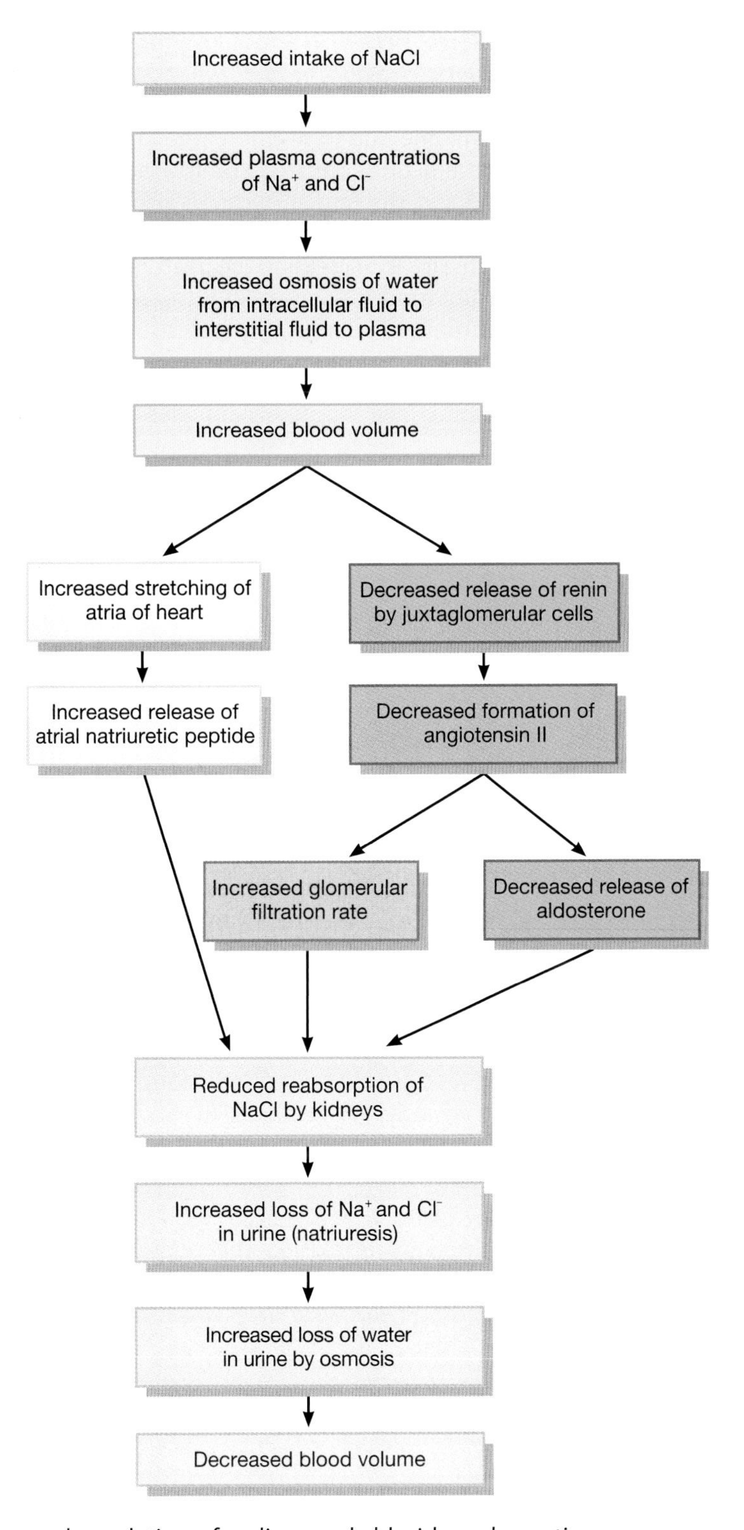


**Figure 2.3** Hormonal regulation of sodium and chloride reabsorption.
*Source:* Tortora and Derrickson 2009, Figure 27.4, p. 1066. Reproduced with permission of John Wiley and Sons, Inc.

pressure. It also inhibits the secretion of both ADH and aldosterone. ANP also has a vasodilating effect on both the arterial and venous blood vessels, which reduces blood pressure (Nair, 2011b).

When there is a reduced fluid volume due to haemorrhage or diarrhoea and vomiting, the amount of blood that returns to the heart is reduced and this is sensed by the baroreceptors (they will be stretched less). In addition, the cardiac output is also reduced and there is a slight reduction in the blood pressure with subsequent reduction in blood supply to the heart and brain. So, the body initially compensates by increasing the heart rate in an effort to improve the circulation and simultaneously the respiratory rate also increases to obtain extra oxygen to transport around the body. Fluid from the interstitial spaces also moves into the intravascular compartments and the blood vessels vasoconstrict, which promotes a normal or slight increase in blood pressure that may go unnoticed. Indeed children are able to tolerate a loss of 20–25% of their circulating volume before exhibiting symptoms (Elliott *et al.*, 2010). However, a keen observer will also notice that the pulse pressure (the difference between the systolic and diastolic pressures) reduces with a reduced urinary output and capillary refill time. Unfortunately, if the cause for the fluid loss is not corrected promptly, decompensation occurs with subsequent hypotension and cell death (Batchelor and Dixon, 2009).

### *Angiotensin*

In response to a reduced circulating blood volume and blood pressure, the afferent arterioles which supply the kidneys with blood are less stretched and the enzyme renin is produced by the juxtaglomerular cells in the kidney. Renin acts to convert angiotensinogen, produced by the liver, into angiotensin 1. This in turn is converted to angiotensin 2 by angiotensin-converting enzyme (ACE), which is formed in the lungs and nephrons. Angiotensin 2 stimulates the secretion of aldosterone and causes vasoconstriction of the afferent arterioles, which increases the blood pressure (Figure 2.4).

### *Aldosterone*

Aldosterone is a mineralocorticoid produced by the cortex of the adrenal glands. It maintains the sodium and chloride balance by regulating their reabsorption by the nephron and the excretion of potassium (Figure 2.5). Since water retention is closely linked to sodium reabsorption, aldosterone is also involved in regulating the volume of circulating blood and the blood pressure. When the amount of potassium circulating in the blood rises, aldosterone is secreted, which stimulates the reabsorption of sodium, chloride and water, thereby increasing the circulating blood volume; it also stimulates the excretion of potassium. By reducing the blood supply to the afferent arterioles, aldosterone also reduces the GFR.

### *Antidiuretic hormone*

ADH is also known as vasopressin and is a hormone produced by the hypothalamus and secreted by the posterior pituitary gland. It acts on the distal part of the convoluted tubule and collecting duct of the nephron by increasing the permeability of the cells, so that more water is reabsorbed and less urine produced. In addition, water loss via sweating is reduced and vasoconstriction occurs in the arterioles, which increases the blood pressure (Figure 2.6).

## Thermoregulation

Thermoregulation is a negative feedback system, controlled by the anterior and posterior parts of hypothalamus called the preoptic area (Tortora and Derrickson 2009), enabling the body in health to maintain a core temperature regardless of the ambient temperature by generating

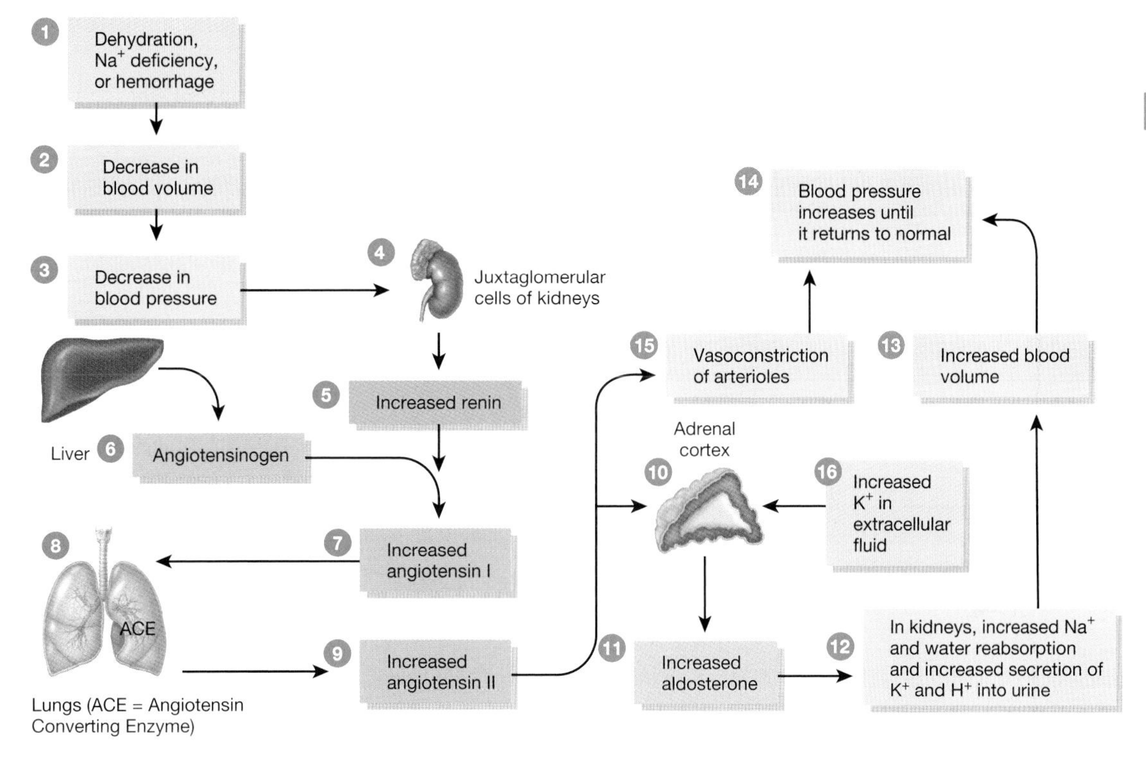


Figure 2.4 Regulation of aldosterone by the renin–angiotensin–aldosterone pathway.
*Source:* Tortora and Derrickson 2009, Figure 18.16, p. 667. Reproduced with permission of John Wiley and Sons, Inc.

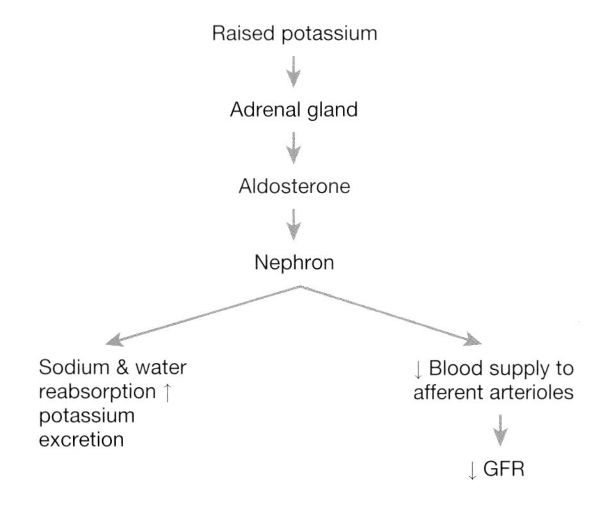


Figure 2.5 Effect of aldosterone on fluid balance.

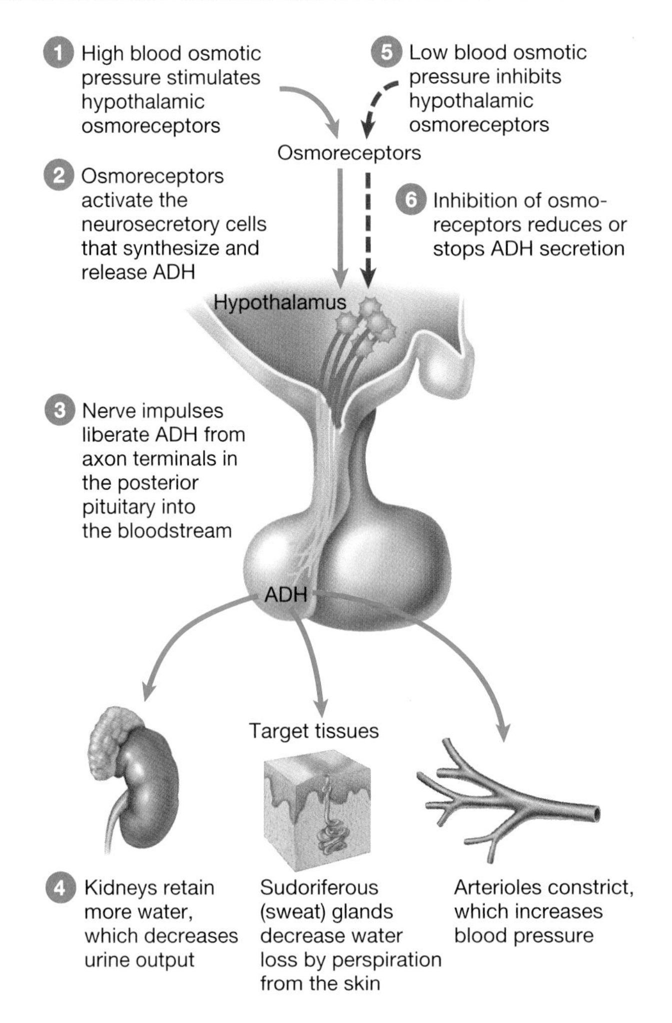


**Figure 2.6** Regulation of the secretion and action of antidiuretic hormone.
*Source:* Tortora and Derrickson 2009, Figure 18.9, p. 657. Reproduced with permission of John Wiley and Sons, Inc.

heat or cooling. The anterior hypothalamus is stimulated by increases in blood temperature and the posterior hypothalamus by decreases in temperature (Kent, 2000).

In hot environments the body adapts by vasodilation of the peripheral circulation and sweating to cool the body. Heat can be lost by radiation, conduction convection and evaporation and the body uses a combination of these to maintain temperature within a narrow range. Older children and young people are usually able to adapt their environment relative to the ambient temperature, such as by removing or adding clothing, drinking extra fluids and exercising. In cold environments the body is able to generate heat through shivering (involuntary skeletal muscle contractions) and vasoconstriction of the peripheral circulation. However babies and young children are unable to shiver and are predominately dependent on non-shivering thermogenesis.

When an infection occurs, the normal response is for body temperature to rise (pyrexia) to assist the cells and enzymes that deal with defending the body against the invasion of pathogens. The immune system releases endogenous pyrogens, which send messages to the hypothalamus to activate the autonomic nervous system (ANS) to increase the temperature. This increase in temperature helps to impede bacterial growth (Cedar, 2012). Sweating is also suppressed, which further increases the temperature. However, in some people this state of pyrexia is dangerous and can result in significant morbidity.

The ANS is not mature until a child is about 5 years old and body temperature regulation in young children is therefore not as precise (Neill and Knowles, 2004). In the older child, receptors in the skin detect changes in the ambient temperature and messages are sent to the hypothalamus so that heat loss or heat production messages can be sent via the ANS to effect actions such as sweating, vasodilation, vasoconstriction and shivering. However newborns have a limited ability to sweat due to the relative lack of sweat glands and are also unable to shiver. It is also worth noting that a baby's head is relatively large in comparison to their body size, being about one quarter of the baby's size in comparison to one eighth in an adult. Thus, a lot of heat and water can be lost via the head in small infants through evaporation.

Sweating enables heat to be lost via evaporation and in the adult up to 4 L/h can evaporate (Cedar, 2012), In addition, some heat is lost via the process of expiration alongside about 0.6 L of water/day in the adult. Vasodilation enables heat to be conducted and radiated away from the skin as the blood vessels widen; with vasoconstriction this process is reduced and the blood supply is redirected to the vital organs such as the brain and kidneys.

Muscular activity due to exercise also generates heat and increases the metabolic rate but this usually requires energy production; however, when shivering occurs the muscles contract rhythmically about 10–20 times/s (Cedar, 2012) generating heat that is transferred around the body.

Newborns are particularly susceptible to ambient temperature changes which can have detrimental effects, since they rely on heat production by brown fat metabolism and have only limited ability to generate heat through this mechanism (non-shivering thermogenesis).

According to Baston and Durward (2010), an acceptable neonatal temperature would range between 36.5 and 37.2 °C. So, a temperature of less than 36.5 °C indicates that the newborn is hypothermic. Newborns have stores of brown fat around their scapula, axillary, adrenal glands and mediastinal areas. Brown fat differs from other adipose tissues in its number of fat vacuoles, mitochondria, glycogen stores, blood supply and sympathetic nervous system supply (Tucker Blackburn, 2013; Potts and Mandleco, 2012). Brown fat can be used to generate heat by thermogenesis, but the process is demanding of oxygen and glucose, which can result in metabolic acidosis and hypoglycaemia. In addition, brown fat also increases the metabolic rate, which further assists with heat production. Brown fat cells start to appear at around 29 weeks of gestation, so premature infants are extremely vulnerable to heat loss due to poor or minimal stores.

# Conclusion

Homeostasis is a complex process which maintains the body in a state of equilibrium. Children differ from adults in many ways as their bodies are still adapting to extrauterine life, enabling them to grow and develop physically into adults.

# Activities

**Now review your learning by completing the learning activities in this chapter. The answers to these appear at the end of the book. Further self-test activities can be found at www.wileyfundamentalseries.com/childrensA&P.**

## Wordsearch

There are several words linked to this chapter hidden in the following two grids. Can you find them? A tip – the words can go from up to down, down to up, left to right, right to left, or diagonally.

# Wordsearch grid 1

| | | | | | | | | | | | | | | |
|---|---|---|---|---|---|---|---|---|---|---|---|---|---|---|
| o | y | t | i | r | a | l | o | m | s | o | r | d | y | h |
| s | p | e | c | i | f | i | c | g | r | a | v | i | t | y |
| m | u | e | o | p | o | l | y | u | r | i | a | u | i | r |
| o | r | n | l | c | i | n | o | t | o | y | h | r | l | e |
| t | i | i | l | a | l | l | u | d | e | m | e | e | i | n |
| i | n | n | e | t | f | e | t | u | l | o | s | t | b | a |
| c | e | i | c | o | a | f | r | a | n | u | r | i | a | l |
| g | o | t | t | n | t | a | e | r | u | i | t | c | e | c |
| r | l | a | i | t | i | t | s | r | e | t | n | i | m | o |
| a | i | e | n | e | l | o | i | r | e | t | r | a | r | r |
| d | g | r | g | e | f | f | e | r | e | n | t | u | e | t |
| i | u | c | d | i | c | a | c | i | r | u | t | e | p | e |
| e | r | i | u | a | i | r | u | t | a | m | e | a | h | x |
| n | i | o | c | v | a | s | o | p | r | e | s | s | i | n |
| t | a | n | t | s | i | s | y | l | a | n | i | r | u | t |

# Wordsearch grid 2

| | | | | | | | | | | | | | | |
|---|---|---|---|---|---|---|---|---|---|---|---|---|---|---|
| g | l | i | g | e | r | e | r | n | c | a | l | y | x | o |
| m | o | i | a | n | o | f | o | s | i | s | o | m | s | o |
| t | o | g | t | o | w | i | k | e | t | u | a | y | t | o |
| t | p | a | r | r | t | y | o | f | o | h | r | l | w | r |
| c | o | l | l | e | c | t | i | n | g | d | u | c | t | n |
| r | f | h | r | t | a | m | y | h | c | n | e | r | a | p |
| e | h | c | u | s | s | b | m | e | d | u | l | l | a | o |
| x | e | t | r | o | c | h | s | p | y | r | a | m | i | d |
| s | n | k | i | d | n | e | y | o | r | n | i | n | e | r |
| i | l | n | e | l | s | u | l | u | r | e | m | o | l | g |
| e | e | s | s | a | t | a | e | l | l | p | t | b | e | d |
| a | n | g | i | o | t | e | n | s | i | n | t | e | s | i |
| f | i | l | t | r | a | t | i | o | n | h | h | i | r | y |
| u | r | e | t | h | r | a | p | e | l | v | i | s | o | u |
| i | t | u | n | o | r | h | p | e | n | t | d | o | o | n |

## Crossword



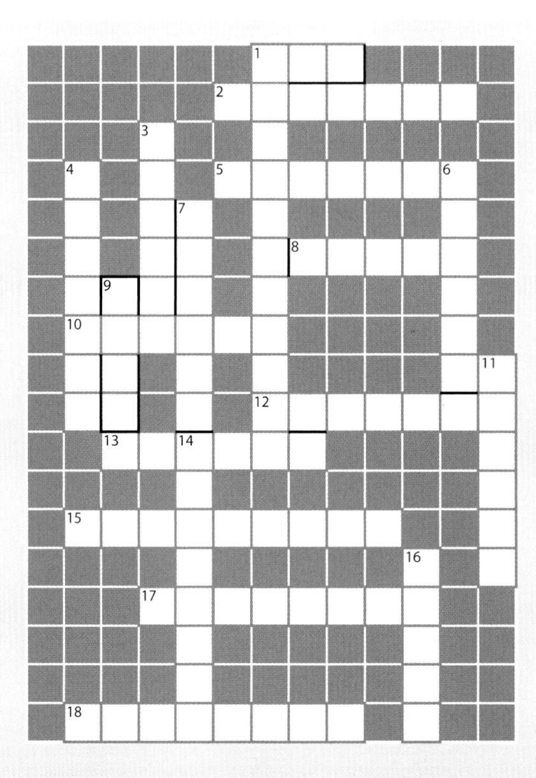


**Across**

1. Released from the posterior pituitary, controls $H_2O$ reabsorption in collecting duct [3, abbrev]
2. Presence in urine is abnormal [7]
5. (First of two words) Beginning of nephron, interfaces with glomerulus [7]
8. Solution containing matter to be filtered by kidneys [5]
10. One of a pair delivering urine to bladder [6]
12. Functional unit of 7 down [7]
13. Organ consisting of millions of 6 down [6]
15. Secreted by tubular cells; altered blood levels of this can cause rhythm problems [9]
17. Secreted by the tubular cells, major component of acid-base balance [8]
18. Kidneys need adequate blood..........to work [8]

**Down**

1. Secreted from the adrenal cortex, controls $Na^+$ absorption and $K^+$ secretion [11]
3. This is convoluted and either proximal or distal [6]
4. (Second of two words; first is 3 down) Beginning of nephron, interfaces with glomerulus [7]
6. Filtered and reabsorbed by the kidney; the most abundant ion in the body [6]
7. Furthest away part of the tubule [6]
9. Normal body waste product excreted by kidney [4]
11. Absence of urine [6]
14. When kidneys fail this may be necessary [8]
16. Negatively charged ion [5]

# Glossary



**Acetyl coenzyme A:** Assists in energy production in the Krebs cycle.

**Adenosine triphosphate:** A compound of an adenosine molecule with three attached phosphoric acid molecules.

**Alanine:** A non-essential amino acid.

**Amino acid:** A basic protein.

**Arginine:** An amino acid synthesized by the body in inadequate quantities in childhood.

**Asparagine:** A non-essential amino acid.

**Aspartic acid:** A non-essential amino acid.

**Baroreceptors:** Specialized receptors found in the aortic arch and the carotid sinus.

**Chemoreceptors:** Specialized receptors that are sensitive to specific chemicals.

**Colloid osmotic pressure:** Osmotic pressure exerted by proteins in the plasma.

**Cysteine:** A non-essential amino acid.

**Cystine:** A non-essential amino acid.

**Cytochrome:** Protein that contains iron and involved in energy transport.

**Diffusion:** Process whereby a substance moves through a membrane without assistance from transport proteins.

**Electron transport chain:** Part of the energy production process, involving the membranes of the mitochondria. Electrons are released in the presence of oxygen and transferred through a series of chemical reactions involving cytochromes.

**Endocytosis:** Process whereby cells take in molecules such as proteins from outside of the cell by engulfing them.

**Exocytosis:** Process whereby vesicles within the cell join with the cell membrane in order to expel their contents from the cell.

**Feedback system:** A monitoring system in which specialist sensor cells detect changes in the levels of oxygen, carbon dioxide, glucose, electrolytes, temperature, urea or water and initiate reactions within the body to maintain a state of equilibrium.

**Flavine adenine dinucleotide (FAD):** A coenzyme involved in transferring energy and in the process is reduced to $FADH_2$.

**Glucagon:** A hormone secreted by the alpha cells of the pancreas that stimulates the breakdown of glycogen, amino acids and fatty acids to provide energy.

**Gluconeogenesis:** The production of energy from amino acids.

**Glutamic acid:** A non-essential amino acid.

**Glutamine:** A non-essential amino acid.

**Glycine:** A non-essential amino acid.

**Glycogen:** A carbohydrate made from glucose.

**Glycogenolysis:** The breakdown of glycogen to produce glucose.

**Glycolysis:** The anaerobic breakdown of glucose to release energy and form pyruvic acid.

**Hepatocytes:** Liver cells.

**Histidine:** An amino acid synthesized by the body in inadequate quantities in childhood.

**Homeostasis:** A state of balance of the internal environment of the body.

**Hormone:** A chemical released into the blood by the endocrine system which has physiological control over the function of other cells or organs.

**Hydrostatic pressure:** The pressure exerted on the blood vessels to force fluid out of the blood vessels; expressed in millimetres of mercury or mmHg; maintained by plasma proteins within the blood vessels.

**Hydroxyproline:** A non-essential amino acid.

**Insulin:** A hormone secreted by the beta cells of the pancreas that facilitates the transfer of glucose into cells and for storage within the liver as glycogen.

**Isoleucine:** An essential amino acid.

**Krebs cycle:** A chain of aerobic reactions that occur within the mitochondria where carbon dioxide and energy as ATP are released.

**Leucine:** An essential amino acid.

**Lipogenesis:** Process whereby triglycerides are synthesized from fatty acids for storage within adipose tissue.

**Lysine:** An essential amino acid.

**Medulla oblongata:** An area of the brain stem involved in the transfer of sensory information from the autonomic nervous system.

**Methionine:** An essential amino acid.

**Mitochondrion:** The energy producing site within the cell.

**Nicotinamide adenine dinucleotide ($NAD^+$):** A coenzyme involved in the transfer of energy; it is reduced to NADH during the process.

**Osmoreceptor:** Specialized cells that are sensitive to pressure or the amount of water present within the body.

**Osmosis:** A process whereby water moves across a semipermeable membrane from an area of high concentration (low solute) to an area of low concentration (high solute).

**Osmotic pressure:** The pressure exerted on a solution.

**Phenylalanine:** An essential amino acid.

**Phosphorylase:** A liver enzyme which initiates glucogenolysis.

**Pinocytosis:** Process whereby cells take in molecules (e.g. proteins) by engulfing them with their cell membrane.

**Proline:** A non-essential amino acid.

**Pyruvic acid:** Formed during glycolysis.

**Serine:** A non-essential amino acid.

**Thermoreceptor:** A receptor that is sensitive to changes in temperature.

**Threonine:** An essential amino acid.

**Transamination:** Process whereby non-essential amino acids are formed from essential amino acids.

**Triglycerides:** Fatty acids with three fatty acid components.

**Tryptophan:** An essential amino acid.

**Tyrosine:** A non-essential amino acid.

**Valine:** An essential amino acid.

## References

Baston, H., Durward, H. (2010) *Examination of the Newborn: A Practical Guide*, 2nd edn. Routledge, London.

Batchelor, S., Dixon, M. (2009) Cardiac. In Dixon, M., Crawford, D., Teasdale, D., Murphy, J. (eds.) *Nursing the Highly Dependent Child or Infant: A Manual of Care*, Wiley Blackwell, Oxford, pp. 76–107.

Cedar, S.H. (2012) *Biology for Health: Applying the Activities of Daily Living*, Palgrave Macmillan, Basingstoke.

Clancy, J., McVicar, A. (2009) *Physiology and Anatomy for Nurses and Health Care Practitioners: A Homeostatic Approach*, 3rd edn. Hodder Arnold, London.

Elliott, B., Callery, P., Mould, J. (2010) Caring for children with critical illness. In: Glasper, A., Richardson, J. (eds) *A Textbook of Children and Young People's Nursing*, 2nd edn. Churchill Livingstone Elsevier, Edinburgh, pp. 691–718.

Higgins, C. (2007) *Understanding Laboratory Investigations for Nurses and Health Professionals*, 2nd edn. Blackwell Publishing, Oxford.

Kent, M. (2000) *Advanced Biology*, Oxford University Press, Oxford.

Lissauer, T., Clayden, G. (2007) *Illustrated Textbook of Paediatrics*, 3rd edn. Mosby Elsevier, Edinburgh.

Lloyd, J., Askie, L.M, Smith, J., Tarnow-Mordi, W.O. (2009) Supplemental oxygen for the treatment of prethreshold retinopathy of prematurity. *The Cochrane Collaboration* [Online]. Available from: http://onlinelibrary.wiley.com/doi/10.1002/14651858.CD003482/pdf Accessed on 2 May 2013.

Mulryan, C. (2011) *Acute Illness Management*, SAGE Publications Ltd, London.

Nair, M. (2011a) The renal system. In: Peate, I., Nair, M. (eds) *Fundamentals of Anatomy and Physiology for Student Nurses*, Wiley Blackwell, Oxford, pp. 446–475.

Nair, M. (2011b) Cells: cellular compartment transport system, fluid movements. In: Peate, I., Nair, M. (eds) *Fundamentals of Anatomy and Physiology for Student Nurses*, Wiley Blackwell, Oxford, pp. 33–61.

Neill, S., Knowles, H. (2004) *The Biology of Child Health: A Reader in Development and Assessment*, Palgrave Macmillan, Basingstoke.

Peate, I., Nair, M. (eds) (2011) *Fundamentals of Anatomy and Physiology for Student Nurses*, Wiley Blackwell, Oxford.

Potts, N.L., Mandleco, B.L. (2012) *Pediatric Nursing: Caring for Children and Their Families*, 3rd edn. Delmar Cengage Learning, New York.

Ramsay, J., Moules, T. (2008) Principles of care-children with altered breathing. In: Moules, T., Ramsay, J. (eds) *The Textbook of Children's and Young People's Nursing*, 2nd edn. Blackwell Publishing, Oxford, pp. 530–541.

Royal College of Nursing (2005) *Perioperative Fasting in Adults and Children: An RCN Guideline for the Multidisciplinary Team*. [Online]. Available from: http://www.rcn.org.uk/__data/assets/pdf_file/0009/78678/002800.pdf Accessed on 17 June 2013.

Summers, K., Teasdale, D. (2009) Renal. In: Dixon, M., Crawford, D., Teasdale, D., Murphy, J. (eds.) *Nursing the Highly Dependent Child or Infant: A Manual of Care*, Wiley Blackwell, Oxford, pp.181–199.

Tortora, G.J., Derrickson, B.H. (2009) *Principles of Anatomy and Physiology*, 12th edn. Wiley, Hoboken, NJ.

Tucker Blackburn, A. (2013) *Maternal, Fetal and Neonatal Physiology*, 4th edn. Elsevier, St Louis.

Watson, R., Fawcett, T.N. (2003) *Pathophysiology, Homeostasis and Nursing*, Routledge, London.

Waugh, A., Grant, A. (eds) (2010) *Ross and Wilson's Anatomy and Physiology in Health and Illness*, 11th edn. Churchill Livingstone Elsevier, Edinburgh.

Wheeldon, A. (2011) The respiratory system. In: Peate, I., Nair, M. (eds) *Fundamentals of Anatomy and Physiology for Student Nurses*, Wiley Blackwell, Oxford, pp. 328–365.

World Health Organization (2013a) *Breast feeding*. [Online]. Available from: http://www.who.int/topics/breastfeeding/en/ Accessed on 20th May 2013.

World Health Organization (2013b) *Childhood Overweight and Obesity*. [Online]. Available from: http://www.who.int/dietphysicalactivity/childhood/en/ Accessed 5th August 2013.

# Chapter 3

# Scientific principles

Peter S. Vickers

## Aim

The aim of this chapter is to introduce you to the basic principles that underpin bioscience in order for you to understand better the chapters that follow, which rely much on a knowledge and understanding of these basic scientific principles.

## Learning outcomes

On completion of this chapter the reader will be able to:

- Describe the levels of organization of a body and the characteristics of life.
- Understand and explain an atom and how it relates to molecules and the ways in which atoms can bind together.
- Describe elements and their characteristics.
- Understand how to read chemical equations.
- List the differences between organic and inorganic substances.
- List the various ways in which we measure things.

## Test your prior knowledge

- What is an atom and what are the three basic components of an atom?
- What are the essential requirements of all organisms (including humans) in order for them to survive and flourish?
- What are elements and which do we need for respiration?
- What are the three types of carbohydrates?
- Explain the difference between organic and inorganic substances in scientific terms.
- What does a pH scale signify in terms of the measurement of acids and alkalis and what is the importance of these for us in the maintenance of human life?
- Can you name two differences between organic and inorganic molecules?
- What does the $\rightleftarrows$ symbol mean in a chemical equation, as in:

  $HCl + NaHCO_3 \rightleftarrows NaCl + H_2CO_3$
- What are the reactants and products in a chemical equation?
- What is the definition of a chemical element?

*Fundamentals of Children's Anatomy and Physiology: A Textbook for Nursing and Healthcare Students*, First Edition. Edited by Ian Peate and Elizabeth Gormley-Fleming.

Companion website: www.wileyfundamentalseries.com/childrensA&P

# Introduction

You wish to look after child patients; in order to do that properly, you will need to understand everything about how a child functions, including psychological and social functioning, how a child's body functions, in both health and sickness. To do that you need to learn about the normal body – how it physically develops and functions (anatomy and physiology) as well as the abnormal (pathophysiology). Learning about anatomy and physiology of the body is much like learning a foreign language – there is new vocabulary, new grammar and new concepts to learn and understand. This chapter introduces you to the basics of this new language in order for you to use this knowledge to help you to understand the anatomy and physiology of the different parts of the body that are explained and discussed throughout this book.

First of all there are two (possibly new) terms to learn and understand: anatomy and physiology.

- **Anatomy** is the study of the **structure** of the body.
- **Physiology** is the study of the **function** of the body.

However, it is important to remember that structure is always related to function because the structure determines the function, which in turn determines how the body/organ is structured – the two are interdependent.

# Levels of organization

The body is a very complex organism that consists of many components from the smallest of them – that is, the **atom** – to the larger organs that are found within the body (Figure 3.1). So, starting from the smallest component and working upwards, we have:

- the atom – for example, hydrogen, carbon;
- the **molecule** – for example, water, glucose;
- the macromolecule (large molecule) – for example, protein, DNA;
- the **organelle** (found in the cell) – for example, nucleus, mitochondrion;
- the tissues – for example, bone, muscle;
- the organs – for example, heart, kidney;
- the organ system – for example, skeletal, cardiovascular;
- the organism – for example, human.

# Characteristics of life

All living organisms have certain characteristics in common. Although the actual mechanics and physiology of these characteristics may differ from organism to organism, they are important for the maintenance of life. These characteristics are:

- **Reproduction:** At the micro-level and the macro-level, reproduction is essential. At the macro-level it is the reproduction of the organism, and at the micro-level it is the reproduction of new cells to maintain the efficiency and growth of the organism.
- **Growth:** Essential for the development of an organism.
- **Movement:** Both a change in position within the environment and motion within the organism itself are parts of movement. This characteristic is essential to allow the organism to seek out nutrition and partners for reproduction, as well as to escape predators.
- **Respiration:** This is important for obtaining oxygen and releasing carbon dioxide (external respiration), as well as releasing energy from foods (internal respiration).

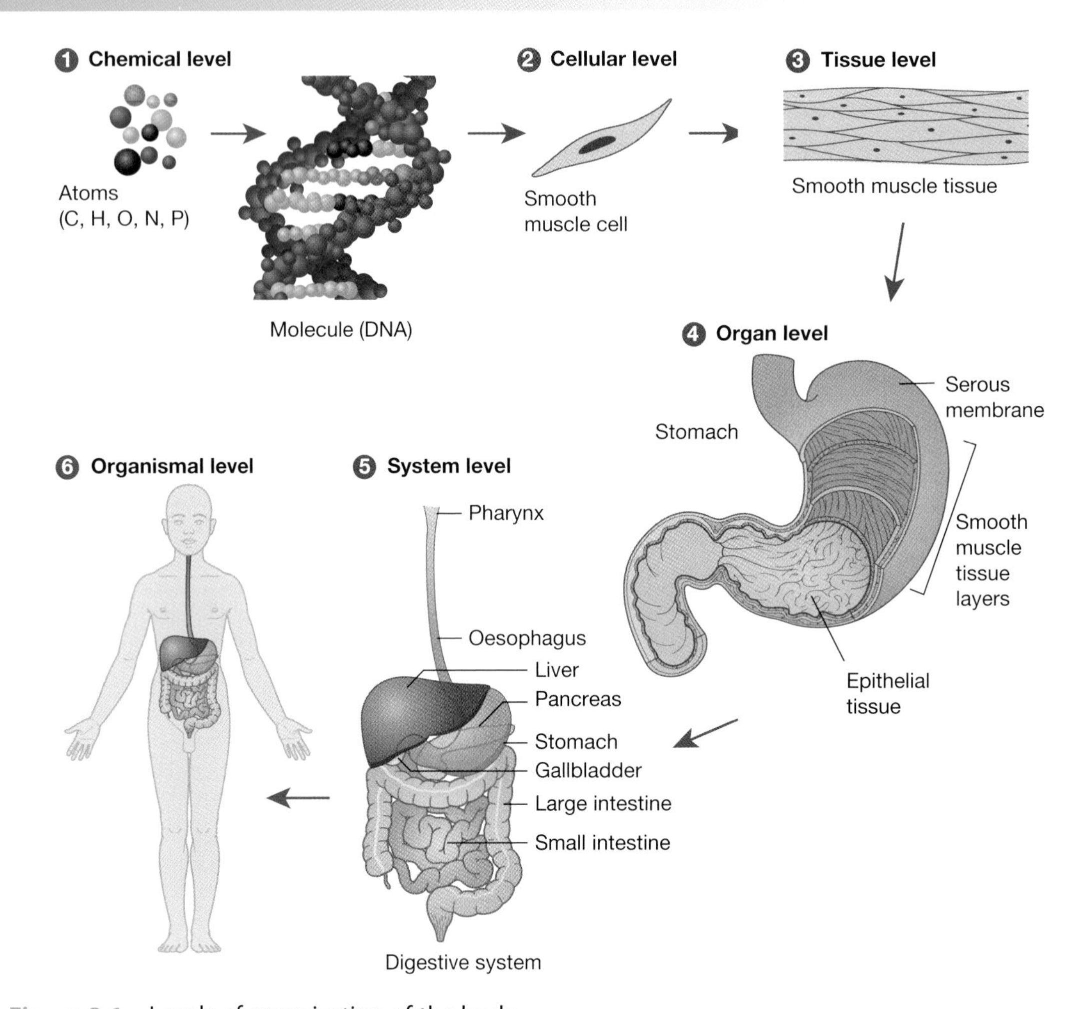


Figure 3.1 Levels of organization of the body.
*Source:* Tortora and Derrickson (2009), Figure 1.1, p. 3. Reproduced with permission of John Wiley and Sons, Inc.

- **Responsiveness:** This allows the organism to respond to changes; for example, in the environment or to other stimuli.
- **Digestion:** This is the breakdown of food substances within the organism, producing the energy necessary for life.
- **Absorption:** The movement of substances (including digested food) through membranes and into body fluids, including blood and lymph, which then carry these substances to the parts of the organism requiring them.
- **Circulation:** The movement of substances, within body fluids, through the body.
- **Assimilation:** The changing of absorbed substances into different substances that can then be utilized by the tissues of the body.
- **Excretion:** The removal of waste substances from the body. Waste substances are either removed because they are of no use to the body or because they are poisonous (toxins) to the body.

# Bodily requirements

There are five essential requirements that all organisms, including humans, require: water, food, oxygen, heat and pressure.

## Water

- Water is the most abundant substance on the surface of the Earth, and the most abundant substance found in the body. At birth, 78% of a baby is water; at 1 year of age this has dropped to 65%. In adult males the figure drops to 60%, and in adult females the figure is 55% (females have more fat than males as a percentage of their body, which accounts for the difference).
- Water is required for the various **metabolic processes** that are necessary to ensure an organism's survival.
- Internally, water is necessary to **transport** essential substances around the organism.
- Water regulates **body temperature** – a human, for example, operates within a very narrow temperature range and has a very small tolerance for temperature change within the body. If the body temperature exceeds this range (either by being too high or too low), then death will occur.

## Food

- Supplies the **energy** for the organism, allowing it to fulfil all the essential characteristics mentioned previously.
- Also supplies the raw materials for these characteristics – particularly for growth.

## Oxygen

- Forms approximately 20% of the air surrounding the organism – essential for the release of energy from the assimilated nutrients.

## Heat

- A form of energy that partly controls the rate at which **metabolic reactions** occur.

## Pressure

- There are two types of pressure required by an organism:
  - **atmospheric pressure**, which is important in the process of **breathing**;
  - **hydrostatic pressure**, which keeps the **blood** flowing through the body.

# Atoms

It is now time to turn to the smallest building blocks of the body (indeed of all matter) – the **atom** (the word 'atom' comes from a Greek word meaning 'incapable of being divided', but now we know that an atom itself is composed of even smaller substances).

## The atom

The atom (Figure 3.2) is made up of:

- **protons**
- **neutrons**
- **electrons**.

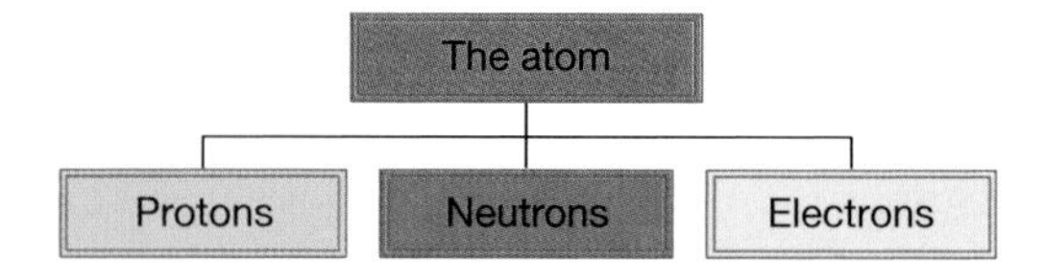


Figure 3.2 The atom.
*Source:* Peate and Nair (2011), Figure 1.2, p. 5. Reproduced with permission of John Wiley and Sons, Ltd.

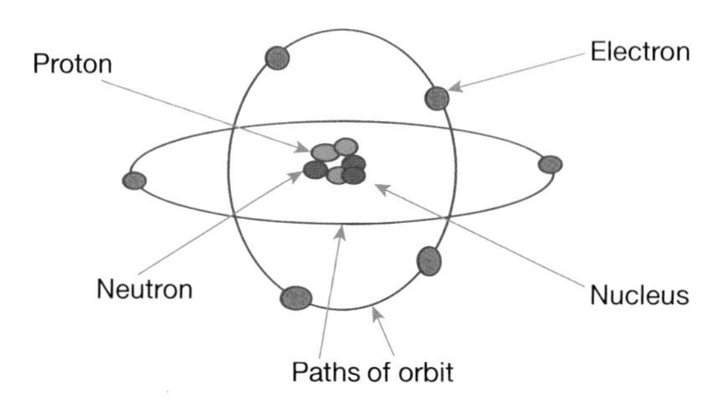


Figure 3.3 Schematic diagram of an atom.
*Source:* Peate and Nair (2011), Figure 1.3, p. 5. Reproduced with permission of John Wiley and Sons, Ltd.

**Protons** and **electrons** carry an **electrical charge**, whereas the **neutron** carries no electrical charge (i.e. it is **neutral** – hence its name). Protons carry a **positive** electrical charge and electrons a **negative** electrical charge. The electrons move rapidly around the nucleus of the atom, which consists of protons and neutrons (Figure 3.3).

Although there are many different types of atoms, they always have the same features – the **paths of orbit**, and the electrons, neutrons and protons – as well as the same characteristics:

- The nucleus is always central.
- The inner **shell** (path of orbit) always has a maximum of two electrons.
- Every atom tries to have eight electrons in its outermost (or valence) shell (the **octet** rule). This may require them to give up, share or take electrons – thus leading to the formations of ions (see later).
- For example, the potassium atom has four shells, and the third shell (counting from the nucleus) has eight electrons.

## Atomic number

All atoms are designated a number; this is known as the **atomic number**. The atomic number of an atom is the same as the number of protons in that atom. For example, a carbon atom has six protons and so the atomic number of a carbon atom is 6. Carbon is very important in bioscience because we are all carbon-based entities. Similarly, the sodium atom has 11 protons, and therefore its atomic number is 11, whilst a chlorine atom has 17 protons and has an atomic number of 17.

Looking in more detail at the atoms in Figure 3.4 illustrates the structure of an actual atom. For example, carbon has six electrons orbiting around a nucleus that consists of six protons and six neutrons. However, having the same numbers of electrons, protons and neutrons is generally not usual because, as explained shortly, whilst it is normal to have the same numbers of elec-

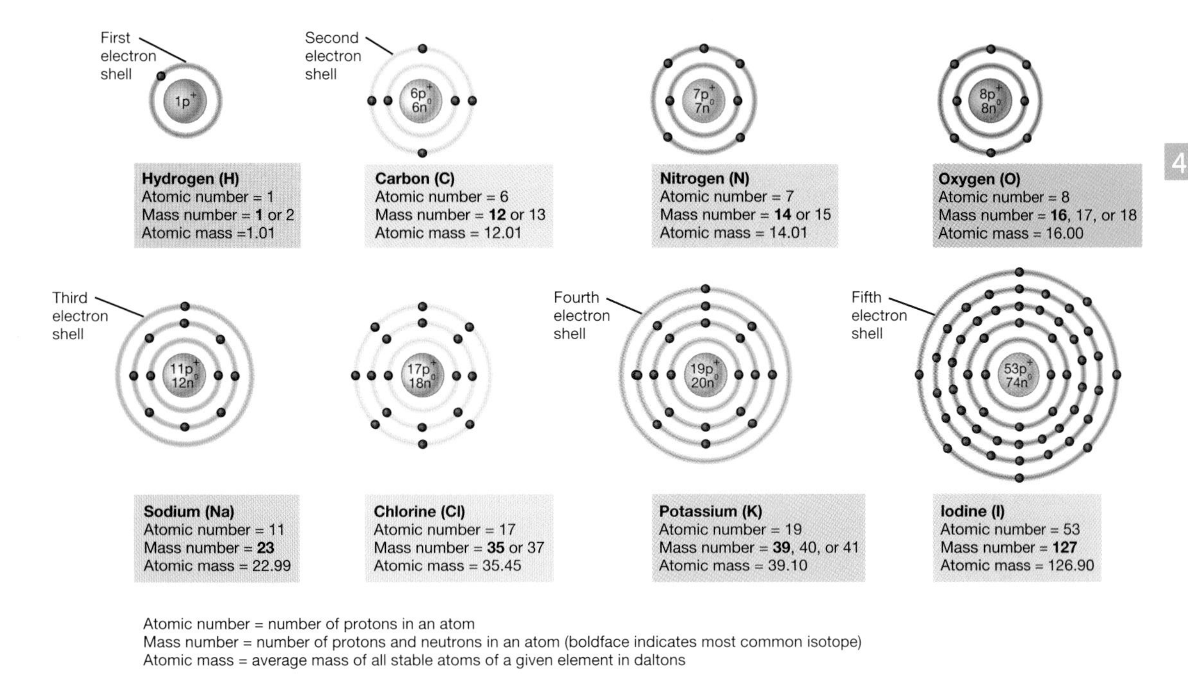


Figure 3.4 Different atoms.
*Source:* Tortora and Derrickson (2009), Figure 2.2, p. 31. Reproduced with permission of John Wiley and Sons, Inc.

trons and protons, the number of neutrons can differ from the numbers of electrons and protons in an atom.

A **basic principle** of the atom is the importance of having equal numbers of electrons and protons in order to maintain electrical neutrality (i.e. an electrically stable state). Because protons carry a positive electrical charge, electrons carry a negative electrical charge and neutrons carry no electrical charge (i.e. they are neutral) it is thus important that an atom has equal numbers of electrons and protons in order to maintain this stable state. To understand how this works, the carbon atom carries six electrons, six protons and six neutrons – the electrical charges of the six electrons and six protons cancel one another out. Consequently, overall the atom is neutrally charged and it is said to be in a state of equilibrium. Looking at the other atoms in Figure 3.4, you can see that this applies to all atoms.

When the number of neutrons differs from the number of protons we have what is called an isotope. For example, carbon normally has six neutrons (carbon-12), but it can have seven (carbon-13) or even eight neutrons (carbon-14). Chemically, the isotopes are identical to the normal carbon with six neutrons, but the elemental masses are different.

## Molecules

This need for the atom to be in equilibrium is the driving force behind the combining of atoms to make **molecules** (the next largest structure). A molecule is an atom or group of atoms capable of independent existence. It contains atoms that have bonded together. Sodium chloride (NaCl) is a molecule containing one atom of sodium (also known as natrium – abbreviated to 'Na') has bonded to one atom of chlorine (Cl). Similarly, the molecule $H_2O$ (which is water) is made up of two atoms of hydrogen (H) bonded to one atom of oxygen (O).

## Ions

When neutral atoms lose or gain electrons they become positively or negatively charged respectively – they are **ions**. For example:

- $Na^+$ (sodium positive)
- $Cl^-$ (chlorine negative).

However, we write the resultant sodium chloride molecule as NaCl because the positive and the negative charges have cancelled each other out.

- Ions that carry a positive electrical charge are known as **cations**.
- Ions that carry a negative electrical charge are known as **anions**.

## Chemical bonds

These are the means by which atoms can bind to one another by losing, gaining or sharing their outer shell electrons with other atoms. Atoms are electrically neutral, but once an atom has unequal numbers of electrons and protons it loses its electrical stability and so becomes an ion. This can be overcome by binding with another ion that also has unequal numbers of protons and electrons. A **chemical bond** is the 'attractive' force that holds atoms/ions together, resulting in the formation of different compounds that are more stable than the original atoms or ions.

The formation of chemical bonds also generally results in the release of **energy** previously contained in the atoms, as shown schematically by

$$A + B \rightarrow AB + \text{energy}$$

The combining power of atoms is known as **valence**; atoms that combine easily have a high valency, whereas atoms that combine poorly have a poor valency. Because the only shell that is important in bonding is the outermost shell, this shell is known as the valence shell (Marieb and Hoehn, 2009).

There are three types of **chemical bonds** occurring between atoms:

- **ionic bonds**
- **covalent bonds**
- **polar bonds/hydrogen bonds**.

### *Ionic bonding of atoms*

Atoms prefer to be in a state of equilibrium, but sometimes an atom that has a stable structure may lose an electron, in which case it becomes unstable. For example, sodium (Na) atoms are atoms that may lose an electron. To become stable again, it must connect with an atom that can accept an electron; for example, chlorine (Cl). When sodium and chlorine atoms are mixed together, one electron of each sodium atom moves to one atom of chlorine (Figure 3.5), thus forming the molecule sodium chloride (NaCl), better known as common salt. This is known as **ionic bonding**, because **ions** are involved.

To summarize, an **ionic bond** is a bond formed between negatively and positively charged ions. These **ions** are attracted to, and stabilize, each other, but they neither transfer nor share electrons between themselves. Consequently, this is more of an electrostatic interaction between atoms rather than a bond between them (Fisher and Arnold, 2004).

### *Covalent bonds*

Unlike ionic bonding, **covalent bonding** involves the sharing of compatible **valence** electrons between atoms. None of the atoms involved in this type of bonding loses or gains electrons.

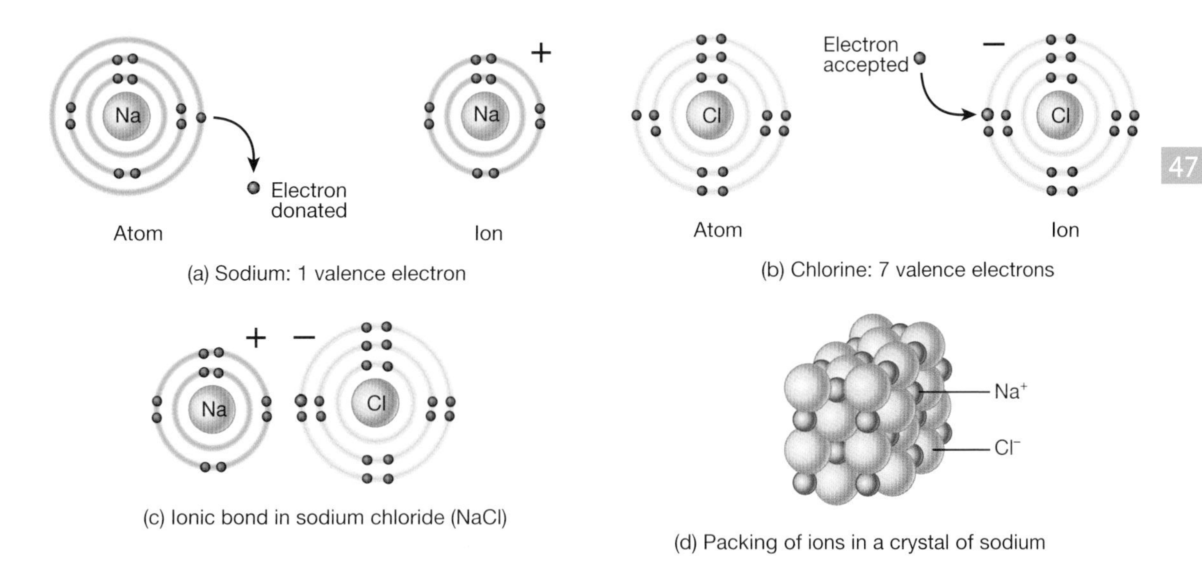


Figure 3.5 Ionic bonding of a sodium and a chlorine atom to form a sodium chloride molecule.
*Source:* Tortora and Derrickson (2009), Figure 2.4, p. 33. Reproduced with permission of John Wiley and Sons, Inc.

Instead, electrons are shared between them so that each of the atoms will have a complete valence shell (i.e. outermost shell) for at least part of the time (Marieb and Hoehn, 2009).

Covalent bonding occurs when two atoms are close to one another, and so an overlapping of the atoms occurs. By overlapping, each atom's electrons are attracted to the other's nucleus (Figure 3.6). This type of bonding does not require positive and negative electrically charged electrons as with ionic bonding. Covalent bonding allows any number of atoms to be bonded together, such as, for example, one carbon and four hydrogen atoms to produce methane, or one oxygen and two hydrogen atoms to produce water.

Types of covalent bonding, depending upon the number of electrons shared between the bonded atoms, are:

1. Single covalent bonds (one electron from each atom shared in the outermost shell); for example, hydrogen molecule.
2. Double covalent bonds (two electrons from each atom shared); for example, oxygen molecule.
3. Triple covalent bonds (three electrons from each atom shared); for example, nitrogen molecule.

## Polar bonds

Sometimes, molecules do not share electrons equally – there is a separation of the electrical charge into positive or negative, known as **polarity**. Because of this separation of electrical charge there is an additional weak bond. However, this bond is *not* between **atoms**, but between **molecules**. As with ionic bonding, polar bonding occurs because of the rule that **opposites attract**. Thus, the small opposite charges from different polar molecules can be attracted to each other. **Polar bonding** (also known as **hydrogen bonding**) generally only occurs with molecules that contain the atom hydrogen (Figure 3.7).

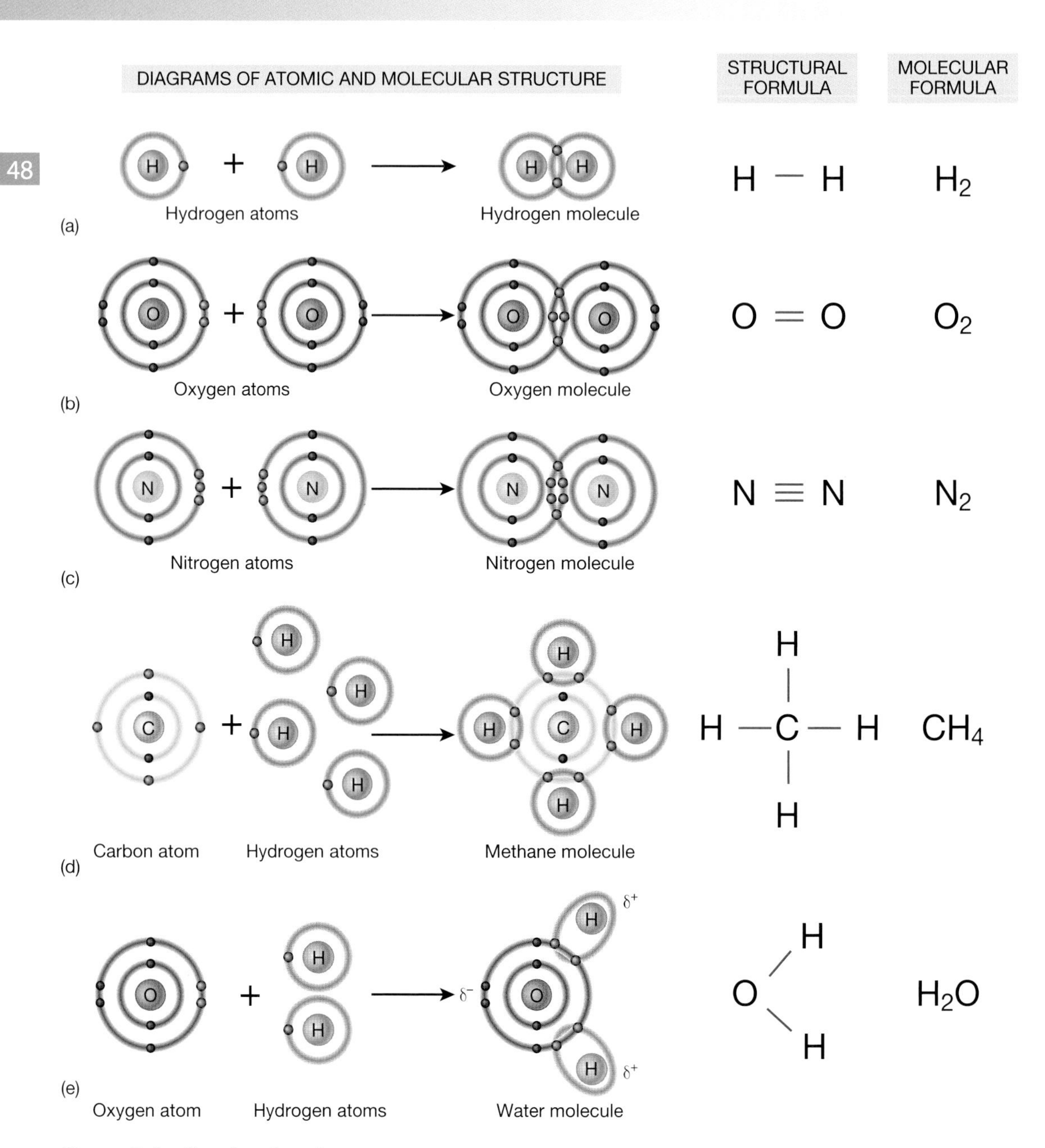


**Figure 3.6** Covalent bond.
*Source:* Tortora and Derrickson (2009), Figure 2.5, p. 34. Reproduced with permission of John Wiley and Sons, Inc.

The fact that molecules can form polar bonds (albeit only weakly) is very important in determining the structure and function of physiologically active substances such as:

- enzymes
- antibodies
- genetic molecules
- pharmacological agents (drugs).

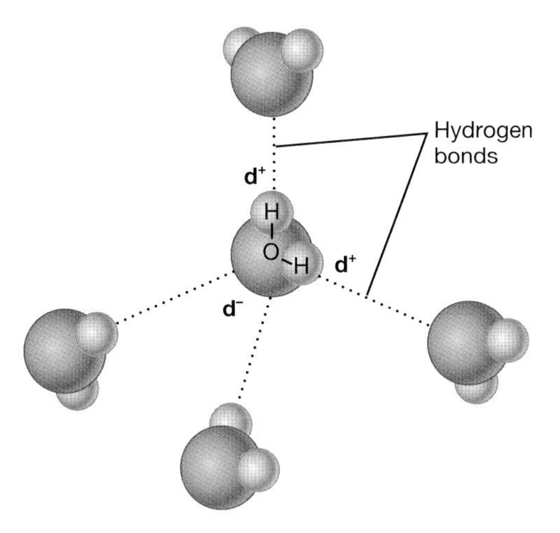


Figure 3.7 Hydrogen bonds and water.
*Source:* Tortora and Derrickson (2009), Figure 2.6, p. 35. Reproduced with permission of John Wiley and Sons, Inc.

DNA is a very good example of the value of molecules only being weakly bonded together, as the bases of DNA are joined by polar bonds. During mitosis (cell division) when the DNA reproduces itself, the bases can easily separate to allow this process (see Chapter 5).

## Electrolytes

**Electrolytes** are substances that move to oppositely charged electrodes in fluids, and are produced following bonding of molecules. If molecules that are bonded together **ionically** (see ionic bonding section) are dissolved in water within the body cells, they undergo the process of **ionization** and become **dissociated** (i.e. separated into ions). These ions are now known as electrolytes.

This only applies to molecules produced by ionic bonding. Molecules produced as a result of other types of bonding are called **non-electrolytes**, and include most organic compounds, such as glucose, urea and creatinine.

Electrolytes are particularly important for three things within the body:

1. They are essential minerals.
2. They control the process of osmosis.
3. They help maintain the acid–base balance necessary for normal cellular activity.

## Elements

A chemical element is a pure substance that cannot be broken down into anything simpler by chemical means. Each element consists of just one type of atom distinguished by its atomic number (which is determined by the number of protons in the nucleus of an atom). If the number of protons in the nucleus of an atom changes, we have a new element. This is different from electrons, in which the number of electrons can change but the atom remains basically the same – although it is now an **ion**. Some common, but very important, examples of elements in the body include:

- iron (Fe)
- hydrogen (H)
- carbon (C)

- nitrogen (N)
- oxygen (O)
- calcium (Ca)
- potassium (K)
- sodium (Na)
- chlorine (Cl)
- sulphur (S)
- phosphorus (P).

As of 2012, there are 118 confirmed chemical elements in the **periodic table**. Of these elements, 98 occur naturally – the remaining 20 having been synthesized. All chemical matter consists of these elements, although new elements of higher atomic number are discovered from time to time – but only as a result of artificial nuclear reactions and are not naturally found in the body.

There are three classes of elements:

1. **Metals** (e.g. Fe, from ferrous).
2. **Non-metals** (e.g. O).
3. **Metalloids** (e.g. arsenic, As).

These three classes of elements all have certain characteristics that define them:

1. **Characteristics of metals:**
   i. They conduct heat and electricity.
   ii. They donate electrons (to other atoms to make molecules).
   iii. At normal temperatures they are all solids – with the exception of mercury (symbol 'Hg').
2. **Characteristics of non-metals:**
   i. They are poor conductors of heat and electricity.
   ii. They accept electrons (from donor atoms).
   iii. They may exist as a solid, a liquid, or a gas.
3. **Characteristics of metalloids:**
   i. They are neither metals nor non-metals – they are sometimes referred to as semi-metals.
   ii. They tend to have the physical properties of metals whilst having the same chemical properties of both metals and non-metals. However, they are not relevant to the biochemistry we are concerned with here, so will not be discussed in this book.

Usually metals bond with non-metals (electron donors with electron acceptors).

The following table lists some examples of metallic and non-metallic elements important for the body:

| Metals | Non-metals |
|---|---|
| Calcium (Ca) | Carbon (C) |
| Potassium (K) | Chlorine (Cl) |
| Sodium (Na) | Nitrogen (N) |
| | Oxygen (O) |
| | Sulphur (S) |
| | Phosphorus (P) |

As well as sodium chloride, some other important compounds include:

- calcium bicarbonate ($CaCO_3$)
- potassium chloride (KCl).

An interesting element is hydrogen (H), because it actually has properties of both metals and non-metals. As a consequence of hydrogen bonding, water ($H_2O$) is an example of a substance that, although it is made up of two gases – namely, oxygen (one molecule) and hydrogen (two molecules) – it becomes a liquid once these have bonded together.

### *Properties of elements*

All substances have certain individual properties, particularly in the way that they react (i.e. behave):

- **Physical properties** – these include such characteristics as colour, density, boiling point, melting point, suitability and hardness.
- **Chemical properties** – these include whether or not a substance is a metal or non-metal (or even metalloid), whether it reacts with an acid or an alkali substance, or whether it dissolves in water or alcohol.

## Compounds

A **compound** is a pure substance that is made up of two or more elements chemically bonded together. The properties of a compound are totally different from the individual properties of the elements that make that compound. We have already met one such compound in water ($H_2O$), which is made up of two gases (hydrogen and oxygen), which once combined in the ratio of two hydrogen atoms to one oxygen atom becomes a liquid. Compounds can be broken down chemically, whilst elements cannot. Two, other examples of a compound are:

- salt (NaCl) – one atom of sodium to one atom of chloride;
- carbon dioxide ($CO_2$) – one atom of carbon to two atoms of oxygen.

**Note:** You will have already noticed that after each chemical, there are one or more letters. These are symbols for chemicals. For example, H = hydrogen, O = oxygen, K = potassium, Na = sodium (Na stands for natrium – another name for sodium), P = phosphorus, and so on. It is important that you learn what are the chemicals and chemical symbols that are of most importance in the body, because these are often used in other books and in practice. Also, you need to know that when the chemical symbol for an atom has a small subscript number after it that this denotes there are that number of atoms of that particular molecule. Water ($H_2O$) is made up of two atoms of hydrogen and one atom of oxygen, whilst carbon dioxide ($CO_2$) consists of one atom of carbon and two atoms of oxygen, and calcium carbonate ($CaCO_3$) consists of one atom of calcium, one atom of carbon and three atoms of oxygen.

If a chemical symbol is preceded by a number (e.g. 2H), then this means that there are two atoms present, and this applies for all the atoms in that combination; for example, $2H_2O$ indicates there are four atoms of hydrogen ($2 \times 2$) and two atoms of oxygen. This will be discussed later in this chapter.

## Chemical equations/chemical reactions

Any mention of chemical equations and most non-chemists/non-scientists immediately start to panic, turning the page quickly. Do not to worry, however. Anyone capable of doing simple addition is capable of working through chemical equations. For example, you will have certainly worked through simple mathematical equations, such as:

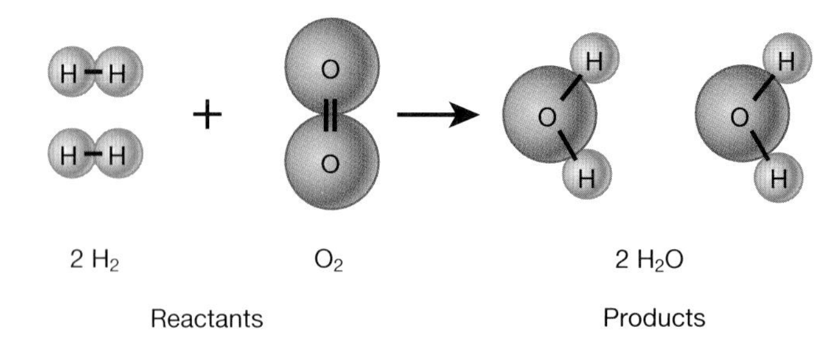


Figure 3.8 Pictorial depiction of the chemical equation (reaction) producing water.
*Source:* Tortora and Derrickson (2009), Figure 2.7, p. 36. Reproduced with permission of John Wiley and Sons, Inc.

- $1 + 1 = 2$
- $2 + 2 = 4$
- $4 + 4 = 8$
- $1 + 1 + 2 + 2 + 2 = 8$
- $1 + 1 + 2 + 2 + 2 = 4 + 3 + 1$

and so on.

Chemical equations work under the same basic principles. When a chemical reaction occurs (which is depicted by an equation), then a new substance is formed, which is called the **product** (as with a mathematical equation). However, in a chemical equation, this new substance (the product) will have different properties from the individual substances involved in the reaction (called the **reactants**).

As discussed earlier, when atoms are combined they form elements or molecules, and symbols are used to describe this process. This look at chemical equations will start with a very simple example, namely the production of water. Two atoms of hydrogen ($H_2$) combined with one atom of oxygen (O) produce one molecule of water ($H_2O$) (Figure 3.8). The chemical equation for this process is

$$H + H + O \rightarrow H_2O$$

hydrogen + hydrogen + oxygen → water

In this equation there are two atoms of hydrogen and one atom of oxygen on the left-hand side, and there are two atoms of hydrogen and one atom of oxygen on the right-hand side. However, because of the chemical reaction, the three atoms of gases have created water – a liquid.

Thus, a chemical equation is just a shorthand way of describing a chemical reaction. Note that the 'equals' sign in a mathematical equation is replaced by an arrow, meaning 'leads to', in a chemical equation. Basically, all chemical equations are as simple as this. There may be more reactants and products, but there are similar basic principles involved in chemical equations as in mathematical equations.

A very important basic principle to remember is that when chemical reactions occur, the amount of each substance must be the same after the reaction has occurred as was present before the reaction (just as in the simple sums at the start of this section). The two sides of a chemical reaction (chemical equation) – the reactants and the product(s) – must balance. In other words, no atoms/molecules are lost in a chemical reaction, they are just organized differently. Another thing to be aware of with chemical reactions is that although the numbers of atoms are the same before and after the reaction, during a chemical reaction, generally something is produced every time; that is, heat/energy.

- In a chemical equation, the reactants and the product may be separated by a single arrow ($\rightarrow$) indicating that the reaction occurs only in the direction that the arrow is pointing.
- Sometimes the reactants and product(s) may be separated by two arrows – one above the other and pointing in different directions ($\rightleftarrows$), indicating that the chemical reaction can be reversed.
- Separation of reactants and the products by an equals sign or by a right/left harpoon ($\rightleftharpoons$) indicates that a state of chemical equilibrium exists. **Chemical equilibrium** is the state in which, during a reversing reaction, both the reactants and the products are present at concentrations that will not change with time (Atkins and de Paula, 2006).

Another important principle to be aware of is that a chemical equation has to be consistent. The elements cannot be changed into other elements by chemical means.

If electrical charges are involved (as occurs with the involvement of **ions**), the net charge on both sides of the equation must be in equilibrium – that is, they must balance.

Now look at the following chemical reaction/equation; although more complicated, the principles just listed and discussed still hold for this:

$$Zn + 2HCl \rightarrow ZnCl_2 + H_2$$

zinc + hydrogen chloride $\rightarrow$ zinc chloride + hydrogen

In this chemical reaction/equation two molecules of hydrogen chloride, made up of two atoms of chlorine ($Cl_2$) and two atoms of hydrogen ($H_2$) – that is, 2HCl – plus one atom of zinc (Zn) have been changed to one molecule of zinc chloride ($ZnCl_2$) plus two atoms of hydrogen ($H_2$). So, the balance between the two sides of the equation in terms of numbers and types of atoms and electrical charge has not been altered.

With this even more complicated chemical reaction/equation, take some time and work out what is happening before reading on:

$$HCl + NaHCO_3 \rightleftarrows NaCl + H_2CO_3$$

hydrochloric acid + sodium bicarbonate $\rightleftarrows$ sodium chloride + carbonic acid

Finally, in this section on chemical equations, we will now look at one of the most important chemical reactions taking place in the body – without which life would not be possible – the production of salt and water. Again, look at it closely, and take some time working it out.

$$HCl + NaOH \rightleftarrows H_2O + NaCl$$

hydrochloric acid + sodium hydroxide $\rightleftarrows$ water + sodium chloride

Both sides of this reversing equation balance out. The two hydrogen atoms bind with the oxygen to give a water molecule, leaving the chlorine and sodium to bind together as sodium chloride (salt).

This is the end of the section on chemical equations. As can be seen, if someone can do simple arithmetical sums, they can understand and work with chemical equations (remembering that the equations are a depiction of chemical reactions that are taking place within the body all the time), so it is really quite easy to work out just what is happening when you see a chemical equation in a book – or even in practice.

## Organic and inorganic substances

All substances in life are classed as **organic** or **inorganic** depending upon their molecules.

**Organic molecules:**



- contain carbon (C) and hydrogen (H);
- are usually larger than inorganic molecules;
- dissolve in water and organic liquids;
- as a group include carbohydrates (sugars), proteins, lipids (fats) and nucleic acids (part of DNA).

**Inorganic molecules:**

- do not generally contain carbon (C);
- are usually smaller than organic molecules;
- usually dissolve in water or they react with water and release ions;
- as a group include water ($H_2O$), oxygen ($O_2$), carbon dioxide ($CO_2$) and inorganic salts.

# Units of measurement

To conclude this chapter introducing certain bioscientific concepts and preparing the reader for the remaining chapters, some brief notes about units of measurement are in order. This is an important section because the ability to identify and understand units of measurement will enhance the understanding of the complex organism known as humans. The accepted arbiter of the weights and measures we use is the International Bureau of Weights and Measures (2006).

A unit is a standardized, descriptive word specifying the dimension of a number. Traditionally, there have been seven properties of matter that have been measured independently of each other:

- **time** – measures the duration that something occurs;
- **length** – measures the length of an object;
- **mass** – measures the mass of an object;
- **current** – measures the amount of electric current that passes through an object;
- **temperature** – measures how hot or cold an object is;
- **amount** – measures the amount of a substance that is present;
- **luminous intensity** – measures the brightness of an object.

In the 1860s, British scientists took the lead in laying down the foundations for a coherent system based on length, mass and time. However, it was not until 1960 that an international system of units was agreed upon and published by most major countries (however, a notable exception to this agreement was the United States of America, along with Burma and Liberia). This new agreed system became known as the Système International d'Unités (or SI units for short). It is a system of units that relates present scientific knowledge to a unified system of units. Tables 3.1–3.3 provide SI units and measurements of weight, volume and length that will be useful as a reference whilst working through this book.

# Conclusion

This chapter has been an introduction to, and overview of, basic scientific principles. Although extremely fascinating in their own right, the real importance of these principles is that they

Table 3.1 The fundamental SI units and other common SI units. *Source:* Peate and Nair (2011), Tables 1.1 and 1.2, p. 21. Reproduced with permission of John Wiley and Sons, Ltd.

| Physical quantity | Unit | Symbol |
|---|---|---|
| *Fundamental SI units* | | |
| Length | metre | m |
| Mass | kilogram | kg |
| Time | second | s |
| Current | ampere | A |
| Temperature | kelvin | K |
| Amount of substance | mole | mol |
| Luminous intensity | candela | cd |
| *Other common SI units* | | |
| Force | newton | N |
| Energy | joule | J |
| Pressure | pascal | Pa |
| Potential difference | volt | V |
| Frequency | hertz | Hz |
| Volume | litre | L |



Table 3.2 Multiples of SI units. *Source:* Peate and Nair (2011), Table 1.3, pp. 21–22. Reproduced with permission of John Wiley and Sons, Ltd.

| Prefix | Symbol | Meaning | Scientific notation |
|---|---|---|---|
| Tera | T | One million million | $10^{12}$ |
| Giga | G | One thousand million | $10^{9}$ |
| Mega | M | One million | $10^{6}$ |
| Kilo | k | One thousand | $10^{3}$ |
| Hecto | h | One hundred | $10^{2}$ |
| Deca | da | Ten | $10^{1}$ |
| Deci | d | One tenth | $10^{-1}$ |
| Centi | c | One hundredth | $10^{-2}$ |
| Milli | m | One thousandth | $10^{-3}$ |
| Micro | μ | One millionth | $10^{-6}$ |
| Nano | n | One thousandth of a millionth | $10^{-9}$ |
| Pico | p | One millionth of a millionth | $10^{-12}$ |
| Femto | f | One thousandth of a pico | $10^{-15}$ |
| Atto | a | One millionth of a pico | $10^{-18}$ |

Table 3.3 Measures of weight, volume, length and energy. *Source:* Peate and Nair (2011), Tables 1.4–1.7, pp. 22–23. Reproduced with permission of John Wiley and Sons, Ltd.

| *Weight* | *Length* |
|---|---|
| 1 kilogram = 1000 grams | 1 metre = $10^{-3}$ kilometre |
| 1 gram = 1000 milligrams | 1 centimetre = $10^{-5}$ metre |
| 1 milligram = $10^{-3}$ gram | 1 millimetre = $10^{-6}$ metre |
| 1 microgram = $10^{-6}$ gram | 1 metre = 39.37 inches |
| 1 pound = 0.454 kilogram/454 grams | 1 mile = 1.6 kilometres |
| 1 ounce = 28.35 grams | 1 yard = 0.9 metre |
| 25 grams = 0.9 ounces | 1 foot = 0.3 metre |
| 1 ounce = 8 drams | 1 inch = 25.4 millimetres |
| *Volume* | *Energy* |
| 1 litre = 1000 millilitres | 1 calorie = 4.184 joules |
| 100 millilitre = 1 decilitre | 1000 calories = 1 dietary Calorie or kilocalorie |
| 1 millilitre = 1000 microlitres | 1 dietary Calorie = 4184 joules or 4.184 kilojoules |
| 1 UK gallon = 4.5 litres | 1000 dietary Calories = 4184 kilojoules |
| 1 pint = 568 millilitres | 1 kilojoule = 0.239 dietary Calories |
| 1 fluid ounce = 28.42 millilitres | |
| 1 teaspoon = 5 millilitres | |
| 1 tablespoon = 15 millilitres | |

underpin all the anatomy and physiology to which you will come in contact throughout the remaining chapters of this book.

As you can now appreciate, particularly if you have never studied science previously, biochemistry, anatomy and physiology are quite complicated, but at the same time very interesting, and not a little exciting. After all, you are learning about your bodies – their structures, functions and how they work. And who is not interested in anything to do with themselves? We are all interested in what is going on in our bodies – in good times and bad times. We want to know what happens when we grow, when we eat and drink to fuel our bodies, when we go to the toilet to rid our bodies of waste matter, when we exercise and when we die. Remember that what is happening to you is also happening to your patients, and so all this is important knowledge and understanding to possess in order to help you when you are looking after your patients and their families.

Think of this chapter as the beginning of a journey – as if you are packing your suitcase ready for your journey – a journey of learning, of self-knowledge and awareness, and of experience, all, of which will lead you to your destination: a good knowledge of the structures and functionings of the human body.

# Activities

**Now review your learning by completing the learning activities in this chapter. The answers to these appear at the end of the book. Further self-test activities can be found at www.wileyfundamentalseries.com/childrensA&P.**

## Crossword

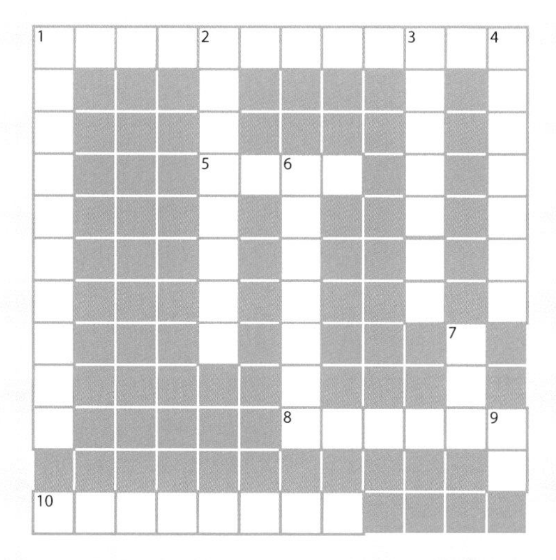


### Across

1. One type of organelle found in the cells.
5. The basis of all life consisting of protons, neutrons and electrons.
8. A very important atom that defines all life on earth and has the atomic number '6'.
10. The name of an atom that has the abbreviation of Cl and the atomic number of 17.

### Down

1. Term used to describe all chemical reactions involved in maintaining the living state of the cells and the organism.
2. A bond between atoms caused by the sharing of electrons between themselves.
3. A characteristic of life that an organism needs to do to obtain oxygen and release carbon dioxide.
4. The study of the physical structures of the body.
6. A chemical substance that contains a carbon molecule.
7. The number of atoms of hydrogen in one molecule of water.
9. Chemical abbreviation of sodium.

## Wordsearch

There are several words linked to this chapter hidden in the square. Can you find them? A tip – the words can go from up to down, down to up, left to right, right to left, or diagonally.



| | | | | | | | | | | | | | | | | | |
|---|---|---|---|---|---|---|---|---|---|---|---|---|---|---|---|---|---|
| Z | H | G | Z | I | S | P | M | M | U | P | I | R | G | L | D | S | M |
| Y | T | T | R | H | R | I | S | J | B | M | B | V | T | M | I | R | L |
| A | C | H | E | M | I | C | A | L | B | O | N | D | U | A | T | W | I |
| V | N | A | T | O | M | K | R | J | C | R | F | I | U | L | M | B | M |
| N | T | N | G | L | R | B | V | U | A | O | R | G | A | N | I | C | Y |
| R | N | S | M | E | M | P | Y | T | Y | Q | K | E | B | G | R | H | G |
| M | I | T | O | C | H | O | N | D | R | I | A | S | A | V | C | L | A |
| I | O | N | Q | U | O | I | J | I | T | M | I | T | R | Q | U | O | S |
| F | G | E | B | L | C | V | M | L | M | C | F | I | D | A | C | R | V |
| M | B | M | W | E | W | D | A | W | V | G | R | O | W | T | H | I | D |
| I | E | E | R | R | G | N | P | L | J | M | F | N | Q | S | Y | N | L |
| P | X | L | F | V | I | A | U | I | E | P | M | F | L | M | U | E | G |
| Q | Z | E | E | E | D | I | V | R | D | N | C | A | K | E | A | J | C |
| U | Y | L | T | C | A | G | D | M | G | K | C | E | M | B | P | M | M |
| E | T | O | P | F | T | M | V | C | J | O | X | Y | G | E | N | L | C |
| R | R | F | M | W | I | R | S | W | D | B | Z | G | F | U | U | P | V |
| P | O | M | B | L | V | W | O | V | E | N | E | R | G | Y | Q | M | D |
| S | O | D | A | K | M | A | J | N | E | S | A | K | P | L | P | S | M |

## Which is the odd one out?

1. (a) carbohydrates
   (b) water
   (c) lipids
   (d) proteins
2. (a) proton
   (b) anion
   (c) neutron
   (d) electron

3. (a) oxygen
   (b) water
   (c) lipids
   (d) carbon dioxide
4. (a) covalent
   (b) equatorial
   (c) polar
   (d) ionic

## Exercise

In the following equation, quantities of sulphur, oxygen and hydrogen combine together to form sulphuric acid.

1. Using the correct symbols can you complete the equation?

$$SO_3 + H_2O \rightarrow ?$$

sulphur trioxide + ? → sulphuric acid

2. What does $H_2O$ stand for?

# Glossary

**Acid:** a chemical substance with a low pH factor (see Chapter 2).

**Acid–base balance:** the relationship between an acidic environment and an alkaline one. This is essential for the maintenance of good health (see pH).

**Alkali:** a chemical substance with a high pH factor – the opposite of an acidic substance.

**Atmospheric pressure:** the force per unit area exerted against a surface by the weight of air above that surface.

**Anatomy:** the study of the physical structures of the body.

**Anion:** ions that carry a negative electrical charge.

**Antibody:** a protein that can recognize and attach to infectious molecules in the body, and so provoke an immune response to these infectious molecules (see Chapter 7).

**Atomic number:** the number of protons to be found in any one atom.

**Atoms:** the basis of all life; atoms are extremely tiny and consist of differing numbers of protons, neutrons and electrons.

**Base:** another name for an alkaline substance.

**Bonds:** the joining together of various substances, particularly atoms and molecules. See chemical bonds, covalent bonds, ionic bonds and polar/hydrogen bonds.

**Cation:** ions that carry a positive electrical charge.

**Chemical bond:** the 'attracting' force that holds atoms together.

**Chemical reaction:** a process by which chemical substances are transformed into something completely different, and usually depicted by a chemical equation.

**Compound:** a 'pure' substance that consists of two or more elements that are chemically bonded together.

**Covalent bond:** a bond between atoms caused by the sharing of electrons between themselves.

**Conductor:** an object or type of material that permits the flow of electric charges in one or more directions.

**Dissociation:** the act of disuniting or separating a complex object into parts.

**Electrolyte:** a liquid or gel that contains ions and can be decomposed by electrolysis; that is, a substance that is able to move to opposite electrically charged electrodes in fluids. Electrolytes affect the amount of water in the body, the acidity (pH) in the body, muscle function and other important processes.

**Electron:** the part of an atom that carries a negative electrical charge. See neutrons and protons.

**Element:** a pure chemical substance consisting of one type of atom that is distinguished by its atomic number. Elements are divided into metals, metalloids and non-metals.

**Enzymes:** proteins, produced by cells, that can cause very rapid biochemical reactions in the body.

**Hydrogen bond:** another name for a polar bond.

**Hydrostatic pressure:** the pressure exerted by a fluid at equilibrium at a given point within the fluid, due to the force of gravity.

**Inorganic substances:** substances that do not contain carbon molecules (e.g. water).

**Ionic bonds:** bonding that takes place when atoms lose or gain electrons, so altering the electrical charge of the atoms.

**Ions:** atoms that are no longer in an electrically stable state (i.e. they are no longer electrically neutral, but are either positively or negatively electrically charged).

**Metabolism:** the term used to describe all chemical reactions involved in maintaining the living state of the cells and the organism.

**Metabolic process:** a set of chemical transformations within the cells of living organisms. These chemical reactions are catalysed by enzymes and allow organisms to carry out their basic functions (e.g. grow and reproduce).

**Mole:** the unit of measurement for the amount of a substance.

**Molecules:** the smallest part of an element or compound that can exist on its own (e.g. sodium chloride).

**Neutral substance:** a chemical substance that is neither acidic nor alkaline.

**Neutron:** the part of an atom that carries no electrical charge (i.e. they are neutral). See electron and proton.

**Organic substance:** a chemical substance that contains carbon molecules (e.g. carbohydrates, lipids).

**Organelle:** structural and functional parts of a cell (see Chapter 4).

**Osmosis:** the movement of water across a semipermeable membrane from an area of low solute concentration to an area of high solute concentration; this allows for equilibrium of solute and water density on both sides of the semi-permeable membrane.

**Periodic table:** a table of all the known chemical elements, organized on the basis of their atomic numbers, and recurring chemical properties.

**pH:** a measure of the acidity or alkalinity of a solution; see acid–base balance and Chapter 3.



**Physiology:** the way in which the bodily structures function.

**Polar bonds:** these bonds occur when atoms of different electromagnetic negativities form a bond. As a consequence, the molecules that are then formed also carry a weak negative electrical charge that allows molecules to covalently bond – just like atoms. They are also known as **hydrogen bonds** because hydrogen molecules generally have to be present for polar bonding to occur.

**Product (chemical reactions):** the new substance formed following a chemical reaction.

**Proton:** the part of an atom that carries a positive electrical charge. See electron and neutron.

**Reactant (chemical reaction):** the individual substances that are involved in a chemical reaction.

**Receptors (of a cell surface):** specialized integral membrane proteins that take part in communication between the cell and the outside world.

**Shell (of an atom):** the name that is given to the orbits of electrons moving around the nucleus (containing protons and electrons) of an atom.

**Valency:** a measure of the strength of the combining power of atoms.

## References


Atkins P., de Paula, J. (2006) *Atkins' Physical Chemistry*, 8th edn, W.H. Freeman, New York, NY.

Fisher, J., Arnold, J.R.P. (2004) *Chemistry for Biologists*, 2nd edn, BIOS Scientific Publishers, London.

International Bureau of Weights and Measures (2006) *The International System of Units (SI)*, 8th edn, BIPM, Paris.

Marieb, E.N., Hoehn, K. (2009) *Human Anatomy & Physiology*, 8th edn, Pearson Educational, Harlow.

Peate, I., Nair, M. (eds) (2011) *Fundamentals of Anatomy and Physiology for Student Nurses*, Wiley–Blackwell, Chichester.

Tortora, G.J., Derrickson, B.H. (2009) *Principles of Anatomy and Physiology*, 12th edn, John Wiley & Sons, Inc., Hoboken, NJ.

# Chapter 4

# The cell

Peter S. Vickers

## Aim

To introduce the student to the cell, which underpins the whole anatomy (structure) and functioning (physiology) of the body. To explore the cell's own structure and functioning in order to gain an understanding of the wonders of this very small, but intensely dynamic and exciting tiny powerhouse of a biological miracle.

## Learning outcomes

On completion of this chapter the reader will be able to:

- Understand the structure of a cell.
- Outline the structure and function of the plasma membrane.
- List and describe the functions of the organelles.
- Explain cellular fluid transport.
- Identify the fluid compartments of the body.

## Test your prior knowledge

- What are the three main parts of a human cell?
- What is meant by a selective permeable membrane?
- Name two important functions of the cell membrane.
- What is the difference between passive and active transport?
- What are the names of the two principal body fluid compartments?
- What do we mean by hydrostatic pressure?
- What is the function of mitochondria in the cell?
- What occurs in osmosis?
- What is the difference between exocytosis and endocytosis?
- Can you name one of the major hormones that regulates fluid and electrolytes in the body?

---

*Fundamentals of Children's Anatomy and Physiology: A Textbook for Nursing and Healthcare Students*, First Edition. Edited by Ian Peate and Elizabeth Gormley-Fleming.

Companion website: www.wileyfundamentalseries.com/childrensA&P

# Introduction

The body is a wonderful piece of biochemical engineering. It functions so well that most of the time we are unaware of the basic functions of the body, such as respiration, digestion and excretion. In the same way, the cells – which combine together to make up the body – are miracles of biochemical engineering. They function so that the body itself can function, and in this chapter we will explore the cells in order to better understand the structure and functioning of the body.



We are all the result of just two cells fusing together in our mother's uterus (the fertilized egg), but by the time that we are born we consist of approximately $2 \times 10^{12}$ cells (2 000 000 000 000 cells), all of them a result of that first fused cell continually dividing over the 9 months before birth. This process of cell division will be explored in Chapter 5 – Genetics. In the average adult human, there are around 10 trillion cells (10 000 000 000 000), although that depends upon height and weight. Actually, we possess none of those cells with which we were born. In fact, all the cells in our bodies are no more than 10 years old. However, all our present cells are exact clones of the original cells we had at birth.

There are many different types of cells in our body, and they differ as regards to their size, shape, colour, behaviour and habitat. However, there are many similarities between our different cells, including, for example, their chemical composition, their chemical and biochemical behaviour, and detailed structure (Vickers, 2009). The correct biological term for a 'cell' is '**cyte**'.

# Characteristics of the cell

A cell is the structural and functional unit of all living organisms (Nair, 2011) – the smallest part of the body that is capable of the processes that define life (Parker, 2007), and the building block of all living organisms. For example, the human body is made up of many millions of cells and is therefore known as a **multicellular organism**. On the other hand, some organisms consist of just one cell and are thus known as unicellular organisms. An example of a unicellular organism is a bacterium. There are many different kinds of cells (Figure 4.1), but all the cells of the human body work together by cooperating in order to give structure, function and life to the body.

Cells are complex structures that, in many ways, carry out very similar processes as do human beings and all animals. They are made up of many different parts, each of which is fundamental to the life and health of the cell.

Most cells are very tiny. In fact, most are **microscopic**, so they can only be seen with the aid of a powerful microscope (Parker, 2007). A typical cell is less than 10–30 $\mu$m (micrometres) in size. However, there is much variation in size amongst cells, ranging in size from about 4 $\mu$m (the granule cell of the cerebellum), to the largest (nerve cells), some of which can be a metre long – although only reaching about 10 $\mu$m in width.

If you look at Figure 4.1, you will see that the cells are of many various shapes and appear quite simple, but in actual fact they are very complex in structure, as can be seen in Figure 4.2.

Inside the cell, many vital chemical activities take place; for example, just like humans, cells respire (breathe), consume (eat), remove waste matter (excrete), grow, **metabolize** (change structures and chemicals into other structures and chemicals – e.g. change food into energy) and reproduce; they also die. In addition, different types of cells carry out differing functions; all these activities (and the very structure and nature of the cell) are programmed into the cell

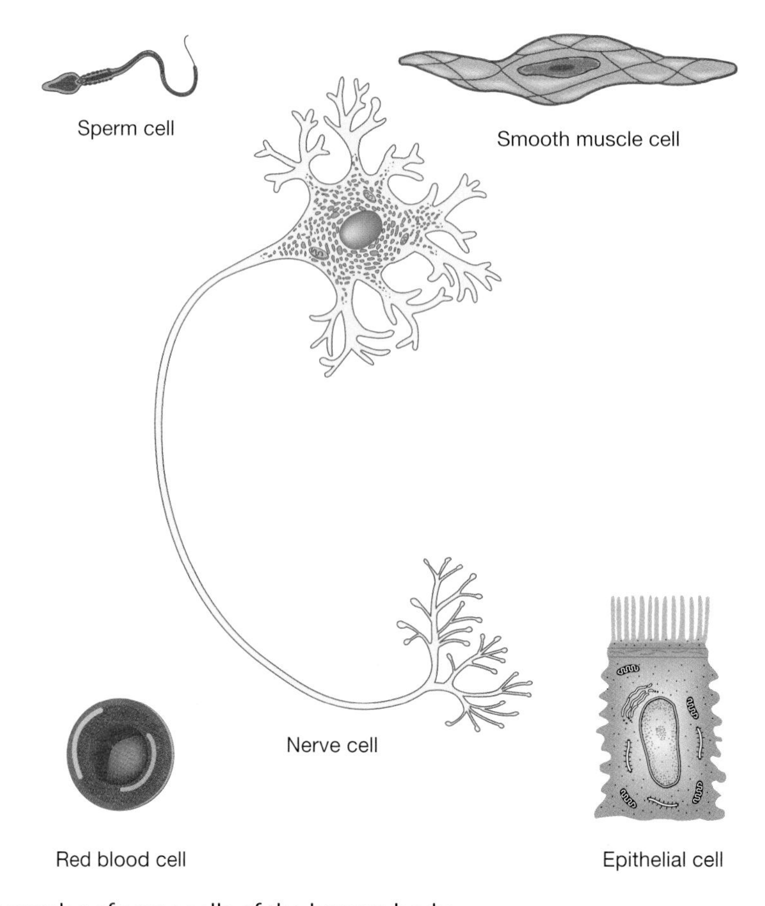


Figure 4.1 Examples of some cells of the human body.
*Source:* Tortora and Derrickson (2009), Figure 3.35, p. 100. Reproduced with permission of John Wiley and Sons, Inc.

by a unique set of instructions that they carry within them. These instructions are found in the cell's **deoxyribonucleic acid** (**DNA**).

## Characteristics of cells

- Cells are active – carrying out specific functions.
- Cells require nutrition to survive and function. They use a system known as **endocytosis** in order to catch and consume nutrients – they surround and absorb organisms such as bacteria and then absorb their nutrients. These nutrients are used for the storage and release of energy, as well as for growth and for repairing any damage to themselves.
- Cells can reproduce themselves, not by means of sexual reproduction but by **asexual reproduction**, in which they first of all develop double the number of **organelles** and then divide, with the same number and types of organelle and structure present in each half. This is known as **simple fission**.
- Cells excrete waste products.
- Cells react to things that irritate or stimulate them; for example, in response to threats from chemicals and viruses.

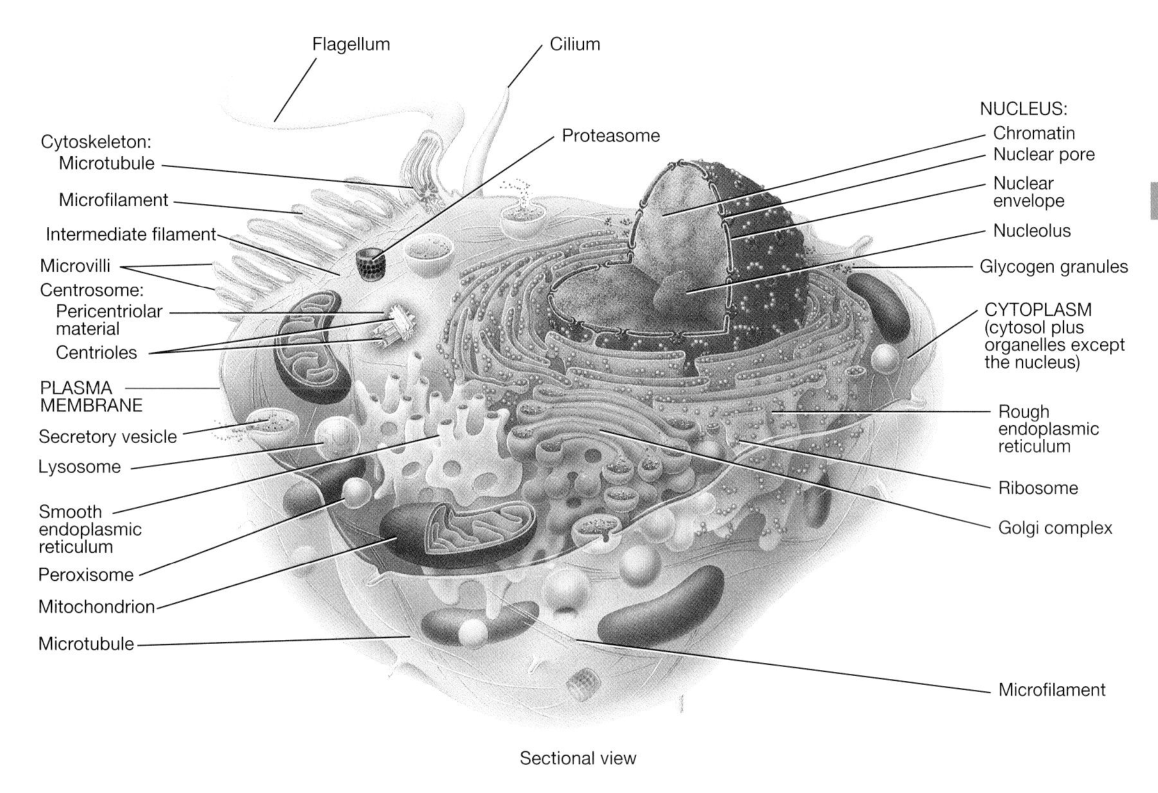


Figure 4.2 Structure of the cell.
*Source:* Tortora and Derrickson (2009), Figure 3.1, p. 62. Reproduced with permission of John Wiley and Sons, Inc.

# The structure of the cell

There are four main compartments of the cell:

- **cell membrane**
- **cytoplasm**
- **nucleus**
- **nucleoplasm**.

Within these compartments are many **organelles**, which are like the cell's internal organs. These organelles perform many functions to keep cells alive and functioning.

## The cell membrane

As can be seen in Figure 4.2, the various structures of the cell are contained within a cell membrane (also known as the **plasma membrane**). This cell membrane is a **semipermeable** biological membrane separating the interior of the cell from the outside environment, and protecting the cell from its surrounding environment. It is semipermeable because it allows only certain substances to pass through it for the benefit of the cell itself. For example, it is selectively permeable to certain ions and molecules (Alberts *et al.*, 2002). Contained within the cell membrane are the **cytoplasm** and the organelles, which include amongst others such organelles as the lysosomes, mitochondria and the nucleus of the cell.

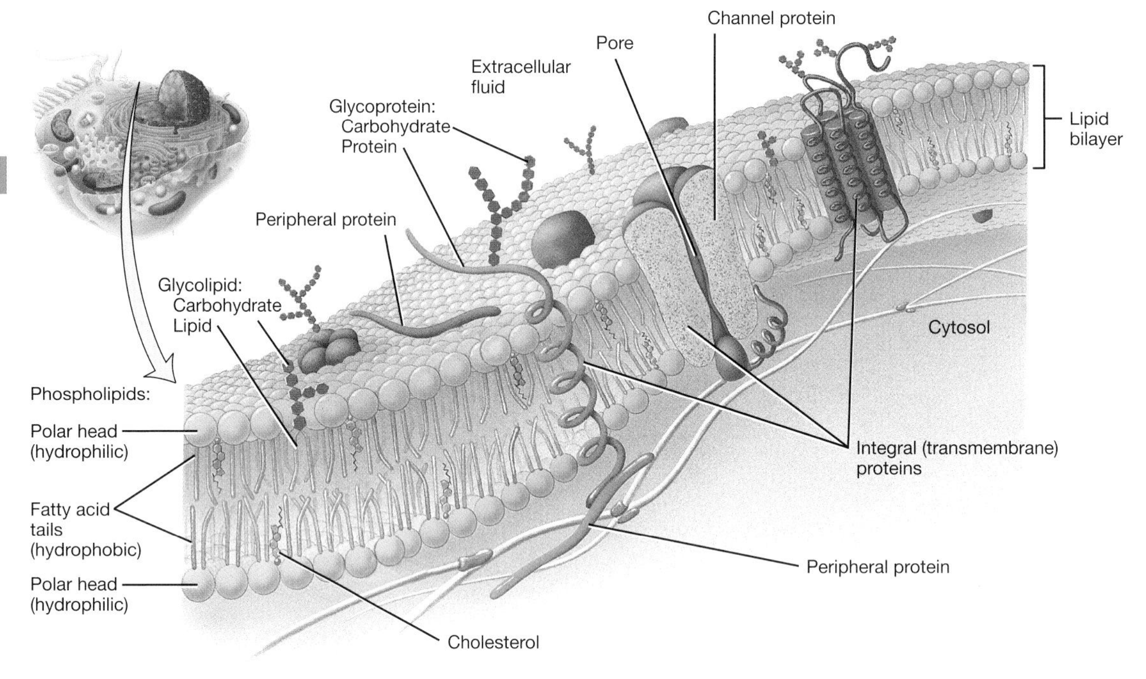


Figure 4.3 Cell membrane.
*Source:* Tortora and Derrickson (2009), Figure 3.2, p. 63. Reproduced with permission of John Wiley and Sons, Inc.

The cell membrane, which can vary in thickness from 7.5 nm (nanometres) to 10 nm (Vickers, 2009) is made up of a self-sealing double layer of **phospholipid** molecules with protein molecules interspersed amongst them (Figure 4.3) A phospholipid molecule consists of a polar 'head' (which is **hydrophilic** – mixes with water) and a tail (made up of non-polar fatty acids, which are **hydrophobic** – they do not mix with water). In the bilayer of the cell membrane, all the heads of each phospholipid molecule are situated facing outwards on the outer and inner surfaces of the cell, whilst the tails point into the cell membrane; it is this central part of the cell membrane consisting of hydrophobic tails that makes the cell impermeable to water-soluble molecules (Nair, 2011). In addition to the phospholipid molecules, the cell membrane contains a variety of molecules, mainly proteins and lipids, and these are involved in many different cellular functions, such as communication and transport. The proteins inserted within the cell membrane are known as **plasma member proteins** (**PMPs**), which can be either **integral** or **peripheral**. Integral PMPs are embedded amongst the phospholipid tails, whilst others completely penetrate the cell membrane. Some of these integral PMPs form channels for the transportation of materials into and out of the cell; others bind to carbohydrates and form **receptor sites** (attaching bacteria to the cell to allow for their destruction – see Chapter 7). Other examples of integral transmembrane PMPs include those that transfer potassium ions in and out of cells, receptors for insulin, and types of neurotransmitters (Nair, 2011). On the other hand, peripheral PMPs bind loosely to the membrane surface, and so can be easily separated from it. The reversible attachment of proteins to cell membranes has been shown to regulate cell signalling, and these cell membrane proteins are also involved in many other important cellular events, such as acting as **enzymes** to **catalyse** cellular reactions through a variety of mechanisms (Cafiso, 2005).

## Fluid mosaic model

According to the fluid mosaic model, biological membranes can be considered as a two-dimensional liquid in which lipid and protein molecules **diffuse** more or less easily (Singer and Nicolson, 1972). Although the lipid bilayers forming the basis of the membranes do form two-dimensional liquids by themselves, the plasma membrane also contains a large quantity of proteins, providing more structure. Examples of such structures are protein-to-protein complexes formed by the **cytoskeleton**.



## Functions of the cell membrane

- It anchors the cytoskeleton – a lattice-like array of fibres and fine tubes integral to a cell's shape.
- It allows the cells to attach to each other and form tissues by attaching to the **cellular matrix**.
- It is responsible for the transport of materials/substances needed for the functioning of the cell organelles.
- By means of its protein molecules, it receives signals from other cells or from the outside environment, which it then converts into messages for organelle response.
- In some of our cells, the membrane protein molecules group together to form enzymes.
- The cell membrane proteins help to transport very small molecules through the cell membrane – but only if the very small molecules are moving from an area of high concentration of these molecules to one of low concentration.

These proteins in the cell membrane have many functions; for example:

- They provide structural support to the cell.
- Some are enzymes – helping to provide chemical reactions.
- Some regulate water-soluble substances through the **pores** in the cell membrane.
- Some are receptors, enabling hormones and other substances, such as neurotransmitters, to play their roles.
- Finally, some of these proteins are **glycoproteins** playing an important role in cell-to-cell recognition; for example, the mucins found in the gastrointestinal and respiratory tracts.

Having introduced the cell membrane, we can now explore some of its functions and roles in more detail.

## Transport systems

There are two processes that are used by the cell to move substances through its membrane: the **passive** and **active transport** systems.

### *Passive transport*

A passive transport system is one in which molecules pass through the cell membrane without the use of cellular energy, but by moving down a concentration gradient. There are four types of passive transport involved in the cell membrane:

- simple diffusion
- facilitated diffusion
- osmosis
- filtration.

Active transport, however, necessitates the use of cellular energy to move substances through the cell membrane. Active transport systems include:

- endocytosis
- exocytosis
- active transport using adenosine triphosphate (ATP).

## *Simple diffusion*



Simple diffusion is the net passive movement of molecules or **ions** due to their **kinetic energy** from an area of higher to one of lower concentration until a state of equilibrium is reached. Diffusion may occur through **selectively permeable membranes**, where large and small lipid-soluble molecules pass through the phospholipid bilayer of the membrane – for example, the movement of oxygen and carbon dioxide between the blood and cells. The rate of diffusion depends upon several factors (Nair, 2011):

- The substance being diffused – gases diffuse rapidly, whilst liquids diffuse more slowly.
- The temperature – the higher the temperature, the faster the diffusion takes place.
- The size of molecules – smaller molecules (e.g. glycerol) diffuse much faster than larger molecules (e.g. fatty acids).
- The size of the surface area of the cell membrane over which the molecule works.
- The solubility of the molecule being transported.
- The concentration gradient.

## *Facilitated diffusion*

As with simple diffusion, this process transports larger molecules across a cell membrane and does not expend cellular energy. Unlike smaller uncharged molecules, which can cross the cell membrane by simple diffusion, larger molecules need to 'hitch a ride' with other proteins in order to cross a selectively permeable cell membrane (Figure 4.4). Among these larger molecules are various sugars, such as glucose, which is insoluble and so cannot penetrate the membrane. In order to cross the cell membrane, the glucose is picked up by a carrier molecule (in this case insulin) and the combined glucose–carrier, which is soluble in the phospholipid bilayer of the membrane, is able to diffuse through the membrane to the inside of the cell where

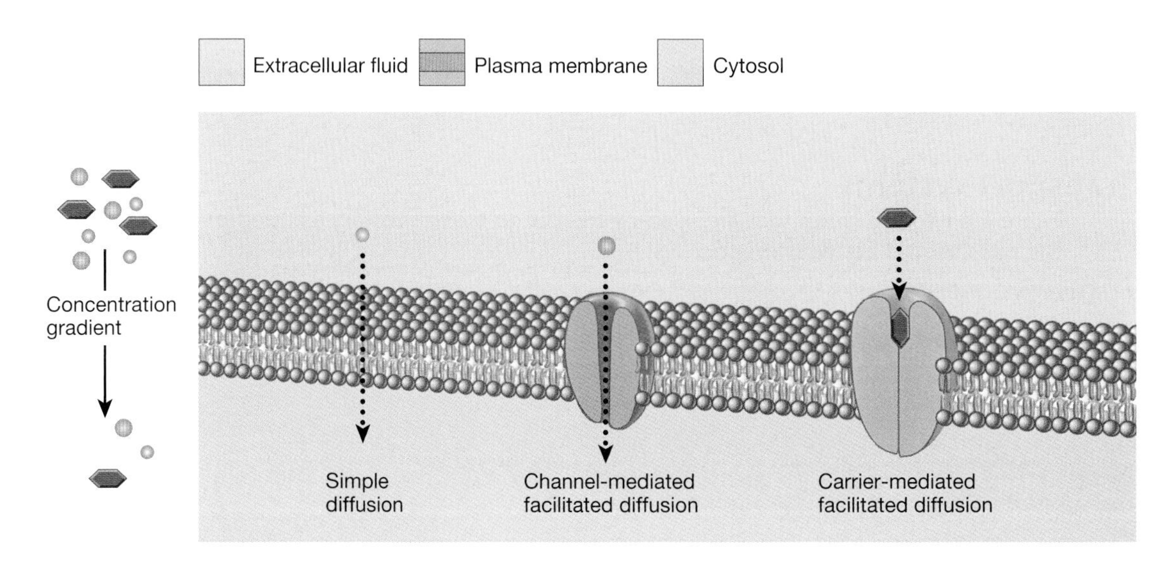


**Figure 4.4** Facilitated diffusion.
*Source:* Tortora and Derrickson (2009), Figure 3.5, p. 67. Reproduced with permission of John Wiley and Sons, Inc.

it releases the glucose to the cytoplasm, thus lowering the blood glucose level by accelerating the transportation of glucose from the blood into the cells. Failure to do this leads to diabetes mellitus.

### *Osmosis*

Osmosis is the net movement of water molecules (due to kinetic energy) across a selectively permeable membrane from an area of higher concentration to one of lower concentration of water until equilibrium is reached. The water molecules pass through channels in the integral proteins within the cell membrane. The passage of water through a selectively permeable membrane generates a pressure called **osmotic pressure**. Osmotic pressure is the pressure needed to stop the flow of water across the membrane and is an important force in the movement of water between various compartments of the body – for example, the kidney tubules.



How concentrated a solution is depends upon the amounts of **solutes** that are dissolved in the water. If there is a high concentration of salt on one side of the cell membrane, then there is less space for water molecules in that same area. In this case, water will move through the cell membrane from the side of the cell membrane with the greater number of water molecules to the area that has fewer water molecules; that is, osmotic pressure (Colbert *et al.*, 2007).

If osmotic pressure rises too much, then it can cause damage to the cell membrane, so the body attempts to ensure that there is always a reasonable constant pressure between the cell's internal and external environments. We can see the possible damage if, for example, a red blood cell is placed in a low concentrated solute, as then it will undergo haemolysis. On the other hand, if it is placed in a highly concentrated solute the result will be a crenulated cell (Figure 4.5). If the red blood cell is placed in a solution with a relatively constant osmotic pressure, it

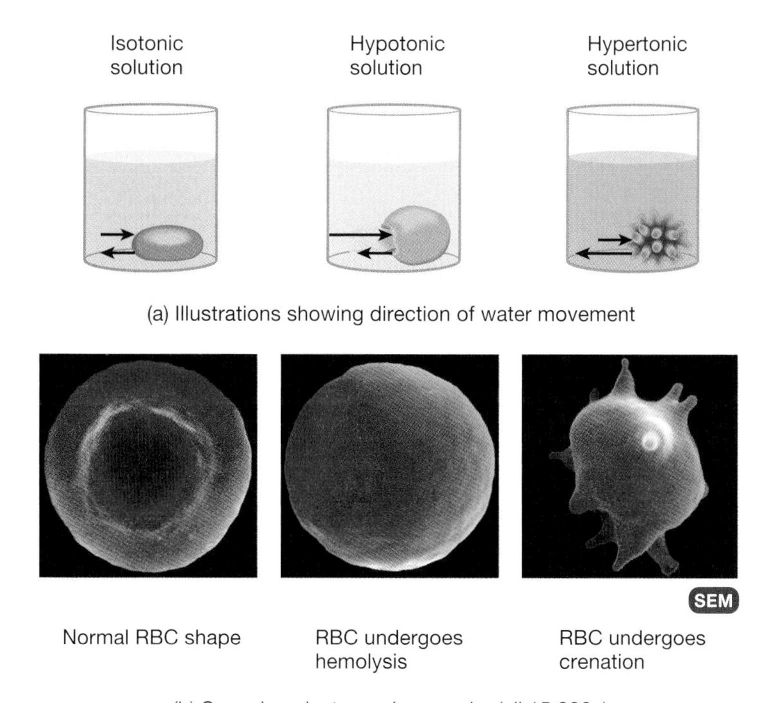


Figure 4.5 Effect of solute concentration on a red blood cell.
*Source:* Tortora and Derrickson (2009), Figure 3.9(a), p. 71. Reproduced with permission of John Wiley and Sons, Inc.

will not be affected because the net movement of water in and out of the red blood cell is minimal (Nair, 2011).

### Filtration

Filtration is the movement of **solvents** and solutes across a selective permeable membrane as a result of gravity or hydrostatic pressure from an area of higher to lower pressure, and this process continues until the pressure difference no longer exists. Most small to medium-sized molecules can be forced through a cell membrane; large molecules or aggregates cannot. Filtration occurs in the kidneys, where the blood pressure forces water and small molecules (e.g. urea and uric acid) through the thin cell membranes of tiny blood vessels and into the kidney tubules. However, the large blood cells are not forced through, but the harmful smaller molecules (e.g. urea) are, and are then eliminated in the urine.

### Active transport

Although facilitated diffusion is the commonest form of protein-mediated transport across the cell membrane, it tends to be overshadowed by active transport. Rather than solutes moving down their concentration gradients to reach equilibrium, in active transport they are actively 'pumped' up a gradient using energy from another source – **ATP**.

An active transport system relies upon the use of cellular energy to move any substances through selective permeable membranes, whereas the passive system does not. This cellular energy occurs as a result of ATP being split into **adenosine diphosphate** (**ADP**) and phosphate (Figure 4.6) ATP is a compound consisting of a base, a sugar and three phosphate ions (hence triphosphate) held together by high-energy bonds, which once broken release a high level of energy and the release of one of the phosphate ions. The phosphate ion that has broken away from the original ATP molecule (leaving behind ADP as the result of the chemical reaction) joins up with another ADP molecule to form a further ATP molecule which will then have energy stored in the phosphate bonds, and so on. This forms the cellular energy that is so important in active transport (Vickers, 2009; Nair, 2011). A typical body cell will expend up to 40% of its ATP for active transport.

Active transport has two advantages over facilitated diffusion:

1. It allows desirable solutes to be accumulated within the cell whilst allowing undesirable ones to be removed.

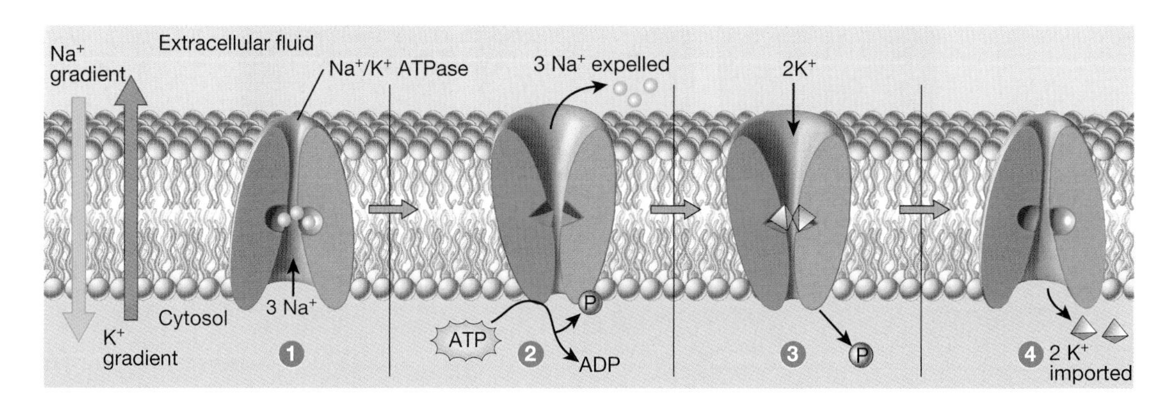


Figure 4.6 Active transport system.
*Source:* Tortora and Derrickson (2009), Figure 3.10, p. 72. Reproduced with permission of John Wiley and Sons, Inc.

2. Much of the energy used in the 'uphill pumping' is conserved, so active transport can provide a way of storing energy, because, by moving back down the concentration gradient, the stored solute can release the energy needed for its accumulation.

Regarding cellular energy, there are four main active transport systems – two pumps and two cotransporters:

1. **The sodium–potassium pump.** The process of moving sodium and potassium ions across the cell membrane involving the hydrolysis of ATP in order to provide the necessary energy for this transport (as briefly described earlier).
2. **The calcium pump.** This pump uses cellular energy to pump calcium ions into the cells of muscles (calcium is crucial for all muscle contraction – including the heart muscle).
3. **Sodium–glucose linked cotransporter.** A family of glucose transporters found in the intestinal mucosa of the small intestine and the proximal tubule of the nephron which contribute to renal glucose reabsorption.
4. **High-affinity hydrogen–glucose cotransporter.** A mechanism for the transfer of glucose and hydrogen ions ($H^+$) from one side of a cell membrane to the other.

To demonstrate how these mechanisms work with regard to integral protein membranes and cellular energy, take the example of glucose. Glucose enters a channel in an integral membrane protein; when the glucose makes contact with an active site in the channel, the energy from the splitting of ATP induces a change in the membrane protein, expelling the glucose through the membrane.

## Endocytosis and exocytosis

### *Endocytosis*

**Endocytosis** is a process by which cells absorb molecules (such as proteins) by engulfing them. It is used by all cells when substances that are important are too large to pass through the cell membrane. Endocytosis involves part of the cell membrane being drawn into the cell along with particles and/or fluids that the cell will ingest. This membrane is then pinched off to form a membrane-bound **vesicle** within the cell (Figure 4.7). The opposite of endocytosis is exocytosis (Marsh and McMahon, 1999).

There are three types of endocytosis:

- **Phagocytosis.** This process results in the ingestion of particulate matter (e.g. bacteria and other cells) from the extracellular fluid (ECF). This process only occurs in certain specialized cells, such as the **neutrophils** and **macrophages** of the blood and immune system.
- **Pinocytosis.** This process, occurring in all cells, results in the engulfing and absorbing of relatively small particles and fluids.
- **Receptor-mediated endocytosis.** This involves the engulfing and absorbing of large molecules, particularly protein, but it also has the important feature of being highly selective because it involves specific receptors that bind to the large molecules in ECF.

### *Exocytosis*

Exocytosis is the removal of unwanted particulate matter from the cytoplasm to the outside of the cell. The process involves the intracellular vesicle with its ingested matter fusing with the cell membrane and discharging its contents into the ECF. The materials excreted by a cell could be a waste product, but it could also function as a regulatory molecule – the cells may communicate with each other through the products that they secrete by means of exocytosis (Figure 4.8).

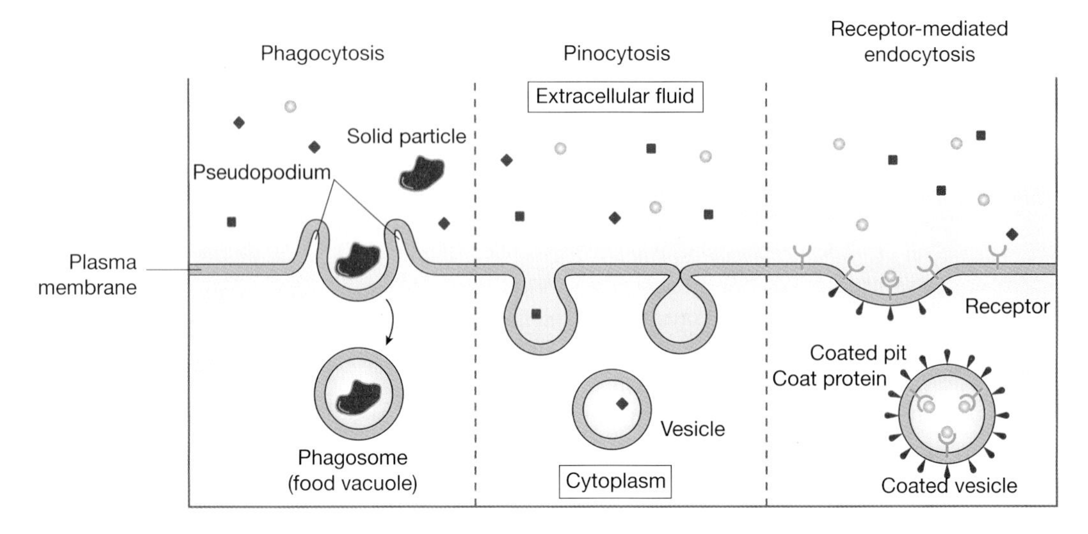


**Figure 4.7** Pinocytosis, phagocytosis and receptor-mediated endocytosis.
*Source:* Peate and Nair (2011), Figure 2.8, p. 42. Reproduced with permission of John Wiley and Sons, Ltd.

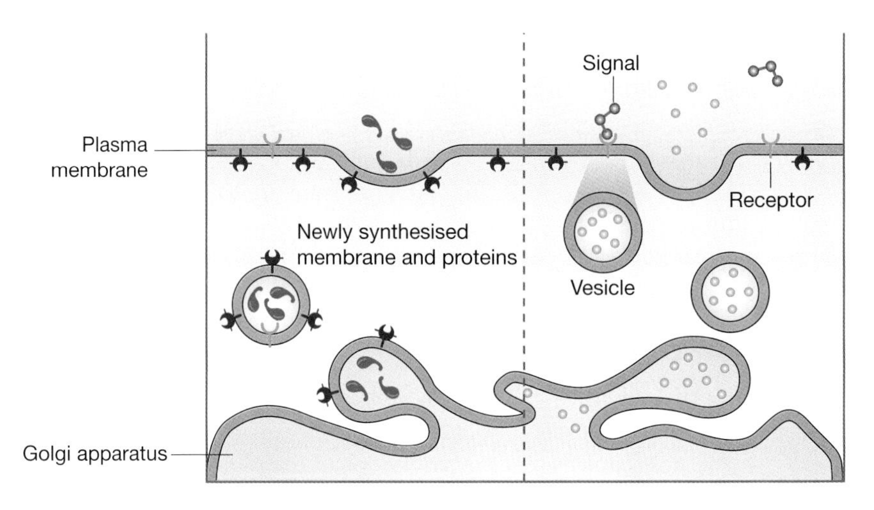


**Figure 4.8** Exocytosis.
*Source:* Peate and Nair (2011), Figure 2.9, p. 43. Reproduced with permission of John Wiley and Sons, Ltd.

Many cells in the body use exocytosis to release enzymes or other proteins that act in other areas of the body, or to release molecules that help cells to communicate with one another. The regulation of glucose is a good example of this process in which the alpha and the beta ($\alpha$- and $\beta$-) **cells** in the **islets of Langerhans** (in the pancreas) secrete the hormones glucagon and insulin respectively. If the level of glucose in the body rises, the $\beta$-cells are stimulated to produce and secrete more insulin through exocytosis. Exocytosis in other cells in the pancreas also releases digestive enzymes into the gut.

Table 4.1 Cellular compartments and their functions. *Source:* Peate and Nair (2011), Table 2.1, p. 36. Reproduced with permission of John Wiley and Sons, Ltd.

| Components | Functions |
|---|---|
| Centrioles | Cellular reproduction |
| Chromatin | Contains genetic information |
| Cilia (pleural) | Moves fluid or particles over the surface of the cell |
| Cytoplasm | Fluid portion that supports organelles |
| Endoplasmic reticulum (rough and smooth) | Many functions, including site for protein transportation, modification of drugs and synthesis of lipids and steroids |
| Glycogen granules | Stores for glycogen |
| Golgi complex | Packages proteins for secretion |
| Intermediate filament | Helps to determine the shape of the cell |
| Lysosomes | Break down and digest harmful substances. In normal cells, some of the synthesized proteins may be faulty – lysosymes are responsible for their removal |
| Microfilaments | Provide structural support and cell movement |
| Microtubules | Provide conducting channels through which various substances can move through the cytoplasm. Provide shape and support for cells |
| Microvilli | Increase cell surface area; site for secretion, absorption and cellular adhesion |
| Mitochondria | Energy-producing site of the cell. Mitochondria are self-replicating |
| Nucleolus | Site for the formation of ribosomes |
| Nucleus | Contains genetic information |
| Peroxisomes | Carry out metabolic reactions. Site for the destruction of hydrogen peroxide. Protects the cell from harmful substances, such as alcohol and formaldehyde |
| Plasma membrane | Regulates substances in and out of a cell |
| Ribosomes | Sites for protein synthesis |
| Secretory vesicles | Secrete hormones, neurotransmitters |

# The organelles

The organelles in the cell are like small 'organs' in cells. Figure 4.2 portrays a diagrammatic representation of the organelles in the cell, which we will now explore. Table 4.1 gives a brief oversight of the cell organelles and their functions.

## Cytoplasm

Although perhaps not, strictly speaking, an organelle, the cytoplasm is a very important part of the interior of a cell. It is the **ground substance** (**matrix**) in which the various cellular components are found; it is a thick, semi-transparent, elastic fluid containing suspended particles and the cytoskeleton. The cytoskeleton provides support and shape for the cell and is involved in the movement of structures in the cytoplasm and the whole cell.

Chemically, cytoplasm is 75–90% water plus solid components – mainly proteins, carbohydrates, lipids and inorganic substances. It is also the place in which some chemical reactions occur. It receives raw materials from the external environment and converts them into usable energy by decomposition reactions (remember ATP). Cytoplasm is also the site where new substances are synthesized for the use of the cell; it is here that chemicals are packaged for transport to other parts of the cell, or to other cells, and it is here that chemicals facilitate the excretion of waste materials.

## Nucleus

The nucleus is like the brain of the cell; however, not all cells have a nucleus. **Prokaryotic cells** do not have a nucleus, whilst most **eukaryotic cells** do. Eukaryotic cells are found in animals and plants, whilst prokaryotic cells are very typical of bacteria. Most human cells have a single nucleus, but some have more than one; for example, some muscle fibres. Other human cells have no nucleus once they have matured; for example, the mature red blood cell.

The nucleus is the largest structure in the cell and is surrounded by a nuclear membrane that has two layers and, like the cell membrane, is selectively permeable. The interior of the nucleus consists of **protoplasm** and is known as **nucleoplasm**; and unless the cell is dividing, there is little variation in the appearance of the nucleoplasm. During division it is possible to observe some spherical bodies known as nucleoli which are responsible for the production of **ribosomes** from ribosomal **ribonucleic acid** (RNA). The RNA is stored in the nucleoli and assumes a function in **protein synthesis**. The nucleoli disperse during cell division and are reformed once two new cells have been formed from the dividing cell.

Inside the nucleus is found the genetic material, which consists principally of **DNA** – when the cell is not reproducing prior to cell division, the genetic material is a thread-like mass called **chromatin**. Just before cell division, the chromatin shortens and coils into rod-shaped bodies – the chromosomes. The basic structural unit of a chromosome is a nucleosome, which is composed of DNA and protein (histones). In humans there are normally 23 pairs of chromosomes in the cells. The exception is sperm and ova, which carry only one copy of each chromosome. DNA has two main functions:

- It provides the genetic blueprint which ensures that the next generation of cells is similar to existing cells.
- It provides the 'plans' for the synthesis of protein in the cell itself, and this information is stored in our genes.

You will learn much more about chromosomes, genes and cell division in Chapter 5.

## Endoplasmic reticulum

The endoplasmic reticulum (ER; Figure 4.9) consists of membranes that form an interconnected series of tubules, vesicles and channels (**cisternae**) dividing the cytoplasm into compartments. There are two types:

- Granular (rough) ER, which is associated with ribosomes, is involved in the synthesis of proteins, and is a 'membrane factory' for the cell.

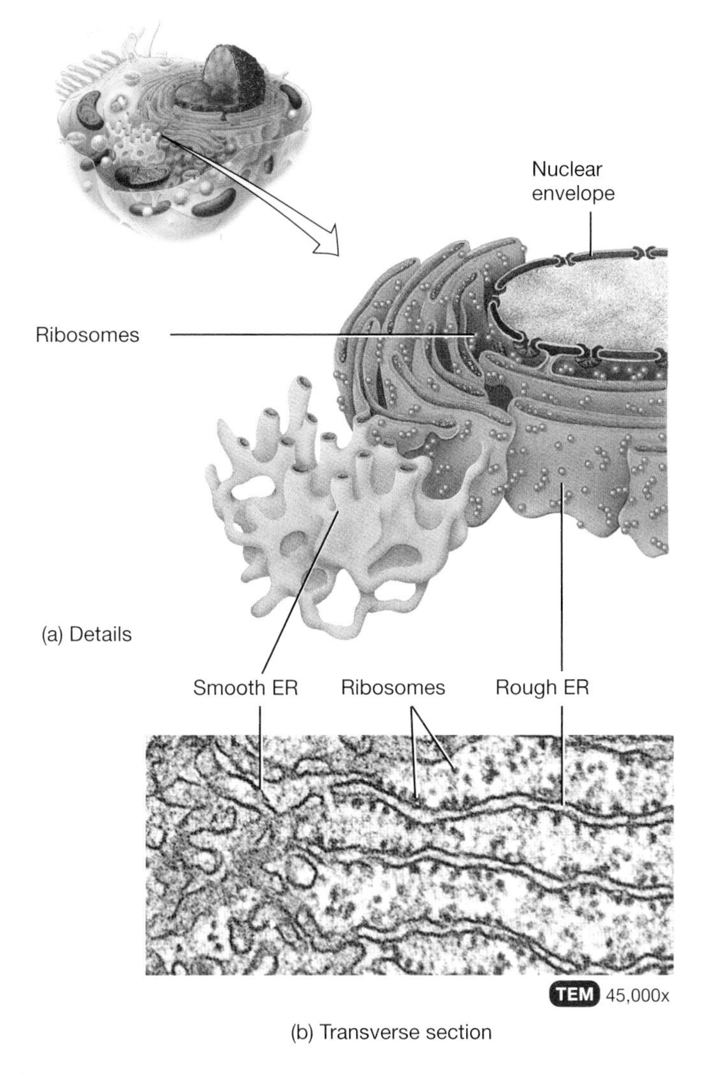


Figure 4.9 Endoplasmic reticulum.
*Source:* Tortora and Derrickson (2009), Figure 3.19, p. 81. Reproduced with permission of John Wiley and Sons, Inc.

- Agranular (smooth) ER is not linked to ribosomes; it is involved in synthesis of lipids (including phospholipids and steroids), metabolizing of carbohydrates, regulation of calcium concentration and detoxification of drugs and poisons.

Ribosomes are tiny particles of RNA that are formed in the nucleus and are involved in the synthesis of proteins needed by the cell (see Chapter 5).

Granular ER is particularly well developed in cells that are involved in the synthesis and export of proteins needed by the body, whilst agranular ER is found extensively in steroid-hormone-secreting cells, such as the cells of the adrenal cortex or the testes. Agranular ER is also present in liver cells, and it is here that it is thought to have an important role to play in drug detoxification.

## The Golgi apparatus

This organelle is a collection of membranous tubes and elongated sacs, which are flattened cisternae, stacked together. It has an important role to play in concentrating and packaging some of the substances that are produced in the cell for use in the cell or for cell secretion to the outside of the cell (exocytosis). Secretory cells have many Golgi stacks, whilst non-secretory cells have relatively fewer stacks.

In addition to being concerned with the transport of materials within the cell, ER contains enzymes that speed up the chemical reactions within the cells and that are important in cell metabolism as well as the alteration or additional processing of proteins destined for export out of the cell. For example, lysosomal enzymes are concentrated in the Golgi complex, surrounded by a membrane, so becoming a vesicle, and then released into the cytoplasm as active lysosomes.

## Lysosomes

Lysosomes are organelles found in the cytoplasm bound to the cell membrane; they contain a variety of enzymes and are originally produced on ribosomes within the cell. Because they contain **hydrolytic enzymes** (enzymes that break down substances) their role is to break down and recycle large organic molecules in the cell. However, it is essential that these enzymes remain in their membranous sacs to remain separated from the cytoplasm; otherwise they could digest the cell itself.

### *Functions*

- Responsibility for the digestion of material taken into the cell by endocytosis, or bacteria that have been drawn into the cell (endocytosis) – see Chapter 7.
- Breaking down of cell components – for example, during the development of the embryo the fingers and toes initially are webbed, but then, before birth, the cells between the toes and fingers are removed by the action of the lysosomal enzymes.
- After birth, the mother's uterus which has grown to accommodate the fetus, and weighing approximately 2 kg at full term, is invaded by phagocytic cells that are rich in lysosomes, whose enzymes reduce the uterus to its normal weight of about 0.5 kg within the space of 3 days or so.
- In cells, some of the synthesized proteins may contain 'errors'; for example, amino acid sequences that do not correspond to their messenger RNA sequences. Lysosomes and their enzymes are responsible for the removal of theses faulty proteins.
- In some human degenerative diseases – for example, juvenile rheumatoid arthritis (Stills disease) – lysosomes break up in macrophage cells and the enzymes that are released can attack living cells and tissues (Watson, 2005).

However, lysosomes, although in some senses 'destructive' organelles, have a highly constructive role to play: they also contribute to hormone production; for example, **thyroxin** – a hormone affecting a wide range of physiological activities, including metabolic rate.

## Peroxisomes

Peroxisomes are organelles similar in structure to lysosomes, but smaller. Although present in most cells, they are found in high numbers in liver cells, containing enzymes related to the metabolism of **hydrogen peroxide** – a substance that is toxic to the cells of our bodies. However, **catalase**, one of the peroxisomal enzymes, breaks down hydrogen peroxide into water and oxygen, and in this way prevents the toxic effects of this substance. Consequently, the role of peroxisome and its enzymes appears to be one of detoxification.

## Mitochondria

Mitochondria are known as the 'power houses' of the cell, generating most of the cell's supply of ATP. They are spherical or rod-shaped organelles within the cytoplasm. As organelles, they have a great variation in shape, size, mobility and numbers – from one per cell to several thousands per cell (Voet *et al.*, 2006). In addition to playing an important role in generating energy in cells, they are involved in other tasks, such as **cell signalling**, **cellular differentiation** and **cell death**, as well as controlling the **cell cycle** and **cell growth** (McBride *et al.*, 2006).

Mitochondria have a double phospholipid bilayer with embedded proteins. The outermost membrane is smooth, while the inner membrane has many folds (**cristae**). These folds increase the surface area available for chemical reactions to occur, leading to internal or cellular respiration. These two membranes divide the mitochondrion into two distinct parts: the intermembrane space (the space between the two membranes) and the mitochondrial matrix (the part enclosed by the internal membrane). Several of the steps in cellular respiration occur in the matrix due to its high concentration of **tricarboxylic acid** enzymes. In addition, it contains enzymes involved in fatty acid oxidation. By using ATP, the mitochondria are able to generate the energy that the cell needs by converting the chemical energy contained in molecules of food. Mitochondria are often found concentrated in regions of the cell associated with intense metabolic activity.

## The cytoskeleton

The cytoskeleton is a lattice-like collection of fibres and fine tubes contained within the cytoplasm (Figure 4.10). Its role is to be involved with the cell's ability to maintain and alter its shape as required. It is very important for intracellular transport (e.g. the movement of vesicles and organelles), as well as for the division/reproduction of the cell. The cytoskeleton has three elements:

1. **Microfilaments** – rod-like structures consisting of the protein actin. In muscle cells, **actin** (which is a thick protein) and **myosin** (a thin protein) are involved in the contraction of muscle fibres. However, in cells other than muscle cells, the microfilaments help to provide support and shape to the cell, as well as cell and organelle movement.

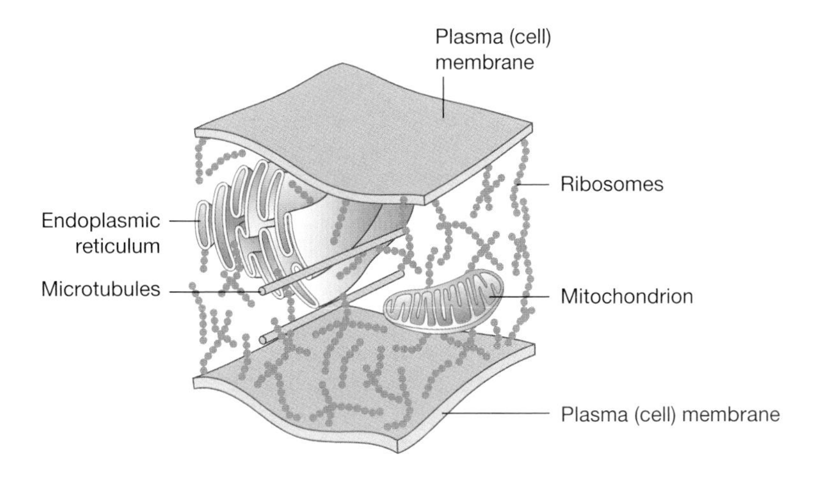


Figure 4.10 The cytoskeleton.

2. **Microtubules** – relatively straight, slender cylindrical structures consisting of the protein **tubulin**. These provide shape and support for the cells, as well as provide conducting channels to allow substances to move through the cytoplasm.
3. **Intermediate filaments** – these help to determine the shape of the cell.



## Cilia and flagella

Unlike all the other organelles, cilia and flagella exist on the outside of the cell. They extend from the surface of some cells and have the capacity to bend – which allows the cell to move.

**Cilia** generally have the function of moving fluid or particles over the surface of the cells. These 'motile' (or moving) cilia are found in the lungs, respiratory tract and middle ear. Owing to their rhythmic waving or beating motion, they help to keep the airways clear of mucus and dirt, allowing us to breathe easily and without irritation. Other non-motile cilia play essential roles in a number of organs by acting as sensory antennae for the cell in order to receive signals from other cells or fluids nearby.

A **flagellum** is a larger structure than the cilia, and the only example in humans is found on sperm, where it acts like a tail to enable sperm to swim to the ovum.

# Fluids and the body

- Sixty per cent of body weight is water.
- Forty per cent of that is intracellular fluid (ICF – inside the cell).
- The other 20% is extracellular (outside of the cell).

## Fluid compartments

Water in the body is to be found in two main body fluid compartments (Figure 4.11):

- **Intracellular compartments (ICF)**
- **Extracellular compartments (ECF).**

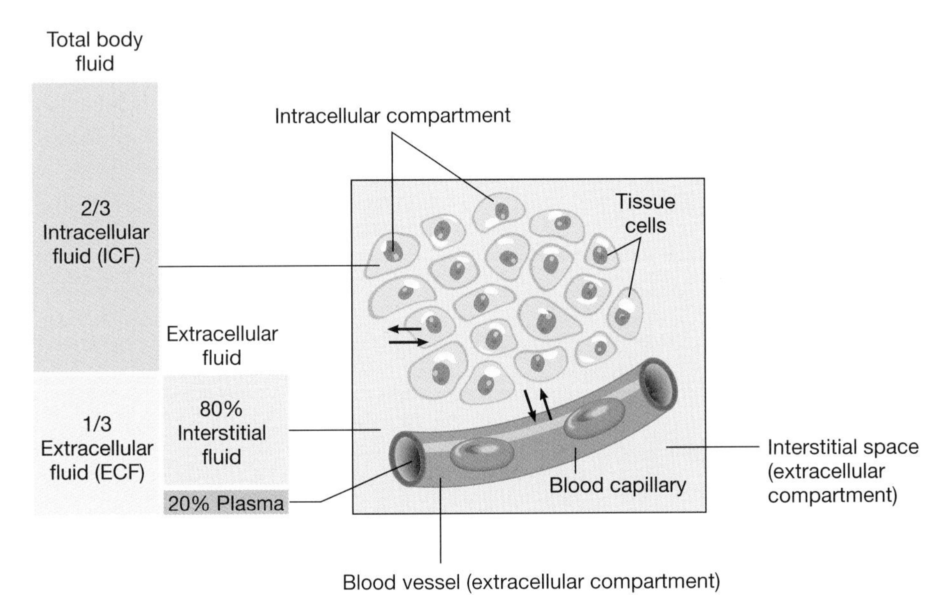


Figure 4.11 Fluid compartments and distribution.
*Source:* Peate and Nair (2011), Figure 2.10, p. 44. Reproduced with permission of John Wiley and Sons, Ltd.

Within these major compartments there are sub-compartments; for example, 80% of the ECF is situated within the **interstitial compartment** of the body, whilst the other 20% is situated within the **intravascular compartments** (i.e. the blood and lymph).

Infants proportionally have much more fluid – both ECF and ICF – than an adult. Total body water is highest in premature and newborn babies, the values ranging from 70 to 83% of body weight (Friis-Hansen, 1954) with proportionally more ECF than adults, whilst at birth the neonate proportionally has three times more water than an adult does. This proportionally large amount of fluid that the young infant requires for health means that the infant is more at risk of serious ill health if they become dehydrated. In a study by Friis-Hansen (1954) the ECF was found to diminish rapidly during the first 6 months of life (from 44% to 30% of body weight), with a gradual decrease during childhood. By the time the infant has reached their first birthday the proportionate body fluids are similar to a young or middle-aged adult (Roberts, 2005).



### Intracellular fluid

This is the internal fluid of the cell consisting mainly of water, dissolved ions, small molecules and large water-soluble molecules (such as protein). It forms the matrix in which cellular organelles are suspended and chemical reactions take place (Kapit *et al*., 2000). In adult humans, the intracellular compartment contains about 25 L of fluid on average – mainly a solution of water, potassium, organic anions and proteins, with other minor constituents. All these are controlled by a combination of the cell membranes and cellular metabolism, and the solutions can differ from cell to cell.

### Extracellular fluid

ECF primarily consists of a solution of water and sodium, potassium, calcium, chloride and hydrogen carbonate, and the ECF pH of 7.4 is very tightly regulated by buffers. Compared with ICF, ECF has lower amounts of protein. Adequate ECF volume (particularly intravascular volume) is essential for normal functioning of the cardiovascular system (Roberts, 2005). There are three sub-compartments within the ECF compartment:

- **Interstitial fluid (ICF)**, which surrounds the cells and makes up about three-quarters of the total ECF, but does not circulate around the body.
- **Plasma**, which circulates around the body as the ECF of the blood and lymph systems, and makes up most of the remaining quarter or so of ECF fluid.
- **Transcellular fluid**, which is a body fluid that is found outside of the cells and is the smallest component of the ECF (just 1–2 L of fluid in an adult). Examples of this type of fluid are mucus, gastrointestinal fluid (digestive juices), cerebrospinal fluid, as well as fluid in the eyes (aqueous humour), joints and bladder urine.

Note that this is mainly a 'virtual' collection of fluid as it is not a pool of fluid in one spot, but consists of different fluids in lots of little places around the body. Plasma is the only major fluid that exists as a true collection of fluid.

## Fluid shift

Fluid shift is the name given to the movement of fluids between the various fluid compartments across semipermeable membranes. This movement is controlled by a combination of two forces:

- osmotic pressure gradients – pressure exerted by the fluid;
- hydrostatic pressure gradients – pressure exerted on a solution to prevent the passage of water into it (Nair, 2011).

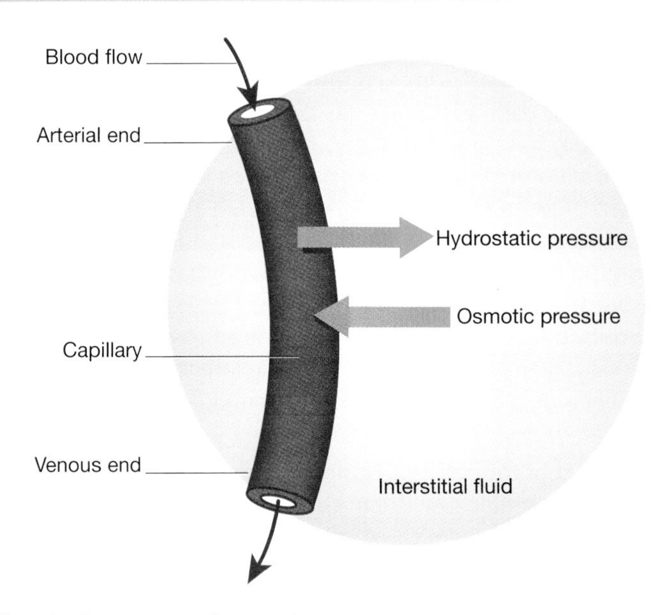


Figure 4.12 Capillary hydrostatic and osmotic pressures.
*Source:* Peate and Nair (2011), Figure 2.11, p. 45. Reproduced with permission of John Wiley and Sons, Ltd.

Water moves from one chamber to another across a semipermeable membrane until the hydrostatic and osmotic pressure gradients balance each other out. Other factors that have an effect on fluid movement between compartments include:

- changes in the concentration of solutes;
- changes in fluid volume.

An example of fluid and solute movement occurs when raised blood pressure (**hydrostatic pressure**) forces fluid and solutes from the arterial end of the capillaries into the interstitial fluid space (Figure 4.12). However, fluid and solutes return to the capillaries at the venous end as a result of osmotic pressure.

## Composition of body fluid

As well as water, body fluid consists of dissolved substances, such as electrolytes (sodium, potassium and chloride), gases (oxygen and carbon dioxide), nutrients, enzymes and hormones (Nair, 2011).

## Functions of body fluid

Water is essential for the body as it:

- Is the major component of the body's transport system – nutrients, oxygen, glucose and essential fats are transported by blood to their target tissues and cells; waste products of cellular metabolism, such as lactic acid and carbon dioxide, are removed by blood. In addition, waste products, such as urea, phosphates, minerals, ketones from the metabolism of fat and nitrogenous waste from the breakdown of protein, are transported out of the body via the kidneys as urine.
- Is needed for the regulation of internal body temperature at around 37 °C. If body temperature rises above the comfort zone of around 37 °C, the blood vessels near the surface of the

skin dilate, releasing some of the heat. The reverse happens if body temperature starts to drop. In addition, sweat glands release sweat (which is 99% water) when the body temperature rises. As the sweat evaporates, heat is removed from the body.

- Provides an optimum medium in which cells can function.
- Provides lubrication for the linings of organs and passages in the body; for example, swallowing is made easier by the oesophagus being lubricated.
- Provides a lubricant for the joints – it is a component of synovial fluid.
- Lubricates the eyes – tears.
- Breaks down food particles in the digestive system.
- Is a component of saliva, providing lubricant to food, and aiding chewing, swallowing and digestion.
- Has a protective function – washes away particles that get in the eyes as well as provides cushioning against shock to the eyes and the spinal cord.
- Is a component of amniotic fluid, which provides protection for the fetus during pregnancy.
- Provides water for chemical reactions that require its presence (Nair, 2011).

Without adequate amounts of water, we would not survive.

## Effects of water deficiency

There are times when our bodies do not have enough water to function properly (**dehydration**) – for example, not drinking enough fluid or fluid lost during ill health. When there is a deficiency of water in our body, many functions of the body are compromised and essential functions cannot be carried out. Some of the problems associated with water deficiency include:

- low blood pressure
- blood clotting
- renal failure
- severe constipation
- multisystem failure
- higher risk of infection
- **electrolyte imbalance**.

## Electrolytes

Electrolytes are chemical compounds that can be acids, bases and salts; they are produced following the ionic bonding of atoms (see Chapter 3). The composition of electrolytes differs between the intracellular and extracellular compartments, and fluid balance is linked to electrolyte balance. See Table 4.2 for a summary of electrolytes and electrolyte functions.

Electrolytes affect:

- the amount of water in the body
- the acidity of blood (pH)
- muscle function
- other important processes.

The body loses electrolytes through sweating or not taking in enough fluids to maintain their presence in appropriate numbers. It is essential that they are replaced by fluids, either by drinking fluids or by an infusion of fluids if drinking is not possible.

Table 4.2 Principal electrolytes and their functions. *Source:* Peate and Nair (2011), Table 2.2, p. 49. Reproduced with permission of John Wiley and Sons, Ltd.

| Electrolytes | Normal values in ECF (mmol/L) | Function | Main distribution |
|---|---|---|---|
| Sodium ($Na^+$) | 135–145 | Important cation in generation of action potentials. Plays an important role in fluid and electrolyte balance. Increases plasma membrane permeability. Helps promote skeletal muscle function. Stimulates conduction of nerve impulses. Maintains blood volume. | Main cation of the ECF |
| Potassium ($K^+$) | 3.5–5 | Important cation in establishing resting membrane potential. Regulates acid–base balance. Maintains ICF volume. Helps promote skeletal muscle function. Helps promote the transmission of nerve impulses. | Main cation of the ICF |
| Calcium ($Ca^{2+}$) | 135–145 | Important clotting factor. Plays a part in neurotransmitter release in neurones. Maintains muscle tone and excitability of nervous and muscle tissue. Promotes transmission of nerve impulses. Assists in the absorption of vitamin $B_{12}$. | Mainly found in the ECF |
| Magnesium ($Mg^{2+}$) | 0.5–1.0 | Helps to maintain normal nerve and muscle function; maintains regular heart rate, regulates blood glucose and blood pressure. Essential for protein synthesis. | Mainly distributed in the ICF |
| Chloride ($Cl^-$) | 98–117 | Maintains a balance of anions in different fluid compartments. Combines with hydrogen in gastric mucosal glands to form hydrochloric acid. Helps to maintain fluid balance by regulating osmotic pressure. | Main anion of the ECF |
| Hydrogen carbonate ($HCO_3^-$) | 24–31 | Main buffer of hydrogen ions in plasma. Maintains a balance between cations and anions of ICF and ECF. | Mainly distributed in the ECF |
| Phosphate – organic ($HPO_4^{2-}$) | 0.8–1.1 | Essential for the digestion of proteins, carbohydrates and fats and absorption of calcium. Essential for bone formation. | Mainly found in the ICF |
| Sulphate ($SO_4^{2-}$) | 0.5 | Involved in detoxification of phenols, alcohols and amines. | Mainly found in the ICF |

Common electrolytes include:

- calcium
- chloride
- magnesium
- phosphorus
- potassium
- sodium.

## Function of electrolytes

Electrolytes are particularly important for the body because they:

1. Form many essential minerals.
2. Control the process of osmosis.

3. Help to maintain the acid–base balance, which is necessary for normal cellular activity.
4. Regulate the movement of fluids between the various fluid compartments.
5. Are essential for the functioning of neurones.

In order for the body to function properly, it must be able to maintain electrolyte levels within very narrow limits. Controlled by signals from hormones, these electrolyte levels are maintained by the movement of electrolytes into, and out of, cells, as required. If the balance of electrolytes is disturbed, and an imbalance occurs, many serious disorders can develop. This can happen if someone:



- becomes dehydrated from problems such as diarrhoea, vomiting, profuse sweating, poor nutrition and poor intake of fluids;
- becomes over-hydrated (water toxicity) through drinking too much water or infusing inappropriate amounts of intravenous fluids;
- takes certain drugs, such as laxatives and/or diuretics;
- has certain medical problems, such as heart, liver and kidney disorders.

Severe dehydration can result in circulatory problems, including tachycardia (rapid heartbeat), as well as problems with the nervous system, leading to a loss of consciousness or shock (Nair, 2011).

With the high proportional fluid body weight of neonates and infants, and the fact that infants and children are more at risk of catching infectious diseases, often accompanied by diarrhoea and vomiting, it becomes apparent that infants and children are at a high risk of developing electrolyte imbalances – either temporarily or permanently.

## Hormones that regulate fluid and electrolytes

There are three principal hormones that regulate fluid and electrolyte balance: antidiuretic hormone (ADH), aldosterone and atrial natriuretic peptide (Thibodeau and Patton, 2007).

### *Antidiuretic hormone*

ADH is a peptide, produced in the hypothalamus, whose main role is to regulate fluid in the body. The target organs for ADH are the kidneys.

### *Aldosterone*

Aldosterone is a steroid hormone produced by the adrenal glands and regulates electrolyte and fluid balance by increasing the reabsorption of sodium and water and the release of potassium in the kidneys, which in turn increases blood volume and blood pressure. It also maintains the acid–base and electrolyte balances.

### *Atrial natriuretic peptide*

This is a hormone secreted by heart muscle cells (**myocytes**) and is a powerful vasodilator. It is involved in the homeostatic control of body water, sodium, potassium and fat (adipose tissue). It is released by myocytes in the atria of the heart in response to high blood pressure and it reduces the water, sodium and adipose loads on the circulatory system, thereby reducing blood pressure.

# Conclusion

This ends this chapter on the subject of cells. Cells are extremely complicated parts of the body, but an understanding of them and their functions is important in allowing us to understand how the human body itself functions.

# Activities

**Now review your learning by completing the learning activities in this chapter. The answers to these appear at the end of the book. Further self-test activities can be found at www.wileyfundamentalseries.com/childrensA&P.**



## Crossword

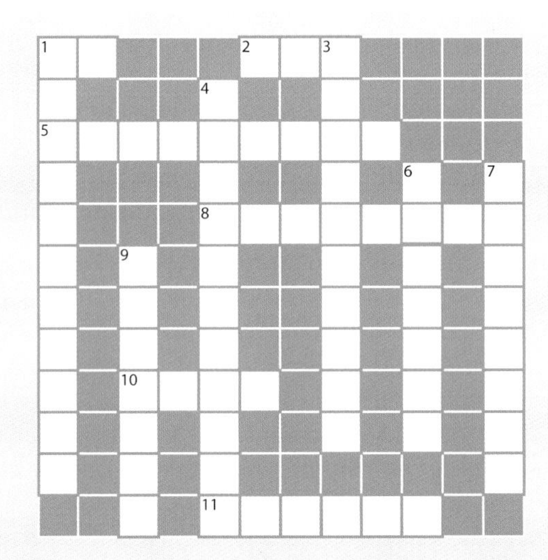


**Across**

1. A site for synthesis of lipids and steroids (abbreviation)
2. A source of energy for use in active transport within the cell (abbreviation)
5. The net passive movement of molecules or ions due to their kinetic energy from an area of higher to one of lower concentration until a state of equilibrium is reached – can be either active or simple
8. These break down and digest harmful substances
10. The focus of this chapter
11. This is a very important part of the interior of a cell, and is also known as 'ground substance'

**Down**

1. A process by which cells absorb molecules (such as proteins) by engulfing them
3. Forms the interior of the nucleus, where it is known as nucleoplasm
4. The interior of the nucleus
6. A substance (e.g. water) that is capable of dissolving another substance
7. A semipermeable biological structure separating the interior of the cell from the outside environment
9. A common electrolyte, and a mineral that is important in the formation of bones

## Fill in the gaps

Fill in the blanks in the sentences below using the correct words from the lists.

1. In order for the________ to function properly, it must be able to maintain ________ levels within very ________ limits. Controlled by signals from ________, these ________ levels are maintained by the movement of electrolytes into, and out of, cells, as required.

Choose from:

active body cell electrolyte electrolyte hormones mitochondrial narrow osmotic passive signals wide

2. Although ________ ________ is the commonest form of protein-mediated transport across the cell ________, it tends to be overshadowed by ________ ________. Rather than ________ moving down their concentration ________ to reach equilibrium, in active transport they are actively '___________' up a gradient using ___________ from another source – ___________ triphosphate (ATP).

Choose from:

active transport diphosphate facilitated diffusion energy protoplasm gradients membrane routes solutes solvents pumped gradient adenosine



## Wordsearch

1. There are several words linked to this chapter hidden in the following square. Can you find them? A tip – the words can go from up to down, down to up, left to right, right to left, or diagonally.

| E | V | L | W | D | H | Y | D | R | O | P | R | J | I | T | T | S | R |
|---|---|---|---|---|---|---|---|---|---|---|---|---|---|---|---|---|---|
| Q | X | O | F | Y | T | L | U | O | P | H | L | E | Q | U | M | E | I |
| O | O | O | S | M | O | S | I | S | N | O | I | E | U | S | I | T | S |
| W | X | T | C | E | L | L | M | E | M | B | R | A | N | E | T | S | I |
| Z | W | Y | Y | Y | R | E | W | Z | T | I | J | L | Z | L | O | W | S |
| O | G | R | T | L | T | B | P | E | L | C | R | L | W | E | C | E | O |
| F | E | P | O | E | G | O | L | G | I | E | D | E | C | C | H | T | T |
| R | A | D | P | C | R | D | S | A | L | I | Q | G | T | T | O | B | Y |
| U | C | I | L | I | A | R | E | I | P | S | O | A | H | R | N | S | C |
| E | E | P | A | E | U | O | P | H | S | Y | D | L | W | O | D | W | O |
| T | L | R | S | F | W | I | G | N | Y | N | I | F | X | L | R | I | N |
| O | L | I | M | X | D | K | O | M | O | T | W | U | J | Y | I | H | I |
| C | T | R | Y | A | K | I | W | T | Y | H | E | Q | U | T | A | T | P |
| F | Y | S | H | Q | T | K | P | E | T | E | K | V | O | E | O | Z | E |
| W | U | C | Q | U | Q | R | I | B | O | S | O | M | E | S | T | W | U |
| Y | L | A | L | Q | E | F | I | I | Z | I | W | E | P | U | I | P | E |
| L | S | O | H | E | Y | K | Y | X | P | S | O | L | V | E | N | T | Y |
| R | S | P | I | K | R | W | I | E | W | G | I | J | T | P | A | G | W |

2. There are several words linked to this chapter hidden in the following square. Can you find them? A tip – the words can go from up to down, down to up, left to right, right to left, or diagonally. Once you have found them, write a brief description of the terms identified.

| | | | | | | | | | |
|---|---|---|---|---|---|---|---|---|---|
| T | N | U | C | L | E | O | P | L | O |
| R | Y | C | O | P | R | O | T | A | R |
| O | L | P | P | R | O | K | E | S | G |
| P | G | P | P | R | O | A | I | M | A |
| S | W | M | R | N | T | R | N | C | N |
| N | U | U | E | I | E | Y | S | Y | E |
| A | Q | P | P | E | T | O | C | T | L |
| R | A | M | U | I | C | L | A | O | L |
| T | N | O | T | E | L | E | K | S | E |
| E | V | I | S | S | A | P | S | H | D |

## Glossary

**Active transport:** the process in which substances move against a concentration gradient by utilizing cellular energy.

**Aggregate:** a collection of items that are gathered together to form a total quantity.

**ATP hydrolysis:** the reaction by which chemical energy that has been stored and transported in the high-energy bonds in adenosine triphosphate is released.

**Cellular matrix:** an insoluble, dynamic gel in the cytoplasm, which is believed to be involved in cell shape determination and movement.

**Chemical reactions:** reactions that involve molecules during which they are altered in some way.

**Concentration gradient:** the gradual difference in the concentration of solutes in a solution between two regions.

**Crenation:** describes a cell's shape as being round-toothed or having a scalloped edge, and occurs with the contraction of a cell after exposure to a hypertonic solution, due to the loss of water through osmosis.

**Cytoskeleton:** cellular scaffolding or skeleton contained within a cell's cytoplasm.

**Equilibrium:** a condition in which all acting influences are cancelled by one another, so leading to a stable and balanced condition.

**Eukaryotic cells:** these are cells that normally include, or have included, chromosomal material within one or more nuclei.

**Glycoprotein:** a molecule that consists of a carbohydrate plus a protein.

**Haemolysis:** the rupturing of red blood cells (erythrocytes).

**Homeostasis:** a stable, relatively constant condition.

**Homogeneous:** the same.

**Hormone:** a chemical released by one or more cells that affects cells in other parts of the organism – essentially a chemical messenger that transports a signal from one cell to another.

**Hydrostatic:** water pressure.

**Kinetic energy:** the energy of movement/motion.

**Metabolism:** this word describes all chemical reactions involved in maintaining the living state of the cells and the organism, and can be conveniently divided into two categories, namely catabolism (the breaking down of molecules in order to obtain energy) and anabolism (which is the synthesis of compounds that the cells need to survive and perform).

**Neurotransmitter:** a chemical that allows the transmission of signals from one neurone to the next across a synapse.

**Osmotic pressure:** the pressure required to prevent the movement of pure water (containing no solutes) into a solution containing some solutes when the solutions are separated by a selective permeable membrane.

**Passive transport:** the process by which substances move on their own down a concentration gradient without using cellular energy.

**Prokaryotic cells:** the opposite of eukaryotic cells; their DNA/RNA is not contained within a discrete nucleus. They are generally small. Bacteria are prokaryotes.

**Selective permeable membrane:** a membrane that allows the unrestricted passage of water, but not solute molecules or ions.

**Solubility:** the property of a chemical substance (the solute) to dissolve in a solvent in order to form a homogeneous solution of the solute.

**Solute:** a substance that is dissolved in another substance.

**Solvent:** a substance (e.g. water) that is capable of dissolving another substance.

**Synthesis:** this word in the context of this subject generally means creation/production/manufacture.

**Vesicle:** a small bubble within a cell, and so considered as a type of organelle. Enclosed by lipid bilayer, vesicles can form naturally; for example, during endocytosis (protein absorption).

## References


Alberts, B., Johnson, A., Lewis, J. *et al.* (2002) *Molecular Biology of the Cell*, Garland, New York, NY.

Cafiso, D.S. (2005) Structure and interactions of C2 domains at membrane surfaces. In: Tamm L. K. (ed.), *Protein–Lipid Interactions: From Membrane Domains to Cellular Networks*, John Wiley & Sons, Ltd, Chichester, pp. 403–422.

Colbert, B.J., Ankney, J., Lee K.T. (2007) *Anatomy and Physiology for Health Professionals: An Interactive Journey*, Pearson Prentice Hall, Upper Saddle River, NJ.

Friis-Hansen, B. (1954) The extra-cellular fluid volume in infants and children. *Acta Paediatrica* **43**: 444–458.

Kapit, W., Macey, R.I., Meisami, E. (2000) *The Physiology Coloring Book*, 2nd edn, Addison Wesley Longman Inc., San Francisco, CA.

Marsh, M., McMahon, H.T. (1999) The structural era of endocytosis. *Science* **285** (5425): 215–220.
McBride, H.M., Neuspiel, M., Wasiak, S. (2006) Mitochondria: more than just a powerhouse. *Current Biology* **16** (14): R551–R560.
Nair, M. (2011) Cells: cellular compartments, transport system, fluid movements. In: Peate, I., Nair, M. (eds), *Fundamentals of Anatomy and Physiology for Student Nurses*, Wiley–Blackwell, Chichester, pp. 32–61.
Parker, S. (2007) *The Human Body Book*, Dorling Kindersley, London.



Peate, I., Nair, M. (eds) (2011) *Fundamentals of Anatomy and Physiology for Student Nurses*, Wiley–Blackwell, Chichester.
Roberts, K.E. (2005) Pediatric fluid and electrolyte balance: critical care case studies. *Critical Care Nursing Clinics of North America* **17**: 361–373.
Singer S.J., Nicolson G.L. (1972) The fluid mosaic model of the structure of cell membranes. *Science* **175** (4023): 720–731.
Thibodeau, G.A., Patton, K.T. (2007) *Anatomy and Physiology*, 6th edn, Elsevier Mosby, St. Louis, MO.
Tortora, G.J., Derrickson, B.H. (2009) *Principles of Anatomy and Physiology*, 12th edn, John Wiley & Sons, Inc., Hoboken, NJ.
Vickers, P.S. (2009) Cell and body tissue physiology. In: Nair, M., Peate, I. (2009) *Fundamentals of Applied Pathophysiology: An Essential Guide for Nursing Students*, Wiley–Blackwell, Chichester, pp. 1–33.
Voet, D., Voet, J.G., Pratt, C.W. (2006) *Fundamentals of Biochemistry*, 2nd edn, John Wiley & Sons, Ltd, Chichester.
Watson, R. (2005) Cell structure and function, growth and development. In: Montague, S.E., Watson R., Herbert R.A. (eds), *Physiology for Nursing Practice*, 3rd edn, Elsevier, Edinburgh, pp. 49–70.

# Chapter 5

# Genetics

Peter S. Vickers

## Aim

To introduce the student to the fascinating and very important subject of genetics, so that a knowledge of genetics will allow them to understand many of the illnesses that have a genetic underpinning.

## Learning outcomes

On completion of this chapter the reader will be able to:

- Understand genes and their importance to our health status.
- Describe the double helix, bases, DNA and RNA.
- Describe the anatomy and functions of a chromosome.
- Understand and describe protein synthesis and cell division.
- Explain the mechanisms involved in inheritance, including Mendelian genetics, and discuss the modes of inheritance – dominant, recessive and X-linked – and their relevance to some childhood disorders.

## Test your prior knowledge

- Name the components of a chromosome.
- What exactly does the double helix consist of, and what is its purpose?
- Can you name the four bases in DNA and their corresponding bases in RNA?
- Which process is involved in genetic knowledge transfer from parents to children, and which from cell to cell?
- Name the stages of mitosis and meiosis.
- What do we mean by Mendelian genetics?
- What is the function of a gene?
- What is the difference between a genotype and a phenotype?
- What do we mean by gametogenesis?
- Discuss the differences between autosomal and recessive genetic disorders.

*Fundamentals of Children's Anatomy and Physiology: A Textbook for Nursing and Healthcare Students*, First Edition. Edited by Ian Peate and Elizabeth Gormley-Fleming.

Companion website: www.wileyfundamentalseries.com/childrensA&P

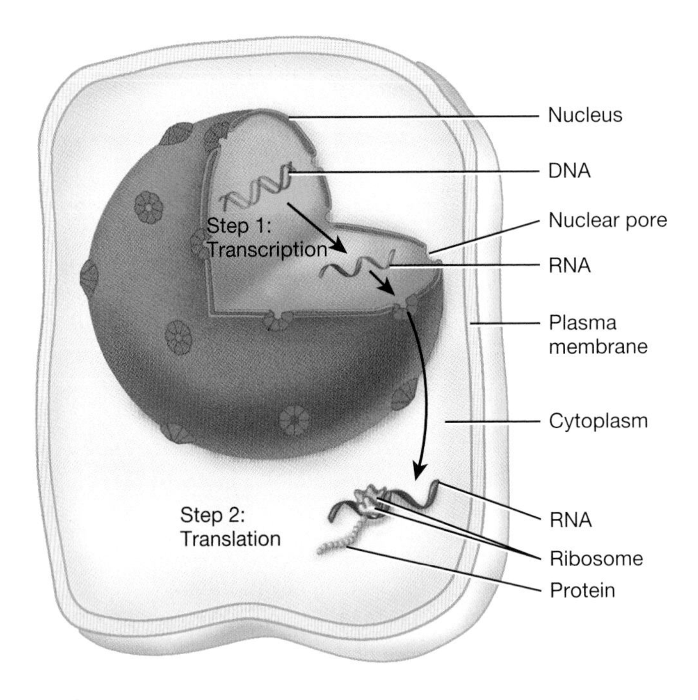


Figure 5.1 The cell nucleus.
*Source:* Tortora and Derrickson (2009), Figure 3.26, p. 88. Reproduced with permission of John Wiley and Sons, Inc.

# Introduction

Genetics is a relatively new field of specialist health care. Although clinical genetic services were first established in the UK in the 1950s, services for affected families were very limited then (Patch and Skirton, 2005). It is a very important subject, because many health problems are linked to genes. Our genes are located within the cell nucleus (Figure 5.1).

# Genes

**Genes** are sections of deoxyribonucleic acid (DNA) carried within **chromosomes**, and these genes contain particular sets of instructions related to our functioning bodies. All the genes that we possess we have inherited from our parents, who in turn inherited theirs from their parents, and so on.

To begin, some technical definitions that will help you to understand genetics.

- **DNA** is the essential ingredient of heredity and comprises the basic units of hereditary material – genes. The ability of DNA to replicate itself is the basis of hereditary transmission, and it provides our genetic code by acting as a template for the **synthesis** of messenger ribonucleic acid (mRNA).
- **Ribonucleic acid (RNA)** and **mRNA** determine the **amino acid** composition of proteins, which in turn determines the function of that protein, and therefore the function of that particular cell.
- **Chromosomes** are a complicated strand of DNA and protein. Each chromosome is made up of two **chromatids** joined by a **centromere**. Each nucleated cell in our body contains, within its genes, all the genetic material to make an entire human being.

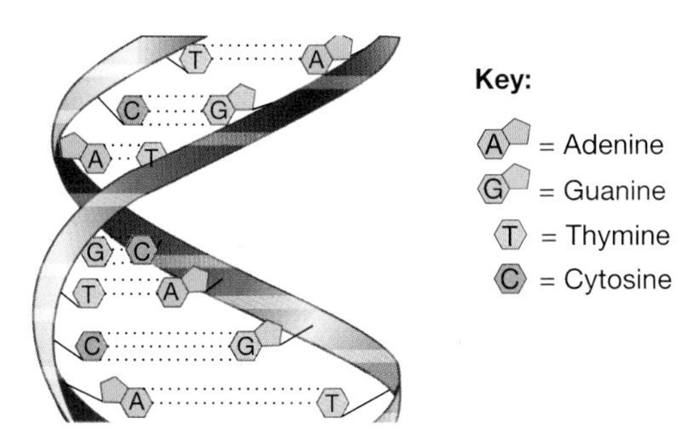


Figure 5.2 A pictorial representation of a portion of the double helix.
*Source:* Tortora and Derrickson (2009), Figure 3.31, p. 94. Reproduced with permission of John Wiley and Sons, Inc.

## The double helix

The double helix structure of DNA was identified in the 1950s by James Watson and Francis Crick (Patch and Skirton, 2005) (Figure 5.2).

The double helix is made up of two strands of DNA. It is a spiral molecule, resembling a ladder, whose rungs are built up of pairs of **bases**, and within the double helix the genetic information is encoded in a linear sequence of chemical subunits, called **nucleotides**.

## Nucleotides

Nucleotides consist of three molecules:

- **deoxyribose**
- **phosphate**
- **base.**

The double helix is like a ladder and consists of two parallel **deoxyribose** and **phosphate** supports (strands) and a series of **bases** that make up the rungs of the ladder.

The **bases** are those elements of the double helix that carry the genetic code. They are arranged in different sequences along the deoxyribose–phosphate strands of the double helix.

## Bases

There are four different bases found in DNA:

- **adenine** (A)
- **thymine** (T)
- **guanine** (G)
- **cytosine** (C).

It is the order of the bases along the length of the DNA molecule that provides the variation that in turn allows for the storage of genetic information.

Look at the drawing of the double helix in Figure 5.2; each strand carries different bases. These bases join together and make the molecule stable. However, these bases do not just pair off haphazardly. The bases are very particular as to which other base they will pair with, and there is a golden rule to remember with this pairing:

- **adenine** (A) always pairs with **thymine** (T);
- **guanine** (G) always pairs with **cytosine** (C).

So, if one half of the DNA has the base sequence

**AGGCAGTGC**

then the opposite side of the DNA will always have the complementary base sequence

**TCCGTCACG**



The bases join together by means of hydrogen/polar bonding, and the individual bases are connected to the deoxyribose of the strands by means of covalent bonds. This is important because hydrogen bonds are not as strong as covalent bonds, and so can separate more easily. Exactly how important this is will become apparent when DNA replication and protein synthesis are discussed (Jorde *et al.*, 2006).

## Chromosomes

A chromosome does not consist only of DNA. Instead, the nuclear DNA (**nucleic acid**) of cells is combined with protein molecules (**histones**). The DNA and histones together make up the **nucleosomes** contained within the cell nucleus. This nucleic acid–histone complex is known as **chromatin**.

However, if we unravelled all the chromatin from every cell in a human adult body, its length would be equivalent to nearly 70 trips from the Earth to the Sun and back, and, on average, a single human chromosome consists of a DNA molecule that is almost 5 cm in length (*The World Book Encyclopedia*, 1966). We only manage to package that amount of DNA and histone molecules in our bodies because it is neatly folded so that it fits into each cell of the body. The chromatin cannot just be pushed into the cell haphazardly – it would never fit and there would be a high possibility of things going wrong (Vickers, 2011). Consequently, the chromosomes twist on one another before being arranged into loops and superloops, until they assume the shape that is recognizable as a chromosome – the X-shape that can be seen in a human cell (Figure 5.3) (Jorde *et al.*, 2006).

Each **chromosome** is made up of two chromatids joined by a **centromere**. Each half of the chromosome is a **chromatid**, and the centromere is where they join near the top of the X (Figure 5.3).

In most humans, each nucleated body cell (i.e. each body cell with a nucleus) has 46 chromosomes, arranged in 23 pairs (Figure 5.4). Of those 23 pairs, one pair determines the gender of the person.

- Females have a matched **homologous** (similar) pair of X chromosomes.
- Males have an unmatched **heterologous** (different) pair – one X and one Y chromosome.
- The remaining 22 pairs of chromosomes are known as **autosomes**. In biology, the word 'some' means body; so autosome means 'self body'. **Autosomes** determine physical/body characteristics – in other words, all characteristics of a person that are not connected with gender.

One of each pair of chromosomes comes from the mother and one comes from the father, and the position a gene occupies on a chromosome is called a **locus**. There are different **loci** for colour, height, hair, and so on. Think of the locus as the address of that particular gene on Chromosome Street – just like your address signifies that that is where you live (Vickers, 2011).

Genes that occupy corresponding loci and code for the same characteristic are called **alleles**. Alleles are found at the same place in each of the two corresponding chromatids, and an allele determines an alternative form of the same characteristic.

Take eye colour as an example. The gene that determines eye colour is found at the same place on each of the two chromatids of one chromosome. One gene will come from the father

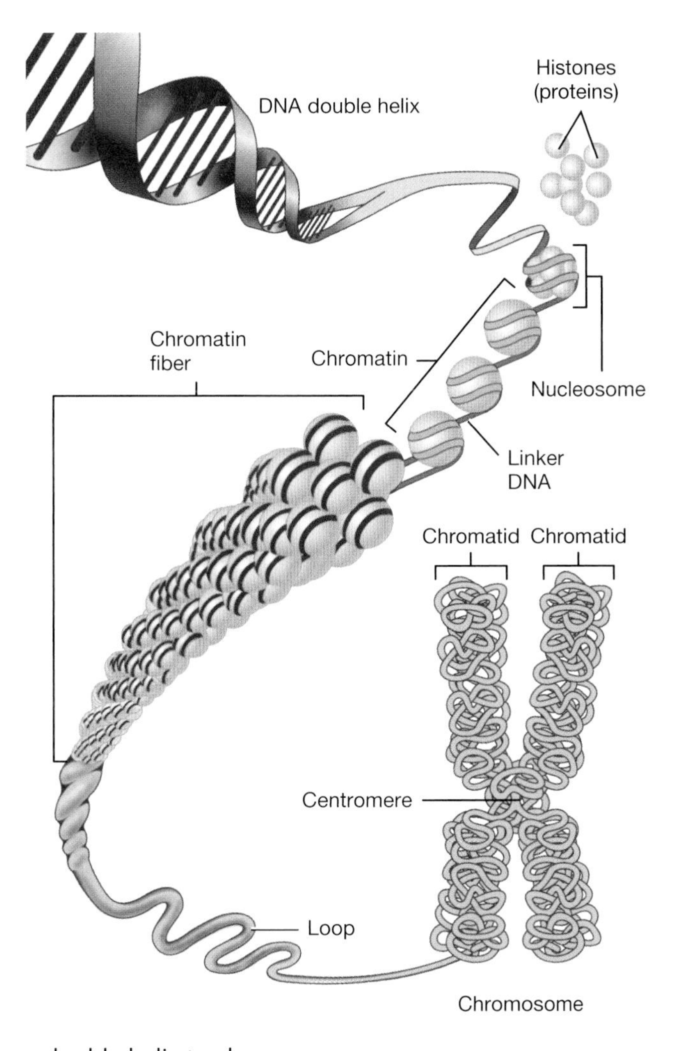


Figure 5.3 DNA from double helix to chromosome.
*Source:* Tortora and Derrickson (2009), Figure 3.25, p. 87. Reproduced with permission of John Wiley and Sons, Inc.

and the other from the mother. If the mother of a child has green eyes and the father has brown eyes, then the child may have green or brown eyes, depending upon factors that will be discussed later in this chapter.

This principle applies to each of a person's characteristics. A person with a pair of identical alleles for a particular gene locus is said to be **homozygous** for that gene, whilst someone with a dissimilar pair is said to be **heterozygous** for that gene.

Two other very important facts to mention about genes are that some genes are **recessive** and some genes are **dominant**.

- A **dominant gene** is one that exerts its effect when it is present on only one of the chromosomes. (The name **genotype** is given to the types of genes found in the body, whilst the name **phenotype** is given to any gene that is manifested in the person.)

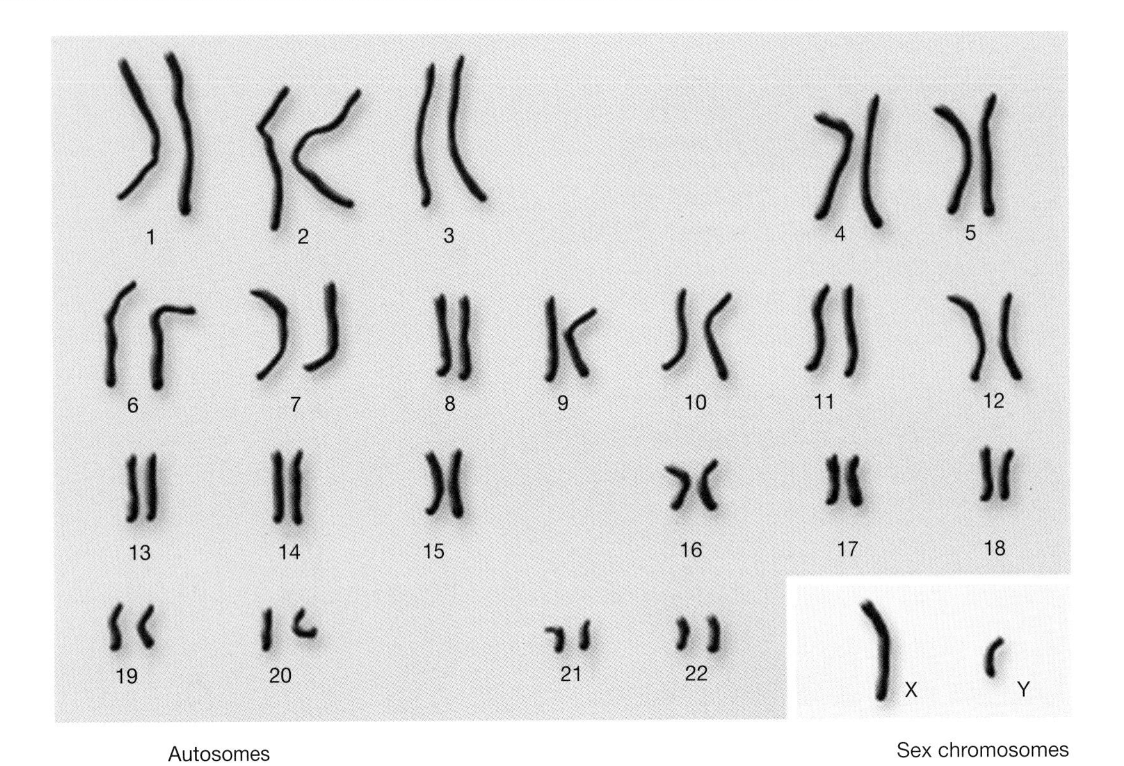


Figure 5.4 Male human chromosomes.
*Source:* Lister Hill National Center for Biomedical Communications (2010), figure on p. 16 (http://ghr.nlm.nih.gov/handbook/basics/howmanychromosomes).

- A **recessive** gene (**genotype**) has to be present on both chromosomes in order to manifest itself (**phenotype**).

This is a very important concept to grasp because of the significance that it has in hereditary disorders.

# From DNA to proteins

As was explained earlier in this chapter, nucleic acids are components of DNA and they have two major functions:

1. The direction of all protein synthesis (i.e. the production of protein).
2. The accurate transmission of this information from one generation to the next (from parents to their children), and, within the body, from one cell to its daughter cells.

## Protein synthesis

Synthesis means 'production'; for example, the production of protein from raw materials. All the genetic instructions for making proteins are found in **DNA**, but in order to synthesize these proteins the genetic information encoded in the DNA has first to be translated.

Initially, all of the genetic information in a region of DNA has to be copied in order to produce a specific molecule of **RNA**.

Then, through a complex series of procedures, the information contained in RNA is translated into a corresponding specific sequence of **amino acids** in a newly produced **protein molecule**.

There are two parts to this procedure: **transcription** and **translation**.



## *Transcription*

In transcription, the DNA has to be transcribed into RNA because protein cannot be synthesized directly from DNA. By using a specific portion of the cell's DNA as a **template**, the genetic information stored in the sequence of bases of DNA is rewritten so that the same information appears in the bases of RNA. To do this, the two strands of the DNA have to separate, and the bases that are attached to each strand then pair up with bases that are attached to strands of RNA. As with the two strands of DNA, the bases of DNA can only join up with a specific base of RNA.

- As with DNA, **guanine** can only join up with **cytosine** in RNA.
- But **adenine** in DNA can only join to **uracil (U)** in the RNA because there is no thymine in RNA.

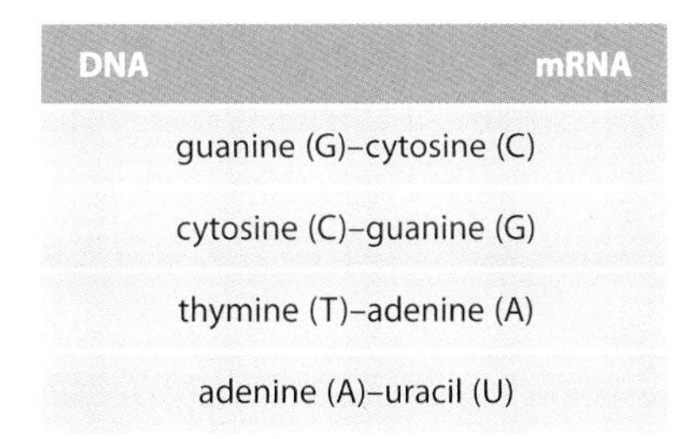

| DNA | mRNA |
|---|---|
| guanine (G)–cytosine (C) | |
| cytosine (C)–guanine (G) | |
| thymine (T)–adenine (A) | |
| adenine (A)–uracil (U) | |

For example:

- if DNA has the base sequence **AGGCAGTGC**
- then mRNA has the complementary base sequence **UCCGUCACG**.

Figure 5.5 shows the way in which the DNA separates and makes more DNA. This same process occurs during transcription, except the new strand with its bases is **RNA** rather than DNA.

**Question**

In the following DNA sequence, what should the RNA bases be?

**C A G C T G C A**

**Answer**

**G U C G A C G U**

Thus, DNA acts as a template for mRNA (Vickers, 2011). However, in addition to serving as the template for the synthesis of mRNA, DNA also synthesizes two other kinds of RNA: **ribosomal RNA (rRNA)** and **transfer RNA (tRNA)**.

- rRNA, together with the **ribosomal proteins**, makes up the **ribosomes**.
- tRNA is responsible for matching the code of the **mRNA** with **amino acids**.

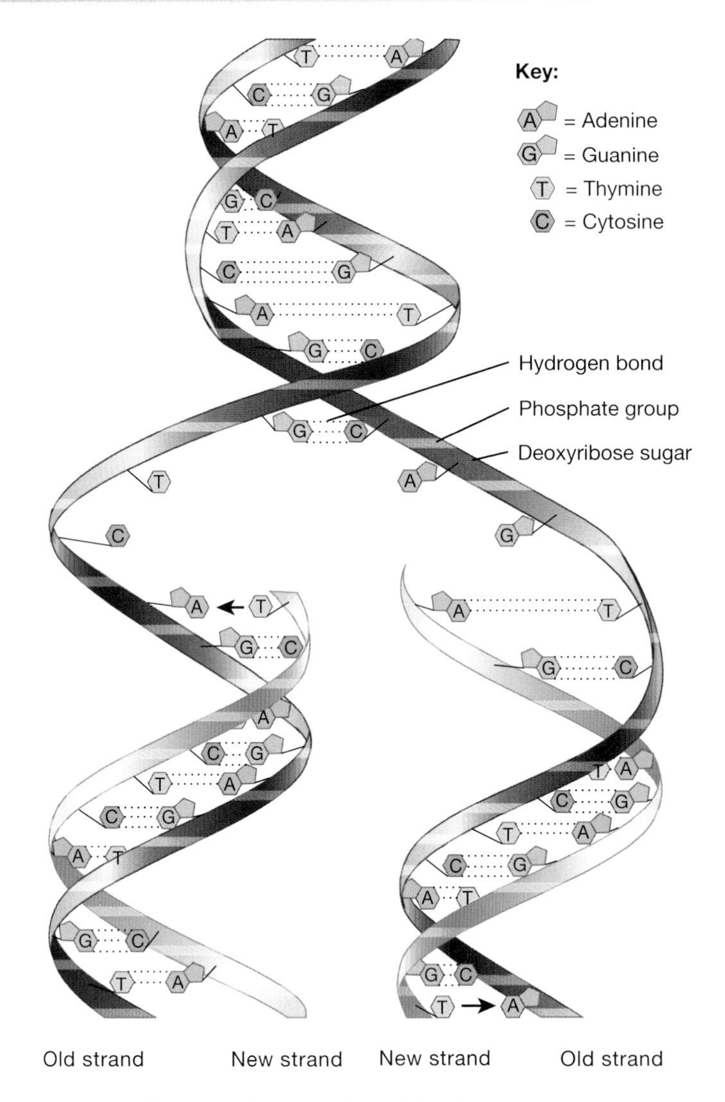


Figure 5.5 The separation of DNA and production of further DNA.
*Source:* Tortora and Derrickson (2009), Figure 3.31, p. 94. Reproduced with permission of John Wiley and Sons, Inc.

Once ready, mRNA, rRNA and tRNA leave the nucleus of the cell and in the cytoplasm of the cell commence the next step in protein synthesis, namely **translation**.

### *Translation*

In genetics, translation is the process by which information in the bases of mRNA is used to specify the **amino acid** of a protein (composed of amino acids). This involves all three different types of RNA as well as ribosomal proteins.

## Key steps of protein synthesis

The key steps of protein synthesis (the production of proteins from DNA) are shown in Figures 5.6 and 5.7.

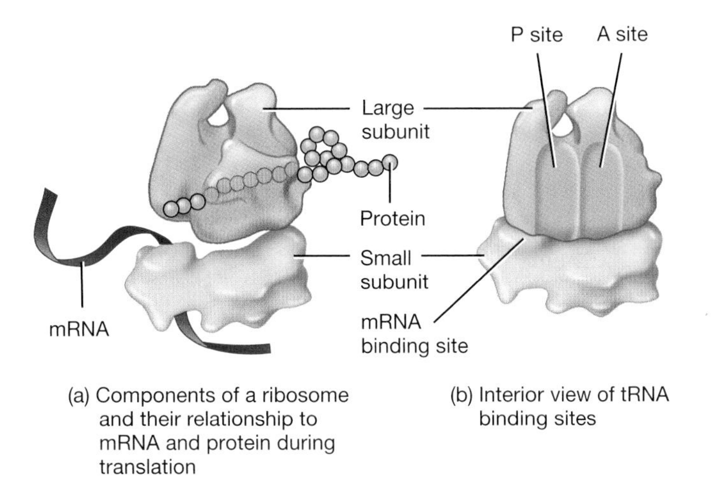


Figure 5.6 mRNA becomes associated with small ribosomal subunit.
*Source:* Tortora and Derrickson (2009), Figure 3.28, p. 91. Reproduced with permission of John Wiley and Sons, Inc.

- In the cytoplasm, a small ribosomal subunit binds to one end of the mRNA molecule. There are a total of 20 different amino acids in the cytoplasm that may take part in protein synthesis, and for each different amino acid there is a different corresponding small tRNA strand of just three bases (known as a **triplet**).
- In protein synthesis (Figure 5.8), a tRNA triplet picks up the selected amino acid known as an **anticodon**.
- One end of the tRNA **anticodon** has receptors that only allow it to couple with a specific amino acid.
- The other end of the tRNA **anticodon** has a specific sequence of bases.
- That tRNA **anticodon** then seeks out the corresponding three bases (**codon**) on the mRNA strand that is bound to the small ribosomal subunit.
- By means of base pairing, the **anticodon** of the specific tRNA molecule attaches to the corresponding **codon** of the mRNA strand (e.g. if the **anticodon** is **UAC**, then the mRNA **codon** will be **AUG**).
- In the process of attaching itself to the mRNA **codon**, the tRNA **anticodon** brings the specific amino acid with it; but this base pairing only occurs when the mRNA is attached to a ribosome.
- Once the first tRNA **anticodon** attaches to the mRNA strand, the ribosome moves along the mRNA strand and the second tRNA **anticodon**, along with its specific amino acid, moves into position.
- The two amino acids that are attached to the two tRNA **anticodons** are joined together by a peptide bond; and once this has happened, the first tRNA **anticodon** detaches itself from the mRNA strand and returns to the body of the cell to pick up another molecule of its specific amino acid.
- Meanwhile, the smaller ribosomal subunit moves along the mRNA strand and the process continues.
- As the correct amino acids are brought into line, one by one the protein becomes progressively larger.

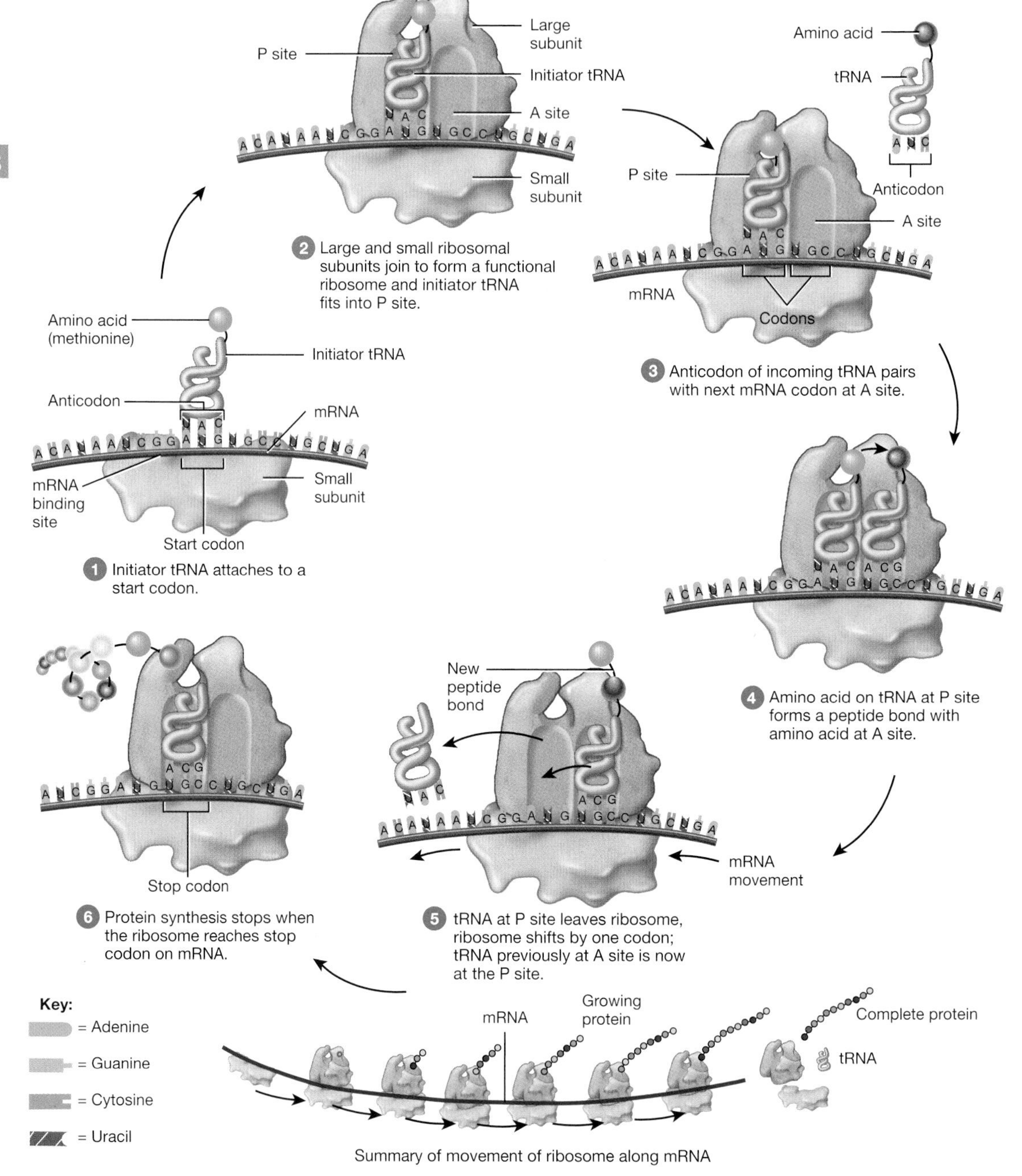


**Figure 5.7** Summary of the movement of the ribosome along mRNA.
*Source:* Tortora and Derrickson (2009), Figure 3.29, p. 92. Reproduced with permission of John Wiley and Sons, Inc.

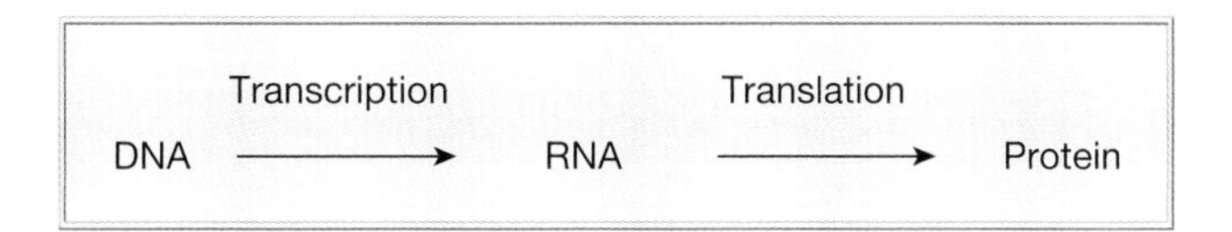


Figure 5.8 Brief summary of protein synthesis.
*Source:* Peate and Nair (2011), Figure 17.8, p. 563. Reproduced with permission of John Wiley and Sons, Ltd.



- As the ribosome moves along the mRNA, and before it completes translation of the first gene into protein, another ribosome may attach to the beginning of the mRNA strand and begin translation of the same strand to form a second copy of the protein. So, several ribosomes moving simultaneously in tandem along the same mRNA molecule permit the translation of a single mRNA strand into several identical proteins simultaneously.
- This process is continued until the protein specified by the mRNA strand (initially specified by the genes on the DNA strand) is complete – the correct number of amino acids has been joined together in the correct order.
- Once the specified protein is completed, further synthesis of amino acids/protein is stopped by a special **codon** known as a **termination codon** (a combination of three bases that signal the end of the protein synthesis process for that particular protein), which effectively blocks any further **codon/anticodon** base pairing, and the assembled new protein is released from the ribosome, whilst the ribosome separates again into its two discrete component subunits (Vickers, 2011).

The process of protein synthesis is extremely rapid, progressing at the rate of about 15 amino acids per second.

So we can now define a **gene** as a group of nucleotides on a DNA molecule that serves as the master mould for manufacturing a specific protein (Jorde *et al.*, 2006):

- Genes average about 1000 pairs of nucleotides, which appear in a specific sequence on the DNA molecule.
- No two genes have exactly the same sequence of nucleotides, and this is the key to heredity.
- The base sequence of the gene determines the sequence of bases in the mRNA.

# The transference of genes

## Introduction

Genetic information is transferred from cells to new cells, as well as from parents to their children. In order for the body to grow, and also for the replacement of body cells that die, whilst ensuring that genetic information is not lost, the cells must be able to reproduce themselves accurately. They do this by **cloning** themselves.

In some organisms, such as the amoeba, this can occur by **simple fission**, where the nucleus in a single cell becomes elongated and then divides to form two nuclei in the same cell, each of which carries identical genetic information. The cytoplasm then divides in the middle between the two nuclei, and so two identical daughter cells result, each with its own nucleus and other essential organelles.

With humans (and other animals), the process of transference of genes (or reproduction of cells carrying genetic information) is much more complicated and is divided into two stages: **mitosis** and **meiosis**.

## Mitosis

In humans, cell reproduction takes place by **mitosis**, in which the number of chromosomes in the daughter cells has to be the same as in the original parent cell.

**Mitosis** can be divided into four stages:

- **prophase**
- **metaphase**
- **anaphase**
- **telophase**.

Before and after it has divided, the cell enters a stage known as **interphase**.

## Interphase

**Mitosis** begins with **interphase**. During this period the cell is actually very busy as it gets ready for replication. If one full cell cycle (Figure 5.9) represents 24 h, then the actual process of replication (**mitosis**) would only last for about 1 h of those 24 h. The rest of the time the cell is producing two of everything, not just DNA, but all the other organelles in the cell, such as the mitochondria. In addition, the cell has to obtain and digest nutrition so that it has the raw materials for this duplication, and also for the energy that will power the various functions of the cell.

During interphase, the chromosomes in the nucleus are present in the form of long threads – they have not yet become **super-coiled**. They need to be in this state so that they can be

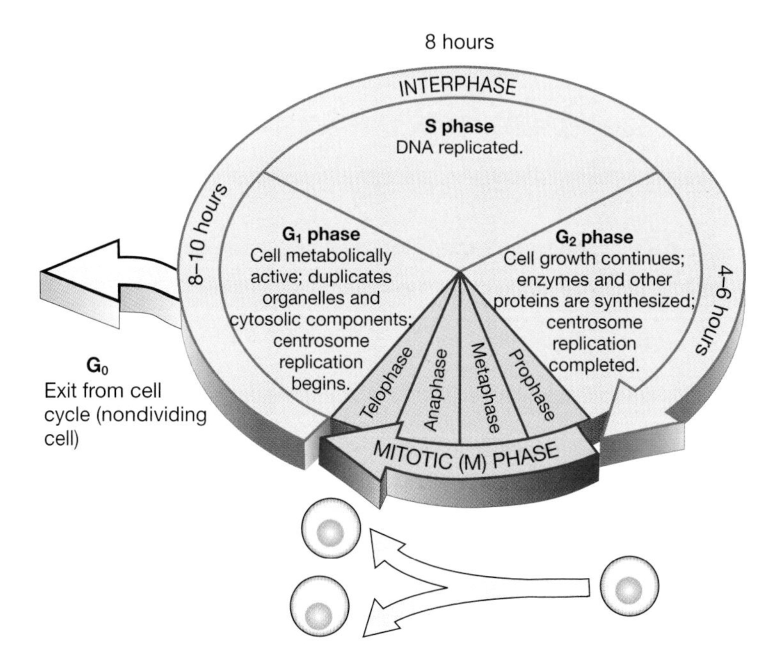


Figure 5.9 The cell cycle.
*Source:* Tortora and Derrickson (2009), Figure 3.30, p. 93. Reproduced with permission of John Wiley and Sons, Inc.

duplicated. During the process of duplication, the cells have to ensure that there will be sufficient genetic material for each of the two daughter cells. The strands of DNA separate and reattach to new strands of DNA. Because of the selectivity of the bases regarding which other base they will join to in this process, an exact replication of the DNA will occur (Figure 5.5).

In addition, during interphase, extra cell **organelles** are manufactured by the replication of existing **organelles**, and the cell builds up a store of energy, which will be required for the process of division.



## Prophase

The **chromosomes** become shorter, fatter and more easily visible. Each chromosome now consists of two **chromatids**, each containing the same genetic information. These two **chromatids** are joined together at an area known as the **centromere**. The two **centrosomes** move to opposite ends of the cell (the **poles**) and are joined together by the **nuclear spindle**, which stretches from pole to pole of the cell. The centre of the cell is now called the **equator**. Finally, the **nucleolus** and **nuclear membrane** disappear, leaving the **chromosomes** in the cytoplasm (Figure 5.10).

## Metaphase

The 46 **chromosomes** (two of each of the 23 **chromosomes**), each consisting of **two chromatids**, move to the **equator** of the **nuclear spindle**, and here they become attached to the **spindle fibres** (Figure 5.10).

## Anaphase

The **chromatids** in each **chromosome** are separated, and one **chromatid** from each **chromosome** then moves towards each **pole** (Figure 5.10).

## Telophase

There are now 46 c**hromatids** at each pole, and these will form the **chromosomes** of the daughter cells. The **cell membrane** constricts in the centre of the cell, dividing it into two cells. The **nuclear spindle** disappears, and a **nuclear membrane** forms around the **chromosomes** in each of the daughter cells. The **chromosomes** become long and threadlike (Figure 5.10).

## Cell division

Cell division is now complete (Figure 5.10), and the daughter cells themselves enter the interphase stage in order to prepare for their replication and division. This process of cell division explains how we grow by producing new cells as well as replacing old, damaged and dead cells.

## Meiosis

Whilst mitosis is concerned with the reproduction of individual cells, **meiosis** is concerned with the development of whole organisms (e.g. human beings).

The reproduction of a human being depends upon the fusion of reproductive cells (known as **gametes**) from each of the parents. These **gametes** are:

- **spermatozoa** (sperm) from the male;
- **ova** (eggs) from the female.

Each cell of the human body contains 23 pairs of **chromosomes** (i.e. 46 in total). It is very important that during the process of human reproduction the cell formed when the gametes fuse has the correct number of chromosomes for a human being (23 pairs). Therefore, each

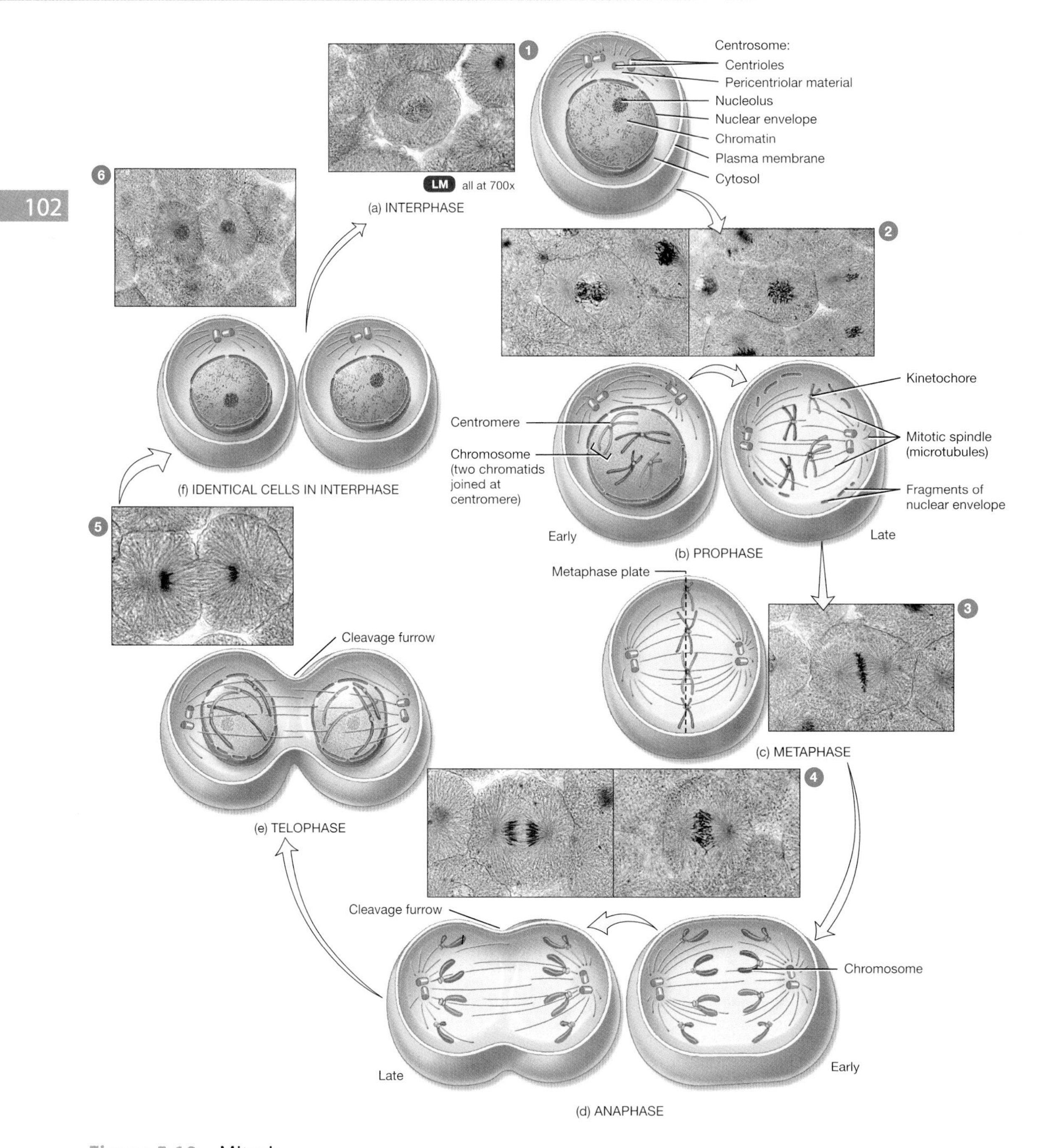


**Figure 5.10** Mitosis.
*Source:* Tortora and Derrickson (2009), Figure 3.32, p. 95. Reproduced with permission of John Wiley and Sons, Inc.

gamete must possess only 23 single chromosomes, because when gametes fuse during reproduction all their chromosomes remain intact in the new life form. If each gamete had a full complement of 46 chromosomes, then the resulting fused cell would possess 92 chromosomes – or four copies of each chromosome rather than the two that a human cell should possess. To prevent this, the gametes only possess one copy of each chromosome, so that the resulting cell would have 46 chromosomes.



Two new terms that describe the number of chromosomes in a cell are **diploid** and **haploid** cells.

1. **A diploid cell** is a cell with a full complement of 46 chromosomes (23 pairs).
2. **A haploid cell** is a cell with only half that number of chromosomes (23 single chromosomes).

Gametes are therefore haploid cells, because they only possess one copy of each chromosome, whilst all other cells of the body are diploid cells.

Gametes actually develop from cells with 46 chromosomes, and it is through the process of **meiosis** that they end up with just 23 chromosomes. Basically, the way that **meiosis** works is that the cells actually divide twice, without the replication of DNA occurring again before the second division.

**Meiosis** can be divided into eight stages (not the four of **mitosis**), and consists of two **meiotic** divisions each with four stages. The stages in each division have the same names (prophase, metaphase, anaphase and telophase) but are appended either with the number I or II. As with **mitosis**, these stages are continuous with one another.

## First meiotic division

This involves the following stages (Jorde *et al.*, 2006):

- prophase I
- metaphase I
- anaphase I
- telophase I.

### *Prophase I*

Prophase I is similar to the stage of prophase in **mitosis**. However, instead of being scattered randomly, the **chromosomes** (consisting of two **chromatids**) are arranged in pairs – 23 in all. For example, the two chromosome 1s will pair up, as will the two chromosome 2s, and so on. Each pair of chromosomes is called a **bivalent**. Within each pair of chromosomes, genetic material may be exchanged between the two chromosomes, and it is these exchanges that are partly responsible for the differences between children of the same parents. This process is called **gene crossover** (Figure 5.11). The important point to remember about meiosis is that the DNA is not replicated during the first meiotic division.

### *Metaphase I*

As in mitosis, the chromosomes become arranged on the spindles at the equator. However, they remain in pairs.

### *Anaphase I*

One chromosome from each pair moves to each pole, so that there are now 23 chromosomes at each end of the spindle.

### *Telophase I*

The cell membrane now divides the cell into two halves, as in mitosis. Each daughter cell now has half the number of chromosomes that each parent cell had.

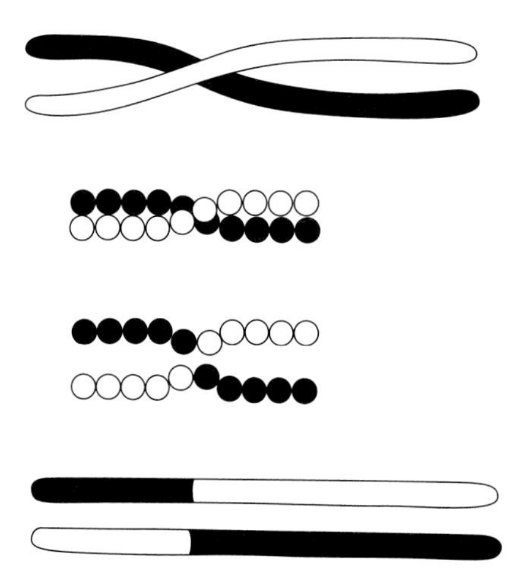



Figure 5.11 Gene crossover.
*Source:* Peate and Nair (2011), Figure 17.11, p. 568. Reproduced with permission of John Wiley and Sons, Ltd.

## Second meiotic division

This involves the following stages:

- prophase II
- metaphase II
- anaphase II
- telophase II.

During the second meiotic division, both of the cells produced by the first meiotic division now divide again.

**Prophase II**, **metaphase II**, **anaphase II** and **telophase II** are all similar to their equivalent stage in **mitosis**, with the exception that the chromosomes are not replicated, before **prophase II**, so there are only 23 single chromosomes in each of the granddaughter cells. That way, when the gametes fuse during reproduction, there are still only 23 pairs of chromosomes per human cell.

Of the 23 pairs of chromosomes, 22 pairs are **autosomal** and one pair consists of the **sex chromosomes**. Remember, autosomal means 'self body' (auto: self; somatic: of the body). In other words, autosomal chromosomes are concerned with the body. On the other hand, the sex chromosomes determine the gender of a person. Male sex chromosomes are designated by the **letter Y**, and female chromosomes are designated by the **letter X**. A male will carry the chromosomes XY (an X chromosome from the mother and a Y chromosome from the father), whilst a female will carry the chromosomes XX (an X chromosome from both the father and the mother).

# Mendelian genetics

## Introduction

So far this chapter has examined the biology of genetics, but now it is time to look at the role of genetics in inheritance. This is very important because, as stated early in the chapter, what we are is designated to a large extent by our genetic makeup – which is inherited from our parents. The *caveat* 'to a large extent' is because as well as being a product of our genes we are also a product of our environment – time, space, relationships, education, and so on.

So how do we inherit our genes from our parents? To understand this we have to return to the 1860s. In Brno, in the Czech Republic, there was a monastery, and in that monastery lived and worked a monk with a very inquiring mind. His name was **Gregor Mendel** and he worked in the monastery gardens where he put his inquiring mind to good use trying to perfect the ideal pea. As part of this work he experimented with cross-breeding. Now, at that time, cross-breeding went on everywhere – on farms and in gardens – and of course humans cross-breed as well. However, what was different about Mendel was that not only did he experiment with cross-breeding different peas, but he also made notes on his experiments and observations. He introduced three novel approaches to the study of cross-breeding – at least these were novel for his time, because no one else was doing this.



1. Not only did he observe, but he experimented and observed.
2. Having observed and experimented he then used statistics.
3. He ensured that the original parental stocks, from which his crosses were derived, were pure breeding stocks.

The phenomena that Mendel discovered/observed were statistical in form: the now-famous ratios made sense only in the context of counting large numbers of specimens and calculating averages. However, the methods for evaluating and statistical data just did not exist then, and were not to be developed for a further 30 years or so. It was only much later that the validity of such an approach would be accepted.

In addition, Mendel's work was carried out not in one of the main centres of science, but at the periphery, and it was obscurely published. In science, as in the rest of life, just who expresses an idea and where they work affects its reception. However, the science of inheritance is now based upon what Mendel discovered all those years ago – but has progressed far beyond Mendel's dreams (Vickers, 2011).

Mendel demonstrated that members of a pair of alleles separate clearly during **meiosis** (remember that alleles are different sequences of genetic material occupying the same gene locus – or place on the DNA, but on different chromosomes). We all have a pair of genes (alleles) at each locus, but because of the process of **meiosis** we can only pass one of those pairs of genes to our child (Figure 5.12).

At the same locus on a chromosome, the father has the two alleles Aa and the mother has the two alleles Bb. When they reproduce, the father can pass either **gene A** or **gene a** (both are at the same locus and are therefore alleles) and the mother can pass on either **gene B** or **gene b** (again both at the same locus). However, each child can only inherit one of **gene A** or **gene a** from the father and one of **gene B** or **gene b** from the mother.

What the child cannot do is inherit both **gene A** and **gene a** or **gene B** and **gene b**. Only one allele from each parent can be inherited by a child. This is known as **Mendel's first law**.

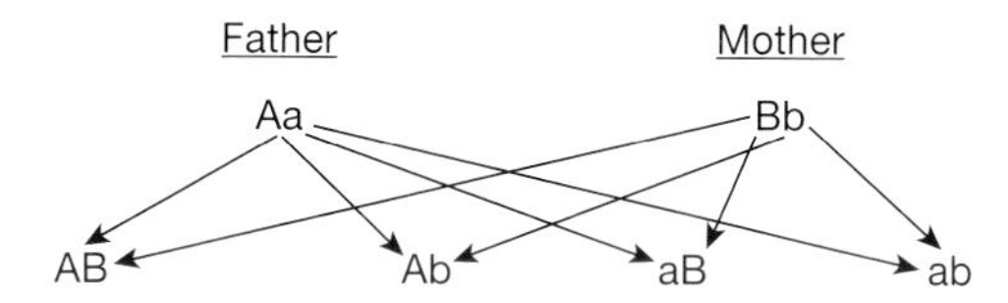


Figure 5.12 Genetic inheritance.
*Source:* Peate and Nair (2011), Figure 17.12, p. 570. Reproduced with permission of John Wiley and Sons, Ltd.

What are the statistical chances of a child inheriting any one of those sets of genes **AB**, **Ab**, **aB** and **ab** from the parents? The answer is 1 in 4 (or 1:4). In other words, there is a 25% chance that any child will inherit one of those pairs of genes from their parents.

So, that brings us to Mendel's second law, which asserts that members of different pairs of alleles sort independently of each other during **gametogenesis** (the production of gametes), and each member of a pair of alleles may occur randomly with either member of another pair of alleles. Note that gametogenesis is the production of **haploid** sex cells so that each carries one-half the genetic make-up of the parents.

This now brings us to the concept of dominant and recessive genes. This has a great bearing on many health disorders that we may encounter, as well as determining such characteristics as eye colour, hair colour, and so on.

## Dominant genes and recessive genes

At each locus, the two alleles (genes) can be either **dominant** or **recessive**. A **dominant** gene is an allele that will be reflected in the **phenotype** (the manifestation of the gene) no matter what the other allele does. Meanwhile, a **recessive** gene is one that will only appear in the phenotype if the corresponding allele is also recessive and has the same characteristic as the first allele.

In genetic representations, dominant genes are usually given capital letters, whilst recessive genes are usually given lower case letters – but not always. Therefore, in Figures 5.12 and 5.13, one gene is dominant and one is recessive.

Look again at Figure 5.12. Suppose two parents had four children and they all had different genotypes (genetic make-up), so that they each were represented by one of the pairs of genes. How many of the offspring would carry at least one **dominant** gene, and how many would carry only **recessive** genes at this locus?

The answer is that three out of the four children (75%) would carry at least one **dominant** gene, and one out of the four children (25%) would carry both **recessive** genes. Of course, in real life, all four children may inherit the same pair of genes at this locus, or maybe two will inherit the same genes. **Mendel's first law** is relevant only in saying that there is a 1-in-4 chance at each pregnancy that the children will carry a certain genotype.

Another example: a man with red hair married a woman with brown hair. As time goes by, they have several children, all of whom have brown hair. Which is the dominant gene for hair colour and who carries it?

The gene for brown hair carried by the mother was the dominant gene in this instance. However, their offspring all married partners with brown hair, but some of their offspring had red hair like their maternal grandfather. How can this be explained?

There are two possible explanations:

1. Some of the children carried the red hair recessive gene from their mother and their partners also carried a recessive red hair gene – this is the most likely explanation.
2. The father was not the genetic father of those children – possible, but not the most likely explanation.

## Autosomal dominant inheritance and ill health

If the dominant gene of one of the parents is the one that causes a medical disorder – for example, Huntingdon's disease or neurofibromatosis – then what will be the risk of any child of those parents having the disease?

The answer is 50% or a 1-in-2 risk of a child having an **autosomal dominant disorder**.

Why is this? Look back at Figure 5.12, and assume that the father (genes **A** and **a**) carries the mutant gene on gene **A**. As a dominant gene is always expressed in the phenotype, then statistically there will be a 50% chance of any child having the disease, because the child could inherit gene **a**. Of course, any child who carries gene **A** will have a 100% chance of having the disease; there is no escaping it.

## Autosomal recessive inheritance and ill health



**Autosomal recessive diseases** occur when both parents are carrying the same defect on a recessive gene at the same locus. Both parents have to carry the defective gene otherwise the child cannot be affected by the disease.

In autosomal recessive diseases, if the child (or parent) only carried the defect on one gene, then they are a **carrier** of that disease and can pass that defective gene on to their children. They in turn could pass it on to their children, who, if they inherit it, would also be carriers, and this situation could continue through many generations until the carrier has children with someone who is also a carrier of that mutant gene. There is then a risk of their children being either a carrier or having the disorder.

So then, what are the risks of:

- A child being a carrier of the recessive gene?
- A child having the disease caused by this mutated/abnormal gene?

To work it out, look again at Figure 5.12. In this case, the lower case letters '**a** and **b**' represent the abnormal recessive gene. As can be seen, both parents carry this abnormal gene; for example, for ***cystic fibrosis*** (CF) – this is a well known disease that is inherited as an autosomal recessive disorder.

If one of the two recessive genes (**a** or **b**) that code for CF is carried by each of the parents, then the chances at each pregnancy of having:

- an affected child are 1 in 4 (or 25%);
- a child who is a carrier are 2 in 4 (or 50%);
- a child who is neither affected nor a carrier are 1 in 4 (or 25%).

Why is this so? Look at Figure 5.12 again.

- Only one child possesses two affected genes (**a** or **b**), and because both affected genes have to be present in order for the disease to appear, then this is one child out of four, or 25%.
- Only one child does not possess an affected gene (**a** or **b**), and so the disease cannot occur; neither can the child be a carrier, because there is no affected gene to be carried. So this is one child out of four, or 25%.
- Two children possess an affected gene, but they also contain an unaffected dominant gene, so they are both carriers.

Whenever there is a dominant gene, then the affected recessive gene cannot be expressed in the **phenotype** (it cannot be manifested) – the dominant gene blocks the action of the affected recessive gene, so two out of four children (or 50%) could be carriers. However, always remember that children who are carriers can pass the affected gene onto their children.

It is important to remember – and stress – that these odds occur for each pregnancy, so you could have four children and have:

- one affected
- two carriers and one unaffected

- four carriers
- three affected and one carrier
- and so on.

Remember that the odds are the same for each child born to those parents (LeMone and Burke, 2008)!



## Clinical application

### Cystic fibrosis

CF affects 1 in 2500 children born in the UK. It is caused by a faulty gene that is transmitted as an autosomal recessive disorder; that is, both parents carry the faulty gene and there is a 1 in 4 chance of each pregnancy leading to a child with CF. There have been over 1,000 different mutations of the CF gene that have been discovered to date. The faulty gene has to be carried by both parents for the child to have CF. It causes thick and sticky secretions in the lungs and the digestive systems, leading to infection, inflammation, lung damage, respiratory failure, malabsorption, malnutrition and poor growth, as well as liver problems, diabetes and potential bowel blockages (Elworthy, 2007). To be able to look after a child and family with this disorder, as well as the physical care of symptoms, the nurse will need knowledge of the underlying genetics in order to be able to counsel the family (and child later on) and will need to be empathetic and understanding of the psychological as well as physical challenges.

## Morbidity and mortality of dominant versus recessive disorders

Autosomal dominant disorders are generally less severe than recessive disorders because if someone carries the affected gene they would have that disorder, whereas with autosomal recessive disorders a person can be a carrier but not have the disease. If autosomal dominant disorders were as severe and fatal as many autosomal recessive disorders, then the disease would die out as all the people with an affected autosomal dominant gene would normally die before being old enough to pass it on to their offspring.

As exception is Huntington's disease, which is a fatal autosomal dominant disorder, but it survives because the symptoms do not usually become apparent until the affected person is in their 30s, by which time they could have passed on the affected gene to their children.

## Clinical application

### Achondroplasia

Achondroplasia is an autosomal dominant condition and is one of the most common genetic causes of disproportionate short stature. It is estimated to occur in 1 in 25 000 births (Bromilow, 2007). The characteristics of a child with achondroplasia are short stature; although the trunk is of a normal length, the limbs are shortened, and there may be possible skeletal abnormalities that can cause medical complications. Children with achondroplasia have a normal IQ and life expectancy, with an expected adult height of approximately 4 feet (1.22 m). Babies with achondroplasia tend to develop motor skills more slowly than other babies do, but they will eventually catch up and achieve development within normal parameters (Bromilow, 2007).

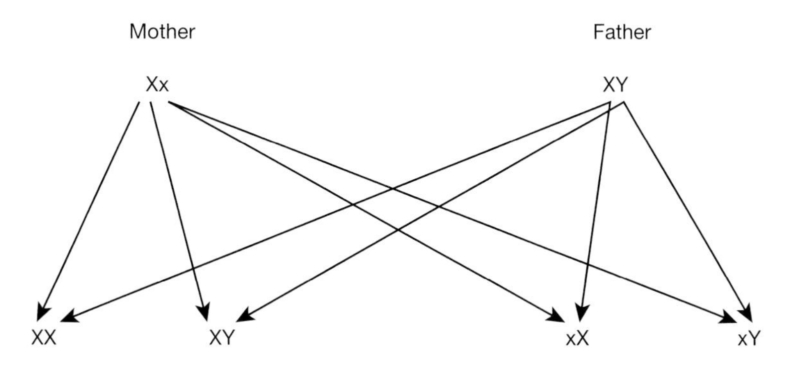




**Figure 5.13** X-linked inheritance.
*Source:* Peate and Nair (2011), Figure 17.13, p. 574. Reproduced with permission of John Wiley and Sons, Ltd.

## X-linked recessive disorders

As well as autosomal inheritance, we can also inherit disorders via the sex chromosomes.

The main role of these chromosomes is to determine the gender of the baby:

- **XX = girl**
- **XY = boy.**

First, look at the possibilities of having a boy or a girl when you decide to have a baby (Figure 5.13). From this you can see that the chances for each pregnancy of a boy or a girl are 50%.

Some disorders are only passed on via the X chromosome. Examples are haemophilia and Duchene muscular dystrophy (DMD). With these disorders, generally only the boys can be affected and only girls may be carriers, though rarely girls can be affected.

## Clinical application

### Duchenne muscular dystrophy

DMD is an X-linked genetic disorder in which the mutated gene is carried on the X chromosome. Consequently, generally only boys will be affected and only girls will be carriers, although, very rarely, girls may be affected. DMD affects about 1 in 3000–4000 live births and is generally diagnosed between 1 and 4 years of age, following a muscle biopsy. It is a progressive and degenerative disorder, and most of the affected boys will die in their late teens or early 20s from heart failure and pulmonary problems.

As with all genetic childhood disorders, the child's nurses will need to be supportive, not only to the child whenever hospitalized, but also to the mother in particular, who may well exhibit feelings of guilt. Following diagnosis and genetic counselling from a specialist genetic counsellor, ongoing genetic counselling is also an important role of the nurse. This condition, as with all X-linked conditions, will require genetic analysis of the X genes in any female siblings of the affected child to detect carrier status and to give counselling (Bromilow, 2007).

If we assume that the lower case **x** is the affected gene for haemophilia, then what is going to happen with our family in Figure 5.13?

- The first child is a girl who does not carry the affected gene, but rather two normal genes, so she is neither a carrier nor affected.
- The second child carries a normal X and a Y, so he is a boy who does not carry the abnormal gene – consequently, he is neither a carrier nor affected.
- The third child is a girl who carries the abnormal gene, but the action of that gene is blocked by the other normal X gene, so she is not affected, but is a carrier.
- The fourth child is a boy who carries an abnormal X gene and a normal Y gene. Unfortunately, the Y gene is unable to block the action of the abnormal gene, so he is a carrier and is also affected.

Consequently, we can say that there is a chance that:

- one out of two girls (50%) will be a carrier;
- one out of two boys (50%) will be affected.



## Clinical application

### Down syndrome

Down syndrome is a chromosomal condition caused by an extra chromosome 21 being present, rather than a condition caused by a faulty gene. There are three types of Down syndrome: trisomy 21, in which each cell contains three chromosomes 21 rather than the normal two (usually linked to advanced maternal age of 35 or over); translocation 21, where a segment of a chromosome 21 is attached to another chromosome (usually inherited from one parent, but can be a spontaneous mutation); mosaic 21, where a fault occurs after fertilization. Pre-natal diagnosis can be obtained from amniocentesis or chorionic villus sampling.

Children with Down syndrome have physical characteristics and medical/developmental problems, and have delayed motor and cognitive skills, but can live for more than 50 years. However, for these children to have long fulfilled lives, they will require special educational provision, physical and social support and therapy, and, of course, effective health care (Wiggins, 2007). There are many degrees of abilities and disabilities that these children possess, and the paediatric nurse must be aware not only of the needs of these children if they require medical/nursing care, but more importantly of the abilities and skills that they possess and develop following skilled parental and professional support.

## Spontaneous mutation

Now to briefly mention another way in which an unusual or abnormal gene can occur in someone and cause genetic disorders. This is known as **spontaneous mutation**. Because of the great speed and precision needed at each replication of DNA in the germ cells, and of protein synthesis, it is possible for mistakes to occur, and so genetic mutations arise.

Finally, there are also the problems of chemical /trauma mutations to consider.

# Conclusion



This has been a rather brief introduction to genetics. Although genetics may appear very complicated, it is a very important subject for you to understand, because not only do our genes make us what we are, but also they can leave us susceptible to certain diseases and have a say in how we respond to treatment for diseases, and how we live our lives, work, develop relationships and, indeed, survive in the world. Paediatric nurses often come across patients who have a genetic disease (because many of the most serious manifest from a very early age, if not from birth); consequently, throughout your career as a nurse, you will need to explain things, not only to children diagnosed with a genetic disease, but also to their families as they struggle to come to terms not only with their child being ill, but also their guilt as they come to terms with the fact that their child is ill because of their genes.

Finally, in recent years, there has been much interest in using genetic therapy to treat illnesses, with varying levels of success. However, probably the most exciting and, to date, successful gene therapy is that used to treat a very few of the many primary immunodeficiency diseases which, unlike secondary immunodeficiencies, have a genetic cause. In the early 2000s, the first successful replacement of a faulty gene in a child with adenosine deaminase deficiency – a rapidly fatal disorder – took place (Hacein-Bey-Abina *et al.*, 2002; Rosen, 2002). Since then this treatment has been used successfully in children with this disorder, and occasionally on children with other severe immune disorders, and research continues to try to improve this technique for other genetic disorders.

# Activities

**Now review your learning by completing the learning activities in this chapter. The answers to these appear at the end of the book. Further self-test activities can be found at www.wileyfundamentalseries.com/childrensA&P.**

## Fill in the gaps

Fill in the blanks in the sentence below using the correct words from the list.

Genes that occupy corresponding ________ and ________ for the same characteristic are called ________, which are found at the same place in each of the two corresponding ________, and each one determines an alternative form of the same characteristic.

Choose from:

autosome, loci, centromere, code, alleles, haploid, diploid, chromatids, amino acid, nuclei.

## Crossword

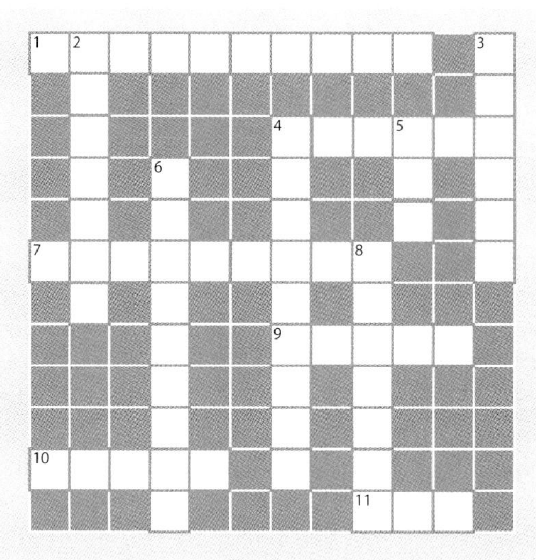


**Across**

1. A complicated strand of DNA and protein – contains our genetic blueprint
4. 'The father of genetics'
7. A triplet of bases on tRNA encoding for amino acids that join with other triplets to produce the appropriate proteins
9. A gene's position on a chromosome
10. The spiral formation of DNA – usually has the prefix 'double'
11. This determines the amino acid composition of proteins, which in turn determines the function of that protein, and therefore the function of that particular cell

**Down**

2. Cell nucleus protein whose role is to package the DNA into nucleosomes
3. The place on the chromosomes where genes which code for the same function are to be found
4. The name given to the type of genetics in which members of a pair of alleles separate clearly during meiosis (named after the first person who worked things out)
5. The essential ingredient of heredity and comprises the basic units of hereditary material
6. These are coded for by genes and can be considered as the building blocks of proteins
8. The name of the membrane around the nucleus

## Wordsearch

There are several words linked to this chapter hidden in the following square. Can you find them? A tip – the words can go from up to down, down to up, left to right, right to left, or diagonally.

| H | I | S | T | O | N | E | P | O | E | Q | U | O | N | C | T | U | Y |
|---|---|---|---|---|---|---|---|---|---|---|---|---|---|---|---|---|---|
| A | N | M | C | Z | T | M | Y | W | M | U | T | A | T | I | O | N | B |
| C | T | G | E | N | E | C | I | R | E | O | U | U | N | C | N | U | E |
| W | E | C | N | Q | I | R | Q | U | N | N | A | W | R | R | I | C | R |
| O | R | Z | T | P | E | O | X | E | D | N | A | C | E | O | L | L | H |
| W | P | E | R | Q | Y | S | V | A | E | I | Z | M | X | I | L | E | H |
| V | H | I | O | N | A | S | E | Y | L | E | P | N | C | L | X | A | R |
| R | A | A | M | I | N | O | P | R | I | C | W | L | C | C | A | R | W |
| X | S | X | E | P | E | V | E | O | A | T | A | B | O | E | O | S | Y |
| H | E | M | R | A | I | E | Z | U | N | C | W | E | D | I | E | P | S |
| L | A | X | E | Q | C | R | T | E | G | T | E | C | O | T | D | I | Y |
| W | O | P | R | W | I | O | P | W | E | P | A | M | N | O | X | N | N |
| O | Y | R | L | C | S | C | A | M | N | I | O | N | E | A | T | D | T |
| A | W | I | E | O | E | R | L | E | E | X | B | I | E | X | N | L | H |
| Z | C | C | M | E | I | O | L | O | T | O | A | Q | U | O | Q | E | E |
| R | A | E | O | W | L | D | E | N | I | N | A | U | G | M | U | A | S |
| L | E | T | O | C | E | T | L | Z | C | P | R | T | P |  | T | S | I |
| E | S | A | H | P | A | T | E | M | S | O | E | W | O | E | M | T | S |

## Exercise

Draw a simple diagram that demonstrates the process of Mendelian genetic inheritance, where the dominant genes are denoted by **A** and **B**, and the recessive genes by **a** and **b**.

**FATHER** **MOTHER**

## Conditions

The following table contains a list of conditions. Take some time and write notes about each of the conditions. You may make the notes taken from text books or other resources – for example, people you work with in a clinical area – or you may make the notes as a result of people you have cared for. If you are making notes about people you have cared for you must ensure that you adhere to the rules of confidentiality.

| Condition | Your notes |
|---|---|
| Achondroplasia | |
| Cri du chat | |
| Fragile X syndrome | |
| Cystic fibrosis | |
| Neural tube defects | |

## Glossary

**Adenine:** one of the four nitrogen–carbon bases of DNA.

**Allele:** the place on the chromosomes where genes that code for the same function are to be found.

**Amino acid:** these are coded for by genes and can be considered as the building blocks of proteins.

**Anaphase:** the stage in cell division where the chromosome separates and moves to the poles of the cells.

**Anticodon:** a triplet of bases on tRNA encoding for amino acids that join with other triplets (or anticodons) to produce the appropriate proteins.

**Autosome:** the name given to chromosomes that are not one of the two sex chromosomes.

**Autosomal dominant disorder:** a medical disorder cause by a faulty dominant gene inherited from one of the parents.

**Autosomal recessive disorder:** a medical disorder cause by a faulty recessive gene inherited from one of the parents.

**Base:** part of the double helix; bases are the code that will eventually lead to the formation of protein.

**Bivalent:** a pair of associated homologous chromosomes formed after replication of the chromosomes, with each replicated chromosome consisting of two chromatids.

**Cell cycle:** the process by which a cell prepares for, and undertakes, cell growth and division.

**Chromatid:** one half of a chromosome.

**Centromere:** the point at which two chromatids become attached to form a chromosome.

**Chromosome:** mixture of DNA and protein – contains our genetic blueprint.

**Codon:** a triplet of bases on mRNA that encodes for a particular amino acid.

**Cytosine:** one of the four nitrogen–carbon bases of DNA.

**Deoxyribose:** a major part of the DNA molecule; deoxyribose is a sugar (ribose) that has lost an atom of oxygen – hence its name.

**Diploid cell:** a cell that contains two sets of identical chromosomes – see also 'haploid'.

**DNA:** abbreviation for deoxyribonucleic acid – present in the double helix.

**Dominant gene:** a gene capable of affecting the body without any help from the recessive gene at the same locus – it dominates the recessive gene.



**Double helix:** two strands of DNA joined together in a spiral formation.

**Equator of cell:** the centre of the cell during cell division.

**Gamete:** a reproductive cell – spermatozoon or ovum.

**Gametogenesis:** the production of gametes.

**Gene:** a unit of heredity in a living organism.

**Gene crossover:** the process at the commencement of meiosis whereby genetic material may be transferred between chromosomes.

**Genotype:** a living organism's genetic makeup – see also 'phenotype'.

**Guanine:** one of the four nitrogen–carbon bases of DNA.

**Haploid cell:** a cell that contains just one set of chromosomes – see also 'diploid'.

**Heredity:** the passing down of genes from generation to generation.

**Heterologous:** means 'different', as opposed to homologous, which means 'same'.

**Heterozygous:** a pair of dissimilar alleles for a particular gene locus.

**Histone:** proteins found in the cell nucleus – their role is to help package and order the DNA into nucleosomes, so making it possible for the chromosomes to be fitted into a cell without becoming tangled.

**Homologous:** means 'same' – see 'heterogeneous'.

**Homozygous:** a pair of identical alleles for a particular gene locus; see also 'heterozygous'.

**Interphase:** the longest stage of the cell cycle during which the cell is growing and preparing for replication.

**Locus:** a gene's position on a chromosome.

**Meiosis:** this is concerned with the development of whole organisms, and is the process by which diploid cells become haploid cells; this ensures that the correct number of chromosomes is passed on to the offspring.

**Mendelian genetics:** the genetics of inheritance.

**Mendel's first law:** only one particular allele from each parent can be inherited by their child.

**Mendel's second law:** during gametogenesis, members of differing pairs of alleles are randomly sorted independently of each other.

**Metaphase:** the stage in the cell cycle when the chromosomes move to the equator of the cell preparatory to separating.

**Mitosis:** the process by which chromosomes are accurately reproduced in cell during cell division.

**mRNA:** messenger ribonucleic acid – it is important in the production of proteins from amino acids.

**Nucleic acid:** a combination of phosphoric acid, sugars and organic bases, nucleic acids direct the course of protein synthesis (production), so regulating all cell activities; both DNA and RNA are nucleic acids.

**Nucleosome:** the basic unit of DNA once it is packaged within a cell's nucleus; it consists of a segment of DNA wound around a histone.

**Nucleotide:** the name for the parts of the DNA consisting of sugar (deoxyribose) and one of the four bases – adenine, cytosine, guanine, thymine; in other words, it is the basis of our genes, and hence of us!

**Ova:** plural of ovum; these are the female reproductive cells – also known as 'eggs'.

**Phenotype:** this describes the expressed features of a living organism as a result of that organism's genotype interacting with the environment.

**Phosphate:** an inorganic molecule that forms part of the double helix.

**Poles of the cell:** the ends of a cell during the stages of cell division.

**Prophase:** the first stage of cell division, during which the chromosomes fold together and become more visible.

**Recessive gene:** a recessive gene requires another recessive gene at the same locus before it can affect the body – it is not 'dominant' over the other gene at that locus (see also dominant gene).

**Ribosomes:** small, bead-like structures in a cell that, along with RNA, are involved in making proteins from amino acids.

**RNA:** ribonucleic acid – transcribed from DNA.

**Spermatozoa:** male reproductive cells.

**Spontaneous mutation disorder:** a medical disorder caused by a new 'fault' that has developed in a gene; that is, neither parent carries a faulty version of that gene.

**Strand:** the long parts of the double helix, consisting of deoxyribose and phosphate.

**Telophase:** the stage in cell division where the cell actually divides and forms two identical daughter cells.

**Termination cordon:** a triplet of bases that acts as the signal to stop the organization of amino acids once the specified protein of that particular DNA/RNA sequence has been produced.

**Thymine:** one of the four nitrogen–carbon bases of DNA.

**Transcription:** the process by which something with which we cannot work is changed into something that we can. In genetics, this is the changing of DNA into RNA.

**Translation:** the process that, in genetics, follows transcription. Translation allows us to understand what a process or a word is; thus, in genetics, the genetic information held in the bases of mRNA is used to specify a particular amino acid that will become part of a specific protein.

**Triplet:** sequences of three RNA bases that code for different amino acids.

**tRNA:** abbreviation of transfer ribonucleic acid; it is important in the production process of proteins from amino acids.

**X-linked recessive disease:** a medical disorder caused by a 'fault' on the X gene (one of the sex genes – the other being the Y gene). Only females can be carriers of these types of diseases, and only males can have these diseases (although there are very rare examples of females also having haemophilia). Haemophilia is an example of such a disease.

# References

Bromilow, G. (2007) Achondroplasia. In: Glasper, A.E., McEwing, G., Richardson, J. (eds.), *Oxford Handbook of Children's and Young People's Nursing*, Oxford University Press, Oxford, pp. 702–705.

Elworthy, S. (2007) Cystic fibrosis (CF): a genetic condition. In: Glasper, A.E., McEwing, G., Richardson, J. (eds.), *Oxford Handbook of Children's and Young People's Nursing*, Oxford University Press, Oxford, pp. 698–701.

Hacein-Bey-Abina, S., Le Diest, F., Carlier, F., *et al.* (2002) Sustained correction of X-linked severe combined immunodeficiency by ex vivo gene therapy. *New England Journal of Medicine*, **346** (16): 1185–1193.

Jorde, L.B., Carey, J.C., Bamshad, M.J., White, R.L. (2006) *Medical Genetics*, 3rd edn, Mosby, St. Louis, MO.

Lister Hill National Center for Biomedical Communications (2010) *Genetics Home Reference: Your Guide to Understanding Genetic Conditions*, http://ghr.nlm.nih.gov/handbook.pdf (accessed 22 July 2014).

LeMone, P., Burke, K. (2008) *Medical–Surgical Nursing: Critical Thinking in Client Care*, 4th edn, Pearson Prentice Hall, Upper Saddle River, NJ.

Patch, S., Skirton, H. (2005) Genetics. In: Montague, S.E., Watson, R., Herbert, R.A. (eds), *Physiology for Nursing Practice*, 3rd edn, Elsevier, Edinburgh, pp. 777–803.

Peate, I., Nair, M. (eds) (2011) *Fundamentals of Anatomy and Physiology for Student Nurses*, Wiley–Blackwell, Chichester.

Rosen, F. (2002) Successful gene therapy for severe combined immunodeficiency. *New England Journal of Medicine*, **346** (16): 1241–1242.

*The World Book Encyclopedia* (1996) Cell, Field Enterprises, Chicago, IL.

Tortora, G.J., Derrickson, B.H. (2009) *Principles of Anatomy and Physiology*, 12th edn, John Wiley and Sons, Inc., Hoboken, NJ.

Vickers, P.S. (2011) Genetics. In: Peate, I., Nair, M. (eds), *Fundamentals of Anatomy and Physiology for Student Nurses*, Wiley–Blackwell, Oxford, pp. 550–584.

Wiggins, S. (2007) Down syndrome. In: Glasper A.E., McEwing, G., Richardson, J. (eds), *Oxford Handbook of Children's and Young People's Nursing*, Oxford University Press, Oxford, pp. 706–707.

# Chapter 6

# Tissues

Peter S. Vickers

## Aim

The aim of this chapter is to introduce you to the tissues of our bodies so that whilst working through this book you will be able to place them within the context of the various systems of the body.

## Learning outcomes

On completion of this chapter the reader will be able to:

- Describe the ways in which the various cells come together to form tissues.
- Understand the structure and function of the various tissues within the body.
- Describe the characteristics and classifications of epithelial tissue.
- Understand the structures and functions of connective tissue.
- List the classifications and functions of muscle tissue.
- Explain, simply, the process of tissue repair.

## Test your prior knowledge

- From what structure are tissues formed?
- Can you name the four main types of tissue in the human body?
- Where is epithelial tissue to be found and what are its six important functions?
- Epithelial tissue is classified in two ways. Can you name them?
- Ground substance is found in connective tissue. Can you say what it is formed of and give any examples of types of connective tissue in the body?
- In which type of tissue would you find neurones?
- What type of tissue is blood?
- What is the common name for adipose tissue?
- What are the two types of tissues that make up the skeleton?
- What are the names of the two types of glands to be found in glandular epithelial tissue?

*Fundamentals of Children's Anatomy and Physiology: A Textbook for Nursing and Healthcare Students*, First Edition. Edited by Ian Peate and Elizabeth Gormley-Fleming.

Companion website: www.wileyfundamentalseries.com/childrensA&P

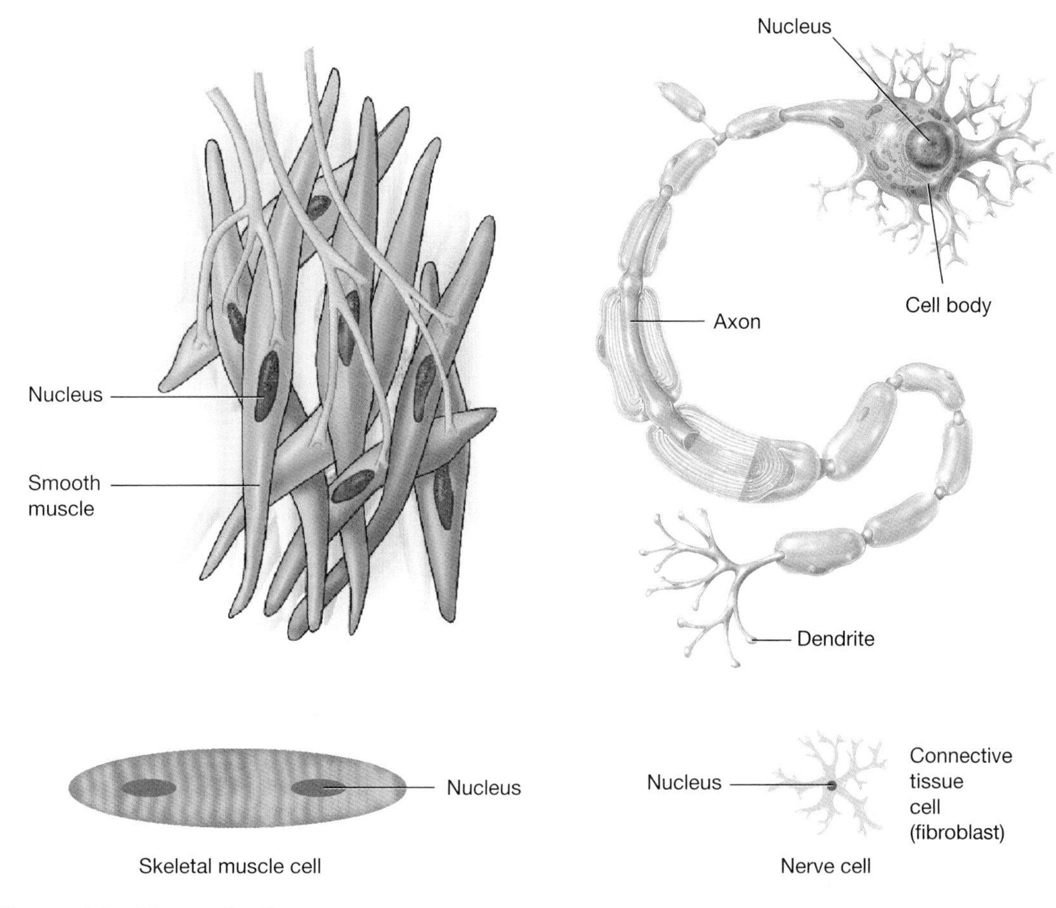


**Figure 6.1** Types of cells.
*Source:* Nair and Peate (2013), Figure 1.12, p. 18. Reproduced with permission of John Wiley and Sons, Ltd.

# Introduction

In Chapter 4 you learned about the cells and their functions. However, cells rarely work alone in our bodies – they organize themselves into groups, and divide and grow together in such a way that they become specialized, for example, muscle cells, skin cells, cells of the lens of the eye, blood cells – Figure 6.1 (Marieb, 2006). Cells that have become specialized in order to perform a common function are grouped together, along with all their associated intracellular material, such as organelles, and so on (Watson, 2005). These groupings are called **tissues**, which are basically groups of cells that are similar in structure and generally perform the same functions (McCance and Huether, 2009). Tissues possess the capability of self-repair.

An example of a type of tissue is muscle; muscle cells have evolved and adapted to perform one particular function – that is, to bring about body movement (Watson, 2005). In the same way, cells that help to control **homeostasis** form nervous tissue.

There are four primary types of tissue, and most organs contain a selection of all four types:

- epithelial
- connective
- nervous
- muscle.

These four primary tissue types then 'interweave to form the fabric of the body' (Marieb, 2006: 85).

Taking the heart as an example (see Chapter 9), the heart contains **muscle** tissue (cardiac muscle – essential for pumping blood around the body). This action is controlled by the **nervous** tissue. The heart is lined with **epithelial** tissue and is supported in the body by **connective** tissue (Wheeldon, 2011).

Put simply (Marieb, 2006):

- **epithelia**l tissue is concerned with 'covering';
- **connective** tissue is concerned with 'support';
- **muscle** tissue is concerned with 'movement';
- **nervous** system is concerned with 'control'.

As the cells need to form themselves into the correct structure for the tissues to be generated, the specialized cells form themselves into tissues in one of two ways.

1. By means of **mitosis**. Cells formed as a result of mitosis are clones of the original cell. Therefore, if one cell with a specialized function undergoes mitosis, and subsequent generations of daughter cells continue to undergo mitosis, the resulting hundreds of cells will all be of the same type and will have the same function – they become tissues. For example, **epithelial cell sheets** (such as **skin**) are formed as a result of mitosis (McCance and Huether, 2009).
2. The second way in which specialized cells can form tissues involves their migration to the site of tissue formation and then they assemble there. This can be seen during the development of the embryo when, for example, cells migrate to sites in the embryo where they differentiate and assemble into a variety of tissues (McCance and Huether, 2009). This process is known as **chemotaxis**, which is 'movement along a chemical gradient caused by chemical attraction' (McCance and Huether, 2009: 33).

# Types of tissues

## Epithelial tissue

There are different types of epithelial tissues that line and cover areas of the body, as well as form glandular tissue. The exterior of the body (the skin) is covered by one type of epithelial tissue, whilst another type of epithelial tissue lines some digestive system organs, such as the stomach, small intestines and the kidneys (Marieb, 2006). Epithelial tissue covers most of the internal and external surfaces of the body. A very important role of epithelial tissue is to act as an **interface** between organs and areas of the body, because almost all of the substances absorbed or secreted by the body must pass through epithelial tissue.

As well as its main function of 'covering', epithelial tissue has six important functions:

- absorption
- protection
- excretion
- secretion
- filtration
- sensory reception.

Not all epithelial tissues carry out the six functions mentioned above; many epithelial tissues only carry out one or two of them. For example, the epithelial tissue found in the digestive system specializes in the absorption of nutrients, whilst the epithelial tissue of our skin provides protection.

Epithelial cells closely bind together to form continuous sheets (e.g. skin), and they possess an **apical** and a **basal** surface:

1. **Apical surface.** The apical surface faces outwards, towards the outside of the body or the organ that it is lining. An apical surface may be smooth, but most have **microvilli** that greatly increase the surface area of the epithelial tissue, so increasing the tissues' ability to carry out the functions of absorption and secretion. In addition, some areas of the body, such as the respiratory tract, possess **villi** (hair-like extensions that are larger than the microvilli), which are able to propel substances along a tract.
2. **Basal surface.** The basal surface, which faces inwards, has a thin sheet of glycoproteins that acts as a selective filter. In this way, the tissue is able to control which substances can enter it.



Epithelial tissue contains many nerve cells, but although epithelial tissue has an excellent supply of nerves, it does not have a blood supply; it obtains its nutrients from blood vessels that are present in the vicinity.

Epithelial tissue is very hardy and tough – which it has to be because it has to cope with a tremendous amount of abrasion, environmental damage and other traumas. This robustness is a result of epithelial cells being capable of dividing and regenerating at great speed, so that damaged cells are rapidly replaced; as long as there is a plentiful supply of nutrient.

Epithelial tissue is classified in two ways.

1. By the number of cell layers
   i. **simple** – where the epithelial is formed by a single layer of cells (Figure 6.2);
   ii. **stratified** – where the epithelium has two or more layers of cells (Figure 6.4).
2. By the shape (Figure 6.3):
   i. squamous;
   ii. cuboidal;
   iii. columnar.

## *Simple epithelia*

Simple epithelial tissue is the type of epithelium that is most concerned with the functions of absorption, secretion and filtration. However, because they are usually very thin, they are not involved in the function of protection.

Simple squamous epithelium rests on a basement membrane (the basement layer), which is a structureless material secreted by the cells and separates epithelial tissue from underlying connective tissue (Marieb, 2006). These basement membranes support a layer of cells (Figure 6.2).

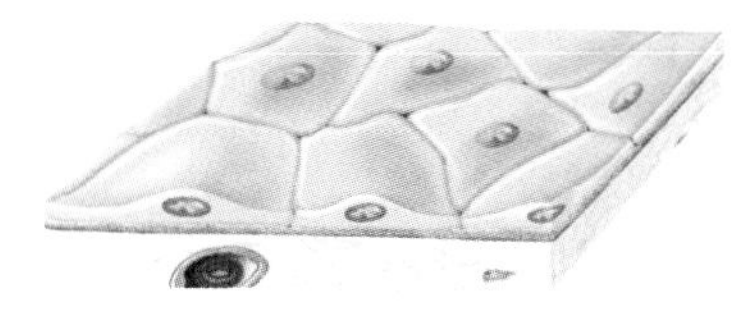

**Figure 6.2** Simple epithelium.
*Source:* Nair and Peate (2013), Figure 1.13, p. 20. Reproduced with permission of John Wiley and Sons, Ltd.

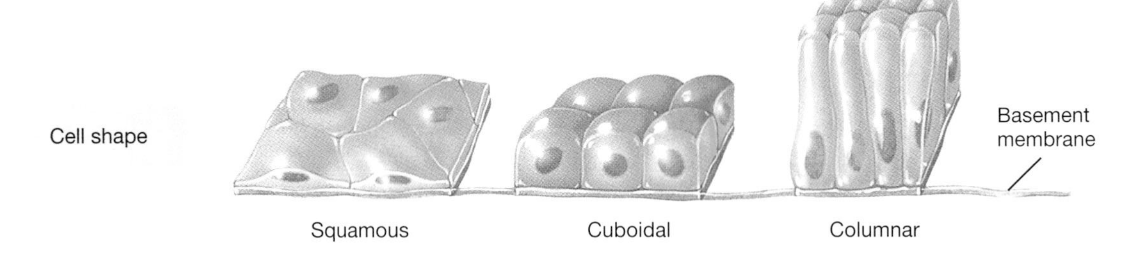


**Figure 6.3** Simple epithelia types.
*Source:* Tortora and Derrickson (2009), Figure 4.3, p. 113. Reproduced with permission of John Wiley and Sons, Inc.

Squamous epithelial cells fit very closely together to give a thin sheet forming the tissue; it is quite often very permeable, so is found where the diffusion of nutrients is essential. It lines the **alveoli** of the lungs and the walls of capillaries. Rapid **diffusion** or **filtration** can take place through this very thin tissue. For example, oxygen and carbon dioxide exchange takes place through the epithelial tissue lining the alveoli of the lungs, whilst nutrients and gases can pass through it from the cells into and out of the capillaries. In addition, simple squamous epithelial cells form serous membranes that line certain body cavities and organs, where it is known as **endothelium**. Endothelium is also found in the kidneys and blood vessels (especially capillaries), where it is ideally situated for the diffusion of nutrients between blood vessels and cells.

**Simple cuboidal epithelial tissue** (Figure 6.3) consists of one layer of cells resting on a **basement membrane**. However, because cuboidal epithelial cells are thicker than squamous epithelial cells, they are found in different places of the body and perform different functions. This epithelial tissue is found in glands, such as the salivary glands and the pancreas, and it forms the walls of kidney tubules and covers the surface of the ovaries (Marieb, 2006).This type of epithelial tissue is concerned particularly with **secretion** and **absorption**.

The third type of simple epithelial tissue, **simple columnar epithelium** (Figure 6.3), consists of a single layer of quite tall cells that, like the other two types, fit very closely together. It can be either ciliated or non-ciliated.

**Ciliated simple columnar epithelial tissue** is found in the areas of the body where the movement of fluids, mucus and other substances is required. The cells of this epithelial tissue line the pathways of the central nervous system and help in the movement of cerebrospinal fluid. They also line the fallopian tubes and assist in the movement of oocytes that have been recently expelled from the ovaries.

**Non-ciliated epithelial tissue** lines the entire length of the digestive tract from the stomach to the rectum and performs two functions depending upon structure. Some of the cells of this type of epithelium have microvilli that greatly increase the surface area for absorption. The other type of non-ciliated epithelial cells are **goblet cells** (they are cup shaped); these release a glycoprotein called **mucin**, which dissolves in water to form **mucus** which lubricates and protects surfaces. Those simple columnar epithelial tissues that line all the body cavities that are open to the body exterior are known as **mucous membranes** (Marieb, 2006).

Another type of simple columnar epithelial tissue also found in the body is **pseudostratified columnar epithelium** – simple columnar epithelial cells that are not all of equal size, giving the illusion that this tissue has many layers, like stratified epithelium. It is commonly found within the lining of the respiratory tract, and also within the lining of the male reproductive system (Wheeldon, 2011).

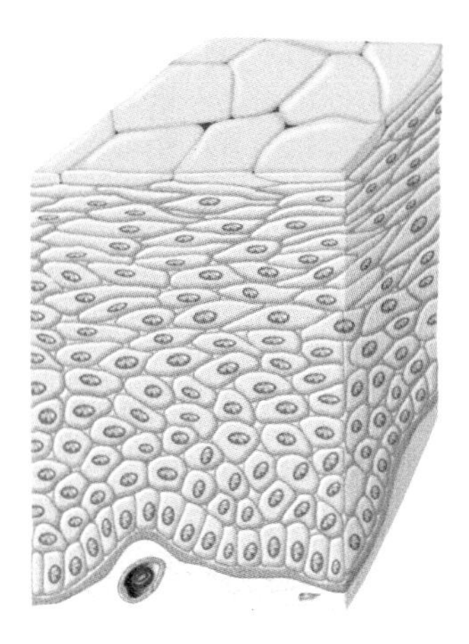

**Figure 6.4** Stratified epithelium.
*Source:* Nair and Peate (2013), Figure 1.14, p. 20. Reproduced with permission of John Wiley and Sons, Ltd.

### *Stratified epithelial tissue*

Stratified epithelial tissue, unlike simple epithelial tissue, consists of two or more cell layers (Figure 6.4). The cells that form this type of epithelium regenerate from the lower levels, with the new cells dividing within the basal layer and pushing the older cells upwards towards the surface. Because these stratified epithelial tissues have more than one layer of cells, they are stronger and hardier than simple epithelia. Thus, a primary function of stratified epithelia is **protection**.

**Stratified squamous epithelial tissue** is the most common stratified epithelium found within the body and consists of several layers of cells (Marieb, 2006). Although this epithelial tissue is called squamous epithelium, it is not made up entirely of squamous cells. The cells at the free edge of the epithelial tissue are composed of squamous cells, whilst those cells close to the basement membrane are composed of either cuboidal or columnar cells. It is found in places that are most at risk of everyday damage, including the oesophagus, the mouth and the outer layer of the skin (Marieb, 2006). Skin stratified squamous epithelial tissue is made stronger and hardier by the presence of **keratin**, which is a very tough, fibrous protein. Stratified squamous epithelial tissue that does not contain keratin lines the 'wet' areas of the body; for example, the mouth, tongue and vagina.

### *Stratified cuboidal epithelial tissue*

Found in the male urethra, the oesophagus and the sweat glands, this epithelial tissue only has two cell layers.

### *Stratified columnar epithelial tissue*

This epithelial tissue is fairly rare in the human body, only being found in the male urethra and the ducts of large glands.

### *Transitional epithelial tissue*

Yet another type of epithelial tissue, **transitional epithelium**, exists in which the basal surface may contain both cuboidal and columnar cells, whilst the apical surface may contain squamous and cuboidal cells. This is a highly modified **stratified** squamous epithelium and forms the lining of just a few organs/structures – all of which are in the urinary system – the urinary bladder, the ureters and part of the urethra. This tissue has been modified so that it can cope with the considerable stretching that takes place with these organs. When these organs or structures are not stretched, the tissue has many layers, with the superficial (top layer) cells being rounded and looking like domes. However, when these organs or structures are distended with urine, the epithelium becomes thinner, the surface cells flatten and they look more like squamous cells. These cells are able to slide past one another and change their shape, so allowing the wall of the ureter to stretch as a greater volume of urine flows through, as well as allowing more urine to be stored in the bladder (Marieb, 2006).



### *Glandular epithelium*

Glandular epithelial tissue is found within the glands of the body. According to Marieb (2006), a gland consists of several cells that make and secrete a particular product.

There are two major types of glands developed from epithelial sheets:

- **exocrine glands**
- **endocrine glands.**

**Exocrine glands** have ducts leading from them, and their secretions empty through these ducts to the surface of the epithelium. Exocrine glands are either unicellular of multicellular. Unicellular glands consist of a single cell type, and **goblet cells** are the main example of such a gland. Multicellular exocrine glands are much more complex and are found in many shapes and sizes. However, all exocrine glands contain an epithelial duct and secretory cells. Some of the exocrine glands are found in the stomach and digestive system and are tubular in shape. Some exocrine glands, such as oil glands within the skin and the mammary glands, are spherical and are known as **alveolar** or **acinar**. Other examples of exocrine glands include the sweat glands, the liver and the pancreas.

**Endocrine glands**, however, do not possess ducts. Instead, their secretions diffuse directly into the blood vessels found within the glands. All endocrine glands, such as the thyroid, the adrenal glands and the pituitary gland, secrete **hormones**.

## Connective tissue

**Connective tissue** is found everywhere in the body and connects body parts to one another. It is the most abundant and widely distributed of all four primary tissue types, and although connective tissues perform many functions and vary considerably in their structure, they have six main functions:

- protection
- insulation
- support
- reinforcement
- binding together other tissues (Marieb, 2006)
- storage sites for excess nutrients (McCance and Huether, 2009).

The most common function of connective tissue is to act as the **framework** on which the epithelial cells gather to form the organs of the body (Figure 6.5).

There are four types of connective tissue:

- connective tissue proper
- cartilage
- bone
- blood.

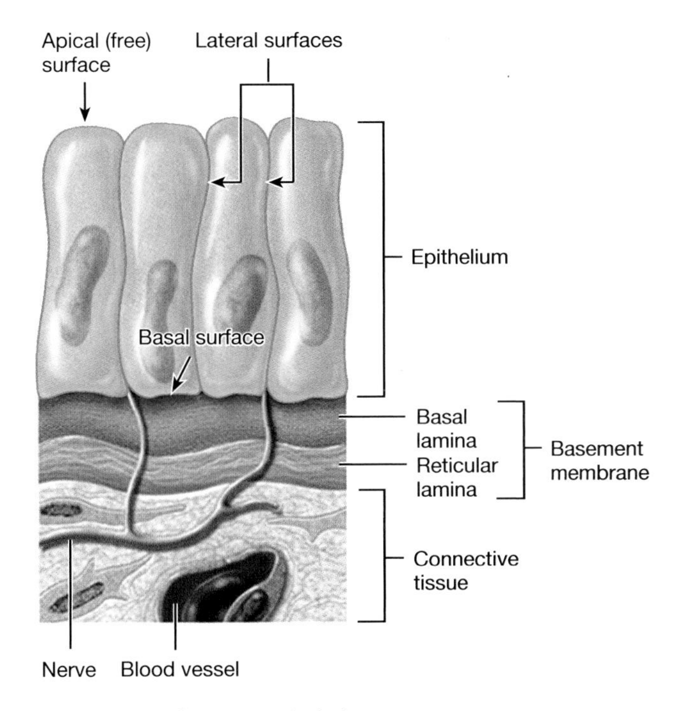


**Figure 6.5** Connective tissue reinforces epithelial tissue.
*Source:* Tortora and Derrickson (2009), Figure 4.2, p. 112. Reproduced with permission of John Wiley and Sons, Inc.

**Table 6.1** The major primary blast cells and their connective tissue type. *Source:* Peate and Nair (2011), Table 4.1, p. 112. Reproduced with permission of John Wiley and Sons, Ltd.

| Connective tissue type | Primary blast cell | Connective tissue cell |
|---|---|---|
| Connective tissue proper | Fibroblast | Fibrocyte |
| Cartilage | Chondroblast | Chondrocyte |
| Bone | Osteoclast | Osteocyte |

Connective tissue is not present on body surfaces. It is generally highly vascular and, therefore, receives a rich blood supply. There are several types of cells to be found in connective tissue:

- **Macrophages** – engulf invading substances, such as bacteria.
- **Plasma cells** – produce antibodies in response to invading substances, prior to the body's immune system destroying them.
- **Mast cells** – produce **histamine**, which promotes **vasodilation** during inflammation.
- **Adipocytes** – fat cells, which, within connective tissue, store fats (triglycerides).
- **White blood cells** – although not normally found in connective tissue in large numbers, they do migrate into such areas during inflammation.
- **Primary blast cells** – these continually secrete ground substance and produce mature connective tissue cells. Each type of connective tissue contains its own particular primary blast cells (Table 6.1).

Blood will be explored in Chapter 8, whilst macrophages, plasma cells, mast cells and white blood cells will be explored in Chapter 7.

There are several common characteristics of connective tissue, and one is that there are few cells in the tissue, but surrounding these few cells there is a great deal of what is known as **extracellular matrix**. This is the intercellular substance of a tissue or the tissue from which a structure develops, and its function is to ensure that connective tissue can bear weight and can withstand the strains, stresses, tensions, traumas and abrasions to which it is constantly subject. The extracellular matrix is composed of **ground substance** and **fibres** and varies in consistency from fluid to a semisolid gel, whilst the fibres, composed of **fibroblasts** – one of the connective tissue cells – are of three types:

- **Collagenous** (white) fibres. These are the most abundant; they are made of the protein **collagen** and are very tough. They are even stronger than steel fibres of the same size (Marieb and Hoehn, 2008).
- **Elastic** (yellow) fibres. These contain the rubber-like protein called **elastin**, which is important in facilitating stretch and recoil, and are found in greater numbers in tissues that have to cope with stretching, such as the skin and the walls of blood vessels.
- **Reticular** fibres. These are much thinner than collagen fibres, and contain bundles of collagen. They provide support and strength to tissues, and are found in large numbers in the 'soft' organs; for example, the spleen and lymph nodes.

**Ground substance** is composed largely of water (interstitial fluid) plus some **cell adhesion proteins** and large **mucopolysaccharide** molecules (called **glycoaminoglycans**). The adhesion proteins serve as a glue that allows connective tissue cells to attach themselves to the fibres, whilst the glycoaminoglycans trap water, giving the ground substance a jelly-like consistency. The change of consistency from fluid to a semisolid gel depends upon the amount of mucopolysaccharide molecules present, with an increase causing the matrix to move from a fluid to a semisolid gel. The ground substance can store large amounts of water, and serves as a water reservoir for the body (Marieb, 2004).

Connective tissue forms a 'packing' tissue around organs of the body and so protects them. It is able to bear weight, and withstand stretching and various traumas, such as abrasions. There is wide variation in types of connective tissue; for example, fat tissue is composed mainly of cells and a soft matrix, whilst bone and cartilage have very few cells but do contain large amounts of hard matrix, which makes them so strong.

There are variations in the blood supply to the tissue. Although most connective tissues have a good blood supply, there are some, like tendons and ligaments, with poor blood supply, whilst cartilages have no blood supply. Thus, cartilages heal very slowly when they are injured – often a broken bone will heal much quicker than a damaged tendon or ligament (Marieb, 2006).

## Types of connective tissues

### *Bone*

Along with cartilage, bones make up the human skeleton. Bone is the most rigid of the connective tissues and is composed of bone cells surrounded by a very hard matrix containing calcium and large numbers of collagen fibres. Because of their hardness, bones provide protection, support and muscle attachment (Marieb, 2006). Bone receives a rich supply of blood and also plays an important role in the production of blood cells.

### *Cartilage*

Cartilage (Figure 6.6) is not as hard as bone, but is more flexible. It has the ability to return to its original shape following stress and movement. This strength and resilience is provided by a

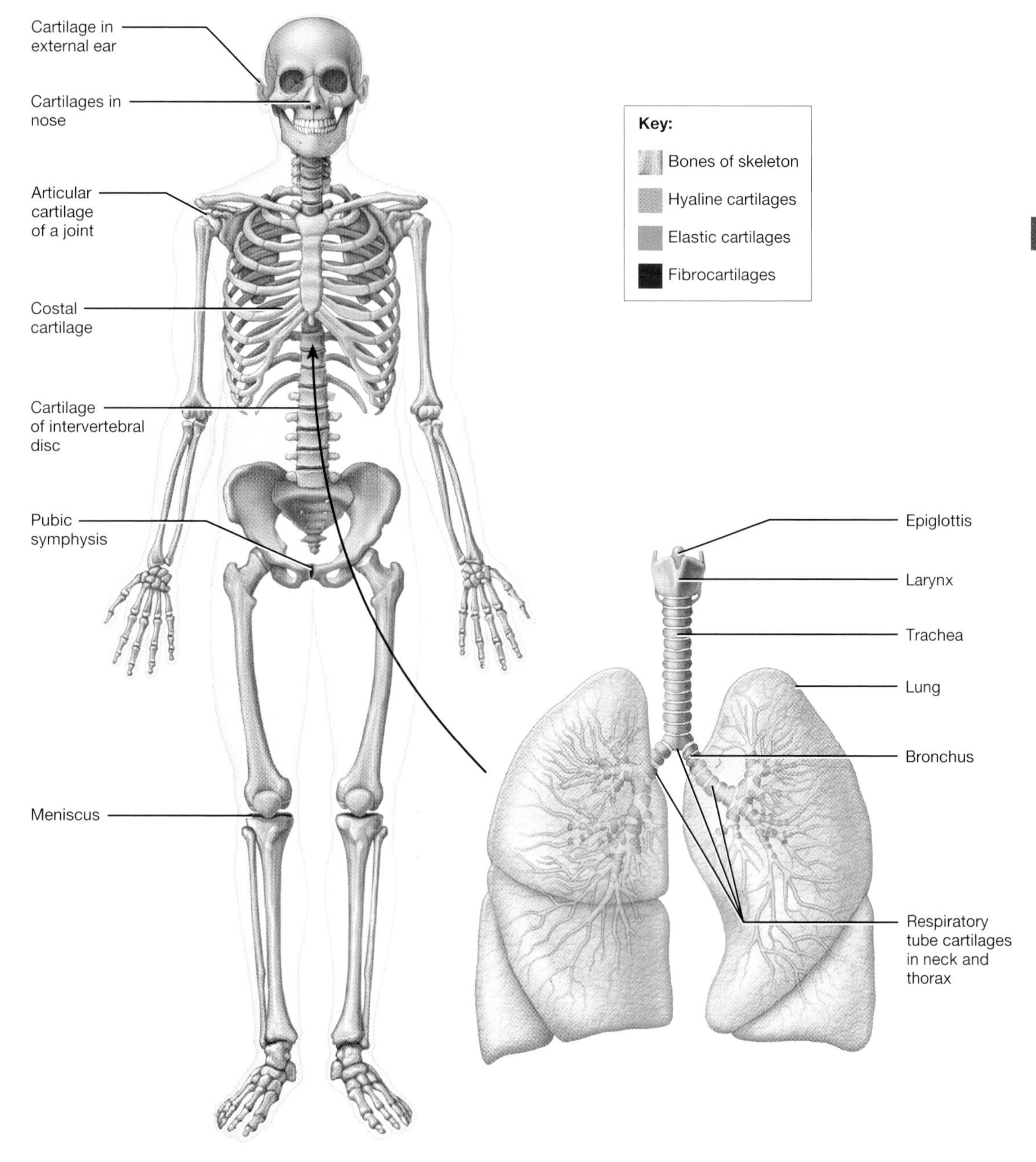


**Figure 6.6** Where cartilage is found in the body.
*Source:* Jenkins and Tortora (2013), Figure 4.6, p. 125. Reproduced with permission of John Wiley and Sons, Inc.

gel-like substance called chondroiton sulphate, which is found in the ground substance of cartilage. Cartilage is surrounded by a layer of dense irregular tissue – the **perichondrium** – which is the only area of cartilage that is served by blood and nervous tissue.

Cartilage is found in only a few places in the body; for example, hyaline cartilage – which supports the structures of the larynx. Cartilage attaches the ribs to the sternum and covers the

**Table 6.2** Types of connective tissue proper – their main constituents, functions and locations. *Source:* Peate and Nair (2011), Table 4.2, p. 114. Reproduced with permission of John Wiley and Sons, Ltd.

| Connective tissue | Main constituent | Functions | Main locations |
|---|---|---|---|
| Loose areolar | Collagen, elastic, reticular fibres | Strength<br>Elasticity<br>Support | Subcutaneous layer beneath skin |
| Loose adipose | Adipocytes | Insulation<br>Protection<br>Energy store | Subcutaneous layer beneath skin<br>Tissue surrounding heart and kidneys<br>Padding around joints |
| Loose reticular | Reticular fibres<br>Reticular cells | Support<br>Filtration | Liver<br>Spleen<br>Lymph nodes |
| Dense regular | Collagen fibres in parallel | Strength<br>Support | Tendons<br>Ligaments |
| Dense irregular | Collagen fibres arranged randomly | Strength | Skin<br>Heart<br>Tissue surrounding bone<br>Tissue surrounding cartilage |
| Dense elastic | Elastic fibres | Stretch | Lung tissue<br>Arteries |

ends of the bones where they form joints (Marieb, 2004). Other types of cartilage include fibrocartilage, which, because it can be compressed, forms the discs between the vertebrae of the spinal column, and elastic cartilage when some degree of elasticity is required; for example, in the external ear.

### *Blood*

'Blood, or vascular tissue, is considered a connective tissue because it consists of blood cells, surrounded by a nonliving, fluid matrix called blood plasma' (Marieb, 2004). Blood is concerned with the transport of nutrients, waste material and respiratory gases (such as oxygen and carbon dioxide), as well as many other substances, throughout the body (see Chapter 8).

### *Connective tissue proper*

Apart from bone, cartilage and blood, all other connective tissue belongs to this class of tissue, which is further subdivided into dense connective tissue and loose connective tissue (Table 6.2). The major difference between the two is that loose connective tissue contains fewer fibres than dense connective tissue does.

#### *Dense connective tissue*

Dense connective tissue contains more collagen or elastic fibres. The dense connective tissue made primarily from collagen is known as either **regular or irregular** collagen fibres, depending upon the organization of the fibres. Dense regular collagen tissue has fibres arranged in parallel rows and is both tough and pliable, whilst dense irregular connective tissue has fibres that are arranged randomly, but closely knitted together. Dense regular connective tissue forms strong, stringy structures such as tendons (which attach skeletal muscles to bones) and the more elastic ligaments (that connect bones to other bones at joints). Dense irregular connective tissue can withstand pressure and pulling forces; it comprises the lower layers of the skin (**dermis**) and the heart, as well as the membranes that surround cartilage and bone. These tissues have collagen fibres as the main matrix element, with many **fibroblasts** (involved in the manufacture of fibres) found between collagen fibres (Marieb, 2006).

### *Loose connective tissue*

Loose connective tissues are softer and contain more cells, but fewer fibres, than other types of connective tissue, with the exception of blood. There are three types of loose connective tissues:



- areolar tissue
- adipose tissue
- reticular tissue.

Areolar tissue is the most widely distributed connective tissue type in the body, and its primary functions are support, elasticity and strength. It is a soft tissue that cushions and protects the body organs that it surrounds, and it helps to hold the internal organs together. It has a fluid matrix that contains all types of fibres forming a loose network – giving it its softness and pliability. It provides a reservoir of water and salts for the surrounding tissues. All body cells obtain their nutrients from this tissue fluid and also release their waste into it. It is also in this area that, following injury, swelling can occur (oedema) because the areolar tissue soaks up the excess fluid just like a sponge does, causing it to become puffy (Marieb, 2004). Areolar tissue is combined with adipose tissue to form the subcutaneous layer, which connects skin with other tissues and organs (Wheeldon, 2011).

### *Adipose tissue*

This tissue, which contains adipocytes and provides insulation, protection and stores energy, is commonly known as 'fat' and is areolar tissue in which there is a preponderance of fat cells. It forms the **subcutaneous** tissue that lies beneath the skin, where it insulates the body and can protect it from the extremes of both heat and cold (Marieb, 2004). It protects some organs, such as kidneys and eyeballs.

### *Reticular connective tissue*

Reticular connective tissue consists of a delicate network of reticular fibres associated with reticular cells (similar to fibroblasts). Its main function is to form a protective internal framework (stroma) to support many free blood cells – mainly the lymphocytes – in the lymphoid organs, such as the lymph nodes, the spleen and bone marrow (Marieb, 2004).

### *Liquid connective tissue*

This is connective tissue that has a liquid extracellular matrix and includes blood and lymph.

## Membranes

Membranes are sheets of tissue that cover or line areas of the body. They contain both epithelial and connective tissues. Membranes consist of an epithelial tissue layer bound to a basement layer of connective tissue (Wheeldon, 2011). There are four major types of membrane:

1. **Cutaneous membrane.** This forms the skin, and consists of an outer stratified squamous epithelial layer sitting on top of a thick layer of dense irregular connective tissue.
2. **Mucous membranes.** These line the external surfaces of body cavities, and include the hollow organs of the digestive tract, the respiratory system and the renal system. All mucous membranes are wet or moist – although not all secrete lubricating mucus (e.g. the mucous membranes of the renal system are wet due to the presence of urine). Most mucous membranes contain stratified squamous or simple columnar epithelium supported by a layer of connective tissue – the lamina propria.

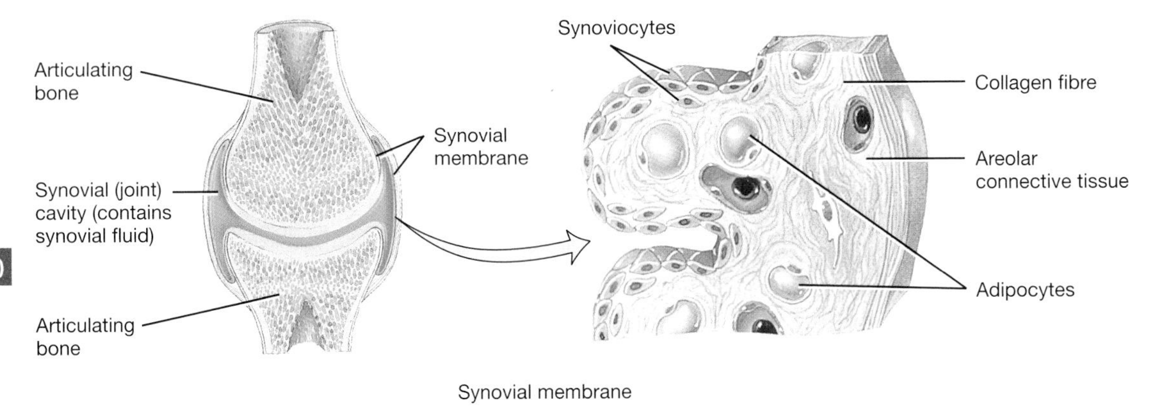


**Figure 6.7** Synovial membranes fill joint cavities.
*Source:* Tortora and Derrickson (2009), Figure 4.7, p. 136. Reproduced with permission of John Wiley and Sons, Inc.

3. **Serous membranes.** These cover internal body cavities and consist of areolar connective tissue covered by a particular squamous epithelium – the **mesothelium** – which secretes **serous fluid** (a watery substance that allows organs to easily slide against one another). Serous membranes consist of an outer (**parietal**) layer and an inner (**visceral**) layer. The **peritoneum**, which lines the organs of the abdominopelvic cavity, is the largest example. Another example is the protective lining of the lungs – the parietal and visceral **pleura** – which glide over one another when the thorax expands on inspiration.
4. **Synovial membranes.** These do not contain any epithelial tissue, and are mainly found in moving joints. They consist of areolar connective tissue, adipocytes, elastic fibres and collagen fibres, and they secrete **synovial fluid** which bathes, nourishes and lubricates the joints of the body. Synovial fluid contains **macrophages** that can help to destroy invading microbes and debris from the joint cavity. These membranes are also found in cushion-like sacs in the hands and feet, and their purpose is to ease the movement of tendons (Figure 6.7).

## Muscle tissue

There are three types of muscle tissues, and these are responsible for helping the body to move, or to move substances within the body. The three types of muscle tissues are:

- skeletal muscle
- cardiac muscle
- smooth muscle.

### *Skeletal muscle*

This is muscle that is attached to bones and is involved in the movement of the skeleton. The structure of the muscle fibres within skeletal muscle gives a striped or **striated** appearance. These muscles can be controlled voluntarily, and form the 'bulk' of the body (the flesh). The cells of skeletal muscle are long, cylindrical and have several nuclei (Figure 6.8). They work by contracting and relaxing, with pairs working antagonistically against each other – that is, one muscle contracts and the opposite muscle relaxes. So, for example, if the muscles in the front of the arm contract and the ones at the back of the arm relax then the arm bends.

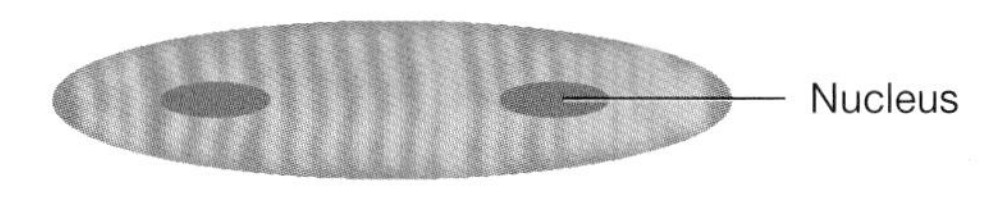


**Figure 6.8** Skeletal muscle cells taken.
*Source:* Nair and Peate (2013), Figure 1.12, p. 18. Reproduced with permission of John Wiley and Sons, Ltd.



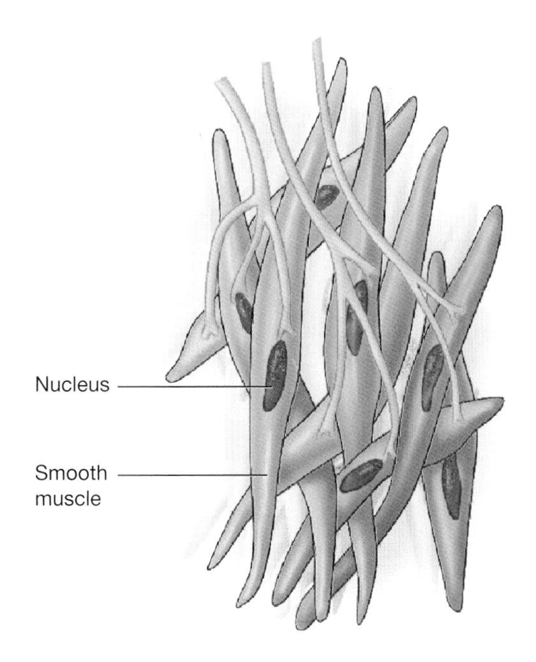


**Figure 6.9** Smooth muscle.
*Source:* Nair and Peate (2013), Figure 1.12, p. 18. Reproduced with permission of John Wiley and Sons, Ltd.

## *Cardiac muscle*

Cardiac muscle is only found in the heart, and it acts as a pump to move blood around the body. It does this by contracting and relaxing, just like skeletal muscles, and it appears striated. However, unlike skeletal muscles, it works in an **involuntary** way – the activity cannot be consciously controlled.

## *Smooth muscle*

Also known as **visceral** muscle, smooth muscle (Figure 6.9) is found in the walls of hollow organs; for example, the stomach, bladder, uterus and blood vessels. Smooth muscle has no striations, and like cardiac muscle it works in an involuntary way. Smooth muscle causes movement in the hollow organs; for example, as smooth muscle contracts, the cavity of an organ becomes smaller (constricted); and when smooth muscle relaxes, the organ becomes larger (dilated). This allows substances to be propelled through the organ in the right direction; for example, faeces in the intestines. Because smooth muscle contracts and relaxes slowly, it forms

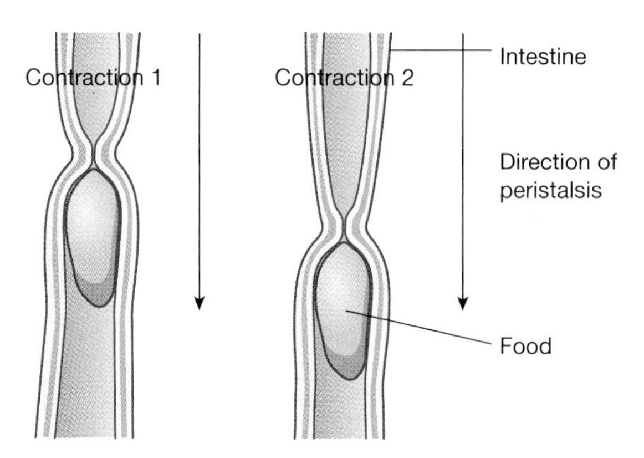


**Figure 6.10** Peristalsis.

a wavelike motion (known as peristalsis) to push, in the case of the intestines, the faeces through the intestines (Figure 6.10).

## Nervous tissue

Nervous tissue is concerned with control and communication within the body by means of electrical signals. The main type of cell that is found in nervous tissue is the **neurone** – the functioning unit of the nervous system –and consists of a **cell body** enclosing a **nucleus** at one end, followed by a long **axon**, and then **dendrites** at the other end (Figure 6.11). All neurones receive and conduct electrochemical impulses around the body. The structure of neurones is very different from other cells. The cytoplasm is found within long processes or extensions (the axon) – some in the leg being more than a metre long. These neurones receive and transmit electrical impulses very rapidly from one to the other across **synapses** (junctions). It is at the synapses that the electrical impulse can pass from neurone to neurone, or from a neurone to a muscle cell. The total number of neurones is fixed at birth, and cannot be replaced if they are damaged (McCance and Huether, 2009).

In addition to the neurones, nervous tissue includes some cells that are known as **neuroglia-supporting cells**. These supporting cells insulate, support and protect the delicate neurones. The neurones and supporting cells make up the structures of the nervous system, namely:

- the brain
- the spinal cord
- the nerves.

# Tissue repair

Tissues repair and replace any cells that have been damaged, have become worn out or have died. Each of the four tissue types is able to regenerate and replace cells that have been damaged as a result of trauma, disease or other events, such as toxins. Not all tissue types are very successful at this process of regeneration and repair, however. Epithelial cells have a great capacity for renewal simply because they are exposed to so much wear and tear in normal living. This is possible because epithelial cells have immature cells – stem cells – that divide and replace lost cells easily. In addition, most connective tissue renews itself easily – apart from cartilage, which, because of a very poor blood supply, takes a long time to repair itself.

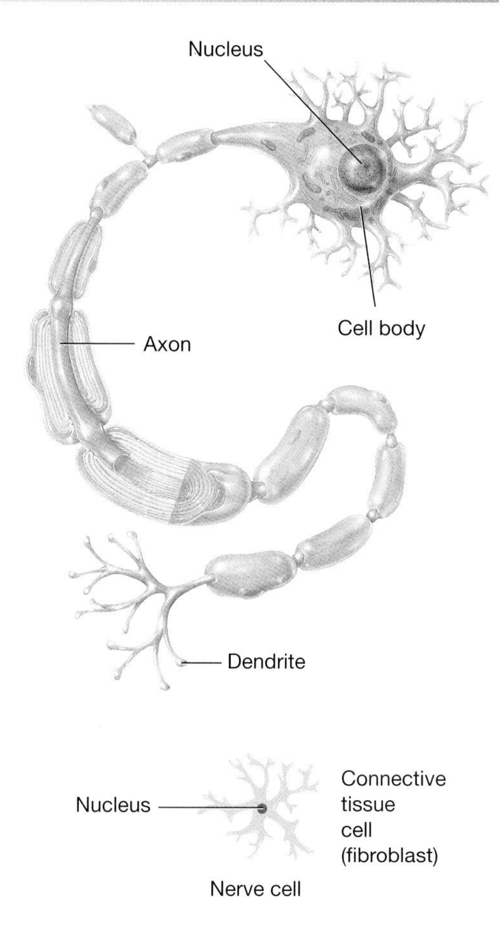


**Figure 6.11** Nerve cell.
*Source:* Nair and Peate (2013), Figure 1.12, p. 18. Reproduced with permission of John Wiley and Sons, Ltd.

Muscle and nervous tissues both have poor regeneration properties, and skeletal and smooth muscle fibres divide very slowly, whilst **mitosis** does not occur in cardiac muscle tissue. Instead, stem cells migrate from the blood to the heart where they divide and produce a small number of new cardiac muscle fibres. Nervous tissue does not normally undergo **mitosis** to replace damaged neurones, and so if nervous tissue is damaged it is lost for ever as a functioning tissue (Wheeldon, 2011).

Inflammation is the body's immediate reaction to tissue injury or damage, because when tissue injury or damage does occur, this stimulates the body's inflammatory and immune responses to spring into action so that the healing process can begin almost immediately (Vickers, 2009). Following damage to a tissue and the instigation of the inflammatory response, the actual repair of damaged tissue involves regeneration – the production and proliferation of new cells (cell division) in the **parenchyma** (tissue and organ cells) or from the **stroma** (supporting connective tissue):

- If **parenchymal** cells are solely responsible for tissue repair, a near-perfect regeneration may well occur.

- If fibroblasts from stroma are involved, new connective tissue (mainly consisting of collagen) is generated to replace the damaged tissue (scar tissue). This process of generating scar tissue is called **fibrosis**, and unlike cells that are regenerated from **parenchymal** cells, scar tissue cells do not perform the original functions of the damaged cells (because they do not have the exact genetic match). So, any organ or structure that contains scar tissue will also have impaired functioning.

In open or large wounds, the process of granulation occurs. **Granulation tissue** is the perfused, fibrous connective tissue that replaces a fibrin clot in healing wounds. Granulation tissue typically grows from the base of a wound to fill wounds of almost any size. During granulation, both parenchymal and stroma cells are active. Fibroblasts provide new collagen tissue to strengthen the area, as well as new blood capillaries sprouting new buds in order to bring the necessary nutrients to the area (Wheeldon, 2011).

# Children and tissue development

Throughout childhood, tissues are developing at different rates in order to fit in with growth and physical needs. The growth and physical/psychomotor and mental/intellectual development that occurs throughout childhood is linked to the growth and development of the various tissues throughout that period.

## Growth

In early childhood, there is a rapid increase in body size for the first 2 years, and then this starts to taper off into a slower growth pattern; but on average, children add 2–3 inches (5–7.5 cm) in height and about 5 lbs (2.25 kg) in weight each year in the early years of childhood. The 'baby fat' apparent in babies begins to decline as the child starts to toddle and then walk, so that children generally become thinner. Girls tend to retain more body fat than the slightly more muscular boys. At the same time, the child's torso – of both boys and girls – lengthens and widens, whilst the spine straightens. By the age of 5 years, toddlers have lost their the top-heavy, bowlegged and pot-bellied physique and are more streamlined with longer legs and flatter stomachs, with body proportions more like those of an adult (Berk, 2008). There are other growth spurts throughout childhood – particularly between the ages of 10 and 16 years. Tissue has to grow and develop throughout childhood in order to allow the child to grow and mature into an adult.

## Bones

At birth, there is very little bone mass in the baby's body, and the bones are softer – they are more cartilaginous – and are much more flexible than the bones of adults. Between the ages of 2 and 6 years, epiphyses (growth centres in which cartilage hardens into bone) emerge in the skeleton. This also occurs during middle childhood (Berk, 2008). An adult has a total of 206 bones which are joined to ligaments and tendons, whilst babies, at birth, have 270 bones. It is not until the age of about 20 years that the bones have finally fused together to leave us with the 206 hard bones of the adult.

## Lymphatic system

This system – linked to the immune system – grows at a constant rate throughout childhood and reaches maturity just before puberty occurs, before it then decreases.

## Central nervous system

This system develops mainly during the first few years after birth. Although brain cell formation is virtually complete before birth, the maturing of the brain, which is not fully developed, continues after birth. This is because the 100 billion or so brain cells have not yet connected and developed into functioning networks. Between birth and the age of 2 years, the brain's weight increases from about 25% of an adult's brain size to about 70%, and then between the ages of 2 and 6 years the brain increases from 70% to 90% of the weight of the adult brain. In addition, the brain also undergoes much reshaping and refining as the brain matures and responds to internal and external stimuli. For example, in some regions, such as the frontal lobes, the number of synapses is almost double the adult value (Berk, 2008).



# Conclusion

Tissues are varied structures that work to make our incredibly complicated bodies. There are four major classifications of tissues: epithelial, connective, muscle and nervous tissues. Each of these tissues plays a part in the total functioning of the body. Epithelial tissue, as well as covering or lining all our structures and organs, is involved in absorption, secretion, protection, excretion, filtration and sensory reception. Connective tissue protects, supports and insulates tissues. Muscle tissue is responsible for movement and posture, whilst nervous tissue is involved in sensory feeling and response. You will learn more about the tissues as you work through this book and look at these structures and roles in situ – within the various systems of the body.

Finally, this chapter has briefly discussed the ways in which tissues are able to regenerate and renew themselves in varying ways and with varying results – with the notable exception of nervous tissue.

# Activities

**Now review your learning by completing the learning activities in this chapter. The answers to these appear at the end of the book. Further self-test activities can be found at www.wileyfundamentalseries.com/childrensA&P.**

## Complete the sentence

1. The main type of cell that is found in nervous tissue is the _____.
2. Nervous tissue does not normally undergo _____ to replace damaged neurone
3. In open or large wounds, the process of granulation occurs using granulation tissue which is perfused, fibrous ______ tissue, which replaces the initial ______ clot.
4. In early childhood there is a rapid increase in ____ ____ for the first __ ____.
5. An adult has a total of ___ bones which are joined to ligaments and tendons, whilst babies, at birth, have ___ bones.
6. Cartilage is found in only a few places in the body; for example, _____ ______ – which supports the structures of the _____.
7. Bone is the most ____ of the connective tissues and is composed of ______ ______ surrounded by a very hard matrix containing ____ and large numbers of _____ _____.

8. _____ _____ produce antibodies in response to invading substances, prior to the body's immune system destroying them.
9. Connective tissue is __ present on the ___ surfaces.
10. The most common function of connective tissue is to act as the ______ on which the_____ _____ gather to form the organs of the body.



## Wordsearch

There are 16 words linked to this chapter hidden in the following square. Can you find them?

A tip – the words can go from up to down, down to up, left to right, right to left, or diagonally.

| f | n | n | f | s | t | e | p | h | e | y | c | a | t | v | e | u | t | n | o |
|---|---|---|---|---|---|---|---|---|---|---|---|---|---|---|---|---|---|---|---|
| i | e | o | d | h | g | e | f | c | m | o | a | y | e | i | y | n | e | s | e |
| b | e | p | i | t | h | e | l | i | a | l | r | e | r | n | s | u | o | l | e |
| r | x | u | r | s | s | d | t | o | e | s | t | m | m | n | r | s | t | b | a |
| o | o | l | e | y | u | o | o | o | t | n | i | i | v | o | t | w | u | v | r |
| b | c | p | h | a | s | f | s | a | r | t | l | n | n | p | b | c | a | e | d |
| l | y | o | m | i | h | s | f | e | s | p | a | n | y | s | h | s | l | b | d |
| a | t | h | s | s | n | o | m | i | r | w | g | e | e | c | c | i | i | j | e |
| s | o | w | h | c | h | i | e | o | d | d | e | x | e | u | e | c | o | d | e |
| t | s | c | a | c | c | e | i | h | e | t | a | p | l | e | v | y | t | g | u |
| a | i | r | r | s | e | g | c | h | e | i | a | a | h | f | i | t | m | i | o |
| l | s | a | a | f | t | e | e | b | r | r | r | i | e | e | t | a | p | b | r |
| g | l | y | c | o | p | r | o | t | e | i | n | e | o | m | c | t | a | e | d |
| g | l | a | n | d | t | h | o | n | f | a | a | o | h | f | e | e | e | m | c |
| i | l | s | t | n | h | r | c | m | o | o | t | a | v | n | n | n | m | p | a |
| h | o | e | e | a | r | h | w | a | a | i | u | b | m | t | n | e | h | m | q |
| n | i | i | u | i | y | f | m | n | r | y | d | t | n | l | o | o | s | t | t |
| t | u | i | c | m | f | i | b | e | s | o | r | d | e | t | c | i | p | o | o |
| c | t | h | a | l | d | k | d | b | k | e | o | e | t | t | s | t | l | l | o |
| n | t | n | i | a | r | l | d | d | m | t | w | a | t | l | a | n | o | t | c |

## Complete the table

Link the connective tissue types, primary blast cells and connective tissue cells from the list below.

| Connective tissue type | Primary blast cell | Connective tissue cell |
|---|---|---|
| | | |
| | | |
| | | |



fibroblast bone chondrocyte
connective tissue proper osteoclast fibrocyte
osteocyte cartilage chondroblast

## Conditions

The following table contains a list of conditions. Take some time and write notes about each of the conditions. You may make the notes taken from text books or other resources; for example, people you work with in a clinical area or you may make the notes as a result of people you have cared for. If you are making notes about people you have cared for you must ensure that you adhere to the rules of confidentiality.

| Condition | Your notes |
|---|---|
| Reynaud's phenomenon | |
| Sarcoidosis | |
| Mesothelioma | |
| Tuberculosis | |
| Scleroderma | |

## Glossary

**Abdominopelvic cavity:** Body cavity that encompasses the abdominal and pelvic cavities. The abdominal cavity contains the stomach, intestines, spleen, liver and other associated digestive organs. The pelvic cavity contains the bladder and some reproductive organs.

**Apical surface:** surface of a body organ that faces outwards, towards the surface.

**Avascular:** structure that does not contain blood vessels.

**Basal surface:** surface that forms the base of a body organ.

**Cartilage:** strong form of connective tissue that contains a dense network of collagen and elastic fibres.

**Diffusion:** the most common form of passive transport of materials – it is the ability of gases, liquids and solutes to disperse randomly and to occupy any space available, so that there is an equal distribution.

**Endocrine gland:** glands that release hormones.

**Endocytosis:** the general name for the various processes by which cells ingest foodstuffs and infectious microorganisms.

**Epithelial tissue:** tissue that lines or covers body surfaces.

**Exocrine glands:** glands that secrete their products externally (e.g. mucus and sweat).

**Exocytosis:** the system of transporting materials out of cells.

**Extracellular fluid:** fluid outside of the cell, but surrounding it.

**Extracellular matrix:** found in connective tissue, this is non-living material that is made up of ground substance and fibres. It separates the living cells found in tissues.

**Fibres:** these are any long, thin structures. The body contains many of them, including nerve fibres and muscle fibres.

**Fibroblasts:** these are the typical cell type of connective tissue. They are responsible for the production and secretion of extracellular matrix materials.

**Gland:** a structure that manufactures a problem (e.g. hormones, mucus, sweat).

**Glycoproteins:** special proteins that contain simple sugar chains, and play an important role in cell-cell communication.

**Goblet cells:** individual cells found in mucosal membranes that produce mucus.

**Ground substance:** the name given to that part of the extracellular matrix (found in connective tissue) that is composed mainly of water, with some adhesion proteins and large polysaccharide molecules.

**Histamine:** a chemical found in some of the body's cells and which causes many of the symptoms of allergies, such as a runny nose or sneezing.

**Hormones:** regulatory chemicals released by endocrine glands for use elsewhere in the body (e.g. thyroxin and insulin).

**Interface:** a point where two systems, subjects, organizations, and so on, meet and interact.

**Innervated:** stimulated by nerve cells.

**Interstitial fluid:** the fluid that surrounds and bathes cells.

**Keratin:** a tough fibrous protein found in skin.

**Macrophages:** white blood cells that specialize in the destruction and consumption of invading pathogens.

**Membrane:** a sheet of tissue that covers or lines an area of the body.

**Microvilli:** hair-like extensions found on the surfaces of cells (singular: microvillus).

**Mitosis:** the process by which cells (other than the gametes) are reproduced by simple division of the nucleus and the cell itself.

**Neuroglia-supporting cells:** these cells are found in nervous tissue and their role is to support the delicate neurones by insulating, supporting and protecting them.

**Neurone:** a nerve cell.

**Parenchyma:** the cells that constitute the part of an organ that is concerned with the function of that organ.

**Oocytes:** female reproductive cells.

**Spleen:** the large lymph organ that is responsible for the production of lymphocytes, and the cleansing of blood.

**Stroma:** the internal framework of an organ.

**Subcutaneous:** underneath the skin.

**Vertebrae:** the disc-shaped bones that make up the spinal column.



# References


Berk, L.E. (2008) *Infants and Children: Prenatal through Middle Childhood*, 6th edn, Pearson Educational, Harlow.

Jenkins, G.W., Tortora, G.J. (2013) *Anatomy and Physiology from Science to Life*, 3rd edn, John Wiley & Sons, Inc., Hoboken, NJ.

Marieb, E.N. (2004) *Human Anatomy and Physiology*, 6th edn, Pearson Benjamin Cummings, San Francisco, CA.

Marieb, E.N. (2006) *Essentials of Human Anatomy and Physiology*, 8th edn, Pearson Benjamin Cummings, San Francisco, CA.

Marieb E.N., Hoehn, K. (2008) *Human Anatomy and Physiology*, 8th edn, Pearson Benjamin Cummings, San Francisco, CA.

McCance, K.L., Huether, S.E. (2009) *Pathophysiology: The Biologic Basis for Disease in Adults and Children*, 6th edn, Mosby, St. Louis, MO.

Nair, M., Peate, I. (2013) *Fundamentals of Applied Pathophysiology: An Essential Guide for Nursing Students*, John Wiley & Sons, Ltd, Oxford.

Peate, I., Nair, M. (eds) (2011) *Fundamentals of Anatomy and Physiology for Student Nurses*, Wiley–Blackwell, Chichester.

Tortora, G.J., Derrickson, B.H. (2009) *Principles of Anatomy and Physiology*, 12th edn, John Wiley & Sons, Inc., Hoboken, NJ.

Vickers, P.S. (2009) Cell and body tissue physiology. In Nair, M., Peate, I. (eds), *Fundamentals of Applied Pathophysiology: An Essential Guide for Nursing Students*. Wiley–Blackwell, Chichester.

Watson, R. (2005) Cell structure and function, growth and development. In: Montague, S.E., Watson, R., Herbert, R.A. (eds), *Physiology for Nursing Practice*, 3rd edn, Elsevier, Edinburgh.

Wheeldon, A. (2011) Tissue. In: Peate, I., Nair, M.(eds), *Fundamentals of Anatomy and Physiology for Student Nurses*, Blackwell, Chichester.

# Chapter 7

# The immune system

Alison Mosenthal

*School of Health and Social Work, University of Hertfordshire, Hatfield, UK*

## Aim

To gain an insight to the function of the immune system and its role in the defence against infectious disease.

## Learning outcomes

On completion of this chapter the reader will be able to:

- Discuss the development of white blood cells in relation to their role in the immune system.
- Describe and discuss the role of the cells, tissues and specialized organs of the immune system.
- Differentiate between innate and adaptive immunity.
- Explain the body's response to infection.
- Discuss the relevance of immunization in the protection of the infant and child from infectious disease.
- Begin to understand how the theory can be applied to practice.

## Test your prior knowledge

- Which blood cells are involved in the immune system?
- Name the microorganisms that can cause infection.
- What barriers does the human body have to prevent infectious organisms gaining entry to the body?
- What are the four classic signs in inflammation?
- What is meant by the term phagocytosis?
- What are the two types of lymphocytes?
- What is the role of antibodies in the immune response?
- What happens when a person is immunized against an infectious disease?
- What is the chain of infection?
- What does the HPV vaccine do?

*Fundamentals of Children's Anatomy and Physiology: A Textbook for Nursing and Healthcare Students*, First Edition. Edited by Ian Peate and Elizabeth Gormley-Fleming.

Companion website: www.wileyfundamentalseries.com/childrensA&P

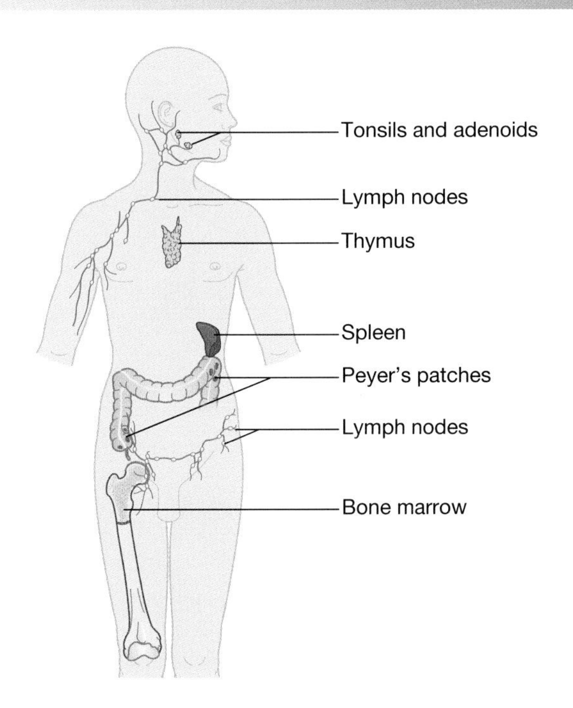




**Figure 7.1** Organs and tissues of the immune system.

# Introduction

Microorganisms such as bacteria and viruses surround us, and our bodies constantly have to protect themselves from the invasion of these organisms, which have the potential to infect and cause harm. The human body has evolved and developed many defence mechanisms to fight infection, and the study of the immune system enables us to understand the processes that are involved in the recognition and prevention of infection.

In this chapter we will examine the generalized (**innate immunity**) and specialized (**acquired** or **adaptive immunity**) responses of the body to infection, their ability to interact and protect the individual, with particular emphasis on how this develops in childhood, and the implications for nursing neonates and children with infections.

Figure 7.1 outlines the organs of the immune system.

# Blood cell development

Blood cells develop by a process of **haemopoiesis** from a multipotent stem cell that is a precursor cell for developing into the different blood cells (MacPherson and Austyn, 2012). In the developing fetus these cells are found in the spleen, liver and bone marrow, but after birth the principal site for haemopoiesis is in the bone marrow.

Figure 7.2 shows the process of haemopoiesis, where the multipotent stem cell divides to give two different precursor cells: one develops into a myeloid stem cell and the other develops into a lymphoid stem cell which goes on to produce the white blood cells that are part of the immune system.

The lymphoid stem cell produces the **T cell** and **B cell** lymphocytes and Natural Killer cells. The myeloid stem cell differentiates to produce the macrophages (the monocytes and the tissue

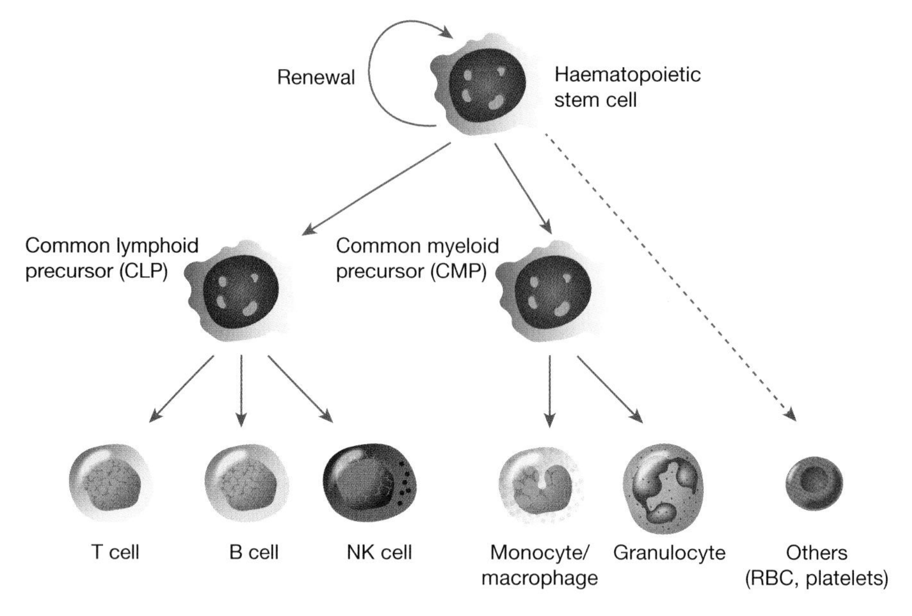


**Figure 7.2** Haemopoiesis. Adapted from MacPherson and Austyn (2012).

macrophages) and the granulocytes (the neutrophils, eosinophils and basophils). It can also be seen that from the stem cell erythrocytes (red blood cells) and megakaryocytes which produce platelets are also produced. It is the white blood cells that this chapter will concentrate on, as they are essential for the immune system. Red blood cells, however, play a significant part in transporting oxygen to other cells of the immune system, and the platelets will be discussed in relation to their role in inflammation.

Following their initial development in the bone marrow from stem cells, these white blood cells mature in different places in the body as part of the immune system.

## Clinical application

Stem cells can be given as part of the treatment in certain cancers such as leukaemia, where high doses of chemotherapy are used to destroy the leukaemia cells but at the same time destroy healthy bone marrow stem cells. Stem cells are obtained from bone marrow donation and peripheral blood donation from matched donors. Infusions of stem cells are given after the patient has received chemotherapy.

Stem cells can also be retrieved from umbilical cord blood and stored for future use in a cord blood bank.

# The organs and tissues of the immune system

The immune system is made up of organs and tissues that are connected by the blood and lymph systems. The organs and tissues where lymphocytes are produced are described as the primary lymphoid organs, and where they come into contact with harmful organisms as mature lymphocytes as the secondary lymphoid organs (Nairn and Helbert, 2007).

## The lymphatic system

Lymphocytes circulate continuously between the blood, body tissues and secondary lymphoid organs. Part of the function of the immune system is to concentrate lymphocytes in areas to encounter antigens, and this is achieved by the lymphatic system.

This is a specialized circulatory system (similar to the blood circulation) consisting of lymph vessels and lymph nodes that contain a fluid called lymph. The lymph is formed from plasma leaking from blood capillaries after blood has circulated through the tissues and the exchange of nutrients, waste and gases has occurred. This leaked fluid and any plasma proteins that escape from the blood need to be returned to the vascular system to maintain blood volume. If this does not occur then fluid accumulates in the tissues and this results in **oedema**, and this impairs the ability of the tissue cells to make exchanges with the interstitial fluid and blood.



The peripheral lymphatic system (Figure 7.3) consists of lymphatic vessels and lymphatic capillaries and encapsulated organs that include:

- **spleen**
- **tonsils**
- **lymph nodes.**

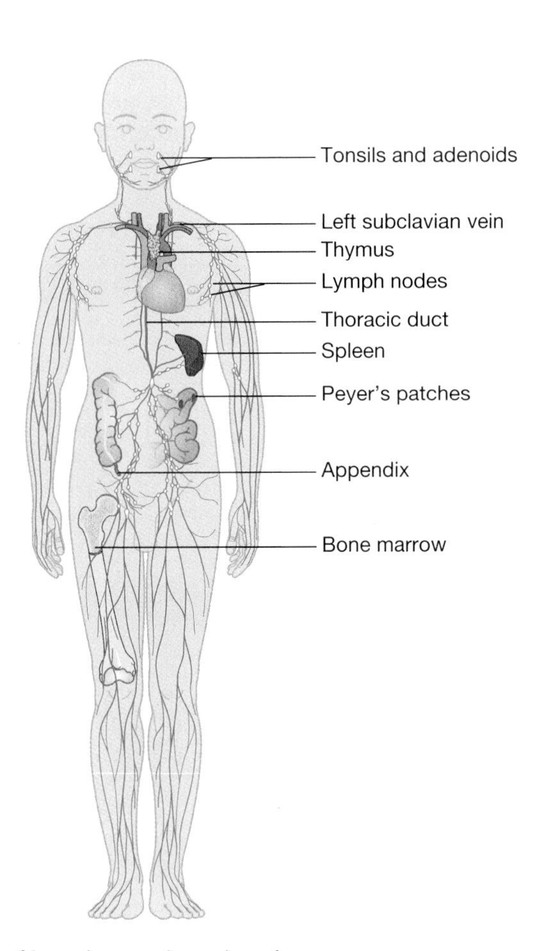


**Figure 7.3** Distribution of lymph vessels and nodes.

As well as the encapsulated organs, the lymphatic system also includes lymphoid tissue that is not encapsulated, and this is found in the gastrointestinal, respiratory and urogenital mucosal areas.

The lymphatic capillaries and vessels join together to form a network throughout the body, and as they become bigger and deeper they form lymph vessels that are located near to major blood vessels. The lymph capillaries are made up of endothelial cells in a single layer, and this allows for movement of molecules to enter the lymph system between capillary walls. Proteins and larger particles, such as bacteria and viruses that are normally prevented from entering the blood capillaries, can also enter the lymphatic system (Marieb, 2012).



The lymphatic system does not have a pump to move the fluid round the system (unlike the heart in the circulatory system). The fluid is moved around the body through the vessels in one direction towards the neck by means of:

- rhythmic contraction of the smooth muscle of the lymph vessels;
- muscle contraction in the upper and lower limbs;
- pressure changes in the thoracic region during breathing.

The lymph returns to the venous circulation by means of two ducts in the thoracic region: the right thoracic duct, which receives lymph from the right side of the body, and the thoracic duct that receives lymph from the rest of the body. These ducts empty the lymph into the right and left subclavian veins respectively.

The primary lymphoid organs have already been mentioned above in the development of the blood cells. In the fetus the primary lymphoid organs are initially the yolk sac where the haemopoietic stem cells are found. By the eighth week of gestation the major sites are the fetal liver, spleen and bone marrow; however, by birth, the production of blood cells, and therefore the production of lymphocytes (**lymphopoiesis**), is in the bone marrow and maturation of the T cell lymphocytes takes place within the thymus gland (Nairn and Helbert, 2007; Figure 7.4).

The secondary lymphoid organs are:

- spleen
- lymph nodes
- lymphoid tissue that lines the respiratory, gastrointestinal and urogenital tracts.

We will now consider the thymus gland and the secondary lymphoid organs in more detail.

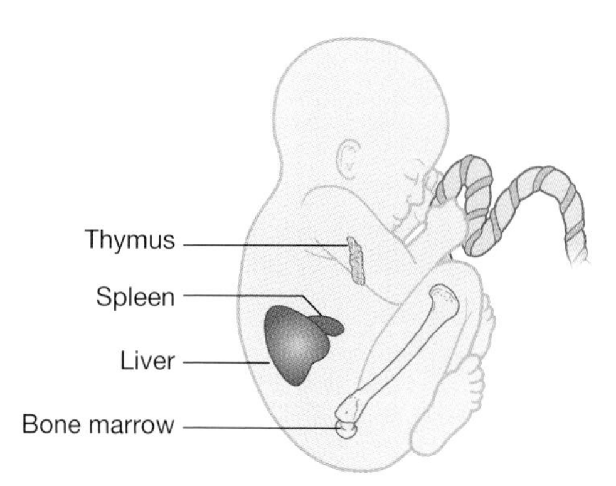


**Figure 7.4** Organs of lymphocyte production in the fetus. Adapted from Nairn and Helbert (2007).

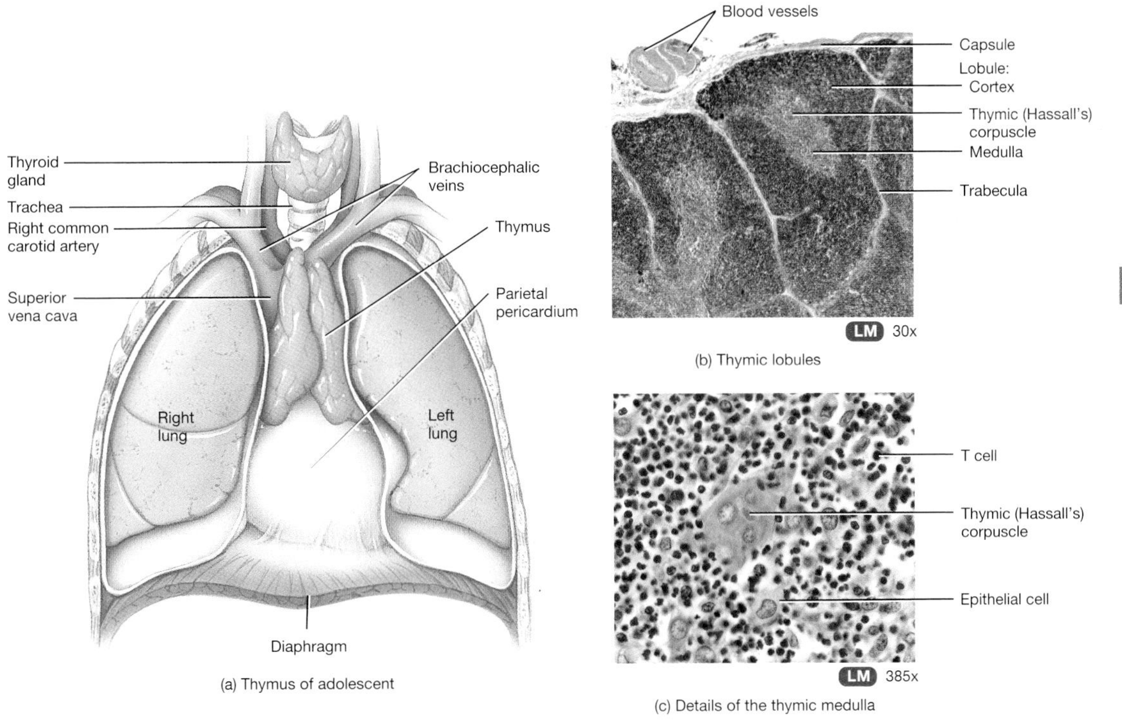




**Figure 7.5** Location of thymus gland.
*Source:* Tortora and Derrickson (2009), Figure 22.5(a), p. 837. Reproduced with permission of John Wiley and Sons, Inc.

## *Thymus gland*

The thymus is a bilobed gland situated in the chest area in the upper anterior thorax just above the heart. Figure 7.5 identifies the location of the thymus gland. It is a relatively large organ in babies and grows until puberty, but it then begins to atrophy so that only a small amount of lymphoid tissue remains in late adulthood (Nairn and Helbert, 2007). T cells, however, continue to develop in the thymus throughout adult life.

The importance of this organ is that lymphocytes formed from the lymphoid stem cell migrate to the thymus gland where they mature into **T cell lymphocytes**. These cells differentiate into their different subclasses and leave the thymus to become peripheral T cells in the secondary lymphoid tissues. Another important aspect of these cells is that they recognize 'self' from 'non-self', where 'self' means cells that originate within the individual. This is particularly important to ensure that an individual's immune system does not destroy 'self' cells and cause an **autoimmune reaction** (Vickers, 2011).

## Clinical application

A child born without a thymus gland with a congenital condition known as **Di George syndrome** may be at risk of opportunistic infections from fungi and viruses.

### *The spleen*

The spleen is situated on the left side of the abdominal cavity behind the stomach and just below the diaphragm. The spleen is the body's largest lymph organ, and within its lymphoid tissue it has phagocytes and lymphocytes that generate an immune response to antigens in blood as it passes through the spleen.

The spleen also has other important functions, as it acts as a filter as blood passes through it, removing cellular debris and dead red blood cells. An important role for the spleen is the destruction of red blood cells, returning some of the breakdown products such as haemoglobin to the liver. With its rich supply of blood vessels it also acts as a reservoir for blood that can be released into the circulation during haemorrhage. It also produces blood cells in the fetus, as discussed above.

## Clinical application

For some children surgical removal of the spleen (splenectomy) may be necessary. This may be due to a haematological disorder such as sickle cell anaemia (where there is abnormal production of red blood cells) or idiopathic thrombocytopaenia (where the spleen is destroying platelets). In some cases it may be due to trauma to the organ.

In the event of a splenectomy other lymphoid tissue can take over the immune function of the spleen. However, a child may be more susceptible to encapsulated bacteria (e,g, *Streptococcus pneumoniae*) and will require protection with pneumococcal vaccines.

### *The tonsils*

The tonsils are small masses of lymphoid tissue found in the mucosa around the pharynx. Their function is to trap bacteria and other foreign bacteria entering the nose and throat. The tonsils are part of the mucosa-associated lymphoid tissue (MALT), and together with lymphoid tissue found in the small intestine (Peyer's patches) and the appendix are also known as the gut-associated lymphoid tissue (GALT). These areas of lymphoid tissue are recognized as providing an important protective barrier against invading microorganisms.

## Clinical application

Tonsils may become enlarged and congested with bacteria, becoming red, swollen and painful. This is known as tonsillitis. This is relatively common in early childhood as the immune system develops, but if there is recurrent infection or the enlarged tonsils make swallowing and breathing difficult then surgery may be considered to remove them – a tonsillectomy.

### *Lymph nodes*

Lymph nodes are found throughout the lymphatic system. They help to protect the body by filtering the lymph as it passes through the lymphatic vessels. Within the lymph nodes are macrophages that engulf and destroy foreign organisms such as bacteria and viruses. This process is known as phagocytosis. The lymph glands are also sites for rapid production of lymphocytes as part of the immune response.

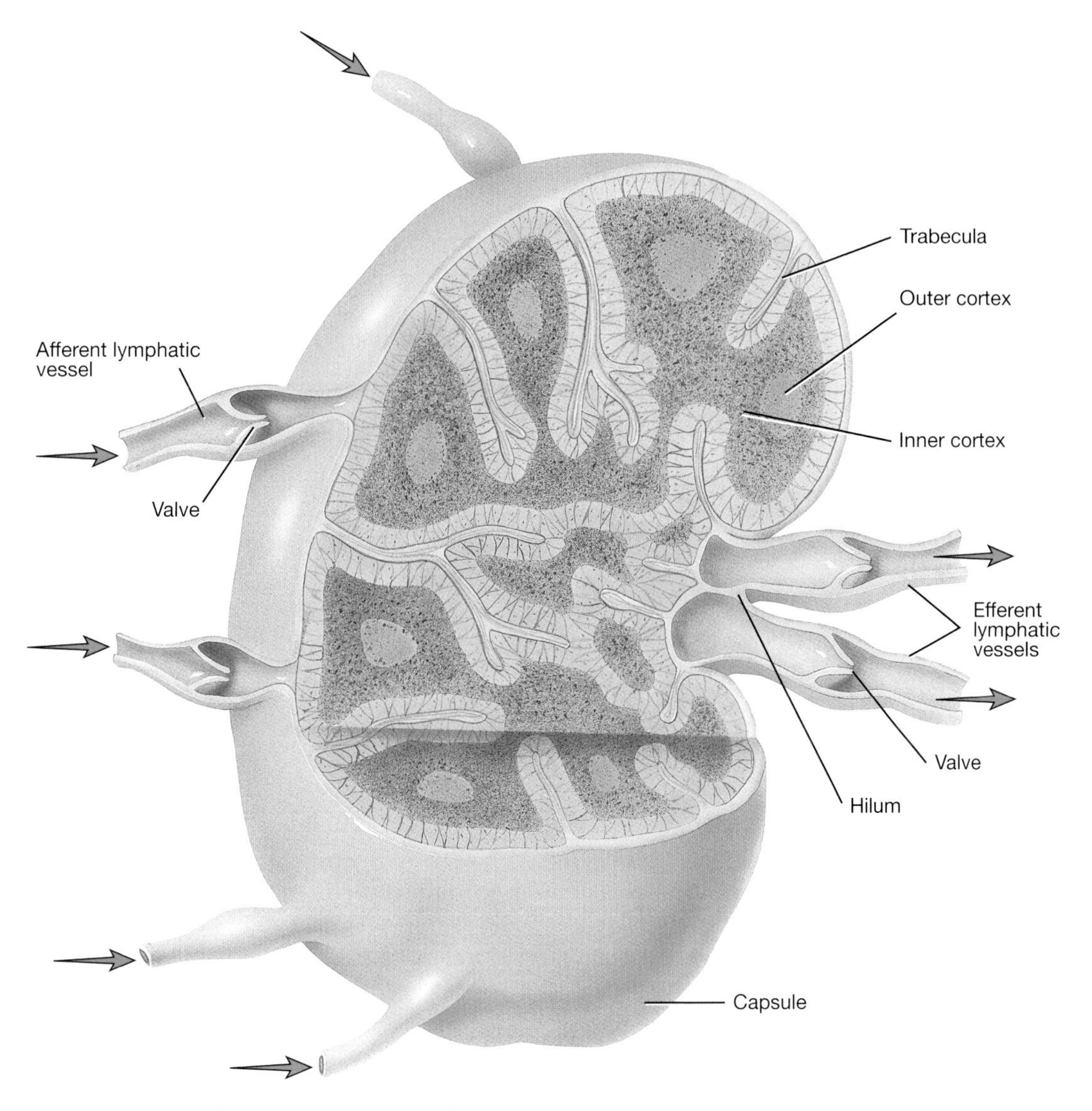


**Figure 7.6** Simplified drawing of a lymph node.
*Source:* adapted from Tortora and Derrickson (2009), Figure 22.6, p. 838. Reproduced with permission of John Wiley and Sons, Inc.

The lymph nodes vary in size but are generally kidney shaped and approximately 2.5 cm long situated within connective tissue. Lymph enters the lymph nodes by the afferent lymphatic vessels. The lymph nodes have a network of cells like a mesh that filters the lymph as it passes through the nodes (Figure 7.6). The lymph node is divided into two regions: an outer cortex and an inner medulla. The cortex contains **B cell lymphocytes** that are organized in lymphoid follicles with germinal centres. These enlarge and proliferate when the B cells encounter their specific antigen. Within the medulla are plasma cells secreting antibodies and macrophages. These macrophages engulf antigens trapped in the meshwork of the connective tissue and also phagocytose dead cells and bacteria. The lymph moves through the lymph node and exits via the efferent lymphatic vessels.

Lymph nodes are generally grouped together varying from two or three to over 100. They can be found either placed deeply within tissues or superficially. The main groups include:

- **cervical nodes,** located in the neck in deep and superficial groups which often become enlarged during upper respiratory infections;
- **axillary nodes,** which are located in the axillae (armpits) and may become enlarged after infections of the upper limbs;
- **tracheobronchial nodes,** which are found near the trachea and around the bronchial tubes;
- **mesenteric nodes,** which are found in the gastrointestinal tract between the two layers of peritoneum that form the mesentery;
- **inguinal nodes,** which are found in the groin area and which may become enlarged with infections in the lower limbs.



## Clinical application

Children will often present with enlarged lymph nodes, particularly in the cervical region, as they fight infections and develop their immune system. The nodes become swollen and tender during an infection and may remain enlarged for several weeks once the infection has cleared.

Part of a physical examination for a child involves the doctor palpating the lymph nodes, which can be difficult if the child is ticklish!

### *Functions of the lymphatic system*

It is apparent that the lymphatic system plays an important part in protecting the individual from infection and is a vital part of the immune system. However, it also has other functions. It is important in maintaining fluid balance by returning tissue fluid back to the circulatory system. It also plays a part in the absorption of fat, as digested fats are too large to be absorbed into the blood through the capillaries in the intestine but are instead absorbed into lymphatic capillaries that line the intestinal tract called lacteals. Fats are then transported in the lymphatic system until the lymph reaches the blood, but this will be discussed in more detail in the digestive system in Chapter 12.

## Types of immunity

It is now time to consider the immune system in more detail. There are two types of immunity: **innate** or non-specific immunity and **acquired** or specific immunity.

Innate immunity is the immunity that is present in all of us from birth, and it provides an effective first-line defence against pathogens such as bacteria, viruses and fungi. It utilizes components that are pre-formed and provides a very fast initial response to all invading organisms and, therefore, is non-specific.

Acquired immunity is found in more developed species such as mammals and humans, and develops during a person's lifetime after encounters with a specific pathogen. It is this ability to respond to a specific organism that provides protection against future encounters with the same organism.

## Innate immunity

There are many parts of the body that act as barriers to provide a first line of defence in addition to the white blood cells that make up the innate immune system. These barriers can be considered as four groups:

- physical barriers
- mechanical barriers
- chemical barriers
- blood cells.



### *Physical barriers*

The natural physical barriers that are first encountered by an invading microorganism are the skin and the mucosal surfaces lining the respiratory, gastrointestinal and urogenital tracts.

The skin protects us from infection and also from mechanical trauma. It has keratin present in the outer layers that makes it difficult for microbes to penetrate. Any damage to the skin, such as a cut or burn, allows entry for microorganisms and the potential for infection.

## Clinical application

A child with eczema where the skin is damaged is likely to be at risk of bacterial and fungal infections.

The skin also has sebum produced from the sebaceous glands that contain antimicrobial chemicals. However, there are potential sites for entry by microorganisms at the orifices to the passages into the body, such as the mouth, nose, urethral opening, vagina and anus. These passages are lined with mucous membranes containing goblet cells producing mucus that traps foreign material in their sticky secretions.

### *Mechanical barriers*

These barriers include the cilia of the respiratory tract, coughing and sneezing, and tears.

- Cilia are tiny hairs found in the mucosa of the upper respiratory tract. These move constantly and help to move secretions and particles away from the internal organs.
- Coughing and sneezing allow the body to remove pathogens and foreign material from the upper respiratory tract by expelling them into the atmosphere.
- Tears act as a mechanical barrier by washing away dirt and microorganisms. They also contain lysosyme which is an enzyme with antibacterial properties. Tears are therefore also a chemical barrier.

### *Chemical barriers*

Body secretions such as tears, saliva, sweat and breast milk provide a chemical barrier as they contain bacterial enzymes such as lysosyme, or **antibodies** such as **immunoglobulin A** (IgA) in breast milk. The acidic environment of gastric secretions, semen and vaginal secretions also inhibits bacteria and in most cases will destroy them.

### *Blood cells*

The blood cells that are involved in the innate immune system are the white blood cells (leucocytes) and the platelets (thrombocytes).

The white blood cells that form the first line of defence against the invading microbes with the barriers described above are:

- neutrophils
- eosinophils
- basophils
- monocytes and tissue macrophages.



The neutrophils are the most abundant of the cells, making up about 60% of white blood cells, and together with the eosinophils and basophils (which make up approximately 1–3% of white cells) contain granules within their cytoplasm. These cells are sometimes called granulocytes. These granules have antimicrobial properties and play an important part in destroying the microbes when they come into contact with them.

These blood cells provide the next line of defence if a pathogen penetrates the physical and chemical barriers of the skin and mucous membranes. The three main activities of these cells are

- phagocytosis
- inflammation
- cytotoxicity.

### *Phagocytosis*

This is the process where a phagocytic cell destroys an invading organism by engulfing and ingesting it and destroying it intracellularly (Figure 7.7) Phagocytosis is also the process that removes old red blood cells and dying cells from tissue remodelling (**apoptosis**).

The two main types of phagocytic cells are macrophages and neutrophils.

The macrophages are derived from monocytes, and when they enter the tissue they develop into macrophages. Macrophages are present in most body tissues and form an important part of the innate immune response. They have receptors that can distinguish between different types of infectious agents, such as bacteria, viruses and fungi. When an infection occurs, neutrophils and monocytes migrate to the infected area and these monocytes enlarge and form macrophages with more phagocytic properties in addition to the resident tissue macrophages. These are sometimes known as **inflammatory macrophages**.

The neutrophils have a very short life span (1–2 days) in contrast to the macrophages, and are produced as mature cells by the bone marrow. Neutrophils are not normally present in the tissues, but they respond very quickly when an infection occurs and migrate quickly to the site of infection. They are highly phagocytic, and once they have engulfed the invading microbe antibactericidal, enzymes such as lysosyme contained within the cytoplasmic granules are released to destroy it.

## Clinical application

When a blood test for a full blood count is taken, an increase in circulating neutrophils can indicate a bacterial infection and inflammation. A child with a low neutrophil count (neutropaenia), as a result of medication given as part of oncology treatments or other haematolological conditions, is at risk of infections.

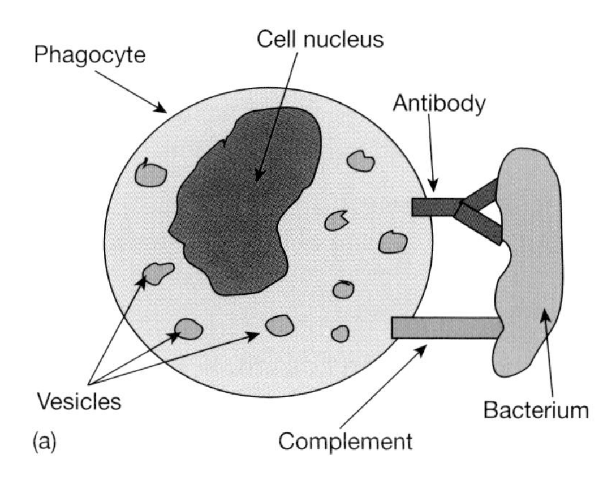


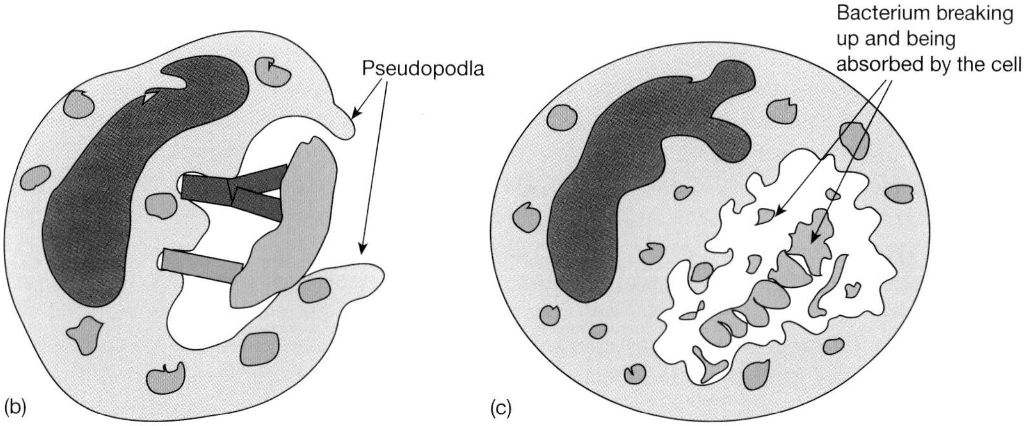


**Figure 7.7** Stages of phagocytosis: (a) stage 1, (b) stage 2 and (c) stage 3.
*Source:* Peate and Nair (2011), Figures 3.5, 3.6 and 3.7, pp. 74–75. Reproduced with permission of John Wiley and Sons, Ltd.

## *Inflammation*

Inflammation is the body's non-specific defensive response to tissue damage or injury and forms an essential part of the innate immune system. This can be caused by:

- trauma;
- infection by pathogens such as bacteria, viruses and fungi;
- irritation by chemicals;
- extreme heat (Vickers, 2011).

The four characteristic signs of inflammation are:

- redness
- pain
- heat
- swelling.

The process of inflammation is a non-specific defence mechanism, and the inflammatory response will occur whether the tissue has been damaged by trauma or by infection. There may also be loss of function in the affected area if the pain and swelling occurs within a joint. The inflammatory response is the means that the body uses to localize the tissue damage and dispose of microbes, toxins and other foreign material. It also attempts to limit the spread of infection and prepare the affected area for tissue repair (Tortora and Derrickson, 2011).

There are three stages in the inflammatory response:

- vasodilation and increased permeability of the blood vessels;
- movement of phagocytes to the site of infection and entering the tissues;
- tissue repair.



When injury occurs, specialized cells called **mast cells** in the connective tissue release the inflammatory mediators **histamine** and **seretonin** that are contained within the cell cytoplasm. These substances cause vasodilation (increase in diameter of blood vessels) and increased permeability of the blood vessels. The dilated blood vessels increase the blood flow to the affected area, causing the redness and heat associated with the inflammatory response. The increased blood flow and the permeability of the blood vessels also allow fluid from plasma in the blood to leak out of the capillaries, and this forms oedema or swelling of the tissue. This tissue fluid increases the pressure on the nerve endings, resulting in the pain of inflammation.

During the inflammatory process there is also activation of various plasma proteins – the **kinins**, the **complement system**, the **clotting system** and the **immunoglobulins**. The kinins help to increase vasodilatation of the blood vessels in the affected area. They can also affect the nerve endings, and this can contribute to the pain associated with inflammation. The complement system consists of proteins that circulate in the blood and become activated when in contact with foreign cells such as bacteria or fungi. They also cause vasodilation and attract the phagocytic cells (neutrophils and macrophages) to the affected area by releasing certain chemicals – a process known as **chemotaxis**. One of the other functions of complement is to enhance the process of phagocytosis by helping the phagocytes to attach to the foreign organism and engulf it. This process is called **opsonization**. Figure 7.8 demonstrates the phases of inflammation.

The clotting factors also arrive at the site of inflammation by migrating through the permeable cell walls of the blood vessels. These proteins are activated and begin to produce fibrin, and this localizes the infected area and helps to trap the invading bacteria, thus preventing the spread of infection. The fibrin also provides the basis for future tissue repair (Figure 7.9).

## Clinical application

During the inflammatory process an accumulation of dead tissue and dead phagocytes collects, and this collection of fluid and dead cells is known as pus. If pus cannot drain out of the affected area it collects and can form an abscess. An abscess may have to be drained surgically before healing can take place.

In some children who have a defect in production of their neutrophils they may present with recurrent abscesses and require further investigation of their immune system.

### *Natural killer cells*

These cells are a form of lymphocyte but are part of the innate immune system rather than the acquired immune system described next. They are present in blood and lymph tissue and they bind to chemical changes on the surfaces of cancer cells and viruses and destroy them by releasing chemicals through the cell membrane.

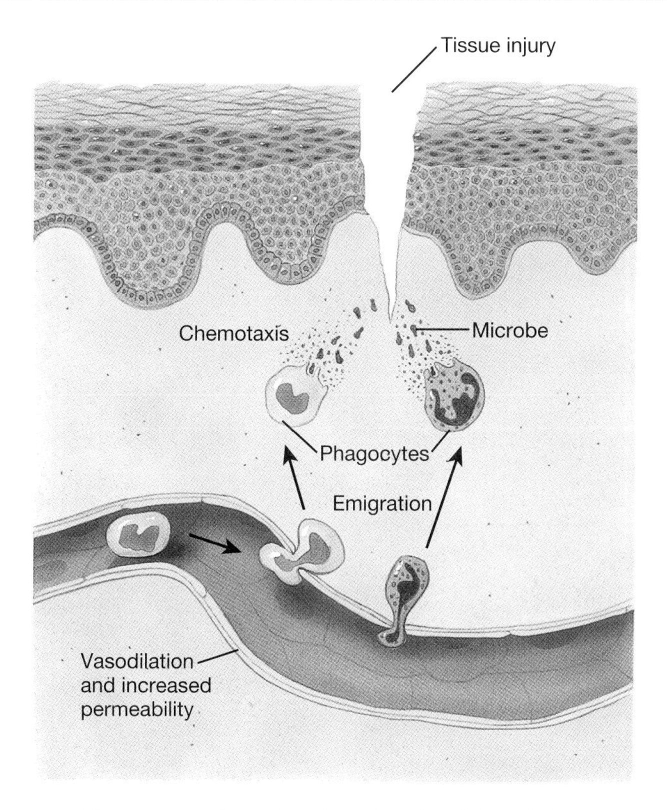


**Figure 7.8** Inflammation and the stages that occur.
*Source:* Tortora and Derrickson (2009), Figure 22.10, p. 845. Reproduced with permission of John Wiley and Sons, Inc.

## The acquired immune system

As we have seen, the innate immune system provides an immediate response to invading organisms such as bacteria, viruses, toxins and foreign tissues. However, the immune system is also able to respond to specific organisms and destroy them. This is by acquired immunity, which is the immunity that we acquire through life – a newborn baby has immunity that has passed from its mother in utero through the placenta and then it starts to develop or acquire its immunity against specific infectious agents as it becomes exposed to them. Acquired immunity is also known as adaptive or specific immunity because this type of immunity provides protection against future exposure to a specific foreign organism.

## Clinical application

We make use of the body's acquired immune system when a baby has its vaccinations, where an injection of the infectious organism that is inactivated (killed) or attenuated (weakened) is given. The baby makes antibodies to the specific organism, and this provides protection when the child next encounters the organism again. This will be discussed in more detail later.

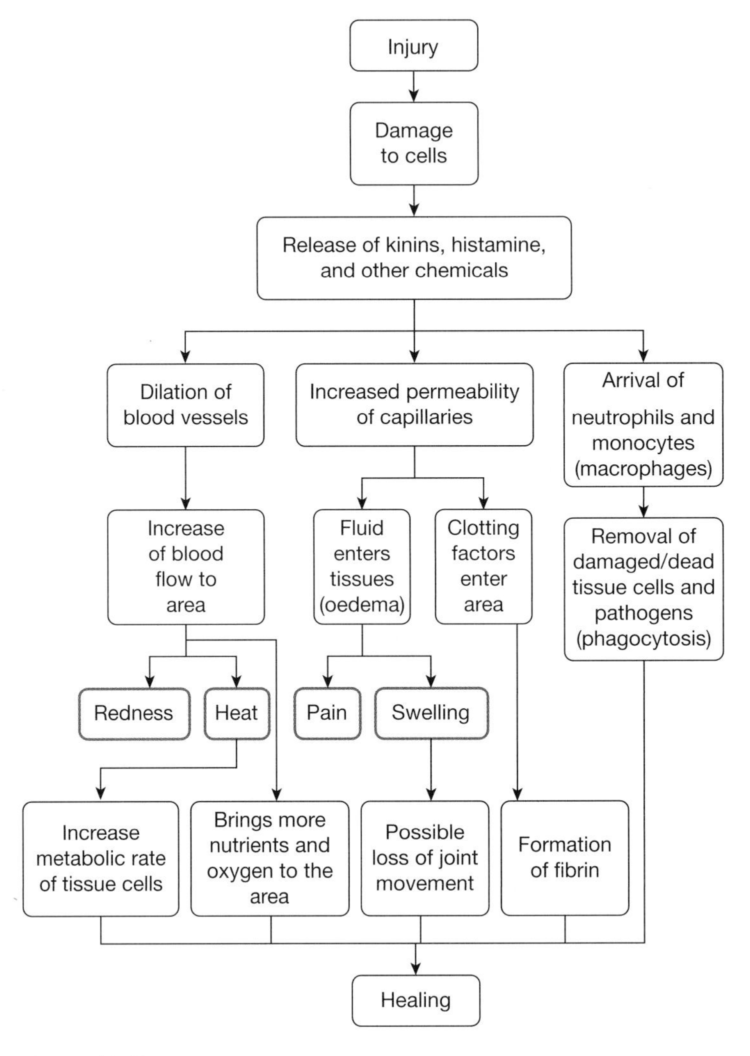


**Figure 7.9** Flowchart of inflammatory events.
*Source:* Adapted from Marieb, 2012.

Acquired immunity involves the lymphocytes, and there are two types of immunity: **cell-mediated** immunity, which involves the T cell lymphocytes, and **humoral** or **antibody-mediated immunity**, which involved the B cell lymphocytes.

## *Antigens*

Before we consider these two types of immunity in greater detail we need to establish what triggers the acquired immune response. Specific immunity depends on the ability of the body to recognize a particular foreign substance. An antigen is any substance that can provoke a response from the adaptive immune system as they are recognized as non-self or foreign. Antigens are usually protein molecules, but they can also be carbohydrates and lipids, and can be found on the surfaces of pathogenic organisms, toxins, cancer cells, transfused blood cells, transplanted tissues, foods and pollens. Lymphocytes have specialized antigen receptors that recognize molecules of pathogenic agents, and these are not present on the cells of the innate immune system.

## *Cell-mediated immunity*

This type of immunity is mediated by T cell lymphocytes. As we discussed earlier, these lymphocytes are produced initially in the bone marrow, but as immature lymphocytes they move to the thymus gland where they undergo a maturation process. During this process they learn to recognize the body's own cells and to differentiate between self and non-self. They also develop the ability to combine with a *specific* antigen. Once the lymphocytes have matured they are described as immunocompetent and account for about 80% of circulating lymphocytes (Taylor and Cohen, 2013). They migrate to the lymph nodes and spleen, where they will encounter an antigen.



The T cell lymphocytes have different functions within the acquired immune system, and this depends on the differentiation that they undergo within the thymus gland. There are four types of T cells:

- **Cytotoxic T cells** These destroy and kill certain abnormal cells, such as virus-infected cells, cancer cells or cells of transplanted tissue. They bind to the foreign cell and release enzymes into the cell that destroy it. They also produce substances that cause the cell to *self-destruct* – a process known as apoptosis. Once the cell destruction has occurred the cytotoxic cell moves on to target another cell.
- **Helper T cells** These are extremely important in the immune response as once they are activated they act on both the innate and the acquired immune systems. They release substances called **cytokines** that stimulate the production of B cell lymphocytes and cytotoxic T cells. They also help to recruit neutrophils and phagocytic cells to the area.
- **Regulatory T cells** These cells *suppress* the activity of both B and T cell lymphocytes and help to stop the immune response once the antigen has been inactivated or destroyed. This helps to prevent overactivity of the immune system.
- **Memory T cells** One of the features of the adaptive immune system is the ability of lymphocytes to recognize an antigen that it has previously encountered. This is described as **immunological memory**. Once an antigen has been destroyed and eliminated, most of the T cells that have been involved with a specific immune response die, but some antigen-specific memory T cells remain. These memory T cells will reactivate rapidly if they encounter the antigen again, which enables the body to respond quickly to future infections by the same pathogen. This is called the **secondary immune response** and will be discussed in detail later in the chapter.

## *Humoral immunity*

Humoral immunity is mediated by the B cell lymphocytes, which, like the T cell lymphocytes, are produced in the bone marrow. However, the B cell lymphocytes mature in the bone marrow before being released into the blood – the term *humoral* refers to body fluids (Taylor and Cohen, 2013).

During this phase the B cell lymphocytes undergo a process of negative selection. This is to ensure that B cell lymphocytes do not develop self-reactivity (the reaction to one's own cells) (Nairn and Helbert, 2007). Only the B cells that respond to non-self antigens survive, and these migrate to the lymph nodes and peripheral lymph tissue where they will encounter the invading antigen.

B cell lymphocytes have specific surface receptors to antigens, and exposure to that antigen stimulates the B cell lymphocytes to grow and multiply rapidly. This production of identical cells

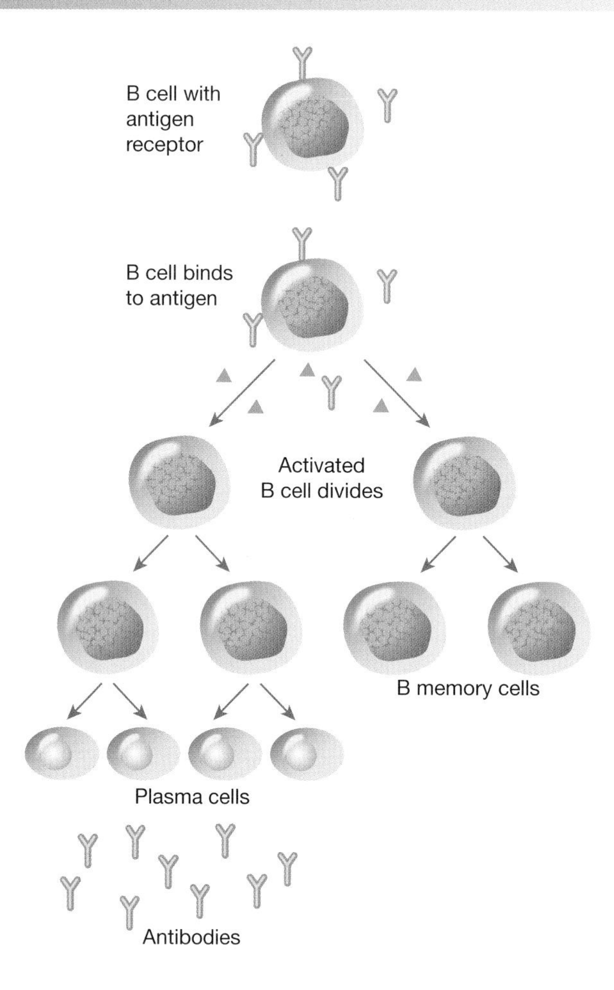


**Figure 7.10** Development of B cell lymphocytes.

from the same cell is called **cloning**. There are two types of mature cells produced: plasma cells, which secrete antibodies, and memory cells. These memory cells do not immediately produce antibodies but circulate in the blood and when they meet the antigen again will rapidly start dividing to produce mature plasma cells.

Figure 7.10 provides an outline of the development of B cell lymphocytes.

# Immunoglobulins (antibodies)

Antibodies that are secreted by plasma cells are also known as **immunoglobulins** (Igs). They are soluble proteins in the blood circulation and they are also found on the surface of B cell lymphocytes (Nairn and Helbert, 2007). Antibodies are produced as a response to a specific antigen, and rather like a lock and key analogy each antibody's structure matches that of the antigen (Tortora and Derrickson, 2011). The antibodies bind with their specific antigen by attaching to receptors on that antigen called **epitopes** and act to disable it in conjunction with other components of the immune system (Vickers, 2011).

There are five different classes of immunoglobulin which vary in molecular size and in their function:

- IgG
- IgA
- IgM
- IgD
- IgE.



## Immunoglobulin G

This is the most abundant of the Igs, forming 75–80% of total serum Ig. It has the lowest molecular weight of all the Igs and is found within both the intravascular and extravascular spaces. It can pass through the placenta from the mother to the fetus, and this provides considerable immune protection for the newborn infant. It also survives longer than other Igs within the blood, and maternal IgG can last for several months within the baby's circulation and tissues. During this time the baby's own acquired immune response is developing and the baby starts to produce its own IgG.

## Clinical application

A baby that is born prematurely has not received sufficient maternal IgG as this occurs mainly in the last trimester of pregnancy, which means that these infants are at severe risk of infections during the neonatal period. Extreme vigilance is required in the nursing care of the preterm infant with infection control practices in hand hygiene and the interventions involved with supporting these babies.

There are four subclasses of IgG, and each of these has slightly different functions. The main functions of IgG are:

- neutralizing bacterial toxins and preventing attachment of some viruses to body cells;
- activation of the complement system, which helps with the breakdown of foreign cells;
- enhancing phagocytosis by binding with the macrophages once antigens have attached to the antibody and coating the cell surface of antigens to make them more susceptible to phagocytosis;
- assisting with the inflammatory response by binding to platelets (Nairn and Helbert, 2007; Tortora and Derrickson, 2011; Vickers, 2011).

## Immunoglobulin A

This Ig is found in the body's secretions such as saliva, tears, sweat, breast milk and nasal secretions. It also found in the mucous membranes of the respiratory, gastrointestinal and urogenital tracts. This presence in the body's secretions provides localized protection from bacteria and viruses. The large amounts that are present in breast milk and colostrum provide protection from gastrointestinal infections in babies.

## Clinical application

Children who have a deficiency in IgA (this is the most common immunodeficiency, occurring in 1:300–1:400 people) may have an increased susceptibility to infections of the respiratory and gastrointestinal tracts. This can result in recurrent ear and chest infections in early childhood.



## Clinical application

Breast feeding is promoted to mothers in the postnatal period, but in neonatal intensive care units breast milk is recognized to improve growth and neurodevelopment and to reduce infections in pre term infants. Mothers are encouraged to express their breast milk for their babies.

### Immunoglobulin M

This is the largest of the Igs in molecular weight and is found in both blood and lymph. It accounts for 5-10% of the total number of Igs but it is the most predominate Ig in the early phase of the immune response. It also causes agglutination of microbes and plays an important part in activating complement (Nairn and Helbert 2007; Vickers 2011; Marieb 2012).

### Immunoglobulin D

There is not a great deal known about this Ig. However, it is found on the surfaces of B cells and is thought to be important in the activation of the B cells.

### Immunoglobulin E

This Ig normally accounts for less than 0.01% of serum Igs and is found on the surfaces of mast cells and basophils. When it binds to an antigen, the activation of the mast cells triggers the release of histamine. This can cause an acute inflammatory response and gives the signs and symptoms of an allergic reaction, such as those seen in asthma and hay fever (Nairn and Helbert, 2007; Vickers, 2011).

## Clinical application

Allergic reactions are hypersensitivity reactions or excessive immune responses to certain antigens such as pollen, dust or foods. The release of histamine causes small blood vessels in the area to dilate (widen) and become more permeable. This accounts for the symptoms of allergy such as watery eyes, runny nose in *hay fever* and itching and reddened skin (hives). Histamine also causes the constriction of smooth muscle, and when this occurs in the walls of the bronchioles it causes the symptoms of *asthma*.

The use of *antihistamine drugs* that block the release of histamine can reduce some of these effects.

# Actions of antibodies

As we have seen, the antibodies have different roles in protecting the body from invading pathogens, but their functions can be summarized as follows:

- **neutralizing the antigen** – by blocking or neutralizing bacterial toxins and preventing attachment of viruses;
- **opsonization of bacteria** – when the microbe is coated with antibody to enhance phagocytosis;
- **agglutination of the antigen** – antibodies have more than one receptor on their surfaces and the attachment of the antigens can be cross-linked causing clumping together which makes phagocytosis easier;
- **activation of complement** – this occurs during the innate immune response as discussed earlier, but the binding of antigen to antibody also triggers this and enhances the phagocytosis and lysis of the cells and the inflammatory response.

Figure 7.11 provides an overview of the cellular and humoral immune responses.

## Clinical application

### Problems with the immune system

There may be times when the immune system does not work properly, and this can result in serious and sometimes life-threatening illness. The most important disorders of the immune system are:

- **Autoimmune disease** – this is when the immune system loses the ability to distinguish self from non-self and the body's immune cells attack the body's own cells. They are generally rare in childhood and can be difficult to diagnose. Some of these conditions include:
  - *juvenile idiopathic arthritis*, which affects joints skin and sometimes the lungs;
  - *glomerulonephritis*, where there is severe impairment of kidney function;
  - *Crohn's disease*, which affects the gastrointestinal tract;
  - *Addison's disease*, which affects the adrenal glands;
  - *type 1 diabetes mellitus*, which affects the pancreas.
- **Primary immunodeficiencies** – these are relatively rare conditions that have a genetic basis affecting the function of one or more components of the immune system, resulting in severe recurrent and unusual infections that can be life threatening. Children usually present with these conditions within the first 2 years of life.
- **Secondary immunodeficiencies** – these immunodeficiencies are acquired and can be related to many factors:
  - *Age of the child* – during the first year of life specific immunity is developing and many children have low levels of Igs, making them more susceptible to infections. Premature babies are at increased risk, as we have already discussed.
  - *Malnutrition* – this is a major cause of immunodeficiency due to protein deficiency and inadequate intake of calories, essential vitamins and mineral nutrients. A child who is unable to take sufficient nutrition orally will need nutritional support by alternative methods of feeding.
  - *Infections* – infections such as human immunodeficiency virus, measles, mumps, congenital rubella, cytomegalovirus, and infectious mononucleosis (glandular fever) can suppress the immune system.

(*Continued*)

- *Cancers* – some cancers. such as leukaemia and lymphomas. can suppress the immune system by preventing the bone marrow from producing white blood cells.
- *Drugs* – certain drugs will suppress the immune system, such as steroids and cytotoxic drugs used in cancer therapies. Some anticonvulsant drugs can also cause antibody deficiency.
- *Burns, trauma and major surgery* – immune suppression occurs, making individuals more susceptible to infections.
- *Protein loss* – severe loss of protein, such as in kidney disease (nephrotic syndrome) and via the gut in severe diarrhoea, may result in lower levels of antibodies.

- **Allergies** – this occurs in susceptible individuals where there is an excessive immune response to an antigen (known as an allergen) which results in the binding of IgE antibodies to mast cells and the release of histamine, as described earlier.
- **Organ and tissue transplantation** – the immune system's normal response to non-self cells is to destroy them, and in transplantation the matching of donor to recipient by tissue-typing technologies reduces the risk of rejection of the transplanted tissue.

## Primary and secondary responses to infection

As discussed earlier, one of the special features of the acquired immune system is its ability to respond to future encounters with a known antigen. This immunological memory is very important as it allows the body to respond effectively to an antigen without having to work out each time it encounters the antigen how to respond to it. The primary response is when the immune system first encounters the antigen, and the secondary response is when it encounters it again.

During an infection when the antigen is new to the body the innate immune system will make an immediate response, but there is an initial delay by the adaptive immune system as specific antibodies are made against the antigen. This is known as the 'lag' phase, and it may take several days to produce sufficient antibodies to be effective. During this phase, IgM is produced first with small amounts of IgG. However, when the antigen is next encountered, the memory cells respond immediately and antibodies are produced very quickly. During this secondary response, large amounts of IgG are produced in much greater quantities, which ensures that the antigen is destroyed effectively (Figure 7.12).

# Immunizations

Protection against infection can be active by activation of the body's immune system or passive by the transfer of antibodies to an individual who does not have them.

## Active immunity

As we have learnt, immunity is acquired by the body's response to a specific disease organism. The reaction to a specific antigen by the T cells and antibodies occurs each time the person is exposed to it and provides immunity to that infection. **Naturally acquired** active immunity occurs after exposure to bacterial and viral infections whereas **artificially acquired** active immunity occurs after receiving a vaccine.

Vaccines contain small amounts of the infective organism that are either *inactivated* (killed) or *attenuated* (weakened). When the vaccination is given the individual will mount an immune response without experiencing the symptoms of the infection during the primary response. When exposed to that infective organism the body will be able to produce antibodies as part

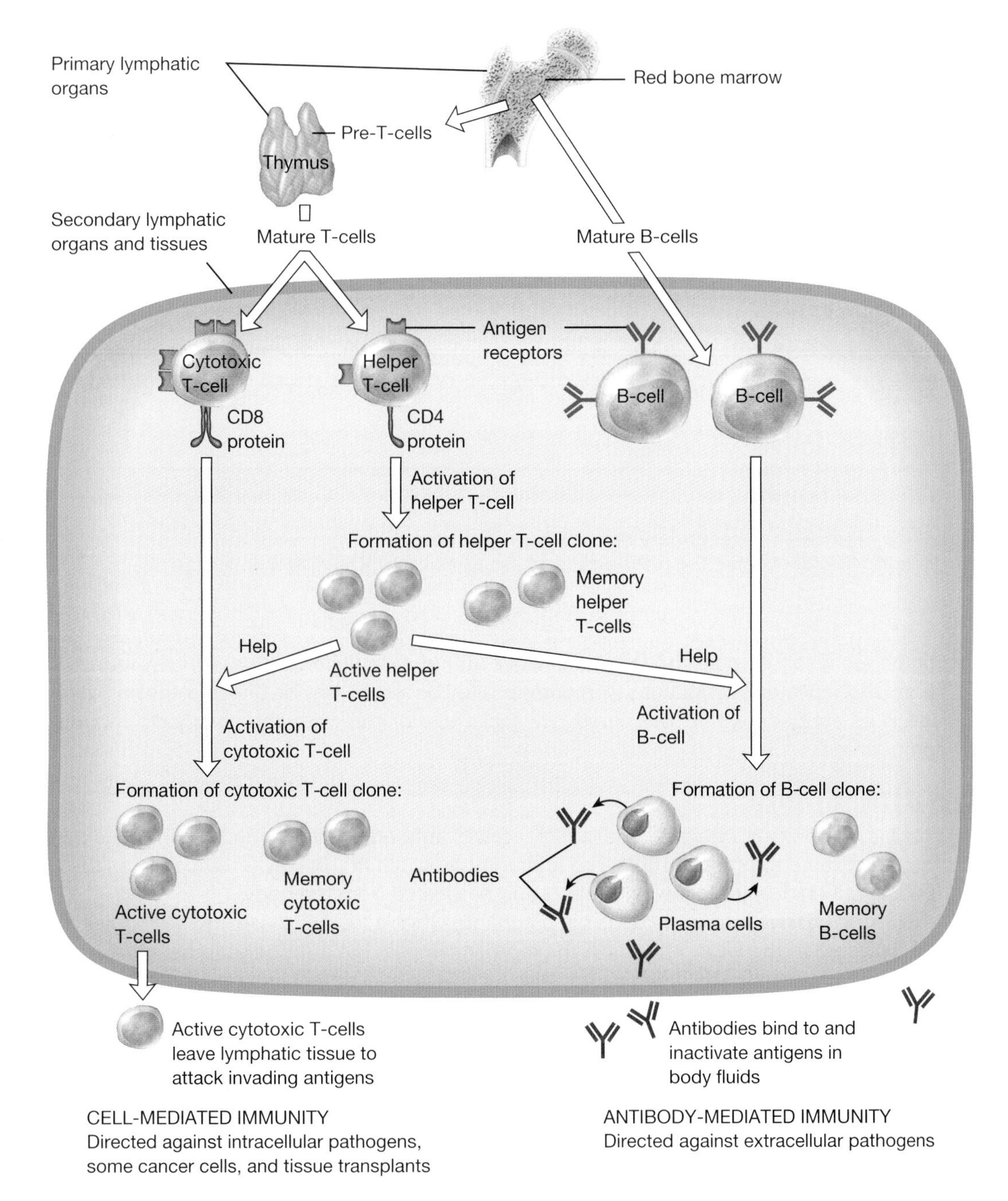


**Figure 7.11** Cellular and humoral immune responses.
*Source:* Tortora and Derrickson (2009), Figure 22.11, p. 847. Reproduced with permission of John Wiley and Sons, Inc.

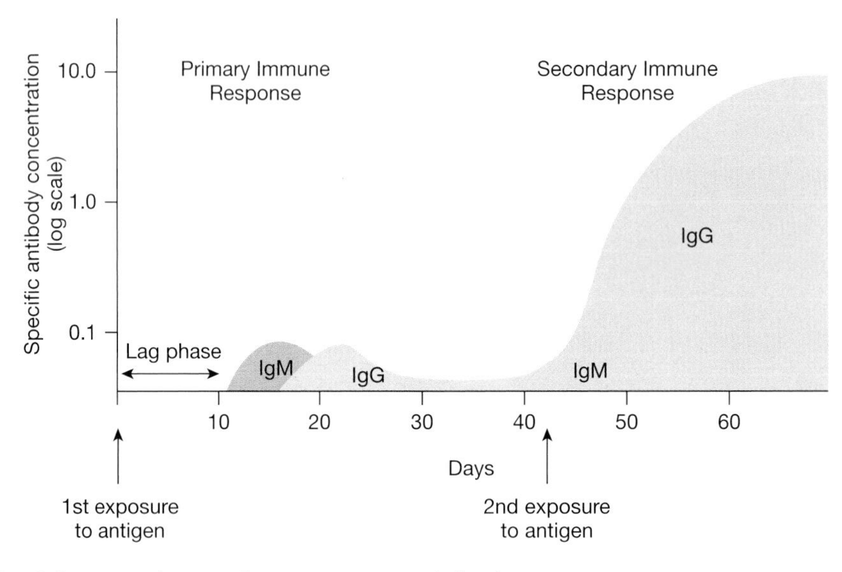

**Figure 7.12** Primary and secondary responses to infection.
*Source:* Peate and Nair (2011), Figure 3.14, p. 89. Reproduced with permission of John Wiley and Sons, Ltd.

of the secondary response to infection. Active immunity acquired in this way may reduce with time, and repeated vaccinations (sometimes called boosters) may be given to ensure adequate levels of antibody within the blood. We make use of this with the childhood vaccination programme.

## Passive immunity

Passive immunity is where the individual receives antibodies directly and the immune system is not activated to produce them. The immunity provided in this way is temporary as the antibodies will gradually decline in numbers and be cleared from the body.

**Natural passive immunity** occurs in pregnancy when a mother passes IgG antibodies across the placenta to the fetus. After birth the mother can pass IgA antibodies to her baby in colustrum and breast milk. An individual can also receive an injection of immune serum to a specific organism if they have been exposed to a large amount of that organism, and this is known as **artificially acquired passive immunity** (Figure 7.13).

## Clinical application

Children undergoing cancer treatment are immunodeficient and are at severe risk of infections. If they are exposed to measles or chicken pox they can receive specific immune serum containing antibodies to these infections to protect them from developing the infection.

Children who have primary immunodeficiency, where they are unable to make antibodies because of a defect in their B cell lymphocytes, receive regular infusions of pooled human Ig to give them protection from a wide range of infections.

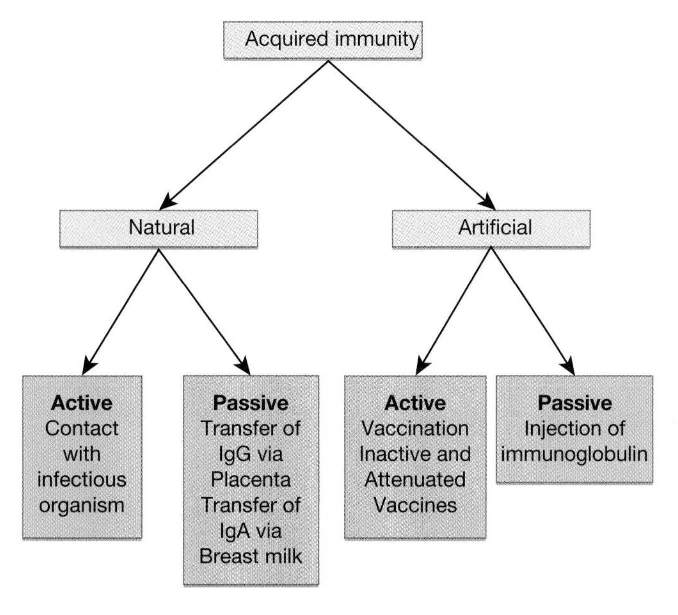


**Figure 7.13** Acquired (specific) immunity. Adapted from Taylor and Cohen (2013).

# Conclusion

The immune system is a complex system with its various components working together to provide the necessary protection from invading pathogens. The immune system develops throughout childhood, and further protection is provided by an immunization schedule that continues to evolve as further vaccines are developed. The immature immune system of the baby and young child has considerable implications for the nursing of sick neonates and children and adolescents with their increased susceptibility to infections.

The study of immunology continues to develop, with research in this field giving us greater understanding of its anatomy and physiology and the disorders that are affected by it.

# Activities

**Now review your learning by completing the learning activities in this chapter. The answers to these appear at the end of the book. Further self-test activities can be found at www.wileyfundamentalseries.com/childrensA&P.**

## Crossword

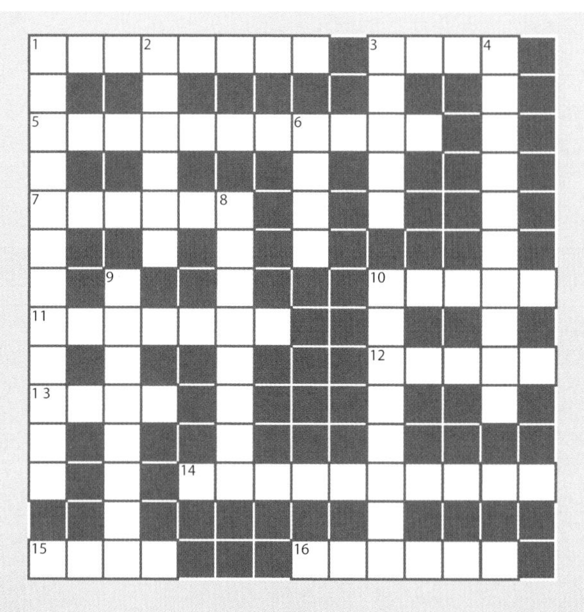


### Across

1. A microorganism that causes disease
3. One of the characteristic signs of infection
5. Happens when antigens become cross-linked on antibodies
7. Accumulation of fluid within the tissues
10. One of the functions of the acquired immune system
11. Lymphatic tissue located within the mucus membrane of the throat
12. Substance produced by bacteria
13. One of the physical barriers in innate immunity
14. White blood cells involved in acquired immunity
15. Cells that release histamine
16. Organ where T cell lymphocytes mature

### Down

1. A function of the macrophages
2. Type of T-cell lymphocyte
3. Lymphoid tissue found in gastrointestinal tract – Peyer's ---------
4. White blood cell involved in phagocytosis
6. Part of lymph tissue
8. Lymph nodes that may become enlarged with infection in the upper arm
9. Something that causes immune response
10. Secreted by B cell plasma cells

## Conditions

The following table contains a list of conditions. Take some time and write notes about each of the conditions. You may make the notes taken from text books or other resources; for example, people you work with in a clinical area or you may make the notes as a result of people you have cared for. If you are making notes about people you have cared for you must ensure that you adhere to the rules of confidentiality.

| Condition | Your notes |
| --- | --- |
| Coeliac disease | |
| Lupus (SLE) | |
| Juvenile idiopathic arthritis | |
| Crohn's disease | |
| Human immunodeficiency virus | |



# Glossary

**Acquired immunity:** immunity that develops during the lifetime of an individual after coming into contact with different infectious organisms.

**Active immunity:** immunity that occurs after exposure to an antigen.

**Antibody:** also known as immunoglobulin. Released in response to a specific antigen.

**Antigen:** foreign substance that provokes an immune response.

**Apoptosis:** death of a cell.

**Autoimmune reaction:** response when antibodies and T cells destroy a person's own tissue.

**B cell lymphocyte:** white blood cell involved in humoral immunity and produces plasma cells that secrete antibodies.

**B cell lymphocytes:** white blood cells that produce antibodies from plasma cells that are derived from these cells. Part of the humoral immunity.

**Bone marrow:** site of production of blood cells.

**Cell-mediated immunity:** acquired immunity that is provided by T cell lymphocytes.

**Complement system:** a group of plasma proteins that when activated enhance phagocytosis and inflammation. They are also involved in opsonization.

**Cytokines:** chemical messenger molecules that affect the behaviour of cells, including those of the immune system.

**Epitopes:** receptors on the cell membrane that allow the antigen and antibody to combine with each other.

**Granulocytes:** White blood cells containing cytoplasmic granules which are involved in the process of phagocytosis.

**Haemopoiesis:** development and production of blood cells.

**Humoral immunity:** acquired immunity that is provided by antibodies secreted from plasma cells produced by B cell lymphocytes.

**Immunoglobulin:** also known as antibody – protein released by plasma cell in humoral immunity.

**Inflammation:** the body's immediate response to tissue damage or injury.

**Innate immunity:** the immunity that is present from birth.

**Kinins:** polypeptides that increase dilation of the arterioles.

**Lymph:** colourless fluid circulating in lymph vessels.

**Lymph nodes:** mass of lymphoid tissue situated within the lymphatic system that filters the lymph and traps antigens to be destroyed by antibodies and other cells of the immune system.

**Lymph vessels:** similar to blood vessels, but transport lymph containing cells from site of infection to the lymph nodes.

**Lymphatic system:** circulatory system that contains lymph vessels transporting lymph and the lymphoid tissue.

**Lymphocyte:** white blood cell that is primarily involved in acquired immunity. Lymphocytes include B and T cells and natural killer (NK) cells (part of innate immunity).

**Macrophages:** develop from monocyte and are involved in phagocytosis within the tissues.

**Natural killer cells:** lymphocytes that are part of the innate immune system and destroy abnormal cells.

**Neutrophils:** white blood cells that are involved in phagocytosis.

**Oedema:** abnormal swelling caused by collection of fluid in tissue or body part.

**Opsonization:** process where bacteria and cells are modified to enhance phagocytosis.

**Passive immunity:** immunity that occurs by the transfer of antibodies to someone who is vulnerable to infection and who may not be able to make antibodies.

**Phagocytosis:** ingestion of particles by cells.

**Primary response:** the immune response that occurs when first exposed to the antigen.

**Secondary response:** the immune response that occurs after exposure to a known antigen, which is a more rapid response and will occur each time the person is exposed to the antigen.

**Spleen:** lymph organ situated on the left side of the abdominal cavity that contains cells of the immune system to fight infections. Also filters blood and destroys red blood cells.

**T cell lymphocyte:** white blood cells involved in cell-mediated immunity as part of acquired immunity.

**Thymus gland:** organ where lymphocytes migrate to mature into T cell lymphocytes.

**Tonsils:** mass of lymphoid tissue situated in the mucosa of the pharynx.

## References


Nairn, R., Helbert, M. (2007) *Immunology for Medical Students*, 2nd edn, Mosby, St Louis, MO.

MacPherson, G., Austyn, J. (2012) *Exploring Immunology*, Wiley–Blackwell, Weinheim.

Marieb, E.N. (2012) *Essentials of Human Anatomy and Physiology*, 10th edn, Pearson Benjamin Cummings, San Francisco, CA.

Peate, I., Nair, M. (eds) *Fundamentals of Anatomy and Physiology for Student Nurses*, Wiley Blackwell, Oxford.

Taylor, J.J., Cohen, B.J. (2013) *Memmler's Structure of the Human Body*, 10th edn, Lippincott Williams and Wilkins, Baltimore, MD.

Tortora, G.J., Derrickson, B.H. (2009) *Principles of Anatomy and Physiology*, 12th edn, John Wiley & Sons, Inc., Hoboken, NJ.

Tortora, G.J. and Derrickson, B. (2011) *Principles of Anatomy and Physiology*, 13th edn, John Wiley & Sons, Inc., Hoboken, NJ.

Vickers, P.S. (2011) The immune system. In: Peate, I., Nair, M. (eds), *Fundamentals of Anatomy and Physiology for Student Nurses*, Blackwell, Oxford, pp. 62–95.

# Chapter 8

# Blood

Peter S. Vickers

## Aim

The aim of this chapter is to introduce the paediatric nursing student to the blood and circulatory system of the child.

## Learning outcomes

On completion of this chapter the reader will be able to:

- Describe the normal composition and properties of blood.
- List the functions of erythrocytes (red blood cells), leucocytes (white blood cells), thrombocytes (platelets) and plasma.
- Explain what is meant by haemopoiesis and how blood clotting occurs.
- Explain the ABO and Rh systems of blood typing.
- Describe the structures of the arteries, veins and capillaries, and list the differences between the arteries and veins.
- Explain what is meant by blood pressure and how it is controlled/regulated.

## Test your prior knowledge

- What are the three main classes of blood cells called?
- What are the components of blood plasma?
- Name the five types of blood vessel.
- What do we mean by blood pressure?
- List the differences between arteries and veins.
- What is the function of red blood cells?
- What is the function of platelets?
- What is the main function of white blood cells?
- Which blood cells are involved in blood clotting?
- How many types of white blood cell are there? Name them.

*Fundamentals of Children's Anatomy and Physiology: A Textbook for Nursing and Healthcare Students*, First Edition. Edited by Ian Peate and Elizabeth Gormley-Fleming.

Companion website: www.wileyfundamentalseries.com/childrensA&P

# Introduction

In this chapter we will be exploring blood and the circulatory system so that myths can be separated from facts.

Without a doubt, blood is truly a miraculous substance – hence, we talk about our **life blood**, because it does give us life, and without it we would die. So, to begin with, some facts about blood:

- Blood is a **viscous** substance – blood is four to five times thicker than water.
- Blood is stickier and heavier than water.
- In the body, blood maintains a temperature of around 38 °C.
- Blood is salty; it contains sodium chloride (NaCl) at a concentration of 0.9%.
- Blood is alkaline and has a pH of 7.35–7.45.
- Blood makes up approximately 8% of an adult's total body weight – but in babies and infants, the proportion is higher.
- In adult males, the average blood volume is 5–6 L, whilst that of an average female is 4–5 L.



Our blood system consists of formed elements (blood cells) and fluid (**plasma**) – see Figure 8.1. The blood system is just one part of the circulatory system, which consists of the blood, the blood vessels, the lymphatic system and, very importantly, the heart.

## Clinical application

### Epistaxis

In children, epistaxis (or nose bleed) is quite common, and most children will have a nose bleed at one time or another, mainly due to trauma, such as an injury to the nose, blowing the nose too hard or too often, or picking the nose. Other causes of epistaxis include insertion of a foreign body into the nose, systemic disease, such as leukaemia, and anticoagulant (anti-clotting) therapy. Most of the time, epistaxis can be simply treated by reassurance, and pinching and squeezing the nose firmly for at least 10 min – do not tilt the child's head backwards, otherwise there is a risk of the child swallowing blood. Alternatively, apply crushed ice (or a bag of frozen peas) to the bridge of the nose, so constricting the blood vessels. If the bleed lasts for more than 20 min, then the child needs to be seen in an accident and emergency department for further investigations and treatment; for example, packing the nose or cauterizing some of the blood vessels (Williams, 2007).

# Composition of blood

Blood consists of red blood cells (**erythrocytes**), white blood cells (**leucocytes**), platelets (**thrombocytes**) and fluid (**plasma**) – Figure 8.1. Plasma consists of water, proteins and other soluble molecules, such as nutrients, hormones and minerals. It makes up 55% of the total blood volume. Erythrocytes account for 45% of the total blood volume, whilst leucocytes and thrombocytes – known as the buffy coat – account for the remaining 10% (Figure 8.2).

The percentage of red blood cells in whole blood is known as the **haematocrit** or the **packed cell volume (PCV)**. The volume of blood in the body is constant – unless there is a physiological problem such as **haemorrhage** (bleeding). However, the haematological values do change according to the age of the person, as can be seen in Table 8.1.

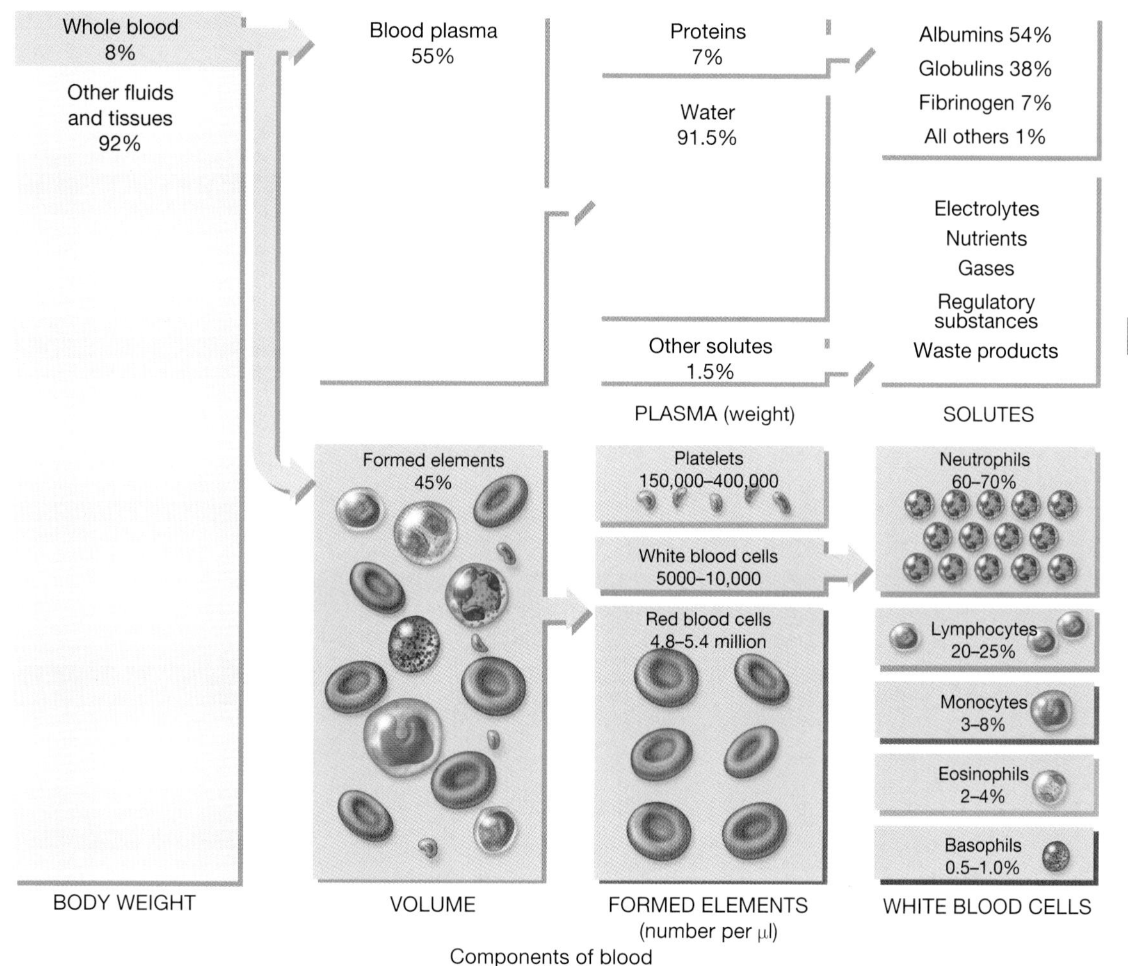


**Figure 8.1** Cells of the blood.
*Source:* Tortora and Derrickson (2009), Figure 19.1, p. 691. Reproduced with permission of John Wiley and Sons, Inc.

Plasma proteins include:

- albumin
- fibrinogen
- prothrombin
- gamma globulins.

They constitute approximately 8% of the plasma in the body, and – with the exception of the gamma globulins (concerned with immunity) – help to maintain the fluid balance in the body, which affects osmotic pressure, as well as increase/decrease the blood viscosity and help to maintain blood pressure (BP).

The normal blood values for children aged between 2 and 12 years are:

- red blood cells – $(3.9–5.03) \times 10^6/\mu L$
- white blood cells – $(5.3–11.5) \times 10^3/\mu L$
- platelets – 150 000–450 000/μL.

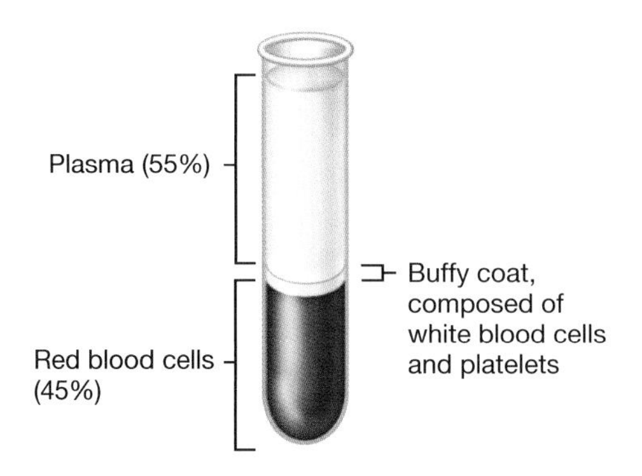




**Figure 8.2** Composition of blood in percentages.
*Source:* Tortora and Derrickson (2009), Figure 19.1, p. 691. Reproduced with permission of John Wiley and Sons, Inc.

**Table 8.1** Haematological values for infants and children.

| Age | Haemoglobin (g/dL) | Packed cell volume (% of blood that is made up of red cells) | Mean cell volume (fL) | Mean cell haemoglobin concentration (%) | White cell count ($\times 10^9$/L) | Neutrophil count[a] |
|---|---|---|---|---|---|---|
| Cord blood | 13.5–20 | 50–56 | 110–128 | 29.5–33.5 | 9–30 | |
| 2 weeks | 14.5–18 | 50–55 | 100–120 | 30–34 | 6–15 | |
| 6 months | 10–12.5 | 33–38 | 80–96 | 32–36 | 6–15 | 2.3–6.9 |
| 1–5 years | 10.5–13 | 36–40 | | | 6–15 | 2.3–6.9 |
| 5–10 years | 11–14 | 37–42 | | | 5–15 | 2.0–8.0 |
| 10–15 years | 11.5–14.5 | 38–42 | | | 4–13 | 2.0–8.0 |

[a]See section on white blood cells for more details on neutrophil count.

Normal plasma in humans has an osmolality (measure of the body's electrolyte–water balance) of 285–295 milliosmol/kg. Osmolality is important in enabling blood cell survival. A raised osmolality (above 600) would cause the red blood cells to crenate (shrivel up) and die, and a reduced osmolality (below 150) would cause haemolysis (rupture) of the erythrocytes.

The specific gravity of blood is 1.045–1.065, compared with 1.000 for water, whilst the normal **pH** of blood ranges from 7.35 to 7.45 (Nair, 2009).

# Functions of blood

There are three main functions of blood:

- Transportation – including the removal of waste products from cellular functions and metabolism.

- Regulation:
  - maintaining body temperature
  - maintaining acid–base balance
  - regulation of fluid balance.
- Protection against infection.

## Transportation

Blood vessels form a huge interlinking network of transportation routes within the body. They carry:



- Oxygen ($O_2$) from the lungs to the tissues, and carbon dioxide ($CO_2$) from the tissues to the lungs for removal by exhalation. $O_2$ is transported by **haemoglobin** (Hb) in red blood cells and as a dissolved substance in blood plasma. Most of the $CO_2$ is transported by bicarbonate ions in the blood plasma.
- Nutrients, such as glucose and amino acids, from the gastrointestinal tract (stomach and intestines) to the cells so they can carry out their cellular functions (Nair, 2011).
- Waste products of metabolism; for example, urea and uric acid – transported by blood for elimination.
- Hormones – from the endocrine glands to other cells in the body where they are needed.
- Enzymes – secreted by some organs to other parts of the body for cellular function (Nair, 2011).
- Heat – from various body cells.

## Regulation

### *Regulation of body temperature*

Heat is produced during the process of cellular metabolism, and blood is essential for dispersing and distributing this heat. Normal body temperature is regulated through the heat absorbing and cooling properties of blood's water content. If the body gets too hot, by dilating the capillaries, blood flow increases to the skin, aiding the removal of excess heat by convection and radiation. If the body is cold, heat is conserved by constricting the capillaries, reducing the blood flow to the skin, and so reducing heat loss.

### *Regulation of acid–base balance*

The acid–base balance is the homeostasis of body fluids at a normal arterial blood pH (7.35–7.45) – pH affects cell membrane structure, enzyme activity and structural proteins in both their structure and optimal range of function. Therefore, the pH of the blood and body tissues needs to be regulated, using various buffers (substances present in blood that help to regulate and maintain the body's pH); for example, bicarbonate ions ($HCO_3^-$). They are chemicals that 'mop up' any excess hydrogen ions ($H^+$) and hydroxide ions ($OH^-$) freely circulating in the blood. If these ions were not neutralied, then either a state of **acidosis** (too many circulating free $H^+$ ions) or a state of **alkalosis** (too many free circulating $OH^-$ ions) would occur.

### *Protection against infection*

Blood also helps to protect the body from damage due to injuries and infection by:

- preventing blood loss through the clotting mechanism;
- preventing invasion by infectious microorganisms and their toxins.

Table 8.2 The normal value of Hb at different ages.

| | |
|---|---|
| Newborn | 14–24 g per 100 mL |
| Infant – 1 week old | 15–20 g per 100 mL |
| Infant – 1 month of age | 11–15 g per 100 mL |
| Child | 10.5–13.3 g per 100 mL |
| Adult male | 14–18 g per 100 mL |
| Adult female | 12–16 g per 100 mL |



# Constituents of blood

## Blood plasma

Water makes up approximately 91% of blood plasma. Solutes make up the other 10% – mainly proteins (albumin, fibrinogen globulin and prothrombin), with 0.9% being inorganic salts. Blood plasma is a pale yellow-coloured fluid, with an adult total volume of 2.5–3 L (Table 8.2 and Figure 8.1), and contains 50–70 mg of protein per millilitre, of which:

- 70% is albumin – 35–50 mg/mL
- 10% is gamma globulin G (IgG) – a main constituent of the immune system.

**Inorganic salts** include:

- sodium – at a concentration of 135–145 mmol/L (millimoles per litre)
- potassium – 3.5–5 mmol/L
- calcium – 2.1–2.7 mmol/L
- phosphate – 0.7–1.4 mmol/L
- chloride – 98–108 mmol/L
- hydrogen carbonate – 23–31 mmol/L.

**Organic substances** make up 0.1% of the plasma constituents, such as:

- fat
- glucose
- urea
- uric acid
- amino acids.

In addition, it has been estimated that plasma may contain as many as 40 000 different proteins, but to date only about 1000 of these have been identified (Nair, 2011).

## Plasma proteins

Blood plasma contains 50–70 mg of protein per millilitre, of which:

- albumin (35–50 mg/mL) makes up approximately 70%;
- gamma globulin (5–7 mg/mL) makes up approximately 10%.

These plasma proteins form three major groups:

- **Albumin** Albumin is synthesized in the liver, and maintains plasma osmotic pressure as well as blood viscosity. Albumin acts as carrier molecules for other substances, such as hormones and lipids. It is found in interstitial fluid, is the most abundant and smallest plasma protein and can pass through blood capillaries.
- **Globulins** Globulins (alpha, beta and gamma) make up approximately 36% of total plasma protein. Alpha and beta globulins are synthesized by the liver, and transport lipids and fat-soluble vitamins around the body to their target cells, whilst gamma globulins are involved in immunity.
- **Fibrinogen** 4% of plasma proteins are fibrinogen – essential for blood clotting.



### *Functions of plasma proteins*

Plasma proteins (Nair, 2011):

- Provide the intravascular osmotic effect – for the maintenance of fluid and electrolyte balance.
- Contribute to the viscosity of blood.
- Are carrier molecules for insoluble proteins and transport them around the blood system.
- Are a protein reserve for the body.
- Aid blood clotting and wound healing.
- Are part of the inflammatory response.
- Protect the body and tissues from infection.
- Maintain the acid–base balance.

## Water

Water constitutes approximately 90–91% of blood plasma and helps to maintain homeostasis. It is the medium where chemical reactions between the intracellular and extracellular areas occur and contains solutes (e.g. electrolytes) whose concentrations change to meet the needs of the body (Nair, 2011).

## Blood cells

The process by which blood cells are formed is known as **haemopoiesis**. After birth, the site of haemopoiesis is mainly the bone marrow of all bones; but later, haemopoiesis only occurs in the marrow of specific bones.

Bone marrow is **myeloid tissue** – a mixture of fat and blood-forming cells. All blood cells are formed from a single type of unspecialized cell – the **stem cell**, also known as a multipotent or pluripotent cell, because it has the potential to develop into several different types of blood cell (Nair, 2011). When a stem cell divides, it initially becomes an immature blood cell (**haemocytoblast**). Whilst still in the bone marrow, haemocytoblasts mature into one of two types of immature cells – either **myeloid** or **lymphoid** stem cells – which then further mature and divide (Figure 8.3).

- Myeloid stem cells develop initially into immature red or white blood cells, or immature platelets (known as **proerythroblasts**, **myeloblasts**, **monoblasts** and **megakaryoblasts**). Proerythroblasts then develop into **erythrocytes** (red blood cells), myeloblasts into **eosinophils**, **neutrophils** and **basophils**, monoblasts into **monocytes** and **macrophages**, and megakaryoblasts into **thrombocytes** (platelets).
- The lymphoid stem cells initially develop into lymphoblasts before becoming **lymphocytes** and **plasma cells**.

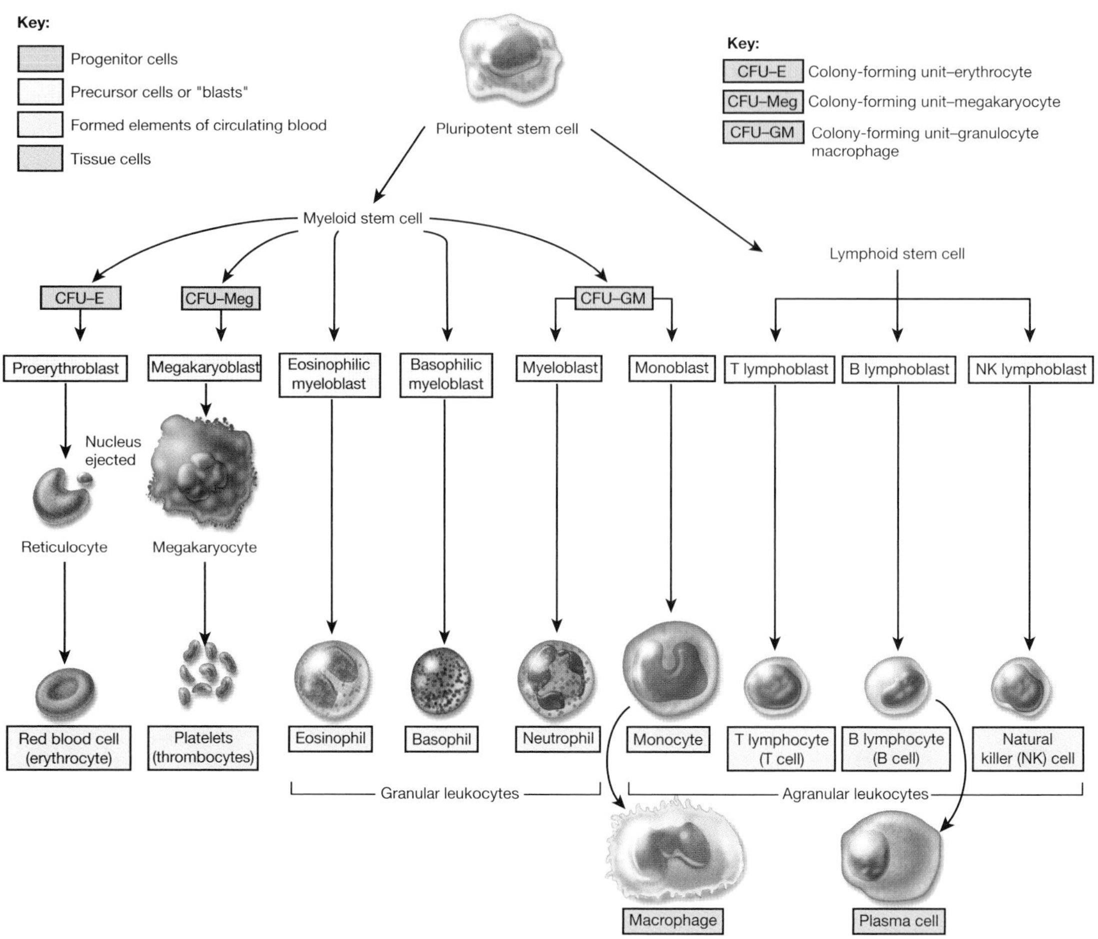

**Figure 8.3** The development of blood cells.
*Source:* Tortora and Derrickson (2009), Figure 19.3, p. 694. Reproduced with permission of John Wiley and Sons, Inc.

## *Red blood cells*

**Erythrocytes** are the most abundant blood cells in our bodies and are **biconcave** flattened discs (Figure 8.4). The biconcave shape allows for a larger surface area for the diffusion of gas molecules ($O_2$ and $CO_2$) that pass through the red blood cell membrane to combine with the Hb molecules within the cell.

They average 8 μm in diameter and do not possess a nucleus. Immature cells, on the other hand, do have a nucleus, but by the time they are fully mature they have lost it, and it is this loss of a nucleus that causes the particular shape of the erythrocyte.

There are approximately 4–5.5 million red blood cells in each cubic millimetre of blood. The biconcave shape of an erythrocyte is maintained by a network of proteins (called spectrin) within its cytoplasm, and this structure allows the erythrocyte to change shape as required during its passage through blood vessels. Red blood cells start to lose their capacity to deliver $O_2$ and lose the flexibility required to squeeze through the smallest capillaries to deliver $O_2$ to the tissues after 21 days.

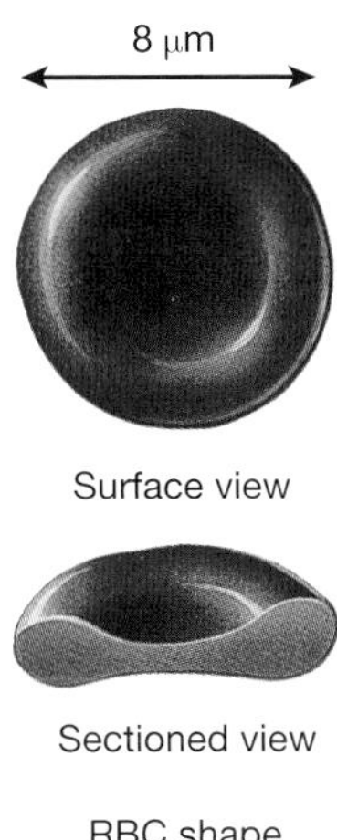


**Figure 8.4** Red blood cells.
*Source:* Tortora and Derrickson (2009), Figure 19.4(a), p. 696. Reproduced with permission of John Wiley and Sons, Inc.

## Clinical application

### Anaemia

In children, anaemia is the reduction of red blood cells or of Hb concentration due to injury or disease. There are three main types of anaemia: hypoproliferative anaemia, which is the result of defective blood cell production (e.g. iron-deficiency anaemia); anaemia caused by bleeding, leading to a loss of blood cells; and haemolytic anaemia, caused by excessive destruction of red blood cells (e.g. sickle cell anaemia). The treatment is to try to treat the underlying cause and so reverse the anaemia; for example, stop bleeding, introduce iron into the diet (Price, 2007a).

## *Haemoglobin and oxygen transport*

The membrane of a red blood cell encloses both the cytoplasm and the red, $O_2$-carrying-protein known as **Hb**, which constitutes 30% of an erythrocyte's total weight and is responsible for the red colour of blood. Note that it is only from the onset of adolescence that Hb levels differ according to gender (Table 8.2).

Hb is extremely important because it combines with the respiratory gases $O_2$ and $CO_2$ – it forms **oxyhaemoglobin** in order to transport $O_2$ around the body to the tissues where it is required, and it forms **carbaminohaemoglobin** to carry $CO_2$ (a waste gas from metabolism) from the tissues to the lungs, where it is expelled. It facilitates the exchange of gases at the alveolar junction in the lungs, Each Hb molecule consists of four atoms of iron, and each iron atom can attach to, and transport, one molecule of $O_2$. As there are approximately 250 million Hb molecules in one erythrocyte, one erythrocyte is able to transport approximately 1 billion molecules of $O_2$.

Erythrocytes take up the $O_2$ molecules from the inspired air that is present in the lungs, bind these $O_2$ molecules to Hb molecules and carry them to the tissues that require the $O_2$. The erythrocytes move into the capillaries, where the $O_2$ molecules diffuse out of the erythrocytes, move across the capillary wall and enter into the cells of the tissues. As the $O_2$ level in erythrocytes increases, they become bright red, and when the $O_2$ content drops they become a dark bluish-red colour. This change in colour can be seen in the tissues – particularly the skin. When it is well-oxygenated, skin is pink, but it becomes bluish when in need of $O_2$. This is a very important sign for the nurse to look for in a patient.

On the erythrocyte's return journey, $CO_2$ is released from tissue cells into the erythrocyte, where it combines with the Hb, is transported back to the lungs and released as expired air. However, although about 23% of $CO_2$ is transported bound to the Hb, most of the $CO_2$ that we exhale is transported in the blood plasma as $HCO_3$ (carbonic acid).



The life span of a red blood cell is approximately 120 days. By that time, the plasma membrane of the cell becomes fragile and dysfunctional, and the plasma membrane and other parts of the worn-out red blood cell are removed from the circulation by **macrophages** (white blood cells) and broken down (**haemolysis**) in the liver and spleen (Figure 8.5):

- The globin (protein) is broken down into its constituent amino acids, which are reused to produce further proteins.
- The iron is separated from haem and is stored in muscles and the liver, and reused – this time in the bone marrow to make new red blood cells.
- Haem is converted into bilirubin, transported to the liver and eventually secreted in bile.

**Bilirubin** is converted by bacteria in the large intestine into **urobilinogen**, and some of this is reabsorbed into the bloodstream where it is converted into a yellow pigment (**urobilin**), which is excreted from the body in urine.

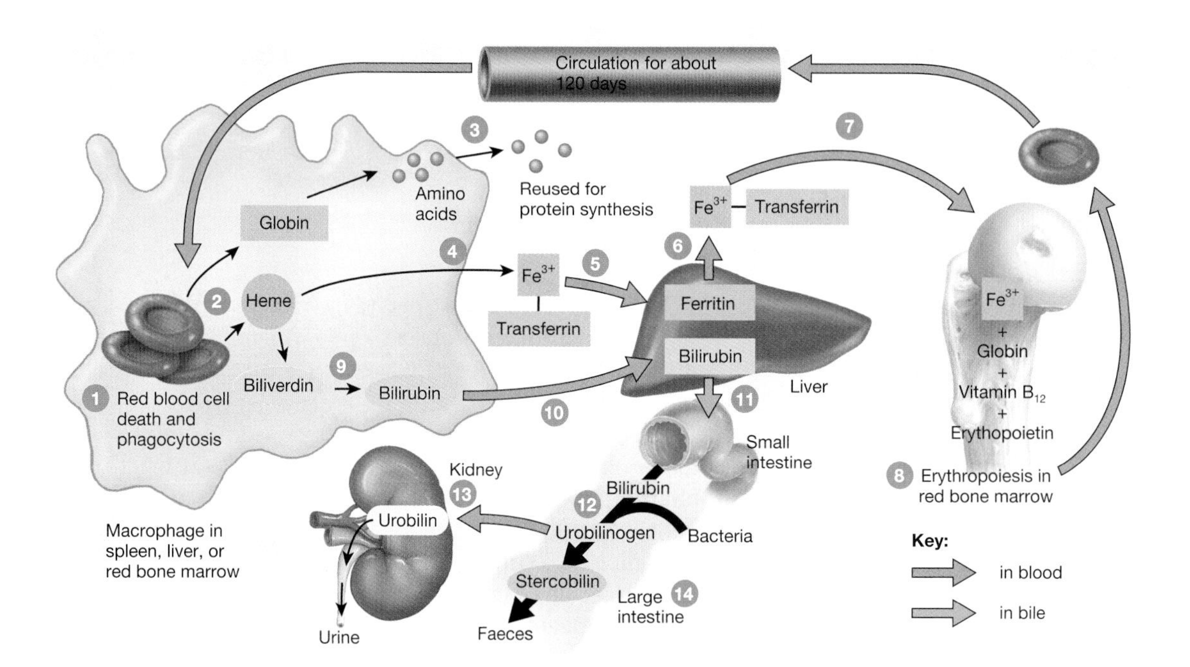


**Figure 8.5** Destruction of the red blood cell.
*Source:* Tortora and Derrickson (2009), Figure 19.5, p. 697. Reproduced with permission of John Wiley and Sons, Inc.

## Erythropoeisis

Production of erythrocytes (**erythropoeisis**) is controlled by **erythropoietin** – a hormone produced by the kidneys and transported by the blood to the bone marrow, where the red blood cells are produced (**synthesized**). In addition to the other substances required for the synthesis of erythrocytes, essential substances include:

- iron
- folic acid
- vitamin $B_{12}$.



Erythropoeisis is a homeostatic mechanism controlled by a negative feedback mechanism (Figure 8.6) and which is triggered by either an increased tissue need for $O_2$ (e.g. during exercise) or a reduced supply of $O_2$ for the body's cells. This reduced supply of $O_2$ can be the result of a reduced number of erythrocytes or other problems with the body's functioning; for example, anaemia, where, although there may be sufficient erythrocytes circulating, they are unable to pick up the $O_2$ molecules. Whatever the cause, the amount of $O_2$ delivered to the tissues is

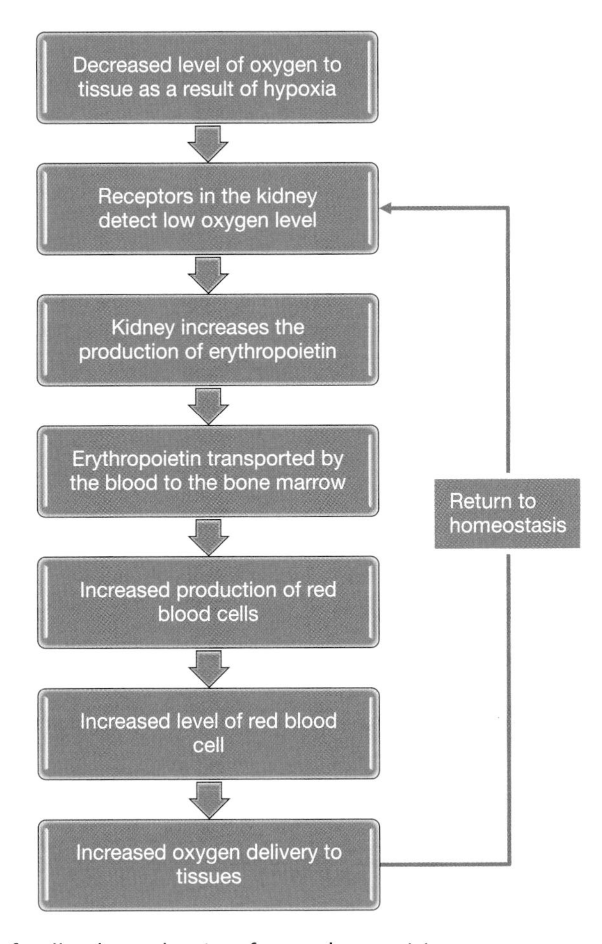


**Figure 8.6** Negative feedback mechanism for erythropoeisis.
*Source:* Peate and Nair (2011), Figure 12.7, p. 377. Reproduced with permission of John Wiley and Sons, Ltd.

deemed insufficient for the body's needs – and so a state of **hypoxia** (low $O_2$ levels) or **anoxia** (no $O_2$ present in the tissues) occurs.

## *White blood cells*

These are known as (**leucocytes**: *leucos* = white and *cyte* = cell). Unlike erythrocytes, leucocytes have nuclei and other organelles. They are able to move out of blood vessels and into tissues. They have a variable life span of a few days to several years. For further information on white blood cells see Chapter 7.



## Clinical application

### Leukaemia

Leukaemia is an oncological (cancer) problem and denotes an absence, or an absence of function, of the white blood cells (leucocytes). There are several different types, such as acute myeloid leukaemia (AML) and acute lymphoblastic leukaemia (ALL), depending upon the particular white blood cell that is not functioning properly. All of them are fatal unless treated successfully. The main treatment is cytotoxic chemotherapy (drugs that destroy the cells of the body – in this case aimed at the cancerous white blood cells). Signs and symptoms include tiredness and weakness, bruising and bleeding, pain, susceptibility to infections, and weight loss. However, it is important to note that the treatment itself can cause many very unpleasant side effects (Langton, 2007; Marriott, 2007; Pritchard, 2007).

## *Bleeding and platelets*

**Platelets** (**thrombocytes**) are small blood cells produced in the bone marrow from **megakaryocytes**. Fragments of the megakaryocytic cytoplasm break off in the bone marrow and become surrounded by a piece of cell membrane to form platelets, which have a life span of 5–9 days. Once platelets have formed, they enter the blood circulation.

Platelets have no nucleus and average 2–4 μm in diameter. Their function is to repair damaged blood vessels and to begin a chain reaction resulting in **blood clotting**. They stick to other proteins, such as **collagen** in the connective tissues, and form platelet plugs that seal the holes in damaged blood vessels and release chemicals that help in blood clotting. Low circulating numbers of platelets can lead to excessive bleeding following damage to blood vessels, whilst high numbers of circulating platelets can lead to increased blood clots (**thrombosis)** occurring, causing many problems, including heart attacks, deep vein thrombosis and pulmonary embolisms.

## *Haemostasis*

Haemostasis is a sequence of responses that can occur in order to stop bleeding (**haemorrhage**) from the smaller blood vessels. It plays an important role in homeostasis and has three main components:

- **vasoconstriction**
- **platelet aggregation**
- **coagulation**.

## *Vasoconstriction*

**Vasoconstriction** occurs as a result of **vascular spasm**, which causes the smooth muscle of the blood vessel wall to contract. This constricts the small blood vessels, preventing the blood from

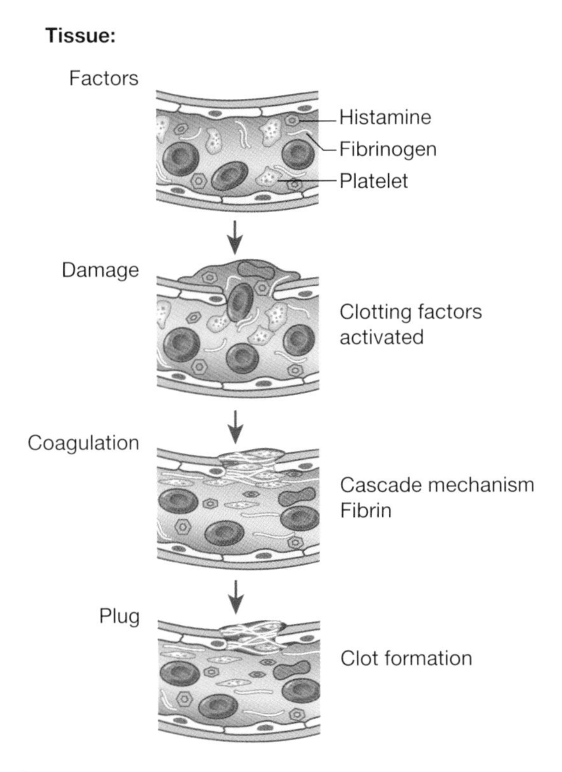


**Figure 8.7** Platelet plug formation.

flowing through them. Vasoconstriction is caused by the sympathetic nervous system acting to restrict blood flow and may last from several minutes to several hours.

Platelets release **thromboxanes** – vasoconstrictors that can cause **hypertension** (high BP) and facilitate platelet aggregation (the massing/clumping of platelets at a site).

### *Platelet aggregation*

Actions involved in platelet aggregation are (Figure 8.7):

- Platelets adhere to the connective tissue of the damaged blood vessels.
- Platelets release adenosine diphosphate, thromboxane and other chemicals, including prostaglandin, serotonin enzymes and calcium ions, so platelets stick to each other, forming a tight clump of platelets.
- Platelets are very effective in preventing blood loss.

### *Coagulation*

**Coagulation** is the process of blood clotting. A platelet plug consists of a network of fibrin threads and platelets. The coagulation procedure involves various proteins, known as **coagulation factors**, essential for clotting to take place. Most are synthesized in the liver, whilst some are obtained from our diet. Clotting is a complex process in which the activated form of one of the coagulation factors catalyses the activation of the next factor in the clotting sequence (or cascade).

Coagulation occurs when the damage to the blood vessel is so extensive that platelet aggregation and vasoconstriction cannot stop the bleeding. The process of coagulation involves the clotting factors displayed in Table 8.3.

**Table 8.3** Plasma coagulation factors.

| Factor | Alternative names |
|---|---|
| I | Fibrinogen |
| II | Prothrombin |
| III | Tissue factor (thromboplastin) |
| IV | Calcium ions |
| V | Proaccelerin<br>Labile factor<br>Accelerator globulin |
| VII | Serum prothrombin conversion accelerator<br>Stable factor<br>Proconvertin |
| VIII | Antihaemophilic factor<br>Antihaemophilic factor A<br>Antihaemophilic globulin |
| IX | Christmas factor<br>Plasma thromboplastin component<br>Antihaemophilic factor B |
| X | Stuart factor<br>Power factor<br>Thrombokinase |
| XI | Plasma thromboplastin antecedent<br>Antihaemophilic factor C |
| XII | Hageman factor<br>Glass factor<br>Contact factor |
| XIII | Fibrin stabilizing factor<br>Fibrinase |

- Platelets in the damaged tissues release **thromboplastin**, triggering the clotting mechanism. In addition, the thromboplastin interacts with other blood protein clotting factors (Figure 8.8).
- **Thromboplastin** converts **prothrombin** (in blood plasma) to **thrombin**.
- **Thrombin** acts, in the presence of **calcium ions**, to change **fibrinogen proteins** (in plasma) into **fibrin**.
- **Fibrin**, in the presence of clotting factor XIII (**fibrinase**), forms a mesh of strands, traps erythrocytes and forms a blood clot.

There are many different clotting factors in blood plasma (Table 8.3), and all need to act together to produce a blood clot. The absence of any of these clotting factors can lead to serious disease; for example:

- an absence of factor VIII (antihaemophilic factor) will lead to **haemophilia**, where the body is unable to produce blood clots in response to tissue damage, and bleeding can continue unabated;

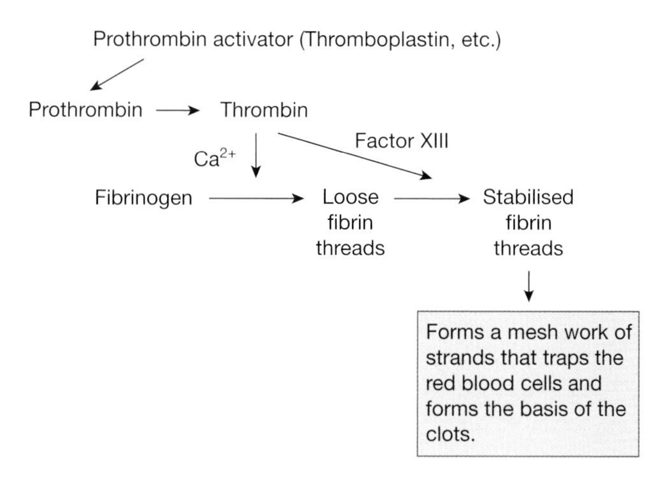


**Figure 8.8** Blood coagulation, simplified.



- an absence of factor IX (Christmas factor) leads to a disorder known as **Christmas disease** – which is similar to haemophilia.

Two pathways have been identified in the formation of a blood clot: intrinsic and extrinsic pathways. The intrinsic pathway is activated when the inner walls of a blood vessel are damaged, whilst the extrinsic pathway (acting much quicker than the intrinsic pathway) is activated when the blood vessels rupture, leading to tissue damage occurring.

## Clinical application

### Haemophilia

Haemophilia is a genetic bleeding disorder caused by defects in the clotting ability of blood due to deficiencies in factor VIII (known as haemophilia A, the most common type at 85%) or factor IX (haemophilia B – also known as Christmas disease – at 15%). The abnormal gene is carried on the X chromosome, and so girls can be carriers, but only boys can have haemophilia. Symptoms include bleeding that is difficult to stop, bruising, joint pain due to bleeding into the joint and muscle pain. It is a lifelong disease with no known cure, and so is managed by palliative treatment, particularly intravenous replacement of the missing blood clotting factors (Price, 2007b).

## Blood groups

All humans belong to one of four blood groups. A **blood group** (blood type) is a classification of blood based on the presence or absence of inherited **antigens** on the surface of red blood cells and their corresponding antibodies found in blood plasma. The antigens found on red blood cells are known as **agglutinogens**, and there are two types of agglutinogen, A and B; thus, **the ABO system** includes four groups, **A, B, AB**, and **O** ('O' signifies an absence of agglutinogens).

In blood group A, red blood cells have A agglutinogens on their membranes, blood group B will have B agglutinogens on the cell membranes. The third blood group, AB, will have both A

**Table 8.4** Blood groups and who they can donate to and receive from.

| Blood type | Agglutinogens on cell membrane | IgM antibodies in plasma | Can donate blood to | Can receive blood from |
|---|---|---|---|---|
| A | A | Anti-B | A and AB | A and O |
| B | B | Anti-A | B and AB | B and O |
| AB | A and B | None | AB | A, B, AB and O |
| O | Neither | Anti-A and anti-B | A, B, AB and O | O |



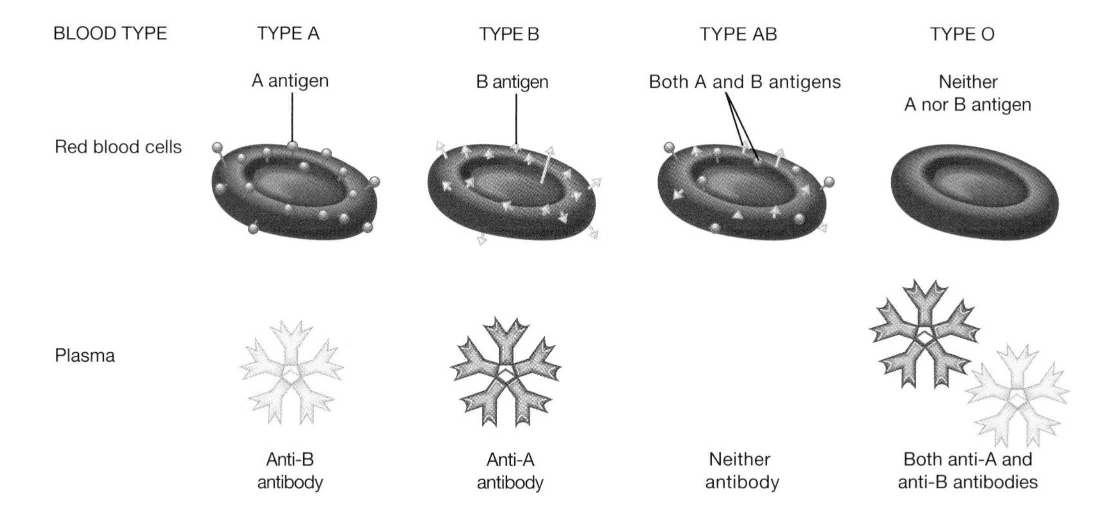


**Figure 8.9** ABO blood groups.
*Source:* Tortora and Derrickson (2009), Figure 19.12, p. 709. Reproduced with permission of John Wiley and Sons, Inc.

and B agglutinogens on the cell membranes and blood group O will have red blood cells that possess neither of these agglutinogens (Table 8.4). People with blood group A will have anti-B antibodies, blood group B will have anti-A antibodies, blood group AB have no anti-A and anti-B antibodies, whilst blood group O have anti-A and anti-B antibodies (Hoffbrand *et al.*, 2006). Because of this direct link between antigens and antibodies, it is important to ensure that a patient requiring a blood transfusion receives the correct blood type, otherwise the patient's own antibodies can destroy the transfused blood, and may lead to a fatal immune reaction, known as **agglutination** – the clumping of red blood cells (Figure 8.9).

There is one other antigen to consider: antigen D, the Rhesus (Rh) factor system. Rh antigens can be present in each of the four blood groups. Not everyone has the Rh antigen on the membrane of their red blood cells.

D antigen is the most significant Rh antigen. It is the most likely to provoke an immune system response. Usually, people who are D-negative do not to possess any anti-D antibodies. However, D-negative individuals can produce anti-D antibodies following an event that can sensitize the blood; for example, a fetomaternal transfusion of blood from a fetus in pregnancy (when blood can cross over from fetus to mother and vice versa, or occasionally a blood transfusion with

D-positive red blood cells. In blood, the presence or absence of the Rh antigens is signified by the plus or minus sign; for example, the A− group does not have any of the Rh antigens, whereas the A+ group would have Rh antigens. Approximately 85% of the UK population are Rh positive and 15% are Rh negative.

The four blood types (ABO system) and Rh groups are not equally distributed amongst the population, nor are they equally distributed between ethnic groups (Jorde *et al.*, 2006).

# Blood vessels



Blood circulates around the body inside blood vessels which form a closed transport system.

There are three major types of blood vessels:

1. **Arteries**, which carry blood away from the heart, via the lungs (where it picks up $O_2$ molecules), and then carry the newly oxygenated blood around the body to oxygenate the cells.
2. **Capillaries**, which enable the exchange of water, nutrients and essential chemicals between the blood and the tissues.
3. **Veins**, which carry deoxygenated blood from the capillaries back to the heart, for the whole cycle to begin again (Figure 8.10).

Apart from the pulmonary artery and the umbilical artery, all the arteries carry oxygenated blood, and all the veins, apart from the pulmonary and umbilical veins, carry deoxygenated blood. Smaller arteries – arterioles – link the large arteries carrying oxygenated blood to the

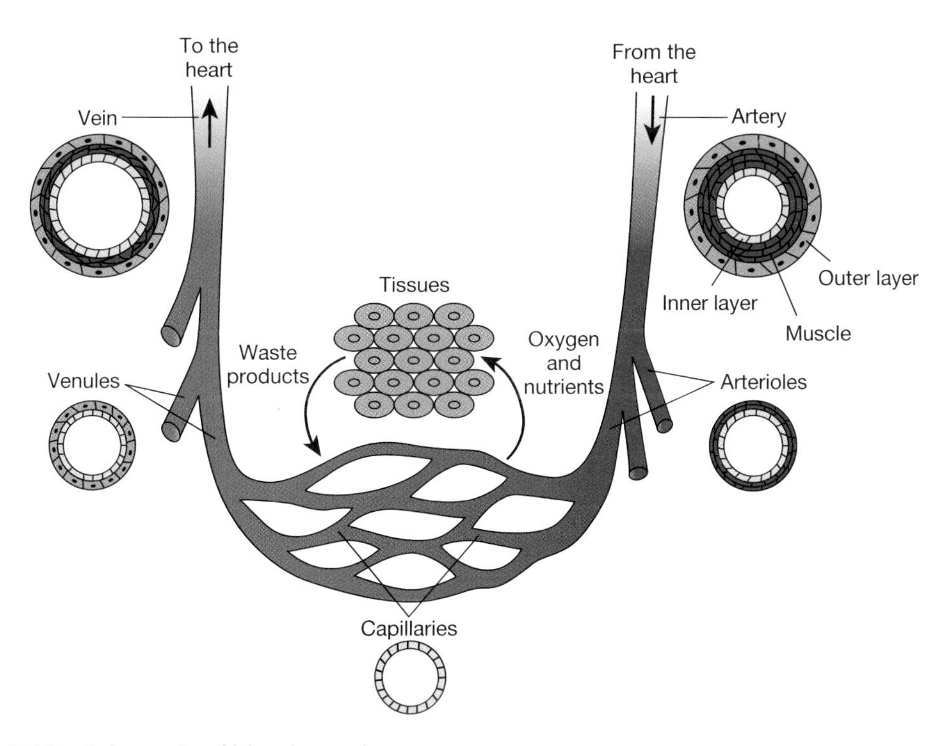


**Figure 8.10** Schematic of blood vessels.
*Source:* Peate and Nair (2011), Figure 12.15, p. 386. Reproduced with permission of John Wiley and Sons, Ltd.

## Box 8.1 Sequence of vessels from arteries to veins.

**Oxygenated blood** **Deoxygenated blood**

Artery → arteriole → capillary → tissue cell → capillary → venule → Vein

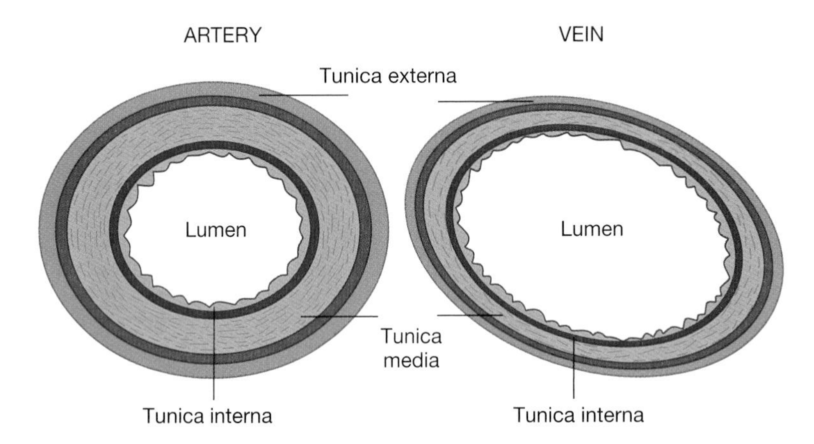


**Figure 8.11** Basic structure of an artery and vein.
*Source:* Peate and Nair (2011), Figure 12.17, p. 388. Reproduced with permission of John Wiley and Sons, Ltd.

capillaries, and, similarly, venules take deoxygenated blood from the capillaries and transfer it to the large veins as shown in Box 8.1.

The capillaries, which are tiny, thin-walled blood vessels (forming the microcirculatory system), transport and deliver, or carry away, nutrients, gases (including $O_2$ and $CO_2$), water and electrolytes, act as a bridge between the arteries/arterioles and veins/venules.

# Structure and function of blood vessels

## Structure

Blood vessels consist of walls and a lumen (Figure 8.11). Except for the walls of the microscopic capillaries, the walls of all the blood vessels have three coats – tunicas (Figure 8.12):

- tunica interna
- tunica media
- tunica externa.

### *Tunica interna*

The lining of the vessels is smooth, allowing for the easy flow of blood through the vessel. However, there are some differences between the different blood vessels:

- arteries have mostly elastic tissue;
- veins have hardly any elastic tissue;
- capillaries have no elastic tissue.

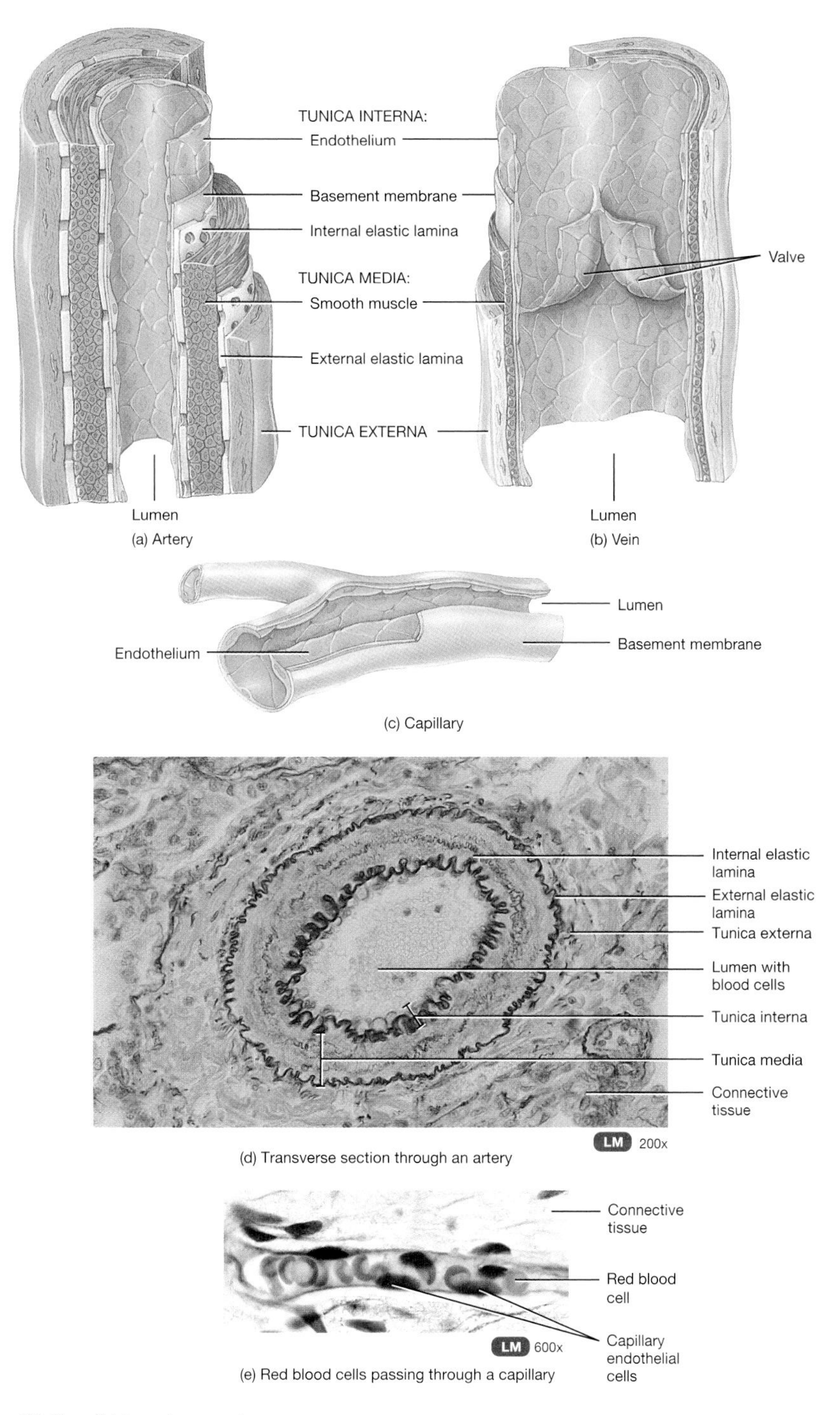


**Figure 8.12** Walls of blood vessels.
*Source:* Tortora and Derrickson (2009), Figure 21.1(a–c), p. 762. Reproduced with permission of John Wiley and Sons, Inc.

### Tunica media

The tunica media consists of elastic fibres and smooth muscle, allowing for vasoconstriction to occur, so enabling blood flow and BP to change when required. This layer is supplied by the sympathetic branch of the autonomic nervous system. When the nerves of the blood vessel are stimulated, the vessel walls contract, leading to a narrowing of the lumen and an increase of pressure within the blood vessel. There are variations in the tunica media depending upon the type of blood vessel:

- arteries have a thickness that varies according to the size of the vessel;
- veins have a thin layer only in these vessels;
- capillaries do not possess a tunic media.

### Tunic externa

The tunica externa consists of collagen fibres, and the thickness varies depending upon the type of blood vessel. The collagen enables the blood vessel to anchor itself to nearby organs, which gives the blood vessel both support and stability:

- arteries are relatively thick;
- veins are relatively thick;
- capillaries are very thin and delicate.

## Functions of blood vessels

### Arteries and veins

Arteries and veins have similar layers within their walls. However, there are some very clear differences between them, linked to their roles within the circulatory system (Table 8.5).

The aorta is the largest artery in the body, and oxygenated blood leaves the heart through the aorta into other arteries. Blood is pumped around the body in the arteries by the action of the heart forcing the blood through the various chambers, and also by the action of gravity. However, the effect of the heart pump is dissipated as the blood flows through the very tiny capillaries, and so has no effect on the blood as it returns via the veins to the heart. The effects of gravity generally work against blood flow through the veins as the blood is flowing 'up' the

**Table 8.5** Differences between arteries and veins. *Source:* Peate and Nair (2011), Table 12.3, p. 388. Reproduced with permission of John Wiley and Sons, Ltd.

| Arteries | Veins |
|---|---|
| Transport blood away from heart | Transport blood to the heart |
| Carry oxygenated blood, except the pulmonary and umbilical arteries | Carry deoxygenated blood, except pulmonary and umbilical veins |
| Have a narrow lumen | Have a wider lumen |
| Have more elastic tissue | Have less elastic tissue |
| Do not have valves | Do have valves |
| Transport blood under pressure | Transport blood under low pressure |

body. Instead, blood is forced through the veins by muscular contractions. Veins have one-way valves in them which prevent the blood from flowing backwards. The heart receives the deoxygenated blood via the inferior and superior vena cavae – the largest veins in the body.

### *Capillaries*

These are very small blood vessels, approximately 6 μm in diameter. Their walls are only one endothelial cell thick, allowing the exchange of materials, such as molecules of $O_2$, water and nutrients into the surrounding tissue fluid by means of diffusion. In the same way, waste products of metabolism (e.g. $CO_2$ and urea) can pass into them. Because the lumens of capillaries are so small, blood cells have to change shape to pass through them, and also have to do so in single file – although the older the red blood cell, the less ability the red blood cell has to change shape and squeeze through the smallest capillaries.



There are networks of capillaries in most of the organs and tissues of the body (Figure 8.13).

## Clinical application

### Sickle cell disease

Sickle cell disease is an autosomal recessive disorder in which the Hb of the blood is abnormal (shaped like a sickle) and has a short life span, and is common in people of African-Caribbean descent, but also found in people from the Mediterranean, Middle and Far East and parts of India. From the age of 3–6 months, symptoms can occur at any age and can differ in severity. Symptoms are many, and include acute vaso-occlusive (blocking of blood vessels) events, which are very painful and can lead to many severe complications, such as painful swelling of hands and feet, fatigue/shortness of breath, haematuria (blood in the urine), acute chest problems, infections, and renal, liver and cardiac problems. Management is aimed at trying to manage complications and haematopoietic cell transplantations for those with very severe complications (Kelsey, 2007).

## Blood pressure

BP, sometimes referred to as arterial BP, is the pressure exerted by circulating blood within a blood vessel (Nair, 2011). Arterial BP is highest closest to the heart and it decreases as the blood moves through the blood circulatory system. During each beat of the heart, BP varies between a maximum (**systolic pressure**), when the heart is squeezing the blood out, and a minimum (**diastolic pressure**), when the heart is filling with blood. It is the differences in mean BP that are responsible for blood flow from one location to another in the circulation. The rate of mean blood flow depends upon the resistance to the blood flow from the blood vessels. Arterial BP in the circulation is mainly due to the pumping action of the heart, whilst for venous BP, gravity affects BP via **hydrostatic** forces; for example, during standing, along with the valves in the veins. Breathing and the contraction of skeletal muscles also influence the BP in veins (Caro, 1978).

Three regulatory forces combine to regulate BP:

- **neuronal** – via the autonomic nervous system;
- **hormonal** – including, but not exclusively, adrenaline, noradrenaline and renin;
- **autoregulation** – via the renin–angiotensin system.

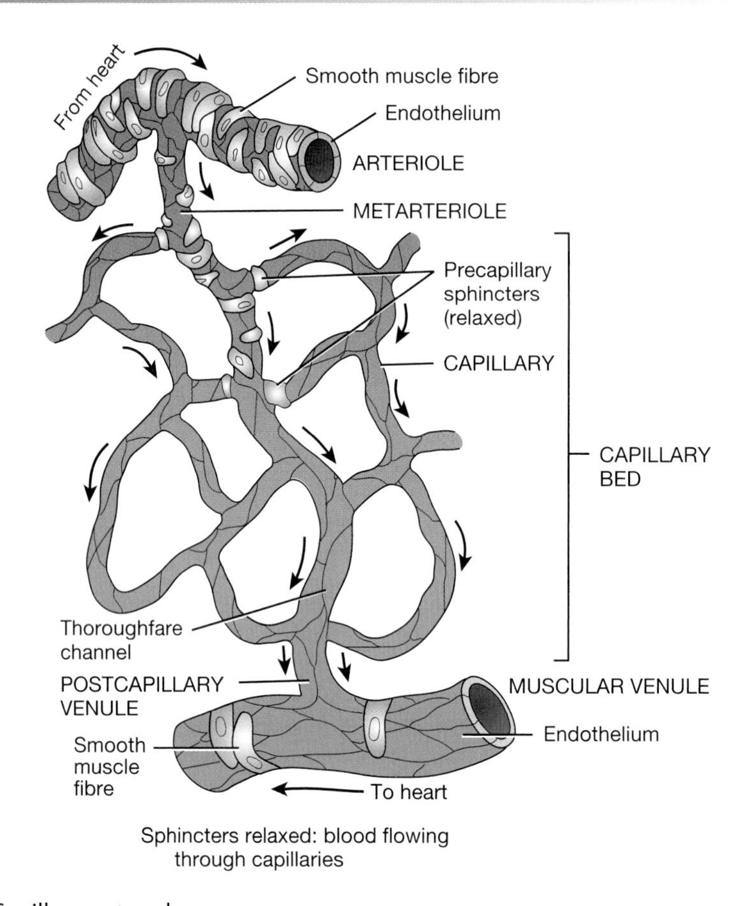


**Figure 8.13** Capillary networks.
*Source:* Tortora and Derrickson (2009), Figure 21.3(a), p. 765. Reproduced with permission of John Wiley and Sons, Inc.

## Physiological factors regulating blood pressure

There are several physiological factors that have an effect on BP (Nair, 2011):

- **Cardiac output** – volume of blood pumped out by the heart in 1 min. This is a function of **heart rate** (the number of heart beats in a minute) and **stroke volume** (the volume of blood – in millilitres – pushed out by the heart with each beat of the heart).
- **Circulating volume** – volume of circulating blood supplying tissues.
- **Peripheral resistance** – resistance provided by the blood vessels.
- **Blood viscosity** – resistance of blood flow, which is provided by plasma proteins and other substances circulating in the blood; that is, the thickness of blood.
- **Hydrostatic pressure** – the pressure exerted by the blood on the wall of the blood vessel.

## Control of arterial blood pressure

It is essential to maintain BP within the large systemic arteries to ensure that there is adequate blood flow to the tissues, and this is maintained and regulated by:

- **Baroreceptors** – situated within the arch of the aorta and the carotid sinus, these are receptive to pressure changes within the blood vessel. When BP increases, signals are sent to the **cardioregulatory centre** (CRC) in the **medulla oblongata** (the brain stem). The CRC increases parasympathetic activity to the heart, reducing heart rate and inhibiting sympathetic activity to the blood vessels. This causes **vasodilation** (widening of the blood vessel), reducing BP. If BP falls, the CRC increases sympathetic activity to the heart and the blood vessels. This increases the heart rate, causing **vasoconstriction** (narrowing of the blood vessels), so reducing the available space for the blood), resulting in an increased BP.
- **Peripheral chemoreceptors** – these consist of aortic and carotid bodies. Both regulate blood $O_2$, $CO_2$ and pH.
- **Circulating hormones** – these include antidiuretic hormones, helping to regulate the volume of circulating blood, and thus have an effect on BP.
- **The renin–angiotensin system** – this is a group of related hormones acting together to regulate BP. The renin–angiotensin system, working with the kidneys, is the body's most important long-term BP regulation system.
- **The hypothalamus** – this responds to stimuli such as emotion, pain and anger, and stimulates sympathetic nervous activity, so affecting BP.



## Normal blood pressure measurements at different ages of children

BP is measured in millimetres of mercury (mmHg).

In children, when considering BP, it has to be taken into account that age, height and gender has a significant effect on BP (Table 8.6).

It is recommend that children have their BP checked regularly, beginning at 3 years (NIH, 2007). The upper limit for normal systolic pressure in children aged between 3 and 5 years is

**Table 8.6** BP (mmHg) by gender and age for the 95th centile.

| Age | Boys | | Girls | |
|---|---|---|---|---|
| | Systolic | Diastolic | Systolic | Diastolic |
| 3 years | 104–113 | 63–67 | 104–110 | 65–68 |
| 4 years | 106–115 | 66–71 | 105–111 | 67–71 |
| 5 years | 108–116 | 69–74 | 107–113 | 69–73 |
| 6 years | 109–117 | 72–76 | 108–114 | 71–75 |
| 7 years | 110–119 | 74–78 | 110–116 | 73–76 |
| 8 years | 111–120 | 75–80 | 112–118 | 74–78 |
| 9 years | 113–121 | 76–81 | 114–120 | 75–79 |
| 10 years | 114–123 | 77–82 | 116–122 | 77–80 |
| 11 years | 116–125 | 78–83 | 118–124 | 78–83 |
| 12 years | 119–127 | 79–83 | 120–126 | 79–82 |

from 104 mmHg to 116 mmHg, depending on height and gender, whilst the upper limit for diastolic pressure will range from 63 to 74 mmHg.

# Conclusion

Blood is essential for life, and this chapter has looked at how blood is formed and how it is used by the body to maintain the body systems.

Three roles within the body are fulfilled by the blood system:



- transportation – of $O_2$, $CO_2$, hormones, nutrients and organic waste products;
- regulation – of acid–base balance and heat;
- protection – against infections.

Blood is carried around the body in a network of blood vessels – with varying sizes of diameters, the largest being the veins and arteries, whilst the smallest are the microscopic capillaries that interact with tissues so that gases, nutrients and waste products of metabolism can be exchanged between the blood system and the tissues. The main driving forces for the movement of blood around the body are the heart, which pushes the oxygenated blood through the arteries, and muscular contractions, which force deoxygenated blood through the veins.

# Activities

**Now review your learning by completing the learning activities in this chapter. The answers to these appear at the end of the book. Further self-test activities can be found at www.wileyfundamentalseries.com/childrensA&P.**

## Make notes about each of the following

- Fanconi anaemia
- Idiopathic thrombocytopaenia purpura (ITP)
- Thalassaemia

## Which is the odd one out?

1. (a) ABO system
   (b) HCO system
   (c) Rh system
   (d) Circulatory system
2. (a) Protection against infection
   (b) Oxygenation of tissues
   (c) Removal of carbon dioxide from the tissues
   (d) Protection against injury

3. (a) Blood cell
   (b) Vein
   (c) Arteriole
   (d) Capillary

## Complete the sentence

From the lists below, fill in the missing words in the sentences:

1. Vasoconstriction occurs as a result of ________ spasm which causes the ________ muscle of the blood vessel _______ to contract, which in turn constricts the small ________ vessels. This process is a result of the __________ nervous system restricting blood flow.

   rough entire blood vascular cell wall smooth muscular antagonistic sympathetic

2. The aorta is the largest ______ in the body and ______ blood leaves the ______ through it.

   liver vein heart deoxygenated kidneys artery arteriole oxygenated whole

3. Blood pressure is maintained by means of ______ which are found in the arch of the ______ and the carotid sinus. When blood pressure increases, this sends signals to the cardioregulatory centre, which increases ______ activity to the heart, reducing heart ______ and inhibiting _______ activity to the blood vessels.

   parasympathetic hormones beat heart aorta rate sympathetic intuitive baroreceptors molecules



## Wordsearch

There are 14 words linked to this chapter hidden in the following above. Can you find them? A tip – the words can go from up to down, down to up, left to right, right to left, or diagonally. Also note that some words are abbreviations.

| | | | | | | | | | |
|---|---|---|---|---|---|---|---|---|---|
| C | S | W | A | N | C | A | B | Y | S |
| B | L | O | O | D | D | L | P | L | T |
| X | A | O | R | T | A | B | E | E | N |
| T | M | U | T | I | F | U | O | U | I |
| T | S | D | A | T | E | M | I | C | E |
| P | A | S | H | N | I | I | V | O | V |
| L | L | E | C | I | E | N | A | C | H |
| E | P | Y | T | S | J | U | G | Y | I |
| M | H | O | R | M | O | N | E | T | S |
| B | A | P | L | A | T | E | L | E | T |

## Crossword

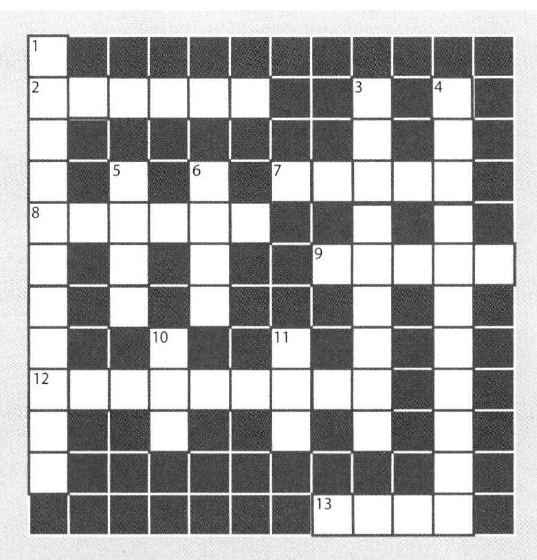


**Across:**

2. The name of the large blood vessels that carry blood from the heart
7. The type of IgM antibody found in the plasma of people with blood group type B
8. The crucial gas in blood that is carried by arteries
9. A classification of blood based on antigens on the surface of red blood cells – important in blood transfusions
12. The element in blood that carries oxygen, etc. Around the body – note there are two words in this answer.

**Down:**

1. A conjugated protein that gives blood its characteristic colour
3. The smallest type of artery
4. The correct term for blood clotting
5. The biological term for a cell
6. The large blood vessels that return deoxygenated blood from the tissues to the heart
10. The chemical formula for carbon dioxide (note that there is a number in this)
11. The type of blood cell that carries oxygen around the body

## Conditions

The following table contains a list of conditions. Take some time and write notes about each of the conditions. You may make the notes taken from text books or other resources; for example, people you work with in a clinical area or you may make the notes as a result of people you have cared for. If you are making notes about people you have cared for you must ensure that you adhere to the rules of confidentiality.

| Condition | Your notes |
| --- | --- |
| Leukaemia | |
| Sickle cell anaemia | |
| Haemophilia | |
| Vitamin K deficiency | |
| Disseminated intravascular dissemination | |



# Glossary

**Agglutinogen:** process by which red blood cells adhere to one another.

**Aorta:** largest artery in the body – emerges from the right ventricle of the heart.

**Antibody:** protein produced in response to the presence of some 'foreign' substances.

**Antigen:** 'foreign' protein against which antibodies are produced.

**Arteries:** blood vessels that carry blood away from the heart.

**Arterioles:** small arteries.

**Baroreceptor:** neurone that senses changes in pressure – either air, blood or fluid pressures.

**Blood pressure:** force exerted by the blood against the walls of blood vessels due to the force caused by the contraction of the heart.

**Bilirubin:** pigment found in bile as a result of the destruction of red blood cells.

**Blood groups:** classification of blood based on the type of antigen found on the surface of the red blood cell.

**Coagulation:** changing from a liquid to a solid; the formation of a blood clot.

**Carbaminohaemoglobin:** a combination of carbon dioxide and haemoglobin.

**Catalyse:** a substance that influences a chemical reaction.

**Chemotaxis:** movement of a cell or organism that is guided by a specific chemical concentration gradient.

**Cortex:** the outer portion.

**Diastolic blood pressure:** the lower blood pressure reading.

**Erythrocyte:** red blood cell.

**Erythropoeisis:** the process by which red blood cells are produced.

**External respiration:** the exchange of oxygen and carbon dioxide between the environment and respiratory organs – lungs.

**Haematocrit:** a measure of the percentage of red blood cells to the total blood.

**Haemocytoblast:** a cell in bone marrow that gives rise to blood cells and platelets.

**Haemoglobin:** an iron-containing protein found in red blood cells and which transports oxygen around the body.

**Haemolysis:** the disintegration of red blood cells.

**Haemopoiesis:** the formation of red blood cells.

**Haemorrhage:** bleeding.

**Haemosiderin:** an insoluble form of tissue storage iron.

**Inorganic salts:** salts that do not contain carbon, and therefore are not living; for example, sodium, potassium and calcium.

**Internal respiration:** metabolic process during which cells absorb oxygen and release carbon dioxide.



**Oxyhaemoglobin:** a combination of haemoglobin and oxygen carried in red blood cells.

**Packed cell volume (PCV):** the ratio of the volume occupied by packed red blood cells compared with the volume of the whole blood as measured by a haematocrit.

**pH:** a measurement of how acidic or basic (alkaline) a substance is, and ranges from 0 to 14.

**Plasma:** fluid portion of blood.

**Platelet:** a type of blood cell involved in blood clotting.

**Stem cell:** a cell that can divide and differentiate into different specialized cell types and can also self-renew to produce more stem cells.

**Systolic blood pressure:** the higher blood pressure reading.

**Thrombocyte:** another name for a platelet.

**Tunica externa:** membranous outer layer of a blood vessel.

**Tunica intima:** the inner lining of a blood vessel.

**Tunica media:** middle muscle layer of a blood vessel.

**Urobilinogen:** a product of bilirubin breakdown.

**Viscous:** having a thick, sticky consistency between a solid and a liquid – having a high viscosity.

## References


Caro, C.G. (1978) *The Mechanics of the Circulation*, Oxford University Press, Oxford.

Hoffbrand, A.V., Moss, P.A.H., Pettit, J.E. (2006) *Essential Haematology*, 5th edn, Blackwell, Oxford.

Jorde, L.B., Carey, J.C., Bamshad, M.J., White, R.L. (2006) *Medical Genetics*, 3rd edn, Mosby, St. Louis, MO.

Kelsey, J. (2007) Sickle cell disease. In: Glasper, A., McEwing, G., Richardson, J. (eds), *Oxford Handbook of Children's and Young People's Nursing*, Oxford University Press, Oxford, pp. 570–573.

Langton, H. (2007) Acute myeloid leukaemia (AML). In: Glasper, A., McEwing, G., Richardson, J. (eds), *Oxford Handbook of Children's and Young People's Nursing*, Oxford University Press, Oxford, pp. 608–609.

Marriott, H. (2007) Acute lymphoblastic leukaemia (ALL). In: Glasper, A., McEwing, G., Richardson, J. (eds), *Oxford Handbook of Children's and Young People's Nursing*, Oxford University Press, Oxford, pp. 610–611.

Nair, M. (2009) The blood and associated disorders. In: Nair, M. and Peate, I (eds), *Fundamentals of Pathophysiology: An Essential Guide for Student Nurses*, John Wiley & Sons, Ltd, Chichester.

Nair, M. (2011) The circulatory system. In: Peate, I., Nair, M. (eds), *Fundamentals of Anatomy and Physiology for Student Nurses*, Wiley–Blackwell, Chichester.

NIH (2007) *A Pocket Guide to Blood Pressure Measurement in Children*, National Institutes of Health, US Department of Health and Human Services, www.nhlbi.nih.gov (accessed 1 July 2014).

Peate, I., Nair, M. (eds) (2011) *Fundamentals of Anatomy and Physiology for Student Nurses*, Wiley–Blackwell, Chichester.

Price, J. (2007b) Haemophilia. In: Glasper, A., McEwing, G., Richardson, J. (eds), *Oxford Handbook of Children's and Young People's Nursing*, Oxford University Press, Oxford, pp. 574–575.

Price, J. (2007a) Anaemia. In: Glasper, A., McEwing, G., Richardson, J. (eds), *Oxford Handbook of Children's and Young People's Nursing*, Oxford University Press, Oxford, pp. 568–569.

Pritchard, G. (2007) T-cell acute lymphoblastic leukaemia. In: Glasper, A., McEwing, G., Richardson, J. (eds), *Oxford Handbook of Children's and Young People's Nursing*, Oxford University Press, Oxford, pp. 614–615.

Tortora, G.J., Derrickson, B.H. (2009) *Principles of Anatomy and Physiology*, 12th edn, Pearson Benjamin Cummings, San Francisco, CA.

Williams, J. (2007) Epistaxis. In: Glasper, A., McEwing, G., Richardson, J. (eds), *Oxford Handbook of Children's and Young People's Nursing*, Oxford University Press, Oxford, pp. 552–553.

# Chapter 9

# The cardiac system

Sheila Roberts

*Children's Nursing, School of Health and Social Work, University of Hertfordshire, Hatfield, UK*

## Aim

The aim of this chapter is for readers to develop their understanding of the anatomy and physiology of the cardiac system from the early development of the embryonic heart through to the fully functioning heart of the older child.

## Learning outcomes

On completion of this chapter the reader will be able to:

- Describe the flow of blood into, through and out of the heart.
- Explain the differences between the fetal circulation and adult circulation.
- Understand the electrical pathway through the heart.
- Indicate the factors which affect cardiac output.
- Relate the anatomy and physiology of the heart to simple congenital heart defects.
- Describe the factors determining the heart rate.

## Test your prior knowledge

- Name the four chambers of the heart.
- Draw a basic diagram to show the blood flow through the heart.
- Name one of the two fetal adaptations that occur within the heart.
- The placenta has many functions, name two of them.
- Is the location of the heart the same in the baby and adult?
- What is the meaning of the term cyanosis?
- Name three congenital heart conditions.
- What is the name of the major blood vessel supplying blood to the body?
- Which side of the heart is more muscular?
- Arteries or veins carry blood away from the heart?

*Fundamentals of Children's Anatomy and Physiology: A Textbook for Nursing and Healthcare Students*, First Edition. Edited by Ian Peate and Elizabeth Gormley-Fleming.

Companion website: www.wileyfundamentalseries.com/childrensA&P

# Body map

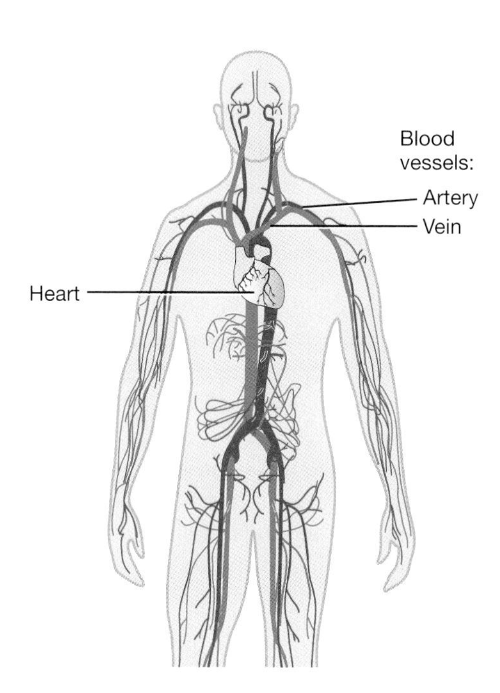




# Introduction

An infant's heart, when the baby is resting, beats at an average rate of 120 times per minute or 7200 times an hour, the equivalent of approximately 173 000 times a day delivering blood to all parts of the body. Growth and development over the following 18 years settles the heart beat to an average of 70 beats per minute, 4200 beats an hour and just over 100 000 a day. The heart pumps a continuous supply of deoxygenated blood to the lungs and oxygenated blood to the rest of the body, providing the much needed oxygen and nutrients to the cells and tissues whilst also removing the waste products. This chapter will provide an overview of the anatomy and physiology of the child's cardiac system to include the:

- fetal circulation
- cardiac adaptations following birth
- structure of the heart
- blood flow through the heart
- electrical pathways through the heart
- cardiac cycle.

# Fetal circulation

The developing embryo has a requirement for an adequate blood supply. Blood begins to circulate within the embryo from about week 3 of gestation and, therefore, it is imperative that the heart begins to develop and function from this early stage. Oxygen and nutrients are provided for the developing embryo from the mother's placenta, but the heart is required to pump these essential elements to the fetus and to remove the waste products, including carbon dioxide, from the fetus back to the placenta. By day 28 of gestation the underdeveloped and primitive heart has four recognizable chambers and has started to pump blood through the embryo. Development continues until the heart of the fetus is fully formed, although there remain structural anomalies that are present until after the baby has been born. These anomalies are a necessary part of fetal development.

As the mother's placenta provides the oxygen required by the developing fetus, the unborn baby's lungs are not required to function until after birth when the baby takes its first breath. The circulatory system of the unborn baby is therefore different from that of an older child or adult. The oxygenated blood enters the fetus through the umbilical vein; it then joins the inferior vena cava and enters the right atrium of the heart. From the right atrium the majority of blood moves directly to the left atrium through the foramen ovale, a valve-like opening in the septum, bypassing the right ventricle. The blood then continues via the bicuspid valve (also known as the mitral valve) into the left ventricle to be pumped via the aorta to the body. In order for the right ventricle to develop, some blood flows through the tricuspid valve into the right ventricle. The blood would then be pumped via the pulmonary artery to the lungs, but in the fetus a second adaptation of the circulatory system, the ductus arteriosus, is present. The ductus arteriosus is a small vessel that connects the pulmonary artery to the aorta, therefore again avoiding the majority of the blood flowing to the non-functioning lungs. Although the blood of the fetus is not oxygenated within the lungs, the lungs themselves require a supply of blood to provide oxygen and nutrients for growth and development.

A further adaptation of the fetal circulation system is the ductus venosus, a continuation of the umbilical vein that allows blood to flow directly into the inferior vena cava bypassing the non-functioning liver. Figure 9.1 provides an overview of the fetal circulation.

# Changes at birth

When a baby is born the circulatory system needs to undergo changes as the umbilical cord is cut, and with it the supply of oxygen and nutrients. A baby will take its first breath within about 10 seconds of delivery; this is probably initiated by the change in environmental temperature and handling of the newborn baby. The first breath and cry inflates the infant's lungs, which prior to delivery have been filled with alveolar fluid, although this begins to decrease with hormonal changes in the mother at the onset of labour. As the lungs inflate, the remaining alveolar fluid is quickly absorbed into the lymphatic system.

With the first breath:

- the pulmonary alveoli open up and fill with air;
- pressure in the pulmonary tissues decreases;
- blood fills the alveolar capillaries;
- pressure in the right atrium decreases as the blood flow resistance to the lungs decreases;
- pressure in the left atrium increases as blood is returned to the heart from the now vascularized pulmonary tissue.

As a result of the increased pressure in the left atrium and the decreased pressure in the right atrium the flap of the foramen ovale will not be able to open and closes to the flow of the blood from the right to the left atria. Blood will now flow from the right atrium to the right ventricle, through the pulmonary artery and to the lungs.

As the umbilical cord is clamped there is increased resistance, and therefore increased pressure, within the circulatory system. The circulatory pressure is now higher than the pulmonary pressure and blood no longer flows through the ductus arteriosus, which starts to constrict and close. The ductus arteriosus is also dependent on a high blood oxygen level in order to close. Bradykinin, a peptide released as the lungs fill with oxygen, works with the increased oxygen

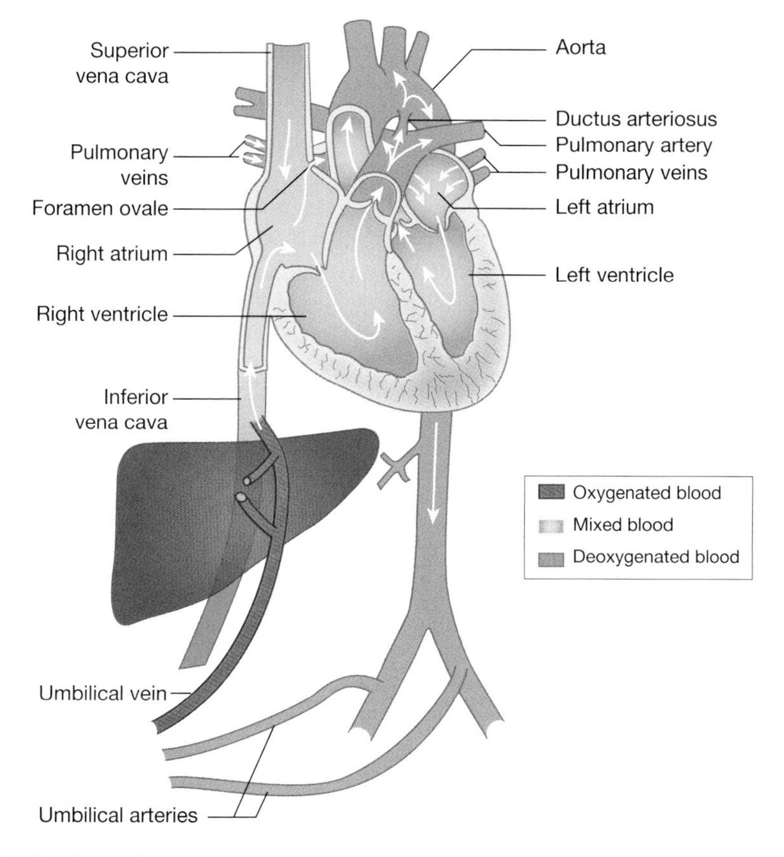


**Figure 9.1** The fetal circulation.



levels achieved from subsequent breaths of the newborn baby, to constrict and close the ductus arteriosus.

The functioning of the ductus venosus closes at birth, whereas structurally it closes between 3 and 7 days of life.

## Clinical application

The umbilical vein remains patent and available for catheterization for up to a week after birth. This is therefore a suitable source of venous access in emergency situations. Once inserted, the umbilical catheter can be used to administer intravenous fluids and medication. As with other forms of central venous access, complications may occur; these include:

- arterial injury – accidental puncturing of an adjacent artery;
- infection;
- thrombosis, specifically in the case of umbilical catheterization hepatic thrombosis;
- air embolism;
- haemorrhage due to vessel perforation.

Once a baby is stabilized, alternative sites are preferable and the umbilical catheter should be removed.

## Clinical application

### Patent ductus arteriosus and transposition of the great arteries

Development of the heart is a complex process, and many defects may arise during that time, the newborn circulation also undergoes several changes. One of those changes is closure of the ductus arteriosus; however, if this does not occur, the ductus arteriosus remains patent (open). An otherwise healthy baby may not initially show any signs or symptoms of having a patent ductus arteriosus, but children's nurses need to understand the changes that may occur.

A patent ductus arteriosus allows some of the oxygenated blood from the aorta to flow back into the pulmonary artery because there is higher pressure in the aorta. The extra blood returning to the lungs increases the pressure within the lungs; hence, the baby may have difficulty in inflating their lungs and, therefore, show signs of shortness of breath. An increase in respiratory effort will require an increased number of calories, but also a potential for poor feeding. The baby may show signs of poor weight gain or even weight loss.



However, in another congenital abnormality the presence of a patent ductus arteriosus is essential for the infant's survival. Transposition of the great arteries is an abnormality where the two major vessels leaving the heart do so from the wrong ventricles. In effect, the aorta is swapped (transposed) with the pulmonary artery (Figure 9.2). Blood returning from the lungs is immediately transported back to the lungs and blood from the body returns directly to the body. Unless other abnormalities are also present, a patent ductus arteriosus is the only way the oxygenated and deoxygenated blood can mix and provide some oxygen to the body. This condition is a cyanotic abnormality; the newborn infant will remain cyanosed (blue).

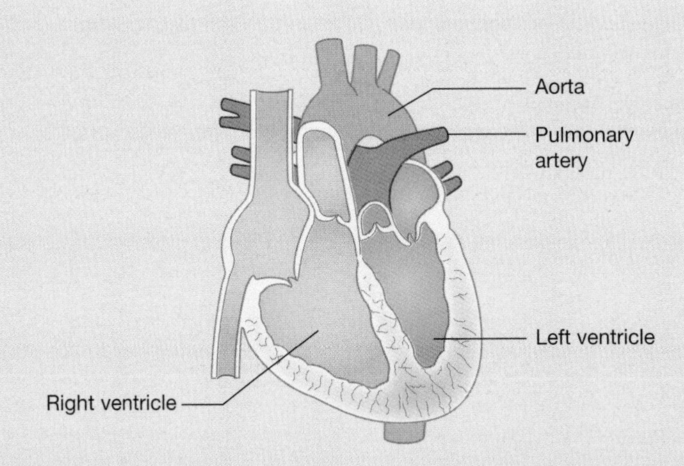


**Figure 9.2** Transposition of the great arteries. The aorta takes blood from the right ventricles and the pulmonary artery takes blood from the left ventricle.

# Position and size of the heart

The heart of an older child is located behind the sternum in the thoracic cavity in a space known as the mediastinum, between the two lungs, with two thirds being on the left side. The apex

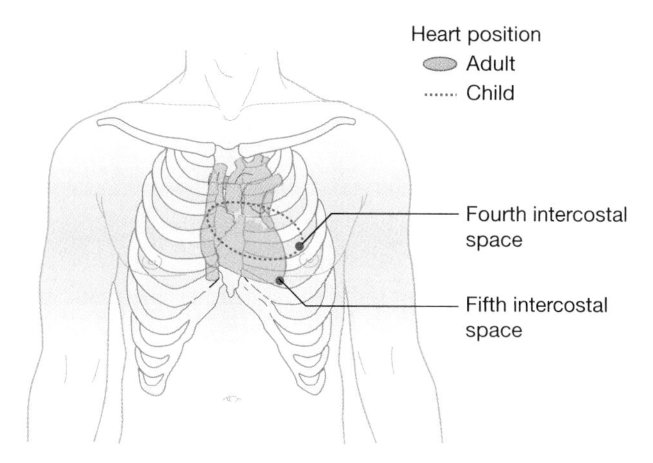




**Figure 9.3** Heart position – adult and child.

(pointed end) is positioned downwards and to the left and on a level with the fifth intercostal space. The size of the heart is approximately the size of the same person's clenched fist, or 0.5% of their body weight. Therefore, a newborn baby weighing 3.5 kg would have a heart weighing approximately 17 g; by the age of 1 year the size will have increased to approximately 50 g. The size and weight will continue to increase until adult size of between 250 and 350 g is reached, with a woman's heart weighing slightly less than a man's heart.

The heart of a newborn baby occupies 40% of the lung fields compared with 30% in adults and has a higher mass as a ratio of body mass (MacGregor, 2008). The heart of an infant lies more transversely than that of the adult, with the apex being near the fourth intercostal space; with lung expansion and over a period of time the heart gets pushed into the oblique position as seen in an adult (Figure 9.3). The heart is suspended in the pericardial sac attached to the aorta and pulmonary artery. This leaves the apex of the heart relatively free; during contraction of the ventricles the apex moves forwards and hits the left side of the chest wall. This characteristic thrust is generally known as the apex beat and can be heard on auscultation (listening) to the chest.

# Structures of the heart

## Heart chambers

The heart is divided into two sides, the left and right, and four chambers: the left and right atria and the left and right ventricles. The two sides of the heart are divided by the septum, a thin partition made of myocardium and covered in endocardium. These four chambers are responsible for pumping blood around the body; therefore, the heart is often referred to as a pump. However, it is actually two pumps:

- the deoxygenated blood pump;
- the oxygenated blood pump.

The first consists of the right atrium and the right ventricle pumping blood to the lungs. Deoxygenated blood is returned to the heart via the superior and inferior cavae and coronary sinus into the right atrium. From there it passes through the right atrioventricular (AV) valve (tricuspid valve) into the right ventricle. The blood enters the pulmonary artery and is pumped to the lungs for oxygenation.

The second pump consists of the left atrium and the left ventricle and pumps blood to the remainder of the body. Oxygenated blood is returned to the left atrium via the pulmonary veins. It passes across the left AV valve (bicuspid valve or mitral valve) and into the left ventricle. The blood is pumped to the remainder of the body via the aorta.



The two atria (singular atrium) are at the top or the base of the heart and are often known as the receiving chambers. The walls of the atria are relatively thin, as their action is to pump blood only into the ventricles. The two atria are separated by the interatrial septum; this prevents blood mixing across the two atria.

- The right atrium receives deoxygenated blood from the lower parts of the body through the inferior vena cava, from the upper body via the superior vena cava and from the heart itself via the coronary sinus.
- The left atrium receives oxygenated blood from the lungs via the four pulmonary veins.

The two atria are separated from the two ventricles by the right and left AV valves. These valves are one-way valves allowing blood to flow into the ventricles as the pressure increases within the atria and close as the pressure increases on contraction of the ventricles. The chordae tendinae prevent the valves opening backwards, and therefore prevent back flow of blood.

- The right AV valve, also known as the tricuspid valve, is made of three flaps or cusps.
- The left AV valve, also known as the biscuspid valve, is made of two flaps or cusps.

The two ventricles have thicker walls than the atria have, in relation to the increased work of pumping that they do, with the left having a thicker wall than the right. Both ventricles contract at the same time, which is after the simultaneous contraction of the atria. The ventricles are separated by the interventicular septum, which, as with the interatrial septum, prevents blood mixing between the two ventricles.

- The right ventricle receives deoxygenated blood from the right atrium, which is then pumped via the right and left pulmonary arteries to the lungs for oxygenation.
- The left ventricle receives oxygenated blood from the left atria, which is then pumped via the aorta to supply the upper and lower body.

Valves guard the entrance to the pulmonary artery and the aorta to prevent blood back flowing into the ventricles when the ventricular muscle relaxes.

- The pulmonary valve lies between the right ventricle and the pulmonary artery and consists of three semilunar (half-moon shaped) cusps and prevents back flow into the right ventricle.
- The aortic valve is also formed of three semilunar cusps and lies between the left ventricle and the aorta to prevent back flow into the left ventricle.

See Figure 9.4 for a diagrammatic representation of the blood flow through the heart.

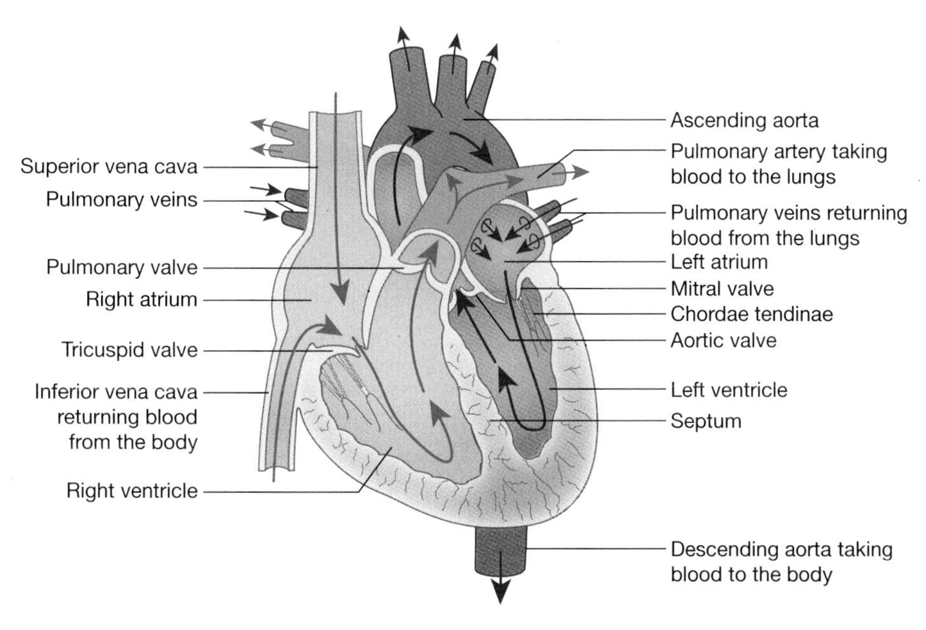


**Figure 9.4** Diagrammatic representation of the blood flow through the heart.



## Clinical application

Another common defect seen in the heart is the ventricular septal defect (VSD). This arises when the wall (septum) between the left and right ventricles does not form correctly, leaving a hole. The VSD is the most common of congenital heart defects, and although there is some association with genetic disorders such as Trisomy 21 (Down's syndrome), the majority are of unknown cause.

VSDs can occur at any point along the ventricular septum:

- **conoventricular** – a hole just below the pulmonary and aortic valves;
- **perimembranous** – a hole in the upper section of the septum;
- **inlet** – a hole in the septum near to where the blood enters the ventricles through the tricuspid and mitral valves;
- **muscular** – a hole in the lower, muscular part of the ventricular septum; this is the most common type of VSD.

Symptoms of and treatment for children with a VSD vary according to the size of the defect. As the pressure within the heart is higher in the left ventricle, blood will be shunted from the left into the right ventricle. Heart defects with a left-to-right shunt are known as acyanotic defects as no deoxygenated blood is ejected from the heart round the body.

Small holes may not cause any problems to the baby; however, larger holes will cause shortness of breath (especially when feeding), dyspnoea, tiredness and potentially poor weight gain if the baby is too tired to feed.

Treatment involves control of symptoms, such as medication to control any cardiac failure and adequate nutritional supplements; in the case of larger holes, open heart surgery to patch the defect may be required.

A VSD may also be seen as part of more complex heart conditions such as tetralogy of Fallot.

## Heart wall

The wall of the heart is made up of three layers:

- the outer layer is the pericardium;
- the middle layer is the myocardium;
- the inner layer is the endocardium.

**The pericardium** is a double-layered (the fibrous pericardium and the serous pericardium) membrane that covers the outside of the heart. The space between the two layers is filled with pericardial fluid and helps to protect or buffer the heart from external jerks and shocks.

- The fibrous pericardium is an elastic dense layer of connective tissue that acts to protect the heart, to anchor it in position and to prevent overstretching, and therefore overfilling of the heart with blood.
- The serous pericardium is further divided into two layers, both of which function to lubricate the heart with serous fluid and prevents friction during heart activity.
  - The outer parietal layer is attached to the fibrous pericardium.
  - The inner visceral layer is known as the epicardium when it comes into contact with the heart.



**The myocardium** is made of specialized cardiac muscle, which is striated and involuntary and makes up the majority of the mass of the heart. It is adapted to be highly resistant to fatigue, but is reliant on a high supply of oxygen to enable aerobic respiration and does not perform well in hypoxic conditions.

**The endocardium** is the inner layer and is a smooth membrane that allows smooth flow of blood inside the heart; it is connected seamlessly to the lining of the blood vessels which enter and leave the heart. The endocardium is thicker in the atria than in the ventricles (Figure 9.5).

## Clinical application

Endocarditis is a very rare condition in children. However, children who have a congenital heart defect have an interruption to the smooth lining of the heart, the endocardium; this gives bacteria a greater opportunity to adhere and multiply, causing an infection of the endocardium – endocarditis. Whilst extremely rare, endocarditis is a very serious disease, often complicated by a delay in diagnosis. Symptoms begin with a gradual onset of malaise and fever, followed by a new or changing heart murmur, petechiae (small red/purple rash) and hepatosplenomegaly (enlarged liver and spleen) (Rudolph *et al.*, 2011).

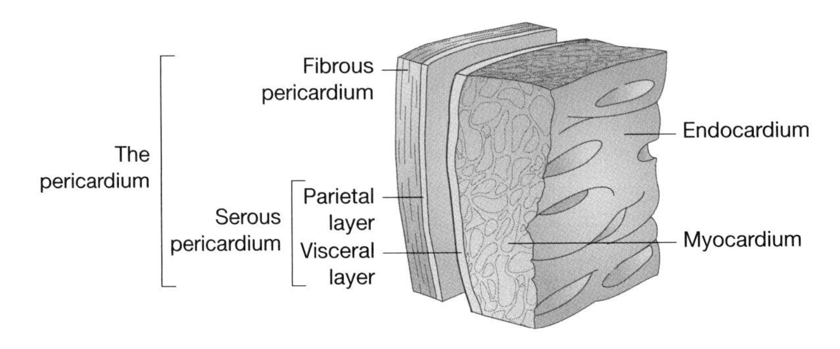


**Figure 9.5** The layers of the heart wall.

# The electrical pathway through the heart

## The sinoatrial node

In order to pump blood round the body the heart needs a source of power in the form of electricity. The sinoatrial (SA) node is a natural pacemaker for the heart, and regularly, according to the needs of the body, will release electrical stimuli. The SA node lies in the wall of the right atrium near the opening of the superior vena cava. The SA node is a mass of specialized cells that are electrically unstable; therefore, they discharge or depolarize regularly, causing the atria to contract (Waugh and Grant, 2010). Although the discharge arises from the SA node in the right atrium it quickly spreads through the myocardium of the right and left atria, allowing contraction to occur. Depolarization is followed by a period of recovery or repolarization, but the instability of the cells immediately triggers a further depolarization, and hence the heart rate is set.



## The atrioventricular node

The electrical impulse then reaches the AV node situated in the wall of the atrial septum near to the AV valves. There is a slight pause allowing the atria to finish contracting and empty the blood into the ventricles and for the AV valves to close. The atria immediately start filling again at the same time as the electrical impulse is spread to the ventricles.

## The atrioventricular bundle

The impulse passes through the AV bundle, sometimes known as the bundle of His, and crosses into the ventricular septum where it divides into the right and left bundle branches. The bundle branches further divide into fine fibres known as Purkinje fibres. The AV bundle, the bundle branches and the Purkinje fibres transmit the electrical impulse from the AV node down to the apex of the heart, where ventricular contraction starts, causing the blood to flow from the ventricles into either the pulmonary artery or the aorta. As the ventricles empty and the valves close, the AV node will repolarize ready for the next discharge of electrical impulse. Figure 9.6 illustrates the electrical pathway through the heart.

Therefore, the stages of a single heart beat are:

- SA node depolarization;
- AV depolarization;
- SA node and AV node repolarization.

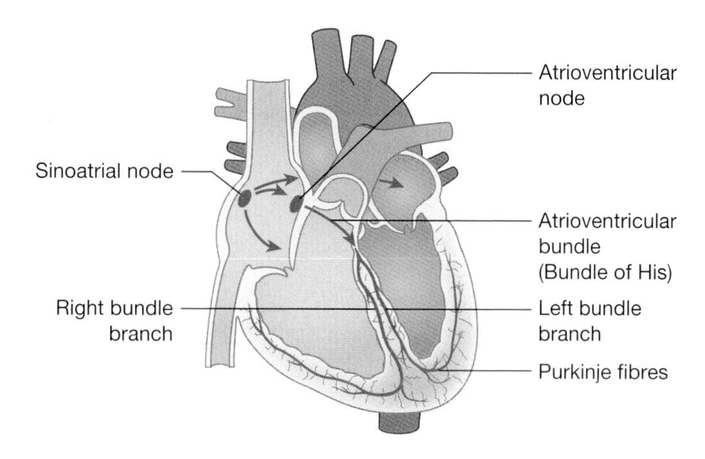


**Figure 9.6** The electrical pathway through the heart.

If the heart beat is 60 beats per minute, each of these stages takes less than a third of a second; this is greatly increased in the young child and babies, where the heart rate may be an average of 120 beats per minute.

# Electrocardiogram

The electrical activity through the heart can be detected by attaching electrodes to the patient and to an electrocardiogram (ECG) monitor. The normal ECG pattern shows five waves, which are known as the P, Q, R, S and T waves.

**P wave** – the P wave shows the electrical impulse at it leaves the AV node and sweeps across the two atria.



**Q, R, S waves or complex** – this represents the spread of the electrical impulse from the AV node through the AV bundle and the Purkinje fibres and also the electrical activity of the ventricular muscle.

The pause between the P wave and the Q, R, S complex is the pause that allows the atria to completely contract before the ventricles begin to contract.

**T wave** – the T wave represents the relaxation of the ventricles (ventricular repolarization).

The atrial repolarization occurs during the ventricular contraction (Q, R, S complex) and therefore cannot be seen on the ECG (Figure 9.7).

## Clinical application

Whilst a pacemaker is more commonly associated with adults, children in a small number of cases may also require a pacemaker. A pacemaker is a small battery-operated device that is surgically fitted under the skin and muscle of the abdomen of younger children or the chest wall in an older child. Wires are connected to the heart muscle and impulses travel from the pacemaker to the heart muscle to stimulate contraction.

Children may need a pacemaker if their SA or AV nodes do not function adequately to ensure the heart beats enough times to provide a sufficient blood flow to the body. Causes of SA or AV dysfunction may be congenital or acquired. Congenital causes may include association with other cardiac disease, whereas acquired may be due to infection or injury (including during surgery).

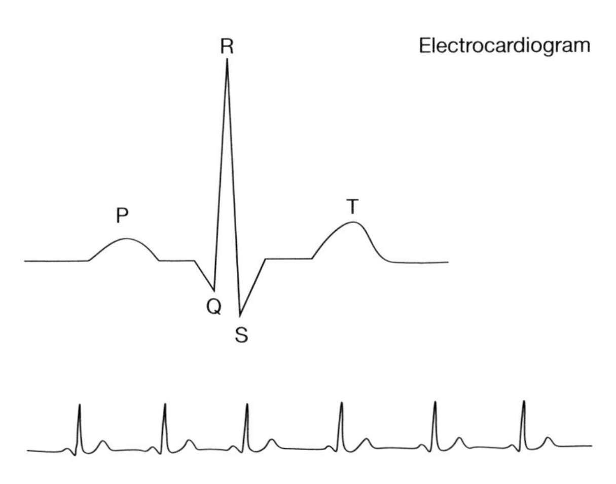


**Figure 9.7** The ECG tracing.

# The cardiac cycle

The cardiac cycle refers to the complete filling and emptying of blood into and out of the heart or from the beginning of one heart beat to the beginning of the next. The mechanism of the cardiac cycle is controlled by:

- systole – the contraction of either the atria or the ventricles;
- diastole – the relaxation of either the atria or the ventricles.

It is important to remember that the two sides of the heart work simultaneously, meaning that the two atria work at the same time followed by the two ventricles. There are various stages involved in the cardiac cycle:



1. Deoxygenated blood enters the right atrium via the superior and inferior vena cavae at the same time as oxygenated blood is entering the left atrium via the four pulmonary veins from the lungs. Up to 80% of the blood will flow passively through the AV valves into the right and left ventricles. The ventricles are relaxed (**ventricular diastole**) and the pressure within the ventricles is low, whereas the pressure in the pulmonary artery and aorta is higher; therefore, at this stage, the pulmonary and aortic valves are closed and the blood is held within the ventricles.
2. The electrical stimulus arising from the SA node triggers a wave of contraction through the myocardium of the atria, causing them to contract. The increased pressure within the atria causes them to complete the emptying of the atria and the filling of the ventricle (**atrial systole**).
3. The brief pause in electrical activity allows the atria to empty completely before the electrical activity continues from the AV node down the bundle of His, the bundle branches and the Purkinje fibres to cause a wave of contraction through the myocardium of the ventricles. At this stage the pressure within the ventricles in now higher than in the atria and so the AV valves close, preventing back flow into the atria. However, the pressure in the pulmonary artery and aorta remains higher than in the ventricles; hence, the pulmonary and aortic valves remain closed. The volume of blood within the ventricles is now constant (**isovolumetric contraction phase**); as the ventricular walls contract, the pressure within the chambers rises; the valves will open when the pressure is greater than that within the pulmonary artery and aorta, allowing blood to flow into the major vessels.
4. As the blood flows from the ventricles they begin to relax (**early ventricular diastole**); the pressure decreases rapidly. As the AV valves remain closed the pressure within the ventricles quickly becomes less than the pressure in the pulmonary artery and the aorta, and again the semilunar valves close (**isovolumetric relaxation phase**).
5. Whilst the ventricles have been contracting and emptying, the atria have been relaxing and refilling (**atrial diastole**). Before the next cycle begins there is a period of complete relaxation (**complete cardiac diastole**) during which time the atria and the ventricles are relaxed and the myocardium recovers.

Whilst this appears to be a long process, assuming a heart rate of 75 beats per minute the average cardiac cycle would take 0.8 s (Marieb, 2001); this is shortened to 0.5 s for a child with a heart rate of 120 beats per minute.

## Heart sounds

The familiar 'lub–dub' sounds that can be heard when using a stethoscope to listen to the heart beat are caused by the closing of the valves. The first, the 'lub', tends to be the louder of the two and is heard when the AV valves close at the beginning of ventricular systole. The second sound, the 'dub', is caused by the closure of the pulmonary and aortic valves at the end of the ventricular systole.

## Blood pressure

The changes in pressure within the chambers of the heart give rise to an individual's blood pressure. Systemic arterial blood pressure is the pressure exerted by the circulating blood on the walls of the blood vessels; it is essential to maintain blood pressure to allow blood to flow to and from the organs of the body. As the left ventricle contracts (ventricular systole), the blood is pushed into the aorta producing pressure within the arterial system; this is the systolic blood pressure. During the complete cardiac diastole the pressure within the arteries is much lower, and this is known as the diastolic blood pressure. The blood pressure is usually measured from the upper arm and expressed as the systolic blood pressure over the diastolic blood pressure in millimetres of mercury (mmHg).



Blood pressure varies from person to person, and differences are seen according to gender, time of day, age and general health. Children's blood pressure is related to their age and height. Within the unborn baby it is the fetal heart that determines blood pressure, not that of the mother, and the pressure increases with gestation.

# Clinical application

### Blood pressure

The arm is the ideal limb for taking a child's blood pressure, although the lower leg may be used in an infant. The arm should be level with the heart and well supported. To ensure an accurate reading, the correct cuff size must be used. The bladder within the cuff must cover 80% of the circumference of the limb, whilst there must also be adequate overlap of the cuff to ensure 100% of the limb is covered (Figure 9.8). The cuff should also cover two-thirds of the length of the limb. Inaccurate cuff size will give rise to inaccurate blood pressure readings. Electronic blood pressure monitors are very sensitive to movement and may not give accurate readings.

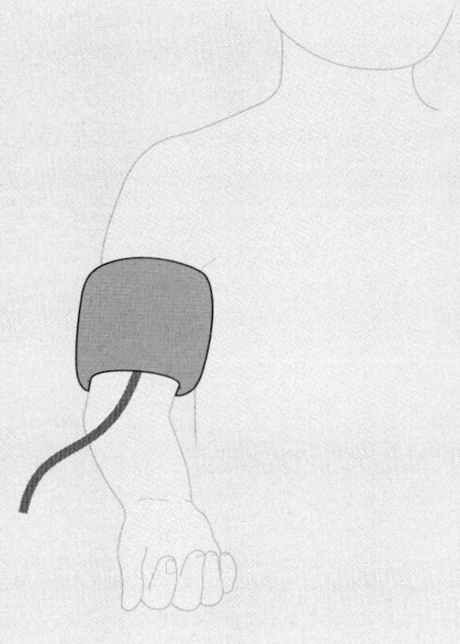

**Figure 9.8** The blood pressure cuff.

# Cardiac output

The amount of blood ejected by the heart in 1 min is known as the cardiac output, whereas the amount of blood ejected from each ventricle on each contraction is the stroke volume. In a healthy adult the stroke volume is estimated at 70 mL; in children, the body surface area is an indication of the stroke volume.

Cardiac output = stroke volume × heart rate

In order for the tissues and organs of the body to receive a supply of oxygen and nutrients an adequate cardiac output is required. Cardiac output can be increased when there is greater demand from the body for oxygen delivery (Farley *et al.*, 2012).

**Stroke volume** is dependent on three factors.



1. **Preload** – the amount of myocardial pressure or distension, sometimes referred to as the stretch, prior to contraction. An increase in the volume of blood, and therefore an increase in the distension of the ventricle, will increase the force of contraction and hence increase the cardiac output. The volume of blood in the ventricles prior to contraction is the **ventricular end-diastolic volume**.
2. **Contractility** – the ability of the cardiac muscle to contract. Increased contractility of the muscle will increase cardiac output.
3. **Afterload** – the resistance against which the ventricles have to pump when ejecting blood (Carson, 2005). This is dependent on arterial blood pressure and vascular tone. More simply, this means a high blood pressure will increase afterload, which in turn decreases stroke volume. A decrease in stroke volume indicates that some blood remains in the ventricles, ultimately lowering the cardiac output.

There are many factors which can affect cardiac output and / or stroke volume. Increasing the heart rate through any means for example exercise or drugs will increase the cardiac output. However an extreme tachycardia may result in the individual being unable to fill the heart adequately and result in reduced cardiac output. A slower heart will allow for time for the heart to fill, increasing the preload and hence increasing cardiac output; however a substantial decrease in heart rate will decrease cardiac output. The sympathetic and parasympathetic nervous system also affects the heart rate. The sympathetic nerve axon works to release norepinephrine which in turn increases the heart rate, whilst the parasympathetic nervous system via the vagus nerve releases acetylcholine which in turn decreases the heart rate.

Stroke volume is also affected by:

- decreased blood volume, which results in decreased venous return to the heart and hence a decrease in preload and decrease in cardiac output;
- cardiac anomalies, such as a VSD, which will affect the filling capacities of the ventricles and hence reduce the preload.

## Clinical application

Children may suffer from a decreased circulatory volume for several reasons.

**Haemorrhage** may occur as a result of trauma or from, for example, a post-tonsillectomy bleed. Any reason causing a substantial loss of blood will result in a decreased venous return. The younger

(*Continued*)

the child, the less circulatory volume available prior to haemorrhage; therefore, small losses can be critical in babies or young children.

A **dehydrated** child will also suffer from a loss of circulatory volume; causes will differ, but may be associated with diarrhoea and vomiting, burns or an inability to take oral fluids.

A decrease in circulatory volume will affect the preload phase of the stroke volume and will result in a decrease in cardiac output.

Treatment needs to include urgent replacement of lost fluid as well as treating the underlying cause, as even a relatively small loss of fluid can have a big impact for infants and young children.



# Factors affecting the heart rate

As already mentioned, the cardiac output is reliant not only on the stroke volume but also on the heart rate. The heart rate is determined either intrinsically or extrinsically. Intrinsic management is the heart's own pacemaker, the SA node, which ensures self-regulation and maintenance of rhythm. Extrinsic factors include the nervous system and hormonal influences.

**Autonomic nervous system** – the balance of the sympathetic and parasympathetic nervous system is essential to the regulation of the heart rate. The sympathetic nerve fibres release norepinephrine when triggered by a stimulus such as exercise, drugs or stress. The autonomic nervous system is located within the medulla oblongata in the lower portion of the brain stem, and two nerves link the medulla oblongata to the SA node: the accelerator nerve (sympathetic system) and the vagus nerve (parasympathetic system).

**Hormones** – sympathetic nervous system activity also causes the release of epinephrine and norepinephrine from the adrenal medulla, the inner portion of the adrenal gland, which is situated above the kidneys. Epinephrine enters the bloodstream and is delivered to the heart where it binds with SA node receptors. Binding of epinephrine leads to further increase in the heart rate.

Thyroxine from the thyroid gland acts to maintain the body's metabolic rate; an increase in thyroxine will increase the heart rate.

**Other factors** that affect the heart rate include:

- changes to the electrolyte balance – for example, an increase in potassium depresses the cardiac function;
- hypoxia and hypercapnia levels, which increase the heart rate;
- exercise, which increases the muscles' requirements for oxygen and nutrients and therefore requires an increased heart rate to deliver the requirements;
- emotions – for example, fear or excitement will stimulate the sympathetic nervous system and ultimately increase the heart rate;
- gender differences – female hearts beat faster than those of males;
- age – babies and younger children have an increased heart rate;
- temperature changes.

# Conclusion

The heart is in effect two pumps: one receives deoxygenated blood from the body and pumps it to the lungs, whilst the second receives oxygenated blood from the lungs and delivers it to the body.

The four chambers of the heart, the left and right atria and the left and right ventricles relax and contract and relax to allow the blood to flow, whilst the valves – the tricuspid and pulmonary valves on the right and the mitral and aortic on the left – ensure the blood is propelled in the right direction and prevent backflow.

Electrical activity arising from the SA node and travelling through the AV node to the bundle of His and finally to the Purkinje fibres provides the stimulus for the heart to beat.

An adequate cardiac output is required to ensure the body receives enough oxygen and nutrients for its needs; this is dependent on the stroke volume and the heart rate.

The heart rate varies according to the needs of the body and is affected by the autonomic nervous system, hormones and other external factors.



# Conditions to consider in relation to the cardiac system

- Congenital heart defects
  - atrial septal defect
  - VSD
  - patent ductus arteriosus
  - transposition of the great arteries
  - tetralogy of Fallot
- Endocarditis
- Cardiac failure
- Supraventricular tachycardia
- Bradycardia.

# Activities

**Now review your learning by completing the learning activities in this chapter. The answers to these appear at the end of the book. Further self-test activities can be found at www.wileyfundamentalseries.com/childrensA&P.**

## Complete the sentences

1. Blood returning from the body enters the heart via the inferior and superior _______ _______, into the _______ atrium across the ___________ valve to the _________ ventricle and then leaves to go to the lungs via the pulmonary _________.
2. Many of the structures within the heart are known by more than one name; complete the following:
   right atrioventricular valve = ___________ __________
   left atrioventricular valve = ___________ __________
   atrioventricular bundle = __________ ___ _________.
3. The structure within the heart which connects the pulmonary artery to the aorta is the ___________ ___________.

4. Connecting the right atrium to the left atrium is the __________ __________.
5. The work of the liver in the unborn baby is carried out by the ____________; therefore, blood bypasses the liver via the __________ ____________.

## True or false?

1. The human heart is referred to as a double circulation?
2. Normal adult circulation would be suitable for a fetus?
3. The sympathetic nerves decrease the heart rate and, therefore, the cardiac output?



## Wordsearch

There are several words linked to this chapter hidden in the following grid. Can you find them? A tip – the words can go from up to down, down to up, left to right, right to left, or diagonally.

| | | | | | | | | | | | | | | |
|---|---|---|---|---|---|---|---|---|---|---|---|---|---|---|
| M | Q | A | T | N | P | L | A | C | E | N | T | A | H | S |
| A | R | T | E | R | Y | Y | A | I | X | O | P | Y | H | T |
| T | U | P | T | U | O | N | Y | H | E | N | I | R | G | R |
| L | U | B | D | U | B | H | W | L | P | V | A | W | B | O |
| M | U | I | D | R | A | C | O | Y | M | N | E | U | M | K |
| J | Y | Y | J | J | T | T | B | K | U | L | N | I | D | E |
| K | C | Q | R | K | S | P | X | L | Q | D | Z | A | N | V |
| S | I | N | O | A | T | R | I | A | L | N | O | D | E | O |
| C | T | F | I | Z | N | M | W | E | J | L | C | N | A | L |
| A | O | D | J | R | E | O | O | A | E | E | T | T | L | U |
| R | N | A | Q | S | Q | F | M | R | M | R | R | T | H | M |
| D | A | M | R | A | H | A | P | L | I | I | A | G | D | E |
| I | Y | K | P | I | T | R | I | C | U | S | P | I | D | Q |
| A | C | O | S | Q | B | J | L | M | K | P | I | P | D | Z |
| C | Y | Z | O | G | W | E | D | I | P | S | U | C | I | B |

## Conditions

The following table contains a list of conditions. Take some time and write notes about each of the conditions. You may make the notes taken from text books or other resources; for example, people you work with in a clinical area or you may make the notes as a result of people you have cared for. If you are making notes about people you have cared for you must ensure that you adhere to the rules of confidentiality.

| Condition | Your notes |
|---|---|
| Congestive cardiac failure | |
| Cardiac arrest | |
| Coarctation of the aorta | |
| Patent ductus arteriosus | |
| Pulmonary atresia | |



## Glossary

**Acetylcholine:** a neurotransmitter.

**Acyanotic:** not cyanosised.

**Air embolism:** air trapped in a blood vessel.

**Alveoli:** tiny air sacs in the lungs for gas exchange.

**Aorta:** major blood vessel supplying blood from the left ventricle to the body.

**Apex:** the lowest part of the heart.

**Atria:** upper two chambers of the heart.

**Atrial systole:** contraction of the atria.

**Atrioventricular bundle:** part of the conducting system of the heart.

**Atrioventricular node:** part of the conducting system of the heart, a small mass of tissue located in the wall of the atrial septum.

**Auscultation:** listening to specific body sounds; for example, the apex beat of the heart.

**Autonomic nervous system:** part of the peripheral nervous system, which acts as a control system.

**Atrium:** singular version of atria.

**Bundle of His:** alternative name for the atrioventricular bundle.

**Chordae tendineae:** cord-like tendons connected to the heart valves to prevent them opening the wrong way.

**Congenital:** something existing at birth.

**Cyanosis:** bluish discoloration of the skin caused by a lack of oxygen.

**Diastole:** relaxation of the atria or the ventricles.

**Ductus arteriosus:** a small vessel connecting the pulmonary artery to the aorta preventing blood flow to the lungs.

**Ductus venosus:** a blood vessel that connects the umbilical vein to the inferior vena cava.

**Electrocardiogram:** tracing showing the electrical activity of the heart.

**Endocardium:** the lining of the inside of the heart.

**Embryo:** an organism in the very early stages of development up to 10 weeks post fertilization in a human.

**Fetus:** a developing baby from 11 weeks to birth.

**Foramen ovale:** a flap between the right and left atria allowing blood flow to bypass the right ventricle.

**Gestation:** the period of time from fertilization to birth.

**Hypoxia:** low levels of oxygen.

**Hypercapnia:** high concentration of carbon dioxide.

**Inferior vena cava:** major blood vessel returning deoxygenated blood from the lower part of the body to the heart.

**Iso-:** remaining constant.

**Lymphatic system:** the transport system within the body.

**Mediastinum:** the area in the chest between the lungs that contains the heart, the windpipe, and the oesophagus.

**Myocardium:** the contractile muscle of the heart.

**Noradrenaline:** a hormone secreted to increase blood pressure and heart rate.

**Pacemaker:** an artificial pacemaker is a battery-operated device that delivers electrical impulses to the heart muscle.

**Patent (persistent) ductus arteriosus:** the ductus arteriosus does not close as expected following birth.

**Pericardium:** the outer layer of the heart.

**Placenta:** an organ attached to the wall of the uterus to provide oxygen and nutrients to the baby.

**Pulmonary:** relating to the lungs.

**Purkinje fibres:** part of the conducting system of the heart.

**Semilunar:** valves shaped like half a moon.

**Sinoatrial node:** a small mass of specialized cells that depolarizes regularly to set the heart beat.

**Superior vena cava:** major blood vessel returning deoxygenated blood from the upper body to the heart.

**Systole:** contraction of the atria or ventricles.

**Tetralogy of Fallot:** congenital heart defect consisting of four anomalies.

**Thrombosis:** a blood clot within a vessel.

**Umbilical catheterization:** using the umbilical vein as a form of central venous access.

**Umbilical cord:** the cord that connects the baby to its mother.

**Ventricle:** lower two chambers of the heart.

# References

Carson, P. (2005) Development of the cardiovascular system. In: Chamley, C., Carson, P., Randall, D., Sandwell, M. (eds), *Developmental Anatomy and Physiology of Children*, Elsevier Churchill Livingstone, Edinburgh.

Farley, A., McLafferty, E., Hendry, C. (2012) The cardiovascular system. *Nursing Standard*, **27** (9), 35–39.

Macgregor, J. (2008) *Introduction to the Anatomy and Physiology of Children: A Guide for Students of Nursing, Child Care and Health*, 2nd edn, Routledge, London.

Marieb, E.N. (2001) *Human Anatomy and Physiology*, 5th edn, Pearson Benjamin Cummings, San Francisco, CA.

Rudolph, M., Lee, T., Levene, M. (2011) *Paediatrics and Child Health*, 3rd edn, Wiley-Blackwell, Chichester.

Waugh, A., Grant, A. (2010) *Ross and Wilson Anatomy and Physiology in Health and Illness*, 11th edn, Churchill Livingstone, Edinburgh.

# Chapter 10

# The respiratory system

Elizabeth Akers

*Great Ormond Street Hospital, London, UK*

## Aim

To understand the anatomy and physiology of the respiratory system and how both physical development and illness can affect its function and the effect this has on the entire body.

## Learning outcomes

On completion of this chapter the reader will be able to:

- Recognize the anatomy of the respiratory system.
- Understand gaseous exchange.
- Understand how pathophysiological processes within the respiratory system affect the child or infant's overall condition.
- Recognize three common respiratory illnesses and understand key aspects of their nursing care.

## Test your prior knowledge

- List the three main regions of the respiratory system in order.
- Describe the main functions of the nose and upper airway.
- Describe the main functions of the lower airway.
- What is the main purpose of respiration?
- Describe in brief terms gaseous exchange.
- What is oxygen used for?
- What waste products are produced in respiration?
- What are three presenting symptoms of respiratory distress?
- List three common respiratory conditions.
- Describe some changes that occur in the respiratory system as the infant grows.

*Fundamentals of Children's Anatomy and Physiology: A Textbook for Nursing and Healthcare Students*, First Edition. Edited by Ian Peate and Elizabeth Gormley-Fleming.

Companion website: www.wileyfundamentalseries.com/childrensA&P

# Body map

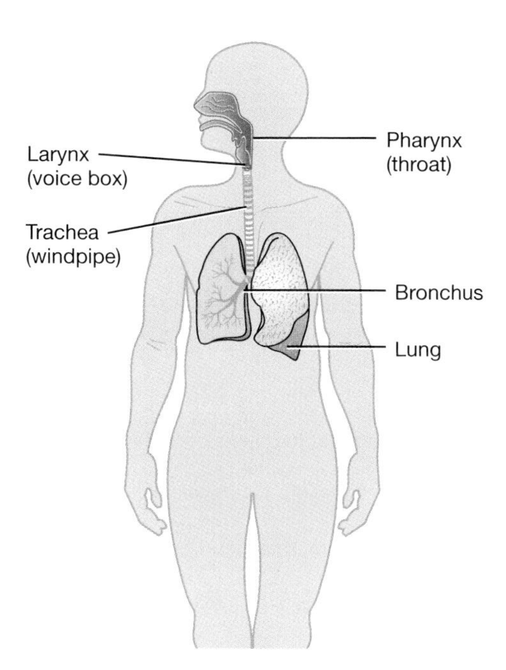




# Introduction

Often we breathe without thinking about it. We rarely think of the process of respiration constantly going on within us, as it is unseen and not usually consciously felt. Of course, without it we would die. All of us at some time have had trouble breathing; we do not take breathing for granted.

Humans need a constant supply of oxygen to sustain life. Respiration, the cycling of oxygen and carbon dioxide between the body and the atmosphere, is a complex process that often takes place without any conscious effort, requiring the coordination of many organs and involving every cell in the body (Winston, 2004).

A good understanding of the respiratory system and its crucial role within the body is essential in nursing. Early recognition of respiratory compromise and subsequent management will prevent the majority of cardiorespiratory arrests in children and infants. Using a standardized assessment tool to assess respiratory function is an essential part of the care of an infant or child and should be incorporated into standard nursing practice (Naddy, 2012). This chapter will describe the anatomy and physiology of the infant and child's respiratory system.

The lungs are relatively immature at birth and continue to develop into childhood. Children have small resting lung volumes and lower oxygen reserves and a higher rate of oxygen consumption, resulting in rapid deterioration when respiratory function is compromised. Infants in particular are increasingly susceptible to respiratory illness with more severe presentations and comparatively high levels of both morbidity and mortality (Tregoning and Schwarze, 2010). Having an awareness of the anatomical differences of the airway allows those caring for infants and children to anticipate complications should they become unwell.

The respiratory system is made up of two parts: the upper and lower respiratory systems; see Figure 10.1 for an overview of the respiratory system and associated structures.

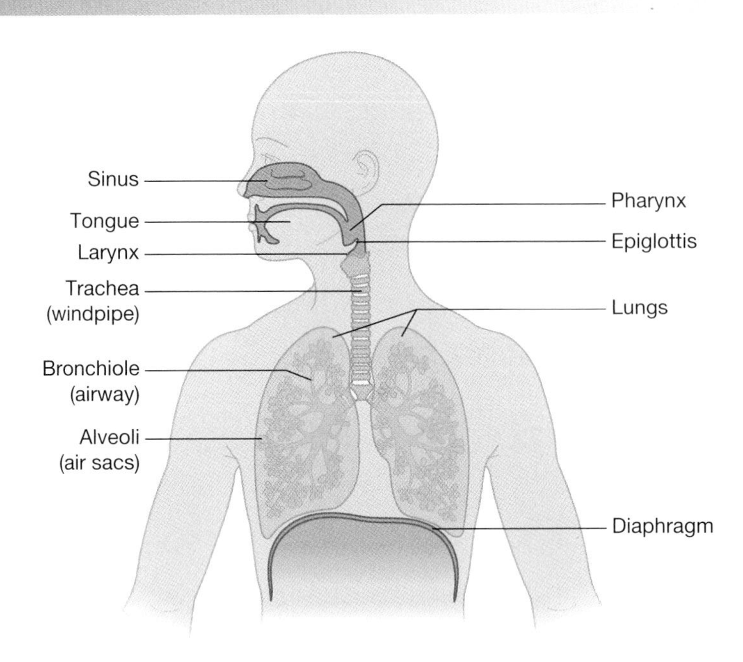




**Figure 10.1** An overview of the respiratory system and associated structures.

The upper respiratory system comprises the nose and oropharynx. The lower respiratory system comprises the larynx, trachea, bronchi, alveoli and the lungs. The primary function of the respiratory system is gas exchange; however, it also carries out other tasks, such as metabolism of some compounds, some filtration of the circulating blood and it can act as a reservoir for blood, but its most significant role is in gaseous exchange. To achieve gaseous exchange, air is conducted through the upper respiratory tract, where it is filtered, warmed and moistened before it moves into the lower respiratory tract where gaseous exchange takes place (Tortora and Derrickson, 2006).

This chapter focuses on development of the respiratory system from birth. However, with the increasing prevalence of infants being born prematurely, it is important to consider the complications to respiratory function that can result from premature birth. In managing the premature neonate, the following strategies have been employed to minimize complications during this fragile time: milder ventilation methods, administration of exogenous surfactant and administration of steroids to the mother. Despite these efforts, the prevalence of pulmonary disease among survivors of prematurity has not decreased. Symptoms of prematurity are being observed well into childhood, including reduced pulmonary function and lung capacity, which suggests that disrupted lung development may be permanent in this group. Not just viewing reduced respiratory function as a complication of prematurity is important in nursing the whole child; premature infants are also at risk of retinopathy of prematurity and neurodevelopmental delay (Buczynski *et al.*, 2012).

## The airway

Figure 10.2 provides a comparison of the adult and child airways.

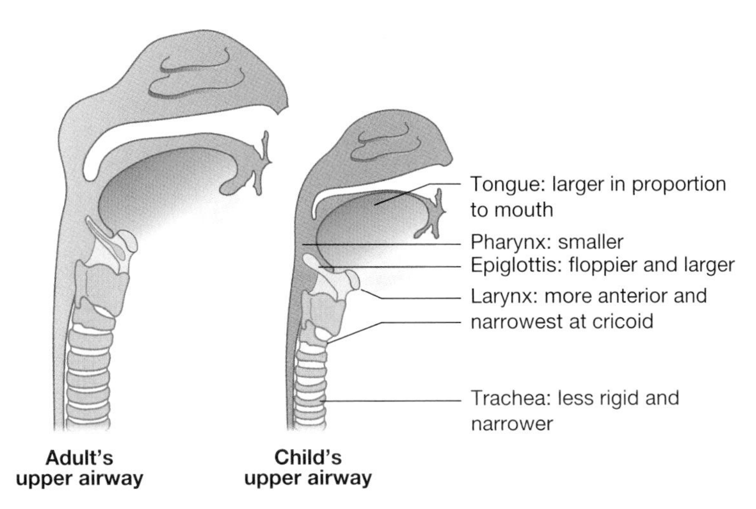


Figure 10.2 The child and adult airway.



## The nose

The functions of the nose are to warm, moisten and filter incoming air, to detect olfactory stimuli (smell) and to modify speech vibrations – providing resonance to the voice. The nose is formed of an external structure, consisting of a supporting framework of bone and hyaline cartilage covered with skin and lined with a mucous membrane. The internal structure is a large cavity in the anterior aspect of the skull and is lined with muscle and mucous membranes. These mucous membranes contain coarse hairs that filter larger particles. The opening of the nose is called the nares or conchae. Ducts from the parasinuses and nasolacrimal ducts also open into the nose. The internal nares are subdivided into the superior, middle and inferior meatuses. This area is also lined with a mucous membrane. The structural arrangement and the mucous membrane help to prevent dehydration by trapping water droplets during exhalation. The olfactory receptors lie in the superior nasal conchae and adjacent septum, called the olfactory epithelium. The space within the internal nose is called the nasal cavity; it is divided by the nasal septum formed primarily of hyaline cartilage. In infants and young children the small nasal passages become easily blocked with secretions, further compromising the airway patency when they are unwell (Tortora and Derrickson, 2006; Stoelting and Miller, 2007, cited in Crawford (2011a)).

## The pharynx

The pharynx (Figure 10.3) can be divided into three anatomical regions: the nasopharynx, the oropharynx and the laryngopharynx. The muscles of the entire pharynx lie in two layers: an outer circular layer and an inner longitudinal layer (Tortora and Derrickson, 2006).

The nasopharynx lies in a posterior position to the nasal cavity and extends to the soft palate. It has five openings: two internal nares, two openings that lead to the auditory or Eustachian tubes (pharyngotympanic) and the opening to the oropharynx. The posterior wall also contains the pharyngeal tonsil. The nasopharynx receives air containing dust trapped by mucus. The lining is made of pseudostratified ciliated columnar epithelium, and the cilia move the mucus down towards the most inferior aspects of the pharynx. The nasopharynx is also linked to the

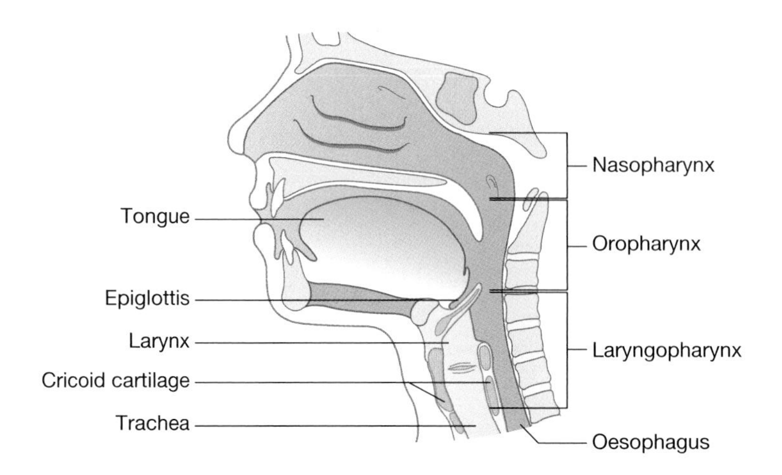




**Figure 10.3** The nasopharynx, the oropharynx and laryngopharynx and associated structures.

middle ear, where exchanges of air take place, ensuring pressure within pharynx and middle ear is equal (Tortora and Derrickson, 2006).

The oropharynx lies in a posterior position to the oral cavity and extends from the soft palate to the level of the hyoid bone. It has only one opening, the fauces (throat) opening from the mouth. This area serves both the respiratory and digestive systems as it is where food, drink and air pass. Two pairs of tonsils (palatine and lingual) are found here. In infants, the tongue, which is large in relation to the oral cavity, can also obstruct the airway when consciousness is impaired, and this needs to be considered in managing the drowsy infant, and also during resuscitation. The laryngopharynx, the inferior region, begins at the level of the hyoid bone. It opens into the oesophagus posteriorly and the larynx anteriorly (Tortora and Derrickson, 2006; Stoelting and Miller, 2007, cited in Crawford (2011a)).

## Clinical application

When a child or infant is found to be unconscious or unresponsive one of the first considerations is hypoxia. To address this immediately, consider an 'airway manoeuvre' by placing the child or infant's head and neck into an appropriate position when applying oxygen.

## The larynx

The larynx is a complex structure that permits the trachea to be joined to the pharynx as a common pathway for respiration and digestion. A key function is protecting the lungs during swallowing. It is also essential in clearance of secretions through coughing and in the production of sound. The infant larynx is positioned higher in the neck than the adult larynx is, located in the region of the first cervical vertebrae (Figure 10.3) (Tortora and Derrickson, 2006).

The larynx, or voice box, is a short passage linking the laryngopharynx with the trachea. It lies in the midline of the neck, anterior to the oesophagus in the region of the third to fourth

(C3–C4) cervical vertebrae in infants, lowering to the fourth to sixth cervical vertebrae (C4–C6) by adulthood. The larynx of an infant is cone shaped at the top with the cricoid cartilage tilting posteriorly. The infant's vocal cords are shorter and the epiglottis is narrower, hanging over the larynx. The axis of both the respiratory and digestive systems allows simultaneous breathing and swallowing in newborns (Tortora and Derrickson, 2006; Savković *et al.*, 2010; Crawford, 2011a,b).

The narrow dimensions of the larynx mean that even a minor obstruction in the infant can be life threatening, unlike in the adult. The narrowest portion of the airway in the older child and adult is the glottic aperture, while the narrowest part of the airway in the infant is the subglottis. A diameter of 4 mm is considered the lower limit of normal in a full-term infant and 3.5 mm in a premature infant. The vocal cords of the neonate are usually 6–8 mm long, increasing to 7–9 mm wide and 11 mm long or approximately one-third the size of an adult. An awareness of the smaller dimensions and less rigid structure of the infant's airway is crucial, as is an appreciation that its large occiput and relatively short neck can result in neck flexion leading to airway compromise when the infant is unwell or when consciousness is impaired (Stoelting and Miller ,2007, cited in Crawford (2011a)).



The wall of the larynx is made of three pieces of cartilage: the thyroid, epiglottis and cricoid cartilages. The arytenoid cartilage, which is paired, is significant owing to their role in changing the position and tension of the vocal folds or true vocal chords. In infants and children, the cricoid ring is a complete ring of cartilage and the narrowest point of the upper airway. The thyroid cartilage (Adam's apple) consists of two fused plates of hyaline cartilage forming the anterior wall of the larynx, giving it, in adults, a triangular shape. This is usually larger in men as it is due to the influence of male hormones during puberty (Tortora and Derrickson, 2006; Savković *et al.*, 2010; Crawford, 2011a,b). Owing to the age-dependent mineralization and ossification changes that take place in the bone and cartilage tissue of the larynx, radiological images should be used with caution as evaluation of this type is difficult in clinical practice if there are concerns about possible aspiration or inhalation (Turkmen *et al.*, 2012).

## Cartilage within the larynx

The epiglottis in infants is an elastic cartilage characterized by thick mucosa and an abundance of serosal glands on both sides. It is proportionally narrower than that of an adult and assumes either a tubular form or the shape of the Greek upper case letter Omega (Ω). The central role of the epiglottis is to protect the respiratory system during swallowing, to prevent food and liquid passing into the airway. During swallowing, the pharynx and larynx move; the pharynx widens to receive the food or liquid and the larynx rises, causing the epiglottis to move down and form a 'lid' over the glottis (Tortora and Derrickson, 2006).

The glottis is made of a pair of folds of mucous membrane, the vocal folds or true vocal chords, and the space between them known as the rima glottis. When small particles of dust, smoke or liquids pass into the larynx a cough is usually triggered to expel the substance. Failure of this mechanism can lead to aspiration and further complications (Tortora and Derrickson, 2006).

The cricoid cartilage is a hyaline cartilage ring forming the inferior wall of the larynx. It is attached to the trachea by the first ring of cartilage known as the cricotracheal ligament. This is the commonly chosen position for tracheostomy insertion. The arytenoid cartilage is a pair of triangular hyaline cartilages located at the posterior, superior border of the cricoid cartilage. Attached to the vocal chords, these contract and move to form vocal sounds. The corniculate cartilages are a pair of elastic cartilages located at the apex of each arytenoid cartilage. They are horn shaped and provide support for the epiglottis. The cuneiform cartilages are a pair of wedge-shaped elastic cartilages anterior to the corniculate cartilages and support the vocal cords and the lateral aspect of the epiglottis (Tortora and Derrickson, 2006).

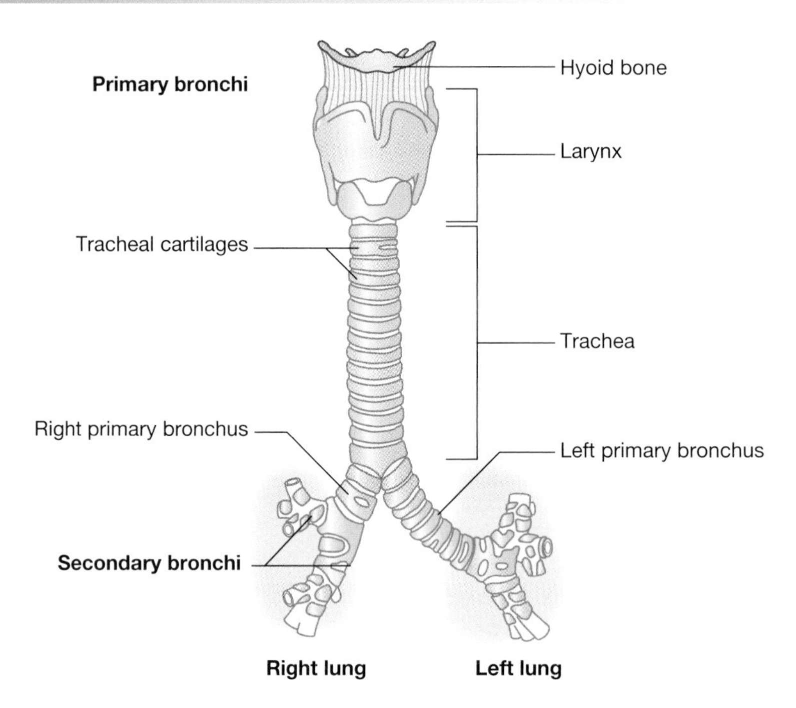




Figure 10.4 The trachea.

## The trachea

The trachea, or windpipe, allows the flow of air to and from the lungs. In an infant the trachea is short, approximately 4–5 cm from the cricoid to the carina; it is narrow and soft and is smaller and less developed than in the older child or adult (Tortora and Derrickson, 2006; West, 2012) (Figure 10.4).

The layers of the trachea are the mucosa, submucosa, hyaline cartilage and adventitia (areolar connective tissue). The tracheal mucosa is lined with an epithelial layer of pseudostratified ciliar columnar epithelium. Transverse smooth muscle fibres, trachealis muscle and elastic connective tissue stabilize the tracheal wall, preventing collapse, especially during inhalation (Tortora and Derrickson, 2006; West, 2012).

## The bronchi

At the base of the trachea the airways divide into the right and left bronchi (Figure 10.4). The point of division is the carina. The mucous membranes of the carina are very sensitive, and stimulation of the carina can trigger a cough reflex. It is important when suctioning a patient that the suction tube is not advanced past the end of the tube as it will then trigger a cough by stimulating the mucosal membrane of the carina. The right main bronchus is more vertical, shorter and wider than the left; as a result of this, an aspirated object or an endotracheal tube that has been advanced too far is more likely to enter the right main bronchus than the left. The bronchi are lined with pseudostratified ciliated columnar epithelium (Tortora and Derrickson, 2006; West, 2012).

## Clinical application

The mucous membranes of the carina are very sensitive, and stimulation of the carina can trigger a cough reflex. It is important when suctioning a patient that the suction tube is not advanced past the end of the endotracheal tube as it will then trigger a cough by stimulating the mucosal membrane of the carina.

An aspirated object or an endotracheal tube that has been advanced too far is more likely to enter the right main bronchus than the left; also, if a patient is not responding well to mechanical ventilation it may be that the endotracheal tube has been advanced too far and entered the right main bronchus, thus only ventilating one side of the lungs.



Distal to the carina, the primary bronchi divide into smaller bronchi – the secondary (lobar) bronchi. These divide into one for each lobe of the lungs, the right side having three lobes and the left having two. The secondary bronchi continue to branch, forming smaller, tertiary bronchi that further divide into bronchioles. Bronchioles continue to branch into terminal bronchioles. This extensive branching appears like an inverted tree and is referred to as the 'bronchiole tree' (Tortora and Derrickson, 2006; West, 2012).

The right bronchus gives rise to three secondary (lobar) bronchi, called the superior, middle and inferior secondary lobar bronchi. The left primary bronchus gives rise to the superior and inferior secondary bronchi. These then give rise to tertiary (segmental) bronchi, of which there are 10 in each lung. Each segment of lung tissue supplied by the tertiary bronchus is called the bronchopulmonary segment. Each bronchopulmonary segment has many smaller compartments called lobules. These are wrapped in elastic connective tissue and contain a lymphatic vessel, an arteriole, a venule and a branch from a terminal bronchiole. Terminal bronchioles subdivide into microscopic branches called respiratory bronchioles, which then subdivide into alveolar ducts (Tortora and Derrickson, 2006; West, 2012).

During exercise, autonomic nervous system activity increases. This leads to a release of adrenaline and noradrenaline, which relax the smooth muscle layer of the lungs. This relaxation of the smooth muscle in the lungs leads to dilatation of the airways. This, in turn, increases the speed at which air reaches the alveoli more quickly and lung ventilation is improved (Tortora and Derrickson, 2006; West, 2012).

## Clinical application

During an allergic reaction, autonomic nervous system mediators such as histamine are released. These cause constriction of the bronchiolar smooth muscle and constriction (tightening) of the bronchioles. Administering adrenaline at this time can be helpful in reversing or limiting the histamine response.

### The alveoli

Around the circumference of the alveolar ducts are numerous alveoli and alveolar sacs (Figure 10.5). The number of alveoli continues to increase until the child is approximately 8 years old

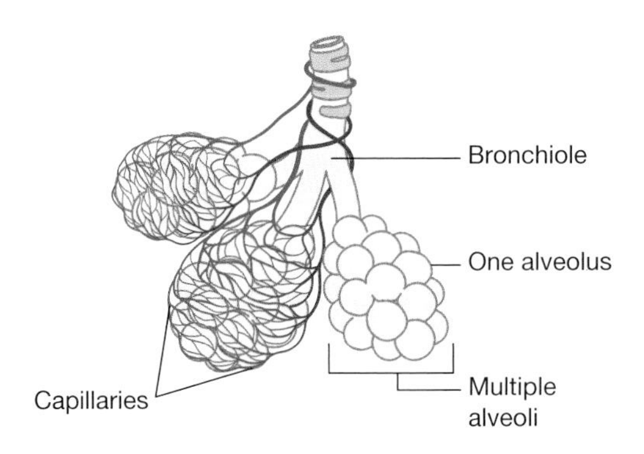


Figure 10.5 Alveoli.



(Crawford, 2011a,b), at which time the lungs contain about 500 million alveoli, thus massively increasing the surface area of the lungs despite the relatively small overall size within the body (West, 2012). Between 3 and 4 years old is considered a critical time for alveoli development, and serious respiratory infections at this age have been linked to adult respiratory disease (Dharmage *et al.*, 2009).

The alveolus is round and is lined by simple squamous epithelium and is supported by a thin elastic membrane. An alveolar sac consists of two or more alveoli sharing a common opening. The walls of the alveoli consist of different types of cells: type I and II cells, and also macrophages and fibrinoblasts. In this region, the most common cells are type I alveolar cells. Type I cells are the main site of gaseous exchange. Type II cells secrete alveolar fluid, which provides a moist environment. The alveolar fluid contains surfactant, which is important in reducing its surface tension, thus reducing the tendency of the alveoli to collapse. In premature infants there is a deficiency of this surfactant, which is supplemented to the neonate as part of their management in an effort to improve respiratory function (Merkus, 2003; Tortora and Derrickson, 2006; Wirbelauer and Speer, 2009).

As the child grows, the number of collateral ventilator channels increases; this means that if an area is blocked or narrowed, the alveoli can be aerated by another channel. By shunting air about within the lungs, gas exchange can happen without a clear connection to the main airway. The cannels of Lambert connect close-lying bronchioles and alveoli, and the pores of Kohn facilitate interalveolar connections (Dixon *et al.*, 2009, cited in Crawford (2011a)).

# The lungs

The lungs are a pair (generally) of cone-shaped organs in the thoracic cavity extending from just above the clavicles to the diaphragm. The lungs continue to develop from birth until the age of 8, at which time they are considered anatomically mature. The convex area of the lungs is broadest at the base, which sits on top of the diaphragm. The surface of the lungs lying against the ribs is called the costal surface and is rounded to 'match' the curve of the ribs. The medial surface of the lungs sits in the mediastinal region of the chest. This region houses the hilum, the point at which the bronchi, pulmonary blood vessels, lymphatic vessels and nerves enter and exit the region. These structures are held together by the pleura and connective tissue. Generally on the left, in the medial region of the thorax is the cardiac notch where the heart lies. Owing to the space occupied by the heart, the left lung is about 10% smaller than the right.

The right lung, although larger, is thicker and broader to accommodate the liver, which sits below it (Tortora and Derrickson, 2006; Crawford, 2011a,b; West, 2012).

The lungs are separated in the mediastinum; this division can serve a protective function, in that if damage to one lung occurs, then the other may remain intact and functional. The two layered pleural membranes enclose and protect each lung. The layer between the thoracic cavity and the lungs is called the parietal pleura; the layer that lines the lungs is called the visceral pleura. Between these layers is the pleural cavity, a small space continuing a lubricating fluid to reduce friction and allow adhesion between the layers; this adhesion is known as surface tension (Tortora and Derrickson, 2006).

## Clinical application



Inflammation of the pleural cavity is known as pleurisy or pleuritis, which can be very painful. Inflammation casing a collection of fluid in the pleural space is known as a pleural effusion. A large volume will compress the area and interfere with normal lung expansion, therefore reducing the effectiveness of the lungs.

## Gas exchange

On the outer surface of the alveoli the arteriole and venules disperse into a network of blood capillaries consisting of a single layer of endothelial cells and basement membrane. These tiny vessels are 'wrapped' around the end of the alveolar sacs, forming a thin barrier allowing the exchange to take place by diffusion across the alveolar and capillary walls, which together form the respiratory membrane. This very thin membrane (about one-sixteenth the diameter of a red blood cell) allows for rapid, passive diffusion of gases.

Whilst being described as a gaseous 'exchange', this is a process of carbon dioxide and oxygen moving across membranes from areas of higher partial pressure to lower partial pressure. The vast number of capillaries near the alveoli and the slow pace of flow ensure that the maximum amount of oxygen is 'collected' by the blood and transported around the body. If there is a decrease in 'availability' of alveoli due to respiratory disease, this decreases the blood's ability to collect oxygen and remove carbon dioxide (Tortora and Derrickson, 2006; West, 2012).

## Oxygen partial pressure and haemoglobin

The higher the partial pressure of oxygen $P_aO_2$, the more oxygen combines with haemoglobin. When haemoglobin is completely converted to oxyhaemoglobin it is 'fully saturated'; when it has only partially bound it is 'partially saturated'. The percentage of saturation of haemoglobin expresses the average saturation of haemoglobin with oxygen; however, it can only bind to a maximum of four oxygen molecules. This is demonstrated in the oxygen–haemoglobin dissociation curve (Figure 10.6).

Partial pressure is the greatest influence on oxygen binding to haemoglobin; however, other factors affect this process (Tortora and Derrickson, 2006; Hammer, 2013):

- **Acidity (↓pH).** As acidity in the blood increases (i.e. pH decreases), haemoglobin's affinity' for oxygen decreases and oxygen dissociates from haemoglobin. Increased hydrogen in the blood leads to haemoglobin unloading oxygen into the blood, and is known as 'compensation'.

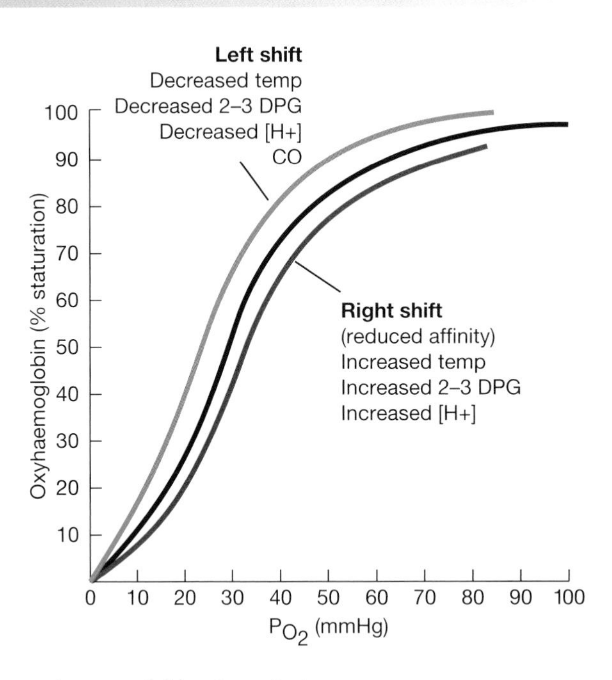


Figure 10.6 The oxygen–haemoglobin dissociation curve.

- **Partial pressure of carbon dioxide.** Carbon dioxide can also bind with haemoglobin, also making oxygen available for the cells and creating an increasingly acidic environment.
- **Temperature.** As temperature increases, the amount of oxygen released from haemoglobin is increased; conversely, when the temperature is lowered, less oxygen is made available for cells.
- **Bisphosphoglycerate (BPG).** This substance is found in red blood cells and decreases haemoglobin's affinity for oxygen, with the greater volume of BPG in the blood directly influencing the amount of oxygen unloading from haemoglobin. BPG levels are influenced by thyroxin, adrenaline, noradrenaline and testosterone, and also by living at high altitude.

## Pulmonary blood flow

The lungs receive blood via the pulmonary and bronchial arteries. Of this, deoxygenated blood to be oxygenated is carried via the pulmonary arteries, and oxygenated blood to perfuse the walls of the bronchi and bronchioles is carried via the bronchial arteries, direct form the aorta. Pulmonary blood vessels are unique in their ability to constrict in response to localized hypoxia. This vasoconstriction within the lungs allows pulmonary blood to be diverted from poorly ventilated areas of the lungs to well-ventilated areas (Tortora and Derrickson, 2006; West, 2012).

As the branching becomes more extensive there are several structural changes of note. The epithelial layer of the primary, secondary and tertiary bronchi changes to a simpler form and becomes nonciliated towards the terminal bronchioles. In this very small region, inhaled particles are removed by macrophages. The cartilage rings disappear in the distal bronchioles; with this there is an increase in smooth muscle encircling the lumen in bands. This can be problematic, as with a muscle spasm, as occurs in an asthma attack, the airway can close off, which is a life-threatening event (Tortora and Derrickson, 2006; West, 2012).

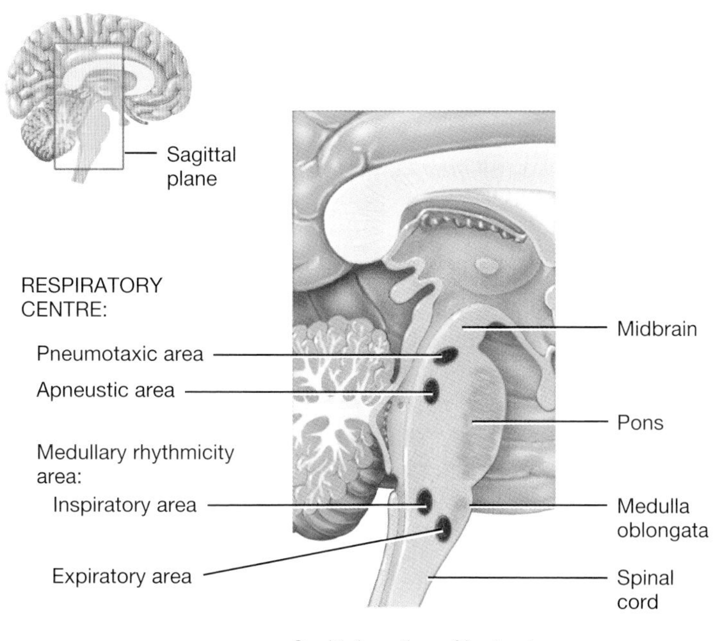


Figure 10.7 Respiratory centre within the brain.
*Source:* Tortora and Derrickson (2006), Figure 23.24, p. 905. Reproduced with permission of John Wiley and Sons, Inc.



## Respiration and the respiratory centre

Respiration is controlled by the respiratory centre within the brain stem (Figure 10.7). This centre is made of three parts: the medullary rhythmicity area in the medulla oblongata, the pneumotaxic area in the pons and the apneustic area, also in the pons. The drive to inhale and exhale is controlled by this area but can be voluntarily controlled. This voluntary control allows us to function under water and to protect our lungs from gases that we know would be harmful. Chemicals can also influence respiration, in particular carbon dioxide and oxygen. Increased levels of carbon dioxide stimulate the respiratory centre into increasing respiratory activity (rapid, deep breathing) in an effort to increase the amount of oxygen in the blood, thus normalizing the balance (Tortora and Derrickson, 2006).

## Mechanism of respiration

The mechanism of respiration is outlined in Figure 10.8.

The most important muscle during the inspiratory phase of respiration is the diaphragm (Figure 10.9). When it contracts, the contents of the abdomen move downwards and the lungs expand vertically and horizontally. During this phase the external intercostal muscles elevate and the ribs move forwards. During expiration, usually a passive process due to the elasticity of the lungs and chest wall, air is expelled, carrying with it carbon dioxide and waste products (Tortora and Derrickson, 2006; Hammer, 2013).

In older children and adults, the position of the ribs and shape of the thorax assist in the work of breathing. In infants, the horizontal position of the ribs, the circular shape of the thorax and the thin chest wall with little muscle-aided stability mean the chest wall is highly compliant;

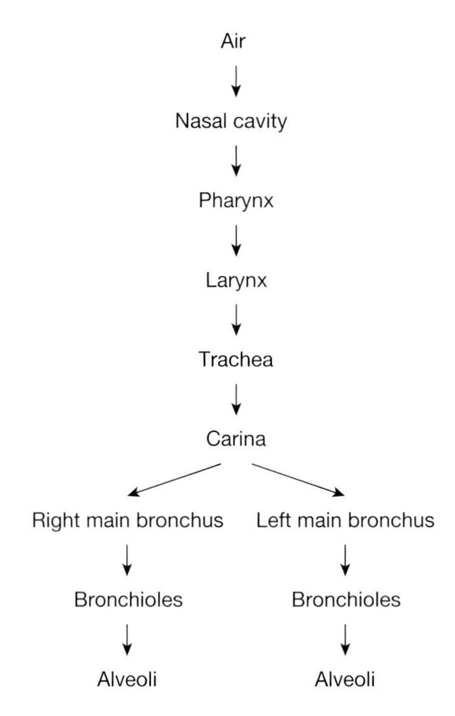




Figure 10.8 The mechanism of respiration. Adapted from Crawford (2011a,b) and Tortora and Derrickson (2006).

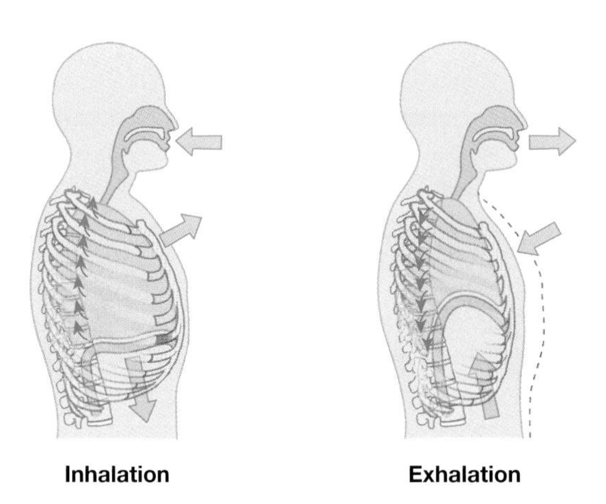


Figure 10.9 Chest movement with inspiration and expiration.

therefore, infants use their abdominal muscles to assist breathing. Also, the flatter diaphragm of the infant makes each contraction (breath) less efficient than in an older child (Tortora and Derrickson, 2006; Hammer, 2013).

**Hypoxia**, a deficiency of oxygen can be described four ways (Tortora and Derrickson, 2006):

1. **Hypoxic hypoxia** – a low partial pressure of oxygen in arterial blood resulting from being at high altitude, an airway obstruction or wet lungs.

2. **Anaemic hypoxia** – too little functioning haemoglobin present in the blood, which may be caused by anaemia, haemorrhage or carbon monoxide poisoning affecting the ability of the haemoglobin to carry oxygen.
3. **Ischaemic hypoxia** – reduced blood flow to tissue.
4. **Histotoxic hypoxia** – adequate oxygen is delivered, but due to the action of a toxic agent the tissue is unable to use it correctly.

## Assessing respiration

Evaluate whether the child is breathing spontaneously and able to maintain their airway first; this is crucial. Seek help if this is not clear or the child cannot maintain its own airway (Hammer, 2013).

Factors involved in assessing respiratory function include the following (Tortora and Derrickson, 2006; Crawford, 2011a,b; Macqueen *et al.*, 2012):



- **Pulse oximetry.** A non-invasive, continuous means of assessing arterial oxygen saturation $S_aO_2$. The pulse oximeter exploits the pulsatile nature of arterial blood flow and 'reads' the amount of oxygen attached to the red blood cell. There are clinical presentations when these readings may not be accurate (e.g. carbon monoxide intoxication), but in most presentations this data can be relied upon.
- **Auscultation.** Listening to the sound of inhalation and exhalation and for any other sounds within the lung fields.
- **Oxygen** requirement and if this is changing.
- **Work of breathing** (see respiratory distress).

## Respiratory distress

Respiratory failure or distress describes the inability to provide oxygen and remove carbon dioxide at a rate matching the body's metabolic demand (Hammer, 2013). All aspects of respiratory function should be used in assessing the child or infant, and care should be taken to consider each aspect, as it may be possible to ascribe an abnormality to its cause; for example, inspiratory stridor indicates upper-airway obstruction (Rojas-Reyes *et al.*, 2009; Crawford, 2011a,b; Hammer, 2013).

- Increased respiratory rate (tachypnoea).
- Cough.
- Grunting in the neonate.
- Nasal flare.
- Continuous, high-pitched musical-like wheeze due to airway turbulence.
- Stridor: inspiratory or expiratory.
- Use of accessory muscles:
  - head bobbing
  - shoulder fixing
  - abdominal breathing
  - uneven chest rise and fall
  - sub- and intercostal recession or in-drawing of the thoracic muscle cage.

## Physiological causes for increased susceptibility to respiratory disease by children and infants

Hammer (2013) discusses a number of factors that are associated with an increased susceptibility to respiratory disease in children, and these include:

- Increased oxygen demand.
- Immature breathing control and risk of apnoea.
- Upper airway resistance, nasal breather, large tongue, small collapsible airway, weaker pharyngeal muscle tone, compliance of upper airway structures.
- Lower airway resistance, small airway size, collapsible airway wall, compliant airways.
- Lung volume – numbers of alveoli, lack of collateral ventilation.
- Efficiency of respiratory muscles – less efficient diaphragm and ribcage with horizontal ribs.
- Endurance of respiratory muscles – increased respiratory rate and more rapid fatigue.

## Oxygen therapy



Oxygen therapy is the administration of oxygen at a concentration greater than in ambient air. It is done with the intention of treating or preventing the symptoms of hypoxia and to reduce the work of breathing. There is a limited evidence base for the use of oxygen in children and infants, including appropriate target saturation ranges, and the recognition of hypoxaemia in very young patients. As such, oxygen should be used with caution (Lamont *et al.*, 2010).

Non-invasive methods of oxygen delivery include (McKiernan *et al.*, 2010; Lee *et al.*, 2013):

- nasal prong (this can be sutured in for stability; therefore, this is a semi invasive procedure)
- nasal cannulae
- high-flow nasal cannulae
- continuous positive airway pressure
- face mask
- head box
- blow-by ('waft').

# Conclusion

The key function of the respiratory system is to supply the blood with oxygen in order for the blood to deliver oxygen to all parts of the body. This occurs through breathing. When we breathe, we inhale oxygen and exhale carbon dioxide. This exchange of gases is how the respiratory system supplies oxygen to the blood.

The anatomy and physiology of the child's respiratory system changes as they develop. The nurse must be aware of these changes as they can impact on the care given. Having an understanding of the anatomy and physiology means you will be able to provide care that is informed, safe and effective.

The various conditions associated with the respiratory tract require skilled nursing care and the ability to monitor the child's condition, noticing sometimes subtle changes that must be reported to the person in charge so that appropriate action can be taken. The nurse must act in a competent and confident manner, instilling confidence in the child and the family.

# Activities

**Now review your learning by completing the learning activities in this chapter. The answers to these appear at the end of the book. Further self-test activities can be found at www.wileyfundamentalseries.com/childrensA&P.**

## Complete the table

Complete the following table using a variety of sources (human and material):

| Complaint | Assessments | Possible causes |
|---|---|---|
| Croup | | |
| Bronchiolitis | | |
| Pneumonia | | |
| Tetralogy of Fallot | | |
| Asthma | | |



## Conditions

The following table contains a list of conditions. Take some time and write notes about each of the conditions. You may make the notes taken from text books or other resources; for example, people you work with in a clinical area or you may make the notes as a result of people you have cared for. If you are making notes about people you have cared for you must ensure that you adhere to the rules of confidentiality.

| Condition | Your notes |
|---|---|
| Pneumonia | |
| Laryngotracheobronchitis | |
| Pneumothorax | |
| Foreign body aspiration | |
| Obstructive sleep apnoea | |

## Glossary

**Asthma:** a common chronic disorder of the airways characterized by exacerbations or attacks. Exacerbation involves inflammation of the airways and airway reactivity causing bronchospasm or contraction of the bronchioles.

**Bronchiolitis:** a viral infection causing fever, nasal discharge and a dry, wheezy cough with fine crackles on auscultation.

**Croup:** a viral infection causing inflammation and oedema of the upper airway mucosa and narrowing of the subglottic region, leading to varying degrees of airway obstruction.

**Hypoxia:** decrease below normal levels of oxygen in inspired gases, arterial blood, or tissue.

**Mild croup:** a barky cough, but no stridor or chest wall recession at rest.

**Moderate croup:** a persistent barking cough, accompanied by stridor and suprasternal and sternal chest wall recession when at rest.

**Respiration:** the cycling of oxygen and carbon dioxide between the body and the atmosphere.

**Severe croup:** significant inspiratory and occasionally expiratory stridor, decreased air entry upon auscultation and evidence of agitation or distress.



## References

Buczynski, B., Yee, M., Paige Lawrence, B., O'Reilly, M. (2012) Lung development and the host response to influenza A virus are altered by different doses of neonatal oxygen in mice. *American Journal of Physiology – Lung Physiology*, **302** (10), L1078–L1087.

Crawford, D. (2011a) Understanding childhood asthma and the development of the respiratory tract. *Nursing Children and Young People*, **23** (7), 25–34.

Crawford, D. (2011b) Care and nursing management of a child with a chest drain. *Nursing Children and Young People*, **23** (10), 27–33.

Dharmage, S.C., Erbas, B., Jarvis, D. *et al.* (2009) Do childhood respiratory infections continue to influence adult respiratory morbidity? *European Respiratory Journal*, **33** (2), 237–244.

Hammer, J. (2013) Acute respiratory failure in children. *Paediatric Respiratory Reviews*, **14** (2), 64–69.

Macqueen, S., Bruce, E.A., Gibson, F. (eds) (2012) *The Great Ormond Street Hospital Manual of Children's Nursing Practices*, Wiley–Blackwell, London.

Lamont, T., Luettel, D., Scarpello, J. *et al.* (2010) Improving the safety of oxygen therapy in hospitals: summary of a safety report from the National Patient Safety Agency. *British Medical Journal*, **340**, c187.

Lee, J.H., Rehder, K.J., Williford, L. *et al.* (2013) Use of high flow nasal cannula in critically ill infants, children, and adults: a critical review of the literature. *Intensive Care Medicine*, **39** (2), 247–257.

McKiernan, C., Chadrick Chau, L., Visintainer, P.F., Allen, H. (2010) High flow nasal cannulae therapy in infants with bronchiolitis. *The Journal of Pediatrics*, **156** (4), 634–638.

Merkus, P.J.F.M. (2003) Effects of childhood respiratory diseases on the anatomical and functional development of the respiratory system. *Paediatric Respiratory Reviews*, **4** (1), 28–39.

Naddy, C. (2012) The impact of paediatric early warning systems. *Nursing Children and Young People*, **24** (8), 14–17.

Rojas-Reyes, M.X., Granados Rugeles, C., Charry-Anzola, L.P. (2009) Oxygen therapy for lower respiratory tract infections in children between 3 months and 15 years of age. *Cochrane Database of Systematic Reviews*, (1), CD005975.

Savković, A., Delić, J., Isaković, E., Ljuca, F. (2010) Age characteristics of the larynx in infants during the first year of life. *Periodicum Biologorum*, **112** (1), 75–82.

Stoelting, R., Miller, R. (2007) *Basics in Anaesthesia*, 5th edn, Elsevier Health Sciences, Philadelphia, PA.

Tregoning, J.S., Schwarze, J. (2010) Respiratory viral infections in infants: causes, clinical symptoms, virology, and immunology. *Clinical Microbiology Review*, **23** (1), 74–98.

Tortora, G.J., Derrickson, B. (2006) *Principles of Anatomy and Physiology*, 11th edn, John Wiley and Sons, Inc., Hoboken, NJ.

Turkmen, S., Cansu, A., Turedi, S. *et al.* (2012) Age-dependent structural and radiological changes in the larynx. *Clinical Radiology*, **67** (11), e22–e26.

West, J.B. (2012) *Respiratory Physiology: The Essentials*, Lippincott Williams and Wilkins, Philadelphia, PA.

Winston, R. (ed.) (2004) *Human*, Dorling Kindersley, London.

Wirbelauer, J., Speer, C.P. (2009) The role of surfactant treatment in preterm infants and term newborns with acute respiratory distress syndrome. *Journal of Perinatology*, **29**, S18–S22.

# Chapter 11

# The endocrine system

Julia Petty

*School of Health and Social Work, University of Hertfordshire, Hatfield, UK*

## Aim

The aim of this chapter is to provide an overview of the anatomy and physiology of the endocrine system, including the vital glands and hormones required for body function. The effects and influence of the endocrine system from prenatal to early adulthood will also be highlighted.

## Learning outcomes

On completion of this chapter the reader will be able to:

- Name the endocrine glands of the body and describe their main functions.
- Name the hormones of the main endocrine glands and how they work.
- Highlight how the endocrine system influences growth and development in a number of important areas.
- Highlight how different hormone levels and endocrine functions vary through development.
- Understand the integration of the endocrine glands and their respective hormones in relation to important major physiological body functions; for example, growth, glucose homeostasis, sex differentiation and maturation of secondary sexual characteristics, and stress.
- Describe and understand a range of endocrine disorders in childhood and the related pathophysiology.

*Fundamentals of Children's Anatomy and Physiology: A Textbook for Nursing and Healthcare Students*, First Edition. Edited by Ian Peate and Elizabeth Gormley-Fleming.

Companion website: www.wileyfundamentalseries.com/childrensA&P

## Test your prior knowledge

- What is the difference between an endocrine and an exocrine gland?
- What is the main function of the endocrine system?
- How can you distinguish the endocrine system from the nervous system?
- What are main glands within the endocrine system?
- Can you name at least one hormone that each of these glands secretes?
- What is a negative-feedback mechanism in relation to how the endocrine system functions?
- Conversely, what is a positive-feedback mechanism in relation to hormone function?
- What are the different *types* of hormones?
- How do endocrine glands *work together* in relation to important body functions such as growth, sexual development and stress?
- What are some of the important developmental changes and influences in hormone release and function throughout childhood?



## Introduction

There are two regulatory systems within the body that are responsible for the transmission of vital messages and the integration of bodily functions: the nervous and endocrine systems. The nervous system works by sending electrical impulses via neurones and neurotransmitters to transfer signals across synapses. The endocrine system comprises a collection of *glands* (Figure 11.1) that secrete a range of different types of hormones directly into the bloodstream which then transports that hormone to take effect in more distant target organs or tissues (Molina, 2013). The unique features of these glands are that they are ductless in nature, have a high vascularity and store their hormones within granules (Rogers, 2012). In contrast, exocrine glands (such as sweat glands, the gall bladder and salivary glands) secrete their hormones using hollow lumen ducts and are less vascular. Endocrine glands are controlled directly by stimulation from the nervous system, as well as by chemical receptors in the blood and hormones produced by other glands (Waugh and Grant, 2010). By regulating the functions of organs in the body, these glands help to maintain the body's homeostasis. Growth and development, sexual development and control of many internal body functions, such as glucose and mineral regulation as well as the stress response are among the many essential physiological processes regulated by the actions of hormones. The anatomy of the endocrine system can be seen in Figure 11.1.

The integrity and health of the endocrine system is essential to maintaining healthy body weight, growth and both physical and emotional development. The endocrine system significantly affects children and teenagers who are experiencing a high rate of development, but different parts of this system play a role as ageing occurs.

The effects of hormones are varied and many. They can be categorized into four broad areas as relevant to the developing child:

- They play a role in the sequential integration of growth and development.
- They contribute to basic processes of reproduction, starting as early as gamete formation, fertilization, stability of the growing fetus, labour and the subsequent adaptation to extra-uterine life.
- They help to control the internal environment throughout life by regulation and homeostasis of many physiological functions.

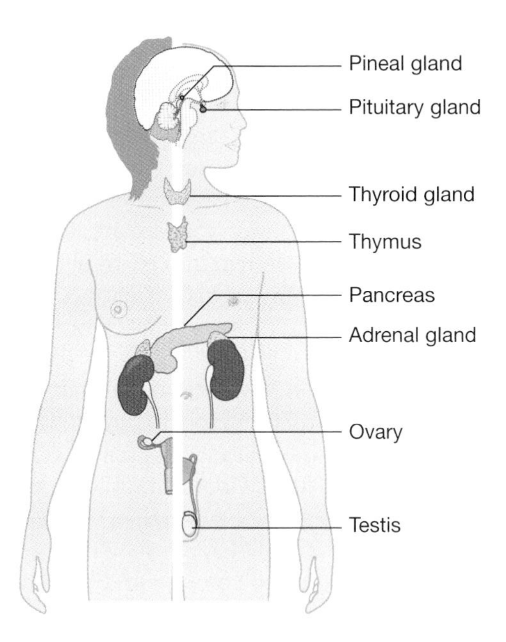


Figure 11.1 The endocrine system.



- They respond to changes in environmental conditions to assist the body to cope with emergency demands, such as stress, trauma, dehydration and temperature imbalances (Waugh and Grant, 2010; Greenstein and Wood, 2011).

# Physiology of the endocrine system

The endocrine system works alongside the nervous system to control many vital functions of the body. The nervous system provides a very fast and narrowly targeted system to act on specific glands and muscles throughout the body. The endocrine system, on the other hand, is much slower acting but has very widespread, long-lasting and powerful effects. Hormones are distributed by glands through the bloodstream to the entire body, affecting any cell with a receptor for a particular hormone (Molina, 2013). While the two systems are different in these specific ways, it is important to remember how they also work together to affect cells in several organs or tissues throughout the entire body leading to many powerful and diverse responses. The two systems work together closely to coordinate their activities in an integrated way. When a gland in the endocrine system releases a hormone, it travels via the bloodstream to a target area where it exerts its effect. Hormones are designed to interact with a specific part of the body so when the hormone arrives at this organ or tissue, a particular action will take place.

## Hormones and how they work

Hormones are classified into two categories depending on their chemical structure and solubility: water-soluble and lipid-soluble hormones. Each of these classes of hormones has specific mechanisms that determine how they affect their target cells.

- **Water-soluble hormones:** Water-soluble hormones include two types; the peptide and amino acid hormones such as insulin, adrenaline, human growth hormone and oxytocin

(Hinson *et al.*, 2010). As their name indicates, these hormones are soluble in water, so they are readily circulated in the blood and do not need a transport protein. As they are unable to pass through the phospholipid bilayer of the plasma membrane, they are therefore dependent upon receptor molecules on the surface of cells. When they bind to a receptor molecule on the surface of a cell, a reaction is triggered inside such as a change in permeability of the membrane or the activation of another molecule. An example here is when molecules of cyclic adenosine monophosphate (cAMP) are triggered to be synthesized from adenosine triphosphate present in the cell in cellular respiration. cAMP is a secondary messenger used for intracellular signal transduction, such as transferring into cells the effects of hormones like glucagon and adrenaline, which cannot pass through the cell membrane.
- **Lipid-soluble hormones:** Lipid-soluble hormones *are* able to pass directly through the phospholipid bilayer of the plasma membrane and bind to receptors inside the cell nucleus. They include the thyroid and steroid hormones, such as testosterone, oestrogens, glucocorticoids and mineralocorticoids. Unlike the water-soluble hormones, lipid-soluble hormones are able to directly control the function of a cell from these receptors, often triggering the transcription of particular genes in the DNA to produce 'messenger ribonucleic acids' that are used to make proteins that affect the cell's growth and function.



For endocrine glands to release their hormones, they can be stimulated by a number of factors:

- The presence of releasing or stimulating hormones from other endocrine glands. The nervous system can control levels of such hormones through the action of the hypothalamus and its releasing tropic hormones. For example, thyrotropin-releasing hormone (TRH) produced by the hypothalamus stimulates the anterior pituitary to produce thyroid-stimulating hormone (TSH). Tropic hormones provide another level of control for the release of hormones. For example, TSH is a tropic hormone that stimulates the thyroid gland to produce triiodothyronine (T3) and thyroxine (T4).
- Environmental factors, such as heat, cold, stress and physical exercise.
- Internal factors or signs from the body; for example, calcium levels, changes in blood glucose, blood pressure and fluid or electrolyte levels.
- Positive-feedback mechanisms, whereby the outcome of the hormonal action then promotes further excretion of that hormone (Hinson *et al.*, 2010; Waugh and Grant, 2010).

Conversely, endocrine function ceases to have an effect when:

- Inhibiting hormones are present. As stated above, the nervous system can control levels of such hormones through the action of the hypothalamus and its inhibiting hormones.
- Environmental factors change.
- Internal homeostasis takes place, bringing the previous abnormal value back to normal range.
- By a negative-feedback mechanism; that is, the high levels of hormone are detected and this leads to the production of that hormone to stop or slow down.

Sensitivity of the target organ or tissue can influence the extent to which endocrine hormones exert their effects, and they do so via receptors (Jameson and De Groot, 2010). Once hormones have been produced by glands, they are distributed through the body via the bloodstream. As hormones travel through the body, they pass through cells or along the plasma membranes of cells until they encounter a receptor for that particular hormone. Hormones can only affect target cells that have the appropriate receptors. This property of hormones is known as specificity. Hormone specificity explains how each hormone can have specific effects in widespread parts of the body (Rogers, 2012). In the presence of large quantities of hormone for an extended period of time, the number of receptors decreases, and this make cells less sensitive ('downregulation') with reduced hormonal control of the cells. The opposite is also the case, in that, if hormones levels are low, 'upregulation' may occur by the number of receptors increasing.

Many hormones produced by the endocrine system are classified according to their effect. A tropic hormone is a hormone that is able to trigger the release of another hormone in another gland. Tropic hormones provide a pathway of control for hormone production as well as a way for glands to be controlled in distant regions of the body. Many of the hormones produced by the pituitary gland are tropic hormones; for example, TSH, adrenocorticotropic hormone (ACTH) and follicle-stimulating hormone (FSH). A similar or alternative term is stimulating – for example, TSH *stimulates* the thyroid gland. Some hormones may stimulate the release of other hormones; for example, growth-hormone-releasing hormone (GHRH) stimulates the release of growth hormone, while gonadotropin-releasing hormone (GnRH) does so for the release of FSH and luteinizing hormone (LH). Conversely, hormones may have an inhibiting effect on other hormones; for example, prolactin-inhibiting hormone (PIH) inhibits prolactin release at such time when this hormone is no longer required.



## Age-related and developmental influences

Some hormone levels also naturally increase and decrease with age. There are a few critical periods within the time in which the endocrine system develops, namely the intrauterine, early postnatal and pre-pubertal periods. Children have a dynamic physiology that is vulnerable because of growth demands but also due to damage during differentiation and maturation of organs and systems. Their needs for energy, water and oxygen are higher because they go through an intense anabolic process.

Neonates in the postnatal period particularly if born preterm, can have altered hormone levels for the first few days of life until full maturity ensues and adaptation to extra-uterine life has been achieved with levels stabilizing following labour and over the first weeks of life. In the main, endocrine organs are fully formed anatomically at term; the endocrine system is overall functional and organized. In infancy and beyond, a healthy child brought up with good nutrition in a sound and stable home environment will grow and develop without problems influenced by both hormonal and environmental influences. Physical growth continues through adolescence as well as the maturation and continued differentiation of physiological functions. Organs grow and their function matures and modifies at different life stages until the end of adolescence. External influences, such as prolonged stress and illness, may, however, interfere with normal endocrine function over time in the vulnerable, growing child. In addition, diseases such as diabetes, an adrenal deficiency or abnormalities with puberty can cause the body to produce an incorrect amount of hormones. These conditions can affect the body as a whole or will affect specific parts of the endocrine system, such as the thyroid or the pancreas depending on the condition. If it appears that a child is not developing properly, cannot manage hunger or they are experiencing fast and unexplained weight gain or growth for example, then there may be a problem with the endocrine system. The clinical implications of compromise to the endocrine system will be discussed in the next section, which covers each gland in turn in relation to the normal anatomy, hormone production and relevant clinical condition that arises if there is dysfunction.

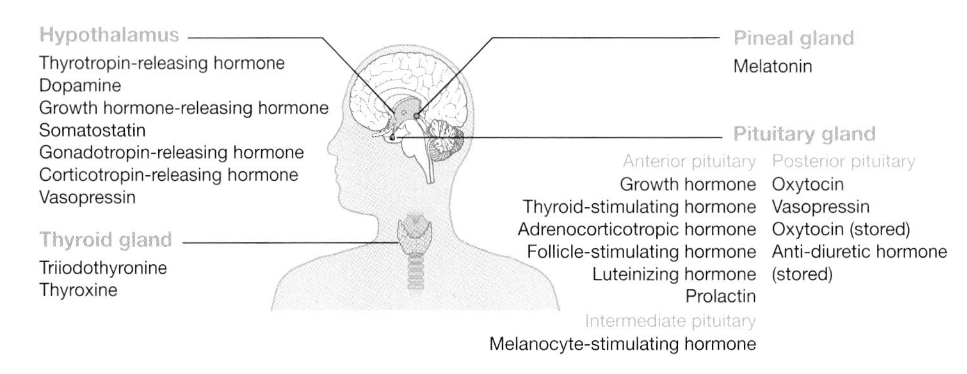


Figure 11.2 The endocrine glands and hormones of the head and neck.



# Anatomy of the endocrine system

All endocrine organs, their hormones and respective functions are now summarized under their respective subheadings. Developmental implications are also discussed for each gland, along with examples of clinical considerations in children.

## Hypothalamus

The hypothalamus is a part of the midbrain located superior and anterior to the brain stem and situated inferior to the thalamus (Figure 11.2). It serves many different functions, such as growth, thermoregulation, control of hunger and thirst, sexual development and regulation of stress defences. The hypothalamus contains special cells called neurosecretory cells – neurones that secrete the hormones seen in Table 11.1.

The hypothalamus is responsible for the direct control of the endocrine system by communication with the pituitary gland. A portal system exists between the two glands that enables hormones to be delivered rapidly and directly from the hypothalamus to the pituitary gland.

Embryonically, this organ develops from the ectoderm layer of the embryo from week 4. Nerve fibres of hypothalamic neurones grow, and then late in the fetal period further cell differentiation occurs when hormones are starting to be produced.

During neonatal life in the period of adaptation to extra-uterine life, the hypothalamus is immature. This can affect specific body functions such as thermoregulation, where the newborn is predisposed to heat loss. Non-shivering thermogenesis, a physiological mechanism for heat generation following birth, is under hormonal control and serves an important function in the early days of postnatal life. Disease relating to hypothalamus pathophysiology is rare but can be as a result of various factors (tumour, hypoxic damage). Any compromise to function will in turn lead to a number of potential problems relating to the specific hormones normally produced within this gland.

## Pituitary gland

The pituitary gland, also known as the hypophysis, is a small pea-sized mass of tissue connected to the inferior portion of the hypothalamus (Dorton, 2000). Situated in a small depression in the sphenoid bone called the sella turcica, the pituitary gland is comprised of two completely separate posterior and anterior components.

Table 11.1 Hormones of the hypothalamus.

| Hormone | Acronym | Function |
|---|---|---|
| Growth hormone (somatotropin)-releasing hormone | GHRH | Stimulates growth hormone (GH) release from the anterior pituitary gland |
| Thyrotropin-releasing hormone | TRH | Stimulates thyroid-stimulating hormone (TSH) release from the anterior pituitary gland |
| Gonadotrophin-releasing hormone | GnRH | Stimulates follicle-stimulating hormone (FSH) release from the anterior pituitary gland<br>Stimulates luteinizing hormone (LS) release from the anterior pituitary gland |
| Prolactin-releasing hormone | PRH | Stimulates prolactin release from the anterior pituitary gland |
| Corticotropin-releasing hormone | CRH | Stimulates adrenocorticotrophic hormone (ACTH) release from the anterior pituitary gland |
| Growth-hormone (somatotropin)-inhibiting hormone | GHIH | Inhibits thyroid-stimulating hormone (TSH) release from the anterior pituitary gland |
| Prolactin-inhibiting hormone | PIH | Inhibits prolactin release from the anterior pituitary gland |
| Antidiuretic hormone (vasopressin) | ADH | Manufactured and then sent to posterior pituitary gland for storage and release. Increases water permeability in the distal convoluted tubule and collecting ducts of the kidney nephrons, promoting water reabsorption and increasing blood volume |
| Oxytocin | OT | Manufactured and then sent to posterior pituitary gland for storage and release. Stimulates uterine smooth muscle contraction during labour and controls the 'let-down' reflex for breast milk ejection within the breasts |



- **Posterior pituitary.** The posterior pituitary gland is not glandular tissue; rather, it is nervous tissue. It is a small extension of the hypothalamus from which the neuronal axons of some of the neurosecretory cells of the hypothalamus extend to. These neurosecretory cells create two hormones that are stored and released by the posterior pituitary (Table 11.2).
- **Anterior pituitary.** The anterior component is the true glandular section of the pituitary gland. The function is controlled by the releasing and inhibiting hormones released from the hypothalamus. The anterior pituitary gland produces seven important hormones. These along with the above two from the posterior side, can be seen in Table 11.2.

Many of the pituitary functions follow a developmental pattern through from prenatal life to puberty. The pituitary gland is also derived from two embryonic sources: the anterior lobe from the oral cavity and the posterior from the base of the brain – that is, neural origin (Greenstein and Wood, 2011).

Table 11.2 Hormones of the pituitary gland.

| Hormone | Acronym | Function |
|---|---|---|
| *Anterior pituitary gland* | | |
| Growth hormone (somatotropin) | GH | Growth and reproduction of body cells; protein anabolism |
| Thyroid-stimulating hormone | TSH | Controls secretion of thyroxine (T4) and triiodothyronine (T3) from the thyroid gland |
| Adrenocorticotrophic hormone | ACTH | Controls secretion of corticosteroids by the adrenal cortex |
| Follicle-stimulating hormone | FSH | In females, initiates the development of ova, maturation of ovarian follicles and induction of secretion of oestrogens by the ovaries. In males, stimulates testes to produce sperm |
| Luteinizing hormone | LH | In females, with oestrogen, stimulates ovulation, formation of corpus luteum, prepares uterus for implantation and breasts to secrete milk. In males, stimulates the testes to produce testosterone |
| Prolactin | PRL | Alongside other hormones, initiates and then maintains milk secretion by the breasts |
| Melanocyte-stimulating hormone | MSH | Stimulates melanin synthesis and release from the melanocytes of the skin and hair |
| Beta-endorphin | | Inhibits perception of pain |
| *Posterior pituitary gland* | | |
| Antidiuretic hormone (vasopressin) | ADH | Made in the hypothalamus and stored for release from the posterior pituitary gland<br>Increases water permeability in the distal convoluted tubule and collecting ducts of the kidney nephrons, promoting water reabsorption and increasing blood volume |
| Oxytocin | OT | Made in the hypothalamus and stored for release from the posterior pituitary gland<br>Stimulates uterine smooth muscle contraction during labour and controls the 'let-down' reflex for breast milk ejection within the breasts |



As for hypothalamic function, pituitary insufficiency due to immaturity may occur in neonatal life, more so in premature infants. However this is temporary and not a disease process. In childhood, certain hormones function in developmental patterns. For example, in childhood, growth hormone levels fluctuate for certain periods throughout the day with bursts at night, particularly when in deep sleep cycles. This hormone then increases in adolescence coinciding with a rise in sex hormone production. Pituitary hormones that result in sexual development and other reproductive functions increase and peak from the pre-pubertal age until puberty is completed. Any disease that influences the pituitary and/or hypothalamus can have significant effects owing to the many functions that they impart.

## Clinical application

### Hypopituitarism

This is the inability of the pituitary gland to provide sufficient hormones, due to inadequate production or an insufficient supply of hypothalamic-releasing hormones. It is usually a mixture of several hormonal deficiencies and can be chronic and lifelong, unless successful surgery or medical treatment of the underlying disorder can restore pituitary function (Schneider *et al.*, 2007).

One example of a problem hypopituitarism can cause is **diabetes insipidus (cranial)**. The pituitary gland produces or releases a reduced amount of antidiuretic hormone (ADH). Signs and symptoms include both excessive thirst and urination. (Khardori and Ulla, 2012).

**Tumours of the pituitary gland** can also be a cause of pituitary dysfunction. They are usually benign and can cause problems by local effects of the tumour, excessive hormone production or inadequate hormone production by the remaining pituitary gland.



## Clinical application

### Growth disorders

**Pituitary dwarfism/poor somatic growth** is a condition characterized by growth that is very slow or delayed early in childhood before the ossification of bone cartilages (Dattani and Preece, 2004) and is caused by insufficient secretion of pituitary growth hormone. The condition begins in childhood, but it becomes more evident during puberty.

# Pineal gland

The pineal gland is a small mass of glandular tissue shaped like a pinecone situated just posterior to the thalamus of the brain. It produces the hormone melatonin (Table 11.3) that helps to regulate the human sleep–wake cycle known as the circadian rhythm. The activity of the pineal gland is inhibited by nervous stimulation from the photoreceptors of the retina. This light sensitivity causes melatonin to be produced only in darkness or low light. Increased melatonin production causes a feeling of drowsiness at night-time when the pineal gland is active.

The pineal gland develops from week 7 to 8 from the ectoderm. Sleep–wake cycles change after birth in the first year or two of age when infants begin to spend less time sleeping during the day and regulate their cycles with the normal pattern of night and day. The pineal gland is large in children relative to adult size but shrinks at puberty; however, the roles of the pineal gland and melatonin in human pubertal development remain unclear. The activity of the pineal gland declines with advancing age.

Table 11.3 Hormones of the pineal gland.

| Hormone | Function |
|---|---|
| Melatonin | Plays a part in the control of circadian rhythm and inducement of drowsiness |

Table 11.4 Hormones of the thyroid gland.

| Hormone | Acronym | Function |
|---|---|---|
| Thyroxine | T4 | Regulates metabolism, stimulates body oxygen and energy consumption, plays a part in growth by promoting protein synthesis and influences the activity of the nervous system |
| Triiodothyronine | T3 | As above |
| Calcitonin | CT | Lowers calcium levels by accelerating calcium absorption by bone osteoblasts (bone-forming cells), which in turn assists bone construction. Inhibits calcium release from bone |



Pineal dysfunction is rare in children, but one consideration relates to the very important topic of sleep and how disruptions to the circadian rhythm may affect the quantity and quality of sleep received. The role of melatonin is one area to consider in this area (Phillips and Appleton, 2004). Disruption of the circadian system can occur as a result of external factors of conditions, such as light at night and crossing meridian time zones (jet lag), but it can also be related to a genetic predisposition or abnormalities that affect the functioning of the retinohypothalamic system, the production of melatonin or rarely, physical damage or tumours of the pineal gland.

## Thyroid gland

The thyroid gland is located at the base of the neck and wrapped around the lateral sides of the trachea. This butterfly-shaped gland produces the three major hormones seen in Table 11.4. Calcitonin is released when calcium ion levels in the blood rise above a certain set point. Calcitonin functions to reduce the concentration of calcium ions in the blood by aiding the absorption of calcium into the matrix of bones. The hormones T3 and T4 work together to regulate the body's metabolic rate. Increased levels of T3 and T4 lead to increased cellular activity and energy usage in the body.

The thyroid gland develops from as early as the third week of gestation from the endoderm layer of the embryo. Follicular cells have developed by the 10th week and by weeks 12–13 are synthesizing thyroglobulin, with T3 being produced by week 16. The fetus has two potential sources of thyroid hormones: its own thyroid and that of it's mother. Human fetuses acquire the ability to synthesize thyroid hormones at roughly 12 weeks of gestation. The net effect of pregnancy is an increased demand on the thyroid gland. Thyroid stimulation is also achieved by chorionic gonadotropin. During this time, blood levels of TSH often are suppressed until after birth. There is also an increased demand for iodine. Nutrition can also control the levels in the body, in that thyroid hormones T3 and T4 require three and four iodine atoms respectively to be produced. Children who lack iodine in their diet will fail to produce sufficient levels of thyroid hormones to maintain a healthy metabolic rate.

Thyroid hormone is critical for normal brain development of the fetus in pregnancy and beyond into childhood. Thyroid disorders may be congenital (present at birth) or develop later in childhood. With proper treatment, most thyroid disorders can be successfully managed in children.

## Clinical application

### Hypothyroidism

Congenital hypothyroidism (CH) can be defined as a lack of thyroid hormones present from birth that, unless detected and treated early, is associated with irreversible neurological problems and poor growth (NHS, 2013). Hypothalamic or pituitary dysfunction accounts also for some cases of CH (Kumar *et al.*, 2005; Willacy, 2011). *Symptoms:* feeding difficulties, lethargy, low frequency of crying, constipation. *Signs:* large fontanelles, myxoedema – with coarse features and a large head and oedema of the genitalia and extremities, macroglossia, low temperature, jaundice – umbilical hernia and hypotonia. The growing child will have short stature, hypertelorism, depressed bridge of nose, narrow palpebral fissures and swollen eyelids.

### Hyperthyroidism

This means a raised level of thyroid hormone. There are various causes. Graves' disease is the most common cause. Hyperthyroidism can produce various symptoms, such as irritability, sleeping poorly, weight loss, tachycardia and a swelling of the thyroid gland (a goitre) in the neck may occur.



## Parathyroid glands

The parathyroid glands are four small masses of glandular tissue found on the posterior side of the thyroid gland (Figure 11.3). They produce the hormone parathyroid hormone (PTH – see Table 11.5), which is involved in calcium ion homeostasis. PTH is released from the parathyroid glands when calcium ion levels in the blood drop below a set point. PTH stimulates the osteoclasts to break down the calcium-containing bone matrix to release free calcium ions into the

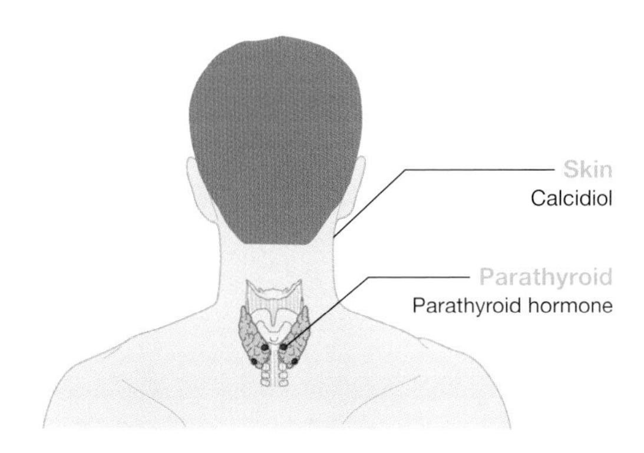


Figure 11.3 The parathyroid glands.

Table 11.5 Hormones of the parathyroid glands.

| Hormone | Acronym | Function |
|---|---|---|
| Parathyroid hormone | PTH | Increases blood calcium level and decreases phosphate level by increasing the rate of calcium absorption from the intestine into the blood. Increases number and stimulates osteoclasts (to break down bone). Increases calcium absorption and phosphate excretion by the kidneys. Activates vitamin D |

bloodstream. PTH also triggers the kidneys to return calcium ions filtered out of the blood back to the bloodstream so that it is conserved.

The parathyroid glands arise from the endoderm from week 5 and start to produce PTH by the seventh week when the bones start to ossify. The principal maternal physiologic modification with respect to calcium metabolism is increasing PTH secretion, which maintains the serum ionic calcium level within its characteristically narrow physiologic limits despite an expanding extracellular fluid volume, increased urinary excretion and calcium transfer to the fetus. The primary feature of perinatal calcium metabolism is the active placental transport of calcium ions from mother to fetus, making the fetus relatively hypercalcaemic. Since none of the calcitropic hormones cross the placenta, hypercalcaemia apparently suppresses either secretion or activity of PTH by the fetus and stimulates fetal calcitonin release, creating an environment (high calcium, low PTH, high calcitonin) favourable to skeletal growth. With birth, the transplacental calcium source terminates suddenly and the serum calcium level declines for 24–48 h, after which it stabilizes and then rises slightly. Neonatal calcium homeostasis suggests multiple influences, including the respective calcitropic hormones and other ions involved, such as magnesium and phosphate. The physiologic mechanisms regulating calcium homeostasis during pregnancy and the perinatal period generally operate very effectively. Thus, aberrations leading to clinically evident disease states are relatively infrequent.



## Thymus gland

The thymus is a soft, triangular-shaped organ found in the chest posterior to the sternum (Figure 11.4). The thymus produces the hormones seen in Table 11.6 called thymosins that help to train and develop T-lymphocytes during fetal development and childhood.

It has two functions: one is the differentiation of primitive lymphocytes and the other is the further proliferation of antigen-stimulating T cells by thymosine.

The thymus gland develops alongside the parathyroid glands between weeks 5 and 8. By week 9, lymphocytes appear in the lymph and blood and by week 16 a large majority of fetal lymphocytes are T cells, illustrating the early presence of an immune system. The T-lymphocytes produced in the thymus go on to protect the body from pathogens throughout a person's entire

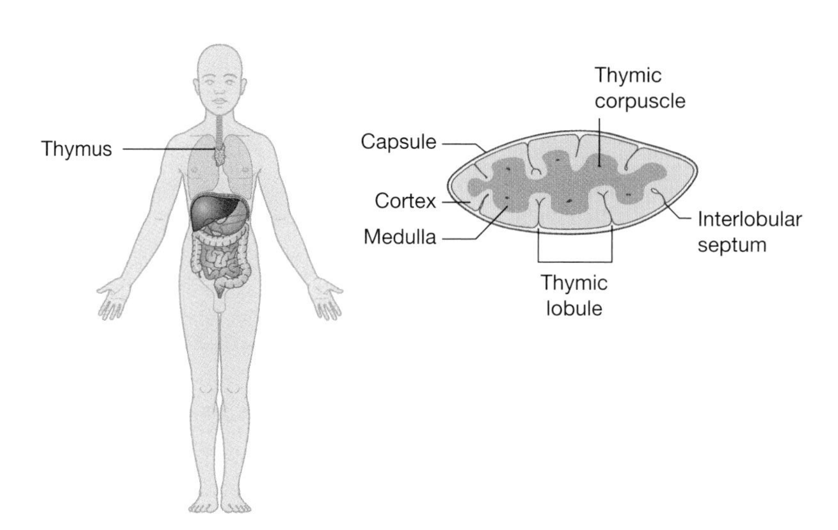


Figure 11.4 The thymus gland.

Table 11.6 Hormones of the thymus gland.

| Hormone | Acronym | Function |
|---|---|---|
| Thymosin | | Integral role in the maturation of T cells produced in the thymus gland as part of the immune system |
| Thymic humoral factor | THF | As above |
| Thymic factor | TF | As above |
| Thymopoietin | | As above |



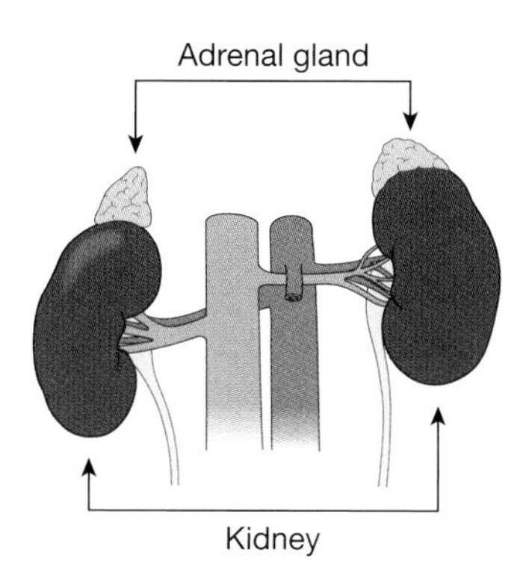


Figure 11.5 The adrenal glands.

life. The thymus is relatively large in children compared with adults and much more active during childhood. It then becomes smaller in size and inactive during puberty and is slowly replaced by adipose tissue throughout a person's life until it becomes quite small in older adults.

Because of this lifetime stockpile of T cells, an adult can have their thymus removed with no ill effects, but a child who loses their thymus is vulnerable to infection (Sauce and Appay, 2011).

## Adrenal glands

The adrenal glands are a pair of roughly triangular glands found immediately superior to the kidneys (Figure 11.5). They are each made of two distinct layers, the outer cortex and inner medulla, each with its own unique functions:

- The adrenal cortex produces many cortical hormones in three classes – glucocorticoids, mineralocorticoids and androgens.
- The adrenal medulla produces the hormones epinephrine and norepinephrine under stimulation by the sympathetic division of the autonomic nervous system.

The hormones from both parts of the adrenal gland can be seen in Table 11.7.

The adrenal glands develop from weeks 5 to 6 from both the mesoderm and ectoderm which continue to migrate and differentiate into the late fetal period. Generally, they are fully functional at birth and are involved in the regulation of many important functions, including the stress response and regulation of sodium and water.

Table 11.7 Hormones of the adrenal glands.

| Hormone | Function |
|---|---|
| Mineralocorticoids (mainly aldosterone) | Stimulates sodium reabsorption in the kidneys so increasing blood levels of sodium and water. Stimulates potassium and hydrogen secretion and subsequent excretion from the kidney |
| Glucocorticoids (mainly cortisol) | Stimulates gluconeogenesis and fat breakdown in adipose tissue, so increasing glucose availability in the blood. Promotes metabolism and resistance to stress by inhibiting inflammatory and immunological responses. Inhibits protein synthesis and glucose uptake in both muscle and adipose tissue |
| Gonadocorticoids (androgens, including testosterone and didehydroepiandrosterone) | Masculinization (virilization) in both males (although relatively small effect compared with androgenic effects of testes) and females |
| *Medulla* | |
| Adrenaline (epinephrine) | Fight or flight response – sympathomimetic – that is, mimics the effects of the autonomic nervous system during stress. For example, boosts oxygen and glucose supply to the brain and muscle by the heart rate and stroke volume increasing while decreasing the flow of blood to and function of organs that are not involved in responding to emergencies |
| Noradrenaline (norepinephrine) | As above |



## Clinical application

### Adrenal insufficiency – Addison's

Adrenal insufficiency leads to a reduction in adrenal hormone release; that is, glucocorticoids and/or mineralocorticoids. Primary insufficiency is an inability of the adrenal glands to produce enough steroid hormones. Addison's disease is the name given to the autoimmune cause. Secondary insufficiency is where there is inadequate pituitary or hypothalamic stimulation of the adrenal glands. Adrenal insufficiency is rare in children. Symptoms can include fatigue and weakness, anorexia, nausea, vomiting, weight loss, dizziness and confusion (Arlt and Allolio, 2003).

The term congenital adrenal hyperplasia encompasses a group of autosomal recessive disorders, which involves a deficiency of an enzyme involved in the synthesis of cortisol, aldosterone or both (Knowles *et al.*, 2011). The clinical manifestations of each form of congenital adrenal hyperplasia are related to the degree of cortisol deficiency and/or the degree of aldosterone deficiency (Speiser *et al.*, 2011). Females with severe forms of adrenal hyperplasia due to deficiencies of 21-hydroxylase, 11-beta-hydroxylase or 3-beta-hydroxysteroid dehydrogenase have ambiguous genitalia at birth due to excess adrenal androgen production *in utero*.

## Pancreas

The pancreas is a large gland located in the abdominal cavity just inferior and posterior to the stomach (Figure 11.6). The pancreas is considered to be a heterocrine gland as it contains both endocrine and exocrine tissue. The endocrine cells of the pancreas make up just about 1% of the total mass of the pancreas and are found in small groups throughout the pancreas called

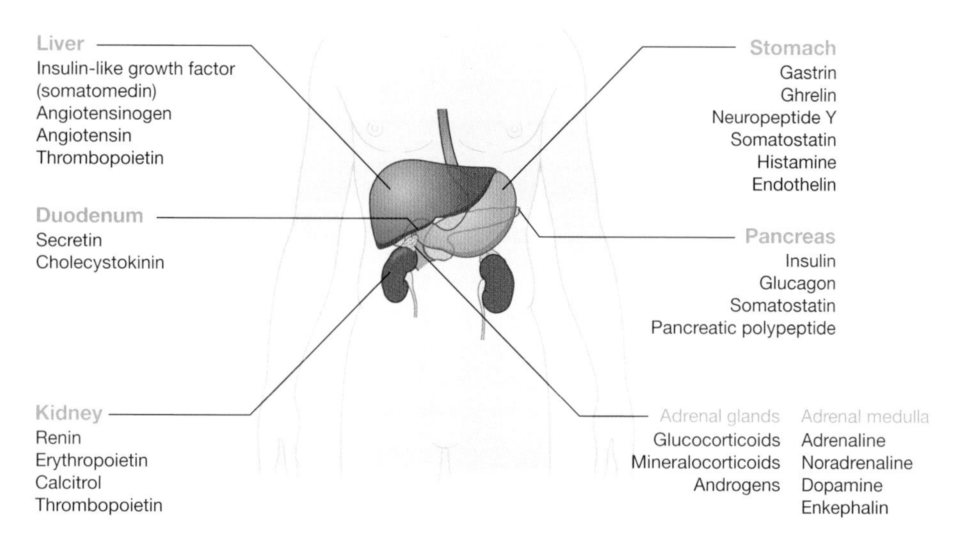


Figure 11.6 The gastrointestinal system and its hormones.



Table 11.8 Hormones of the pancreas.

| Hormone | Function |
|---|---|
| Insulin | Targets cells to take up and use free glucose, to convert glucose to glycogen (glycogenesis), to increase protein and lipid synthesis from glucose and to slow down glycogen breakdown into glucose (glycogenolysis). This decreases blood glucose levels |
| Glucagon | Targets the liver to break down glycogen into glucose and accelerates the conversion of lipids and proteins to form glucose in the liver (gluconeogenesis). This increases blood glucose levels |
| Somatostatin | Inhibits the release of insulin and glucagon and slows absorption of nutrients from the gastrointestinal tract |
| Pancreatic polypeptide | Inhibits the release of somatostatin, inhibits contraction of the gallbladder and inhibits secretion of digestive enzymes from the pancreas |

islets of Langerhans. Within these islets are two types of cells: alpha and beta cells. The alpha cells produce the hormone glucagon, which is responsible for raising blood glucose levels. Glucagon triggers muscle and liver cells to break down the polysaccharide glycogen to release glucose into the bloodstream. The beta cells produce the hormone insulin, which is responsible for lowering blood glucose levels. Insulin triggers the absorption of glucose from the blood into cells where it is added to glycogen molecules for storage. Hormones of the pancreas can be seen in Table 11.8.

The pancreas develops from weeks 10–11 from the endoderm, with insulin appearing in the 10th week and glucagon by the 15th week. The pancreas has a key role to play from birth in digestion and absorption. With regard to its hormonal action however, it is the interplay between insulin and glucagon that is the most important function in relation to glucose homeostasis (McKenna, 2000). Disorders to this vital regulatory process have important health implications for the child as they grow and develop.

# Clinical application

## Diabetes mellitus

Diabetes mellitus is a disease caused by deficiency or diminished effectiveness of endogenous insulin. It is characterized by hyperglycaemia, deranged metabolism and chronic complications. Type 1 diabetes mellitus results from the body's failure to produce sufficient insulin (Dunger and Todd, 2008; NICE, 2010). Type 2 results from resistance to the insulin (Reinehr, 2005; Beckwith, 2010). Gestational diabetes occurs in pregnant women who have high blood glucose levels during pregnancy. Maturity-onset diabetes of the young includes several forms of diabetes with monogenetic defects of beta-cell function (impaired insulin secretion), usually manifesting as mild hyperglycaemia at a young age (Vivian, 2006). Secondary diabetes accounts for a small number of diabetes mellitus and can be due to pancreatic disease and endocrine conditions such as Cushing's syndrome, acromegaly, thyrotoxicosis and glucagonoma, for example.



*Presentation:* polyuria, polydipsia, lethargy, weight loss, dehydration, ketonuria. Diabetes may be diagnosed on the basis of one abnormal plasma glucose (random $\geq$11.1 mmol/L or fasting $\geq$7 mmol/L) (NICE, 2009) in the presence of diabetic symptoms such as thirst, increased urination, recurrent infections, weight loss, drowsiness and, in the worst case scenario, coma (WHO, 2006).

## Gonads

The gonads – ovaries in females and testes in males – are responsible for producing the sex hormones of the body; see Figure 11.7. These sex hormones (Table 11.9) determine the secondary sex characteristics of adult females and adult males.

- **Testes.** The testes are a pair of organs found in the scrotum of males that produce the androgen testosterone in males after the start of puberty. Testosterone has effects on many parts of the body, including the muscles, bones, sex organs and hair follicles.
- **Ovaries.** The ovaries are a pair of glands located in the pelvic body cavity lateral and superior to the uterus in females. The ovaries produce the female sex hormones progesterone and

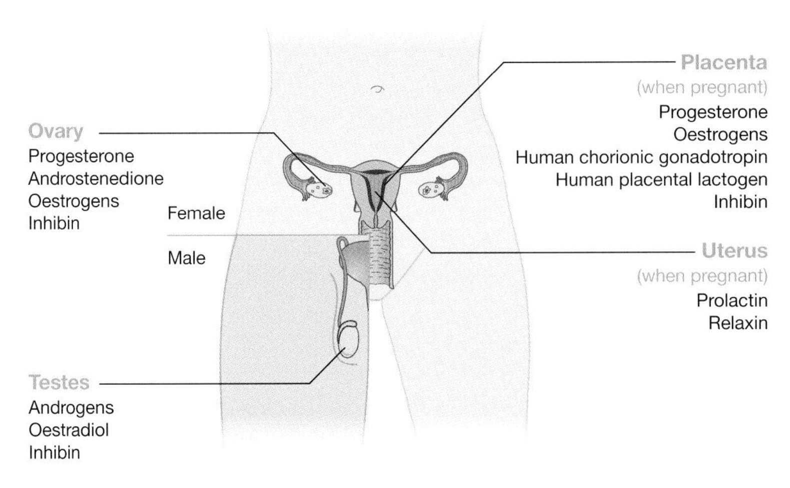


Figure 11.7 The reproductive system and its hormones.

Table 11.9 Hormones of the testes and ovaries.

| Hormone | Function |
|---|---|
| *Testes* | |
| Androgens (mainly testosterone) | Stimulates descent of testes at birth, regulates sperm production and promotes/maintains secondary sexual characteristics. *Includes:* anabolic growth of muscle mass and strength, increases in strength and density of the bones and muscles, including the accelerated growth of long bones during adolescence. Virilization and maturation of sex organs, scrotum formation, deepening of voice and growth of axillary hair. During puberty, testosterone controls the growth and development of the sex organs and body hair of males, including pubic, chest and facial hair |
| Oestradiol | Prevents death of germ cells |
| Inhibin | Inhibits the production and release of FSH |
| *Ovaries* | |
| Oestrogen | Development and maintenance of female sexual characteristics (height and bone formation increase, accelerates metabolism, stimulates endometrial, breast, pubic hair and uterine growth). Together with the gonadotropic hormones of the anterior pituitary gland and progesterone, helps regulate the menstrual cycle and maintain pregnancy. Plays a part in blood coagulation, regulation of sodium and water retention and increase of growth hormone and protein synthesis. Oestrogen also triggers the increased growth of bones during adolescence that lead to adult height and proportions |
| Progesterone | Together with the gonadotropic hormones of the anterior pituitary gland and oestrogen, helps regulate the menstrual cycle and maintain pregnancy. Prepares the endometrium for implantation, prepares breasts for milk production |
| Inhibin | Inhibits the production and release of FSH |
| Relaxin | Relaxes symphysis pubis, helps dilate uterine cervix at the end of pregnancy and plays a role in sperm motility |



oestrogen. Progesterone is most active in females during ovulation and pregnancy, where it maintains appropriate conditions in the human body to support a developing fetus. Oestrogens are a group of related hormones that function as the primary female sex hormones.

In fetal development, the genetic sex of a fetus depends on the nature of the sex chromosomes and this governs the sexual determination of the gonads. The gonads then either produce hormones (if male) or no hormones (if female). The hormonal environment determines the sex of the reproductive tract and developing external genitalia (White and Porterfield, 2013). At birth, the newborn baby may experience temporary changes in the breast and, in girls the vaginal area due to effects of pregnancy hormones. By the third day after birth, breast enlargement may be seen in either sex, which should disappear by week 2. Fluid may also leak from the breast, again ceasing by week 2. Vaginal changes in girls include swollen labia, white discharge and a small amount of bleeding, all a result of oestrogen exposure (Jameson and De Groot, 2010). Such changes will disappear over the early weeks of life.

The most significant changes to sexual maturation in both sexes are seen from the time approaching and during puberty (MacGregor, 2010). Dysfunctions to sexual differentiation and maturity may result from a failure to release the necessary hormones in a timely fashion, leading to precocious (early onset) or delayed puberty (Cesario and Hughes, 2007; Kaplowitz, 2010).

## Clinical application

### Ambiguous genitalia or disorders of sex development

Disorders of sex development are caused by a variety of different conditions, such as congenital adrenal hyperplasia (Knowles *et al.*, 2011; Speiser *et al.*, 2011) and mixed gonadal dysgenesis. *Presentation:* male appearance but with associated abnormalities of genitalia, severe hypospadias with bifid scrotum, undescended testis/testes with hypospadias, bilateral non-palpable testes in a full-term apparently male infant, female appearance but with associated abnormalities of genitalia, clitoral hypertrophy of any degree, non-palpable gonads, vulva with single opening (Hughes *et al.*, 2006; BSPED, 2009; Hutcheson and Snyder, 2012).



## Other hormone-producing organs

In addition to the glands of the endocrine system, many other non-glandular organs and tissues in the body produce hormones, and these are summarized in Table 11.10 (Greenstein and Wood, 2011).

**Table 11.10** Hormones of other organs in the body and their functions.

| Hormone | Acronym | Function |
|---|---|---|
| *Kidneys* | | |
| Renin | | Activates the renin–angiotensin pathway, which regulates sodium and water reabsorption or excretion from the kidney nephrons |
| Erythropoietin | | Stimulates erythrocyte production |
| Thrombopoietin | | Stimulates megakaryocytes to produce platelets |
| Calcitriol | | Increases absorption of calcium and phosphate from the gastrointestinal tract and kidneys, inhibits release of PTH |
| *Heart* | | |
| Atrial-natriuretic peptide | ANP | Increases sodium and water excretion by inhibiting renin secretion by the kidneys and thus secretion of aldosterone by the adrenal cortex |
| *Liver* | | |
| Insulin-like growth factor | | Insulin like effects. Regulates cell growth |
| Thrombopoietin | | Stimulates megakaryocytes to produce platelets |
| Angiotensinogen and angiotensin | | Stimulated by renin – see above; this conversion leads to the release of aldosterone from the adrenal cortex. Vasoconstriction of blood vessels |
| *Bone marrow* | | |
| Thrombopoietin | | As above for liver |
| *Stomach* | | |
| Gastrin | | Secretion of gastric acid |

Table 11.10 (*Continued*)

| Hormone | Acronym | Function |
|---|---|---|
| Ghrelin | | Stimulates hunger and appetite and has a potent effect on growth hormone release from the anterior pituitary gland |
| Neuropeptide Y | NPY | A potent feeding stimulant and causes increased storage of ingested food as fat |
| Somatostatin | | As for pancreas (see Table 11.8) |
| Histamine | | Stimulates gastric acid secretion |
| Endothelin | | Smooth muscle contraction of stomach |
| *Duodenum* | | |
| Secretin | | Stimulates the exocrine portion of the pancreas to secrete bicarbonate into the pancreatic fluid (thus neutralizing the acidity of the intestinal contents) |
| Cholecystokinin | | Stimulates delivery into the small intestine of enzymes from the pancreas and bile from the gallbladder to promote the digestion of protein and fat |
| *Skin* | | |
| Calcidiol | | Inactive form of vitamin D |
| *Adipose tissue* | | |
| Leptin | | Inhibits appetite by counteracting the effects of NPY (see above), counteracting the effects of anandamide, another potent feeding stimulant and promoting the synthesis of $\alpha$-MSH, an appetite suppressant |
| Oestrogen | | See Table 11.9 |
| *Local hormones* | | |
| Prostaglandin | | Plays a role in the normal inflammatory response to local damage – swelling, pain response and raised local temperature protecting the area from any further damage |
| Leukotrienes | | After prostaglandins have taken effect, reduction of inflammation while allowing white blood cells to move into the region to clean up pathogens and damaged tissues |
| *Placenta (when pregnant)* | | |
| Progesterone | | Supports pregnancy: decreases uterine smooth muscle contractility and inhibits onset of labour. Enriches the uterus with a thick lining of blood vessels and capillaries so that it can sustain the growing fetus |
| Oestrogen | | Promotes maintenance of corpus luteum during early pregnancy |
| Human chorionic gonadotropin | HCG | Promotes the maintenance of the corpus luteum during the beginning of pregnancy. This allows the corpus luteum to secrete the hormone progesterone during the first trimester |
| Human placental lactogen | | Modifies the metabolic state of the mother during pregnancy to facilitate the energy supply of the fetus |
| Inhibin | | Suppresses LH |

# Conclusion

This chapter has covered an overview of the endocrine system in relation to the essential knowledge of anatomy and physiology relevant to understanding the clinical application in children's nursing. It is important to understand the glands comprising the endocrine system and how the system works to control and regulate many important physiological body functions. It is clear how many vital functions are controlled by the endocrine system and its' hormones and how these work together closely in an integrated and organized manner to return the body to homeostasis under normal, healthy conditions. Knowing the normal function can help us understand when the system is compromised or does not develop normally and what disease this gives rise to. Endocrine disorders are rare on the whole; nonetheless, if they do occur, they can lead to significant and life-changing consequences for the child and their family that even with treatment, can be prolonged and even permanent. It is essential that children's nurses are clear about this system and the causes of such endocrine disorders should they care for them within the hospital and/or community setting.



# Activities

**Now review your learning by completing the learning activities in this chapter. The answers to these appear at the end of the book. Further self-test activities can be found at www.wileyfundamentalseries.com/childrensA&P.**

## True or false?

1. True or false? Endocrine glands have ducts for the release of hormones into the bloodstream.
2. True or false? The pituitary gland is the master gland of the endocrine system under control of the hypothalamus.
3. True or false? The pituitary gland and hypothalamus are situated within the brain stem.
4. True or false? The kidney, liver, heart and skin have secondary endocrine functions.

## Match the hormone with its function

1.

| | |
|---|---|
| i. Somatotropin | a. Stimulates the adrenal cortex |
| ii. Prolactin | b. Stimulates ovulation formation of corpus luteum in girls |
| iii. TSH | c. Stimulates melanin synthesis from skin |
| iv. ACTH | d. Increases water permeability in the kidney |
| v. Luteinizing hormone | e. Stimulates milk synthesis |
| vi. ADH | f. Stimulates thyroxine from the thyroid gland |
| vii. Melanocyte-stimulating hormone | g. Lactation let-down reflex |
| viii. Oxytocin | h. Growth hormone |

2.

| | | | |
|---|---|---|---|
| i. | Thyroxine | a. | Intake of glucose into cells from blood |
| ii. | Calcitonin | b. | Activates the renin–angiotensin pathway in water control in the body |
| iii. | Insulin | c. | Stress (fight–flight) response |
| iv. | Glucagon | d. | Regulates metabolism and stimulates body oxygen and energy consumption |
| v. | Renin | e. | Promotes resistance to stress by inhibiting inflammatory responses |
| vi. | Cortisol | f. | Water/sodium retention/increases blood pressure |
| vii. | Aldosterone | g. | Gluconeogenesis in liver |
| viii. | Adrenaline | h. | Stimulates osteoblasts |

## Conditions



The following table contains a list of conditions. Take some time and write notes about each of the conditions. You may make the notes taken from text books or other resources; for example, people you work with in a clinical area or you may make the notes as a result of people you have cared for. If you are making notes about people you have cared for you must ensure that you adhere to the rules of confidentiality.

| Condition | Your notes |
|---|---|
| Diabetes mellitus | |
| Diabetes insipidus | |
| Congenital adrenal hyperplasia | |
| Thyrotoxicosis | |
| Pheochromocytoma | |

# Glossary

Source: Adapted in part from De Wood (2011).

**Adrenal cortex:** the outer component of the adrenal gland; secretes cortisol and aldosterone.

**Adrenal medulla:** the inner component of adrenal gland; secretes adrenaline (epinephrine) and noradrenaline (norepinephrine).

**Circadian rhythm:** also known as 'biological clock'; our internal time-measuring mechanism that adjusts according to night and day, seasonally or both in response to environmental cues.

**Adenosine monophosphate (cAMP):** synthesized from adenosine triphosphate in cellular respiration.

**Endocrine system:** system of glands, cells and tissues integrally linked to the nervous system; controls functions through the secretion of hormones and other chemicals.

**Gonad:** primary reproductive organ in which human gametes are produced.

**Hormone:** signalling molecule secreted by the endocrine glands that stimulates or inhibits activities of any cell via the action on receptors. Hormones are transported by the bloodstream.

**Hypothalamic inhibitor:** hypothalamic molecule that suppresses a particular secretion by the anterior lobe of the pituitary gland.

**Hypothalamus:** centre of homeostatic control over the body's internal environment (e.g. water and sodium balance, temperature); influences hunger, thirst, the stress response, sexual differentiation and emotions.

**Local signalling molecule:** a secretion that alters chemical conditions in localized tissues (e.g. prostaglandin).

**Negative-feedback mechanism:** a homeostatic mechanism by which a condition that changed as a result of an activity triggers a response that stops or reverses the change.

**Neurotransmitter:** signalling molecules secreted by neurones that act on cells transmitting signals across synapses (gaps between neurones); they are then rapidly degraded or recycled.



**Pancreatic islet:** clusters of pancreatic endocrine cells.

**Parathyroid gland:** one of four small glands embedded in the thyroid; their secretions influence blood calcium levels.

**Peptide hormone:** a hormone that binds to a membrane receptor, thus activating enzyme systems that alter target cell activity. A second messenger in the cell often relays the hormone's message.

**Pineal gland:** light-sensitive endocrine gland; its melatonin secretions affect internal circadian rhythm.

**Pituitary gland:** endocrine gland that functions closely with the hypothalamus; controls many physiological functions influencing many other endocrine glands. Its posterior lobe stores and secretes hypothalamic hormones. Its anterior lobe produces and secretes its own hormones.

**Positive-feedback mechanism:** homeostatic control that initiates a chain of events that intensify change from an original condition.

**Puberty:** period of development when secondary sexual traits emerge and mature.

**Releasing hormone (hypothalamic):** hypothalamic molecule that enhances or slows secretions from target cells in the anterior lobe of pituitary gland.

**Second messenger:** molecule within a cell that mediates a hormonal signal.

**Steroid hormone:** lipid-soluble hormone made from cholesterol that acts on target cell DNA.

**Thymus gland:** endocrine gland that produces thymosins that help to develop T-lymphocytes necessary for immunity.

**Thyroid gland:** endocrine gland that secretes hormones influencing growth, development and metabolic rate.

**Tropic hormone:** one that influences or stimulates another gland to release it's hormone.

## References


Arlt, W., Allolio, B. (2003). Adrenal insufficiency. *The Lancet*, **361** (9372), 1881–1893.

Beckwith S. (2010) Diagnosing type 2 diabetes in children and young people. *British Journal of School Nursing*, **5** (1), 15–19.

BSPED (2009) Statement on the management of gender identity disorder (GID) in children & adolescents, British Society of Paediatric Endocrinology & Diabetes, http://www.gires.org.uk/assets/Youngsters Treatment/BSPEDStatementOnTheManagementOfGID.pdf (accessed 7 July 2014).

Cesario, S.K., Hughes, L.A. (2007) Precocious puberty: a comprehensive review of literature. *Journal of Obstetric, Gynecologic, & Neonatal Nursing*, **36** (3), 263–274.
Dattani, M., Preece, M. (2004) Growth hormone deficiency and related disorders: insights into causation, diagnosis, and treatment. *The Lancet*, **363** (12), 1977–1987.
De Wood, D. (2011) Glossary Chapter 36 Endocrine system, Biology @ AIPCV, http://aipcvbiology.blogspot.co.uk/2011/09/glossary-chapter-36-endocrine-system.html (accessed 22 July 2014).
Dorton, A.M. (2000) The pituitary gland: embryology, physiology and pathophysiology. *Neonatal Network*, **19** (2), 9–17.
Dunger, D.B., Todd, J.A. (2008) Prevention of type 1 diabetes: what next? *The Lancet*, **372** (9651), 1710–1711.
Greenstein, B., Wood, D. (2011) *The Endocrine System at a Glance*, Wiley–Blackwell, Oxford.
Hinson, J.P., Raven, P., Chew, S.L. (2010) *The Endocrine System: Systems of the Body Series*, 2nd edn, Churchill Livingstone, Edinburgh.
Hughes, I.A., Houk, C., Ahmed, S.F. *et al.* (2006) Consensus statement on management of intersex disorders. *Archives of Disease in Childhood*, **91** (7), 554–563.
Hutcheson, J., Snyder, H.M. (2012) Ambiguous genitalia and intersexuality, Medscape, http://emedicine.medscape.com/article/1015520-overview (accessed 7 July 2014).



Jameson, J.L., De Groot, L.J. (2010) *Endocrinology: Adult and Pediatric*, 6th edn, Saunders, Philadelphia, PA.
Kaplowitz, P. (2010) Precocious puberty, Medscape, http://emedicine.medscape.com/article/924002-overview (accessed 7 July 2014).
Khardori, R. & Ulla, J. (2012) Diabetes insipidus, Medscape, http://emedicine.medscape.com/article/117648-overview (accessed 7 July 2014).
Knowles R.L., Oerton, J.M., Khalid, J.M. *et al.* (2011) British Society for Paediatric Endocrinology and Diabetes Clinical Genetics Group Clinical outcome of congenital adrenal hyperplasia (CAH) one year following diagnosis: a UK wide study. *Archives of Disease in Childhood*, **96**, A27, doi:10.1136/adc.2011.212563.54.
Kumar, P.G., Anand, S.S., Sood, V., Kotwal, N. (2005) Thyroid dyshormonogenesis. *Indian Pediatrics*, **42** (12), 1233–1235.
MacGregor, J. (2010) *Introduction to the Anatomy and Physiology of Children*, 2nd edn, Routledge, London.
Molina, P. (2013) *Endocrine Physiology*, 4th edn, McGraw-Hill Medical, New York, NY.
McKenna, L.L. (2000) Pancreatic disorders in the newborn. *Neonatal Network*, **19** (4), 13–20.
NHS (2013) *Congenital Hypothyroidism: Initial Clinical Referral Standards and Guidelines*, UK Newborn Screening Programme Centre.
NICE (2009). *Type 1 diabetes: diagnosis and management of type 1 diabetes in children, young people and adults*, NICE guidelines update, RCOG Press, London.
NICE (2010) Diabetes – type 1, NICE Clinical Knowledge Summary, http://cks.nice.org.uk/diabetes-type-1 (accessed 7 July 2014).
Phillips, L., Appleton, R.E. (2004) Systematic review of melatonin treatment in children with neurodevelopmental disabilities and sleep impairment. *Developmental Medicine & Child Neurology*, **46** (11), 771–775.
Reinehr, T. (2005) Clinical presentation of type 2 diabetes mellitus in children and adolescents. *International Journal of Obesity*, **29**, S105–S110, www.nature.com/ijo/journal/v29/n2s/full/0803065a.html (accessed 7 July 2014).
Rogers, K. (ed.) (2012) *The Endocrine System*, Britannica Educational Publishers, New York, NY.
Sauce, D., Appay, V. (2011) Altered thymic activity in early life: how does it affect the immune system in young adults? *Current Opinion in Immunology*, **23** (4), 543–548.
Schneider, H.J., Aimaretti, G., Kreitschmann-Andermahr, I. *et al.* (2007) Hypopituitarism. *The Lancet*, **369** (9571), 1461–1470.
Speiser, P.W., Azziz, R., Baskin, L.S. *et al.* (2011) Congenital adrenal hyperplasia due to steroid 21-hydroxylase deficiency: an endocrine society clinical practice guideline. *Archives of Disease in Childhood*, **96**, A27, doi:10.1136/adc.2011.212563.54.
Vivian, M. (2006) Type 2 diabetes in children and adolescents – the next epidemic? *Current Medical Research and Opinion*, **22**, 297–306.
Waugh, A., Grant, A. (2010) *Ross and Wilson Anatomy and Physiology in Health and Illness*, 11th edn, Churchill Livingstone, Edinburgh.
White, B.A., Porterfield S.P. (2013) *Endocrine and Reproductive Physiology*, 4th edn, Elsevier, Philadelphia.
WHO (2006) *Definition and Diagnosis of Diabetes Mellitus and Intermediate Hyperglycaemia*, World Health Organization, Geneva.
Willacy, H. (2011) Childhood and congenital hypothyroidism, Document ID: 1164 Version: 22, www.patient.co.uk/doctor/childhood-and-congenital-hypothyroidism (accessed 7 July 2014).

# Chapter 12

# The digestive system and nutrition

Joanne Outteridge

*Department of Family and Community Studies, Anglia Ruskin University, Cambridge, UK*

## Aim

The aim of this chapter is to explore the role of the digestive system in removing nutrients from the food that a child eats to enable growth and development.

## Learning outcomes

On completion of this chapter the reader will be able to:

- Outline the nutritional requirements of the growing child and identify the food groups necessary to maintain healthy growth and development.
- Label the alimentary canal and identify where each part of the digestive system sits in relation to other organs.
- Discuss how the normal functioning of the digestive system removes the required nutrients from ingested food and eliminates waste products.
- Critically analyse how normal growth, development and family functioning are affected when a child has a disorder of the alimentary canal, or difficulty maintaining adequate nutritional intake.

*Fundamentals of Children's Anatomy and Physiology: A Textbook for Nursing and Healthcare Students*, First Edition. Edited by Ian Peate and Elizabeth Gormley-Fleming.

Companion website: www.wileyfundamentalseries.com/childrensA&P

## Test your prior knowledge

- The digestive system normally has two external openings. Put the following structures in order, through which food passes from the mouth to the anus: *small intestine*; *oropharynx*; *anus*; *stomach*; *mouth*; *rectum*; *oesophagus*; *large intestine*.
- What are the three main food groups that we all need to ingest to enable healthy body functioning?
- What is the difference between mechanical digestion and chemical digestion?
- How many 'milk' teeth do children have?
- What is the process called by which solid food moves through the alimentary canal?
- What is the function of the villi in the small intestine?
- Label the duodenum, jejunum and ileum.



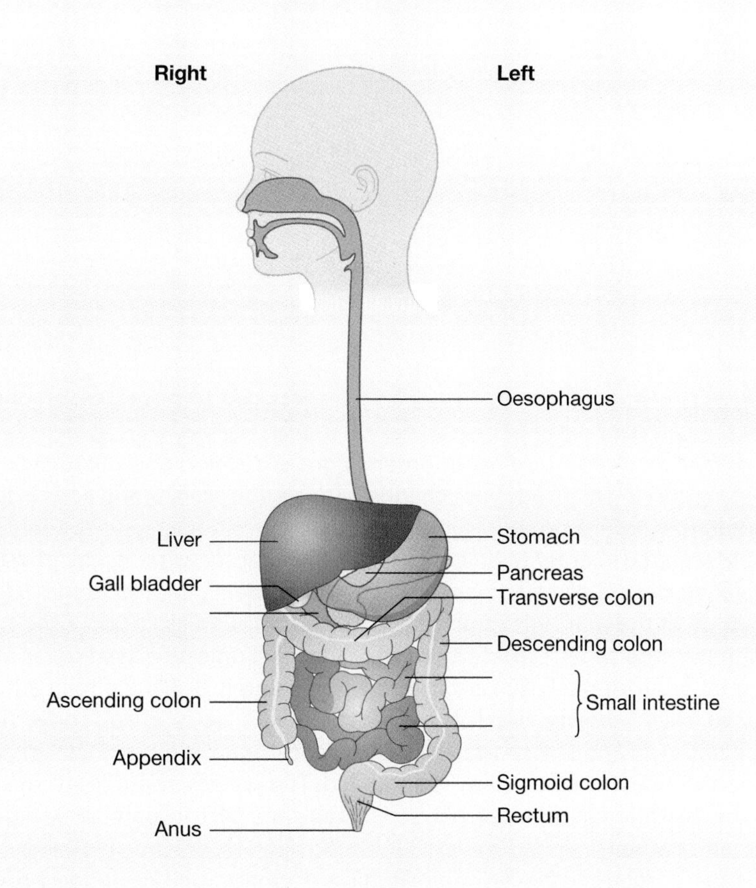


- What is the pH of normal stomach acid? What colour is this on pH paper?
- Define constipation. In a child without an underlying medical disorder, what are the common reasons for constipation, and how may it be diagnosed?
- What are the definitions of vomiting, haematemesis, diarrhoea, steatorrhoea, melaena and meconium?

# Body map

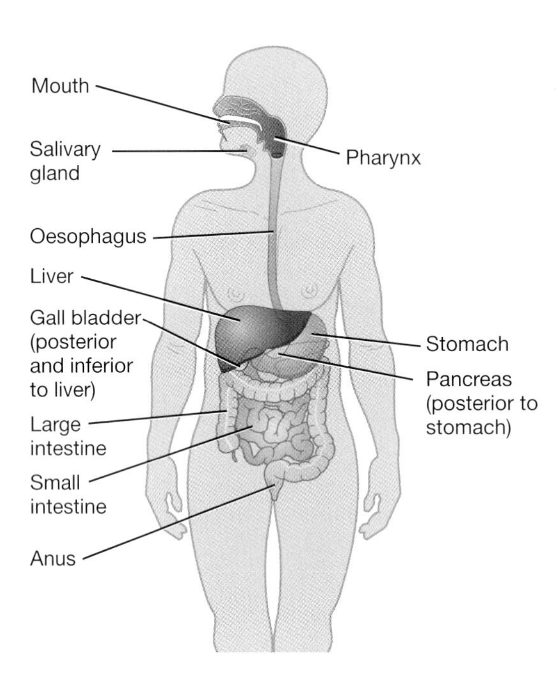




# Introduction

All living organisms need nutrients in order to grow, develop and repair themselves. Growth begins *in utero*, continues throughout childhood and adolescence, and ends in late adolescence or early adulthood. Whilst growth is regulated by the hormonal system, adequate energy and nutrients are required in order to build new cells. This chapter will consider how the fetus, infant, child and young person gains and processes those nutrients in order to grow and develop.

Practically all of the nutrients required are gained through ingestion – the taking in of food or drink through the mouth. These then pass into the digestive system, or alimentary canal (growing from approximately 3 m at birth to approximately 8–10 m long in adults), where they are processed. The necessary nutrients are retrieved from the food and converted into a format that can be removed and transported into the bloodstream and lymphatic system for distribution around the body to where they are needed. The waste products are then eliminated. In addition, the gastrointestinal tract plays an important part in immunity, digesting ingested pathogens and toxins.

This chapter is organized by firstly labelling the normal structures found within the digestive system (Figure 12.1). There will then follow a brief outline of normal fetal development of the digestive system, and an introduction to the more common congenital anomalies of the digestive tract.

There will follow an outline of the functions of the digestive system in relation to the nutrients that the body requires to maintain normal healthy functioning. The structure and function of each part of the digestive system will then be considered in detail, with each structure's role in the digestion of nutrients or elimination of waste clearly stated. Throughout there will be ques-

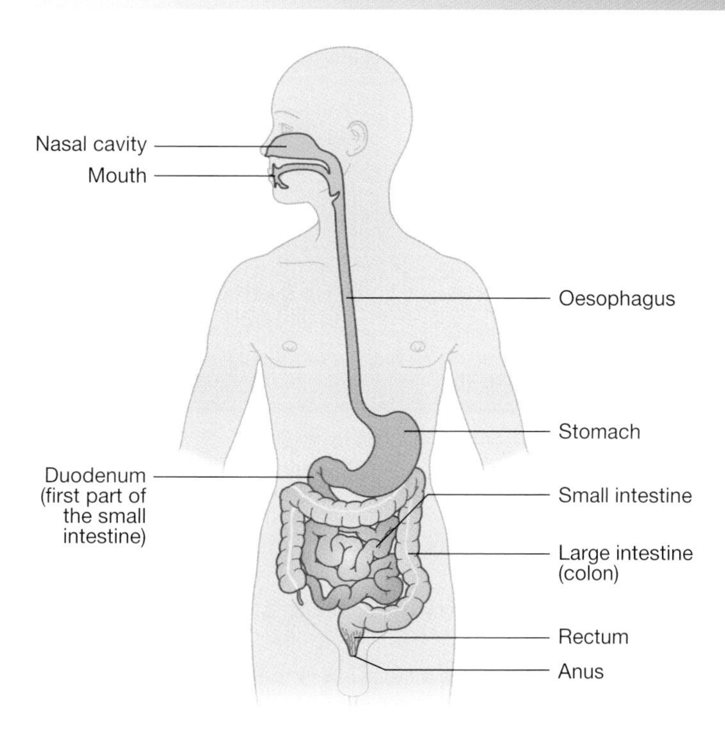


Figure 12.1 The structures associated with the digestion of food.



tions, activities and case studies to help you to process this information, and to gain an understanding to a level that will allow you to apply it to practice.

The common nutrient groups that can only be gained from ingestion of food and drink are carbohydrate, fats and proteins. These are broken down by the processes of mechanical and chemical digestion. Mechanical digestion can be defined as the use of movement to break food matter into smaller particles and to mix food matter with digestive enzymes. Chemical digestion can be defined as the use of the chemicals (acids, enzymes) in digestive juices to break down food matter into its constituent components which can then be absorbed into the blood or lymph.

## Carbohydrates

Carbohydrates contain carbon, hydrogen and oxygen atoms. Large, complex carbohydrates (starches) are metabolized into sugars, which pass from the gastrointestinal tract into the bloodstream. The breaking down of complex carbohydrates takes time, and leads to slow release of sugars, giving a sustainable source of energy. The simplest sugar is glucose ($C_6H_{12}O_6$). This is used by cells for energy. In the presence of insulin (see Chapter 11) it is transported from the blood across the cell membrane. Inside the cell, aerobic respiration takes place (adding oxygen to break down the glucose and release energy – see Chapter 10). The waste products are then carbon dioxide ($CO_2$ – which passes back into the blood for exhalation at the lungs – see Chapter 10) and water (water of oxidation). This can be represented by the equation

$$6O_2 + C_6H_{12}O_6 \rightarrow 6CO_2 + 6H_2O$$

## Clinical application

If refined processed sugars are eaten, they pass into the bloodstream very quickly, as not much further metabolism is needed. This leads to a 'sugar rush', which can be seen as excessive bursts of energy in some children. However, the danger is that, in order to combat this sudden high sugar level, large amounts of insulin are released (instead of the smaller, more steady quantities released when complex carbohydrates are ingested). It is thought that this could contribute to type 2 diabetes later in life (Gross *et al.*, 2004). It also takes time for the released insulin to be metabolized and excreted, leading to feelings of hunger. The child may therefore seek more food, even though they have no nutritional need to do so, which can contribute to obesity. The sequence of events is therefore

High blood glucose
↓
Release of insulin
↓
Normalization of blood glucose
↓
Excess insulin in circulation
↓
Stimulation of hunger



## Fats

Fats are lipid molecules. They pass from the gastrointestinal tract into the lymphatic system (see Chapter 7). Lipids are necessary for the formation of cell membranes (see Chapter 4 and the lipid bilayer), the formation of some hormones, the production of antibodies and recognition of antigens.

## Proteins

Proteins are the building blocks of cells; ingested proteins are broken down into their constituent amino acids. They form the 'head' of the lipid bilayer of cell membranes, and they are necessary for production of DNA and RNA in the nucleus of the cell and during cell division. The source can be animal or vegetable. There are 20 naturally occurring amino acids; if we cannot gain these from the diet, then the liver breaks down the ingested proteins and 'rebuilds' the amino acids that we require (transamination).

Carbohydrates, fats and proteins are all organic compounds because they contain a carbon atom (see Chapter 3). In addition, in order for enzymes and cells to work effectively, inorganic compounds are also needed in the form of vitamins and minerals. Some of these are ingested, but some are synthesized by the body itself (for example, vitamin D is synthesized when the skin is exposed to sunlight).

Table 12.1 outlines many of the vitamins and minerals necessary and their role in normal growth and development, what foods they are found in and what happens if these are not consumed. Greater detail, and the less common deleterious effects of overconsumption, can be found in Clancy and McVicar (2009).

Table 12.1 Vitamins and minerals required for growth and development, the food from which they are derived and the effect of deficit (adapted from Clancy and McVicar (2009) and Tortora and Grabowski (2000)).

| | Role | Food group | Effects of deficit |
|---|---|---|---|
| *Vitamins* | | | |
| A | Maintain epithelial health<br>Bone growth<br>Rhodopsin pigment in rods in eyes | Liver, green leafy vegetables, fish-oils, milk; synthesized in gastrointestinal tract from betacarotene | Atrophy of epithelial cells; drying of cornea<br>Slow bone development<br>Night blindness |
| $B_1$ (thiamine) | Co-enzyme for carbohydrate metabolism<br>Synthesis of acetylcholine (neurotransmitter) | Whole grains, eggs, pork, nuts, liver, yeast | Beri-beri – paralysis of gastrointestinal tract smooth muscle<br>Polyneuritis – degeneration of myelin sheaths |
| $B_2$ (riboflavin) | Co-enzyme in carbohydrate and protein metabolism in cells of the eye, skin, gastrointestinal tract and blood | Liver, beef, veal, lamb, eggs, asparagus, peas, peanuts, whole grains, yeast | Blurred vision, cataracts, dermatitis, intestinal lesions, anaemia |
| $B_3$ (niacin/ nicotinamide) | Co-enzyme for cellular respiration<br>Assists in breakdown of fats and inhibits cholesterol production | Meats, liver, fish, whole grains, yeast, peas, beans, nuts | Pellagra – dermatitis, diarrhoea, psychological disturbances |
| $B_6$ (pyridoxine) | Co-enzyme in amino acid and fat metabolism<br>Assists in antibody production | Liver, salmon, whole grains, yeast, spinach, yoghurt, tomatoes. Synthesized by bacteria in gastrointestinal tract | Dermatitis of mouth, nose and eyes; slowed growth |
| $B_{12}$ (cyano-cobalamin) | Co-enzyme needed in formation of red blood cells<br>Co-enzyme for formation of amino acids<br>Co-enzyme in production of neurotransmitter choline | Meat, liver, kidney, eggs, milk, cheese | Pernicious anaemia, psychological disturbances and nerve degeneration |
| C (ascorbic acid) | Formation of connective tissue<br>Promotes protein metabolism – wound healing | Green leafy vegetables, citrus fruits, tomatoes | Scurvy – poor connective tissue growth, including swollen gums and loose teeth<br>Anaemia<br>Poor growth and healing, and fragile blood vessels |
| D | Absorption of calcium and phosphate from the gastrointestinal tract | Egg yolk, fish liver oils, milk<br>Synthesized under skin in UV light | Rickets – poor growth of long bones in children |
| E | Necessary for nucleotides for cell formation, promoting wound healing, neural function | Green leafy vegetables, whole grains, nuts, seeds | Abnormal membranes and difficulties in muscular functioning |

*(Continued)*

Table 12.1 (*Continued*)

| | Role | Food group | Effects of deficit |
|---|---|---|---|
| K | Co-enzyme necessary for producing clotting factors | Liver, spinach, cauliflower, cabbage | Delayed clotting |
| *Minerals* | | | |
| Sodium | Main component of extracellular fluid<br>Needed for action potentials for nerves and muscles<br>Part of pH buffer system | Salt added to prepared foods; seafoods | Hypovolaemia; poor nerve conduction and muscle contraction |
| Potassium | Main ion in intracellular fluid<br>Needed for action potentials for nerves and muscles | Most foods; high in bananas | Muscle weakness and cramps<br>Cardiac arrhythmias when $<2$ mmol/L |
| Calcium | Needed for growth of bones and teeth<br>Blood clotting<br>Muscle contraction<br>Needed for production of neurotransmitters | Green leafy vegetables, milk, shellfish, egg yolk | Loss of bone density<br>Delayed clotting<br>Poor muscle contractility (especially cardiac)<br>Nerve dysfunction |
| Magnesium | Normal muscle and nerve functioning | Green leafy vegetables, whole grains, peas, peanuts, bananas | Hypertension; muscle weakness |
| Iron | Major component of haemoglobin<br>Needed for intracellular respiration | Meat, liver, egg yolk, whole grains, nuts, beans, pulses | Anaemia |
| Copper | Works with iron for haemoglobin formation<br>Co-enzyme for melanin formation | Liver, fish, whole grains, eggs, beans, asparagus, spinach | Anaemia<br>Decreased growth<br>Cerebral degeneration |
| Chlorine (chloride) | Main component of intracellular and extracellular fluids<br>Necessary for acid–base balance<br>Formation of HCl in stomach | Salt intake in processed foods | Disruption of acid–base balance (rare) |
| Phosphorus (phosphate) | Formation of bones and teeth<br>Part of pH buffer system<br>Nerve conductivity<br>Component of nucleotides, and adenosine triphosphate | Meat, fish, poultry, milk, nuts | Poor growth of bones and teeth |
| Iodine (iodide) | Main component of thyroid hormones | Cod-liver oil, seafood, some vegetables from iodine-rich soil areas | Thyroid deficiency |

## Clinical application

The lack of vitamin D synthesis from ultraviolet light leads to the disease 'rickets'. This has shown resurgence in developed countries, partly as a result of parents avoiding exposing their child to UV light owing to the risks associated with sunburn and basal cell carcinomas. The impact of behaviour on vitamin D synthesis is examined in the article by Shaw and Mughal (2013).

# Fetal development and infant nutrition



Throughout fetal life, nutrition is via the placenta, with nutrients passing from the mother's blood, across the placental membrane, to the fetal blood. In addition, soluble waste products are passed from the fetal blood, across the placental membrane, back into the maternal circulation.

Formation of the gastrointestinal tract begins at day 14 after fertilization with a cavity called the primitive gut. This primitive gut is a single tube, fixed at both ends (the mouth and the anus). As it grows it convolutes, and by week 3 there is differentiation of the foregut, midgut and hindgut (Chamley *et al.*, 2005).

- The foregut becomes the mouth, oesophagus, stomach and duodenum until the bile duct (also the respiratory tract).
- The midgut becomes the duodenum after the bile duct, jejunum, ileum and proximal two-thirds of the colon.
- The hindgut becomes the distal transverse, descending and sigmoid colon, rectum and anal canal (also the urinary bladder and urethra).

At 6–8 weeks' gestation the gut is completely occluded by inner epithelial cells, but it then re-canalizes. This early growth is so rapid that the gut extrudes from the abdominal cavity into the umbilical cord. However, by around 10 weeks it will be withdrawn as the abdominal cavity increases in size; failure of the abdominal wall to close following this leads to an exomphalos. As the primitive gut convolutes, it positions itself around the other developing organs; failure of this process may lead to malrotation of the gut.

As with any tube or cavity, growth *in utero* is dependent on a flow of fluid through it. By 14 weeks there is evidence of peristalsis, and by 16 weeks the fetus swallows about one-third of the total amniotic fluid per hour. This provides about 10% of the protein requirement, and it also stimulates gastrointestinal tract mucosal growth, and liver and pancreatic growth.

Anatomically, the gastrointestinal tract is complete from 24 weeks' gestation. However, digestive enzymes only start to be produced between weeks 24 and 28, glycogen storage in the liver begins at week 31, and the coordination of sucking, swallowing and peristalsis is not present until 34 weeks (Neu, 2007). Postnatal development of the gut continues from growth factors found in human milk. At birth, there is reduced stomach acid, meaning that ingested micro-organisms are not destroyed. This allows them to pass through into the intestines, and is thought to play an important role in development of immunity and recognition of antigens.

As the digestive system grows *in utero*, there will be a small quantity of solid waste product in terms of cellular debris. This passes into the intestinal lumen and mixes with swallowed amniotic fluid and waste products from the liver in terms of breakdown of red blood cells. This forms a small quantity of black–green tarry waste called meconium. After the infant has been delivered and begins to feed, this meconium will pass through the intestines and be excreted; normally within 48 h (Glasper *et al*., 2010).

As with all developing body systems, there is the potential for formation not to follow the normal pattern, and for congenital anomalies to occur. These can affect any part of the digestive system: the oral cavity could form incompletely, causing cleft lip and/or palate; the trachea and oesophagus may not separate completely, causing a tracheo-oesophageal fistula; the abdominal wall may not fully enclose the digestive system, causing a gastroschisis or exomphalos; the outside opening of the anus may not form, causing imperforate anus. A discussion of each of these, and other less common malformations, can be found in neonatal textbooks.

Infant nutrition deserves separate consideration for several reasons. First, infant digestive systems have physical immaturities up until the age of 4 months, which means that they have a limited ability to digest fats and can only digest simple proteins and carbohydrates. In addition, they are developing the coordination necessary in order to take liquid or semi-solid food into the mouth and to swallow this without aspirating it into the respiratory tract or initiating the gag reflex. Initially, a purely liquid diet is recommended; either breast milk or infant formula. Whilst formula milk provides similar nutrition, the profile differs from breast milk, and the non-nutrient benefits cannot be added.



Box 12.1 outlines the features of breast milk over other milk substitutes that make it ideal for infant nutrition.

As the infant develops tongue movements and swallowing coordination, semi-solid foods are introduced (weaning). This is necessary as breast and formula milks do not provide adequate nutrition alone after 6 months of age (Neill and Knowles, 2004). As the infant develops and uses their mouth to explore their world (sucking fingers, putting toys in their mouth), they will be able to tolerate more solid foods. The type of food at this stage also depends on their newly erupting dentition (or 'teething'). Culture, socioeconomic status, religious practices and personal preference of the parents all affect weaning practices. This leads to different families choosing whether to introduce soft foods, hard foods, whether the parent leads by introducing

## Box 12.1 Components of breast milk that make it ideal for infant nutrition.

- High whey-to-casein ratio, meaning that it is more easily digested.
- Low protein concentration – necessary for immature kidneys.
- Increased concentration of fats to provide energy; in particular, long-chain polyunsaturated fats for brain development.
- Carbohydrate is present in the form of lactose. The infant has less amylase to digest this, but breastmilk is high in mammary amylase to assist this digestion.
- Human milk has antimicrobial factors inhibiting anaerobic growth.
- Human milk is a less effective buffer; this means that acids pass into the lumen and acidify the lumen.
- Iron in breastmilk is in the form of lactoferrin, which is easily absorbed.
- Breastmilk contains enzymes, immunoglobulin A, growth factors and some hormones, which help normal growth, development and maturation.

Adapted. *Source:* Wilcox (2004).

a spoon to the baby's mouth, or whether the baby leads by picking up foods itself. The age at which weaning is recommended does change as new research and recommendations emerge. Currently in the UK it is 6 months of age (http://www.nhs.uk/Conditions/pregnancy-and-baby/Pages/solid-foods-weaning.aspx#close).

There are suggestions that late weaning could be a trigger for development of inappropriate immune responses to food (allergies – see Chapter 7). This is because the digestive tract is programmed to accept 'foreign bodies' in terms of food with foreign proteins, and therefore does not develop a response if this is the *first* exposure to that protein. However, in late weaning, the first exposure to these foreign proteins is usually through the skin in the form of proteins found in house dust. This is made worse by the cultural bathing of babies; frequent bathing results in very dry skin and resultant breaches in the epidermis through which the foreign proteins can enter. As the proteins entering through the skin are the infant's first exposure to these proteins in late weaning, there is a hypothesis that this can lead to an immune response in hyper-allergenic individuals. At a later date, when this food substance is then ingested for the first time, an immune response already exists, and the child may show a food allergy. There is currently a major research trial in the UK investigating this hypothesis – see http://www.eatstudy.co.uk/eat-study-info/.



As the infant becomes a toddler, their hand–eye–mouth coordination will develop to allow them to use a spoon by themselves. However, this physical development also coincides with social and emotional development (see Chapter 1). Infants quickly learn that eating is one aspect of their lives over which they can have some control. They can make their carer smile and praise them by eating the food; they can get laughter (or reprimands) by putting hands in food, or food on their face; and they can get a carer's complete attention and cause upset by refusing to open their mouth or spitting out all food and refusing to eat. The way in which carers manage this behaviour can affect a child's future eating habits and perception of sense of control by manipulating these habits. Thus, nutrition in early childhood, and often throughout a person's life, is affected by emotional responses, preferences in taste and environmental triggers as much as it is by nutritional requirements.

Children are growing continuously, and this requires a proportionately large number of calories per kilogram body weight. They also require a high-protein diet in order to build new cells, in addition to replacing worn out cells. Fats are also necessary for the production of new cells and for the production of growth hormones; vitamins A, D, E and K are also only fat soluble. Therefore, it is recommended that all children up to the age of 2 should receive full-fat versions of milk and yoghurts (http://www.nhs.uk/Conditions/pregnancy-and-baby/Pages/drinks-and-cups-children.aspx#close). This can be confusing for parents where societal pressures are to prevent obesity and to consume a low-fat diet.

The process of digestion begins with the sight or smell of food, or sounds associated with predicted food intake. You will have experienced how the smell of food when walking past a hot-food outlet can make you salivate, or how your stomach rumbles if someone else opens a noisy food wrapper when you are hungry. This prepares the alimentary canal for the arrival of food by releasing those digestive juices necessary for processing the predicted intake of food. Cranial nerve X (vagus nerve) is responsible for stimulating the digestive system (see Chapter 15).

The feeling of hunger is triggered by the endocrine system (see Chapter 11). A drop in blood glucose levels causes the alpha cells of the pancreas to release glucagon. This converts the glycogen stored in the liver back into glucose to raise blood glucose back to normal limits (4–7 mmol/L). It also triggers the child to feel hunger, and seek food. This is the point at which the digestive juices containing all of the enzymes necessary to digest the predicted food intake will start to be activated at the sight and smell of food.

## Clinical application

Consider what might happen to this mechanism of controlling hunger if a child is allowed to 'graze' and eat snacks continuously throughout the day? How might this lead to limited food exposure or 'picky eaters'? What are the implications for practice?

For a child to ingest food, they need age-appropriate food to be provided in appropriate quantities, at an appropriate temperature and at regular intervals, as they have a rapid metabolism. Any concerns over a child being under- or overweight need to take these issues into consideration; and although physical reasons need to be considered, safeguarding issues in terms of neglect must also remain a consideration.



# The anatomy and physiology

Each of the structures within the gastrointestinal system will now be discussed. However, in order to function, each needs a blood supply and drainage, and innervation.

Arterial blood supply to the gastrointestinal system is provided by

- the coeliac artery (foregut – oesophagus to proximal duodenum);
- the superior mesenteric artery (midgut – distal duodenum to proximal transverse colon);
- the inferior mesenteric artery (provide the blood supply to the gastrointestinal tract distal transverse colon to anus);
- the marginal artery of the colon (joins the superior mesenteric artery and inferior mesenteric artery).

Venous drainage is via the hepatic portal vein into the liver (Figure 12.2).

The proximity of the lymphatic system to the circulatory system and the gastrointestinal tract means that dissolved nutrients will also pass into the lymph. This is especially important for the movement of fats (see section on small intestine).

Innervation of the gastrointestinal system is via the autonomic nervous system. The network of nerves relays signals to the brain via the spinal cord at a subconscious level, controlling peristalsis and release of mucus and enzymes. Figure 12.3 shows the innervation throughout the digestive tract.

## The mouth

The mouth is a cavity lined with stratified squamous epithelial cells (see Chapter 6). This allows hard items to come into contact with the buccal lining without causing significant trauma. The rate of cellular repair is also high in the mouth. The mouth contains the teeth, the tongue and three pairs of salivary glands: the parotid, the submandibular and the sublingual. The anterior portion of the roof of the mouth consists of the hard palate, and the posterior portion consists of the soft palate from which the uvula hangs. The palatine tonsils sit on the posterior lateral walls of the oral cavity (Figure 12.4).

The mouth is responsible for both mechanical and chemical digestion. Mechanical digestion begins with chewing and the teeth (see Figure 12.5). Chewing is controlled by cranial nerve V (the trigeminal nerve).

The groups of cells or 'buds' required for tooth development form in the fetus, and all 20 primary teeth are usually present under the gums at birth. The time at which these teeth erupt

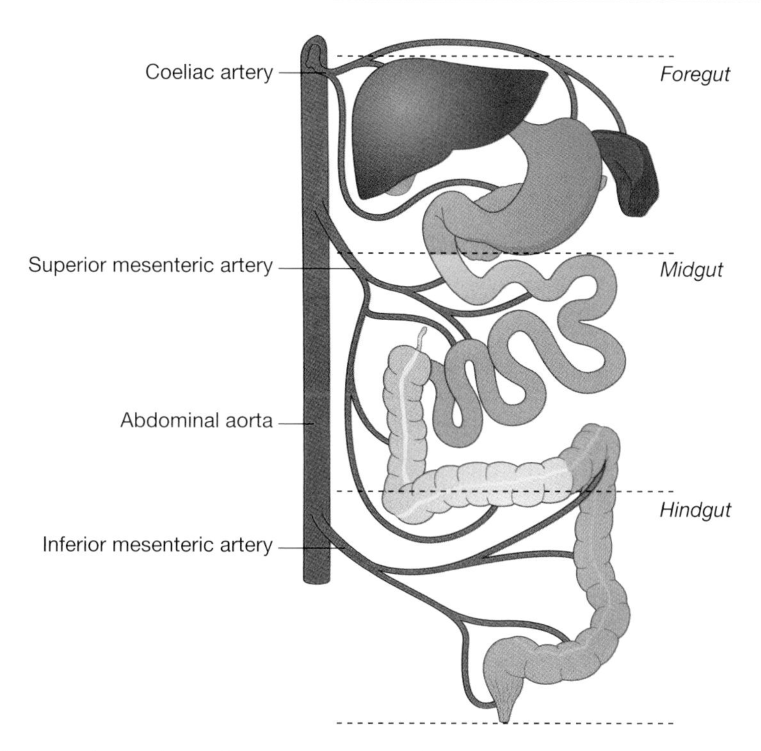


Figure 12.2 The blood supply to the gastrointestinal system.



through the gums depends on the individual child; some will be born with teeth, whilst others will have no teeth until around 1 year of age (Sajjadian *et al.*, 2010). However, the teeth do erupt in a typical pattern:

lower incisors followed by upper incisors

↓

followed by first molars

↓

followed by canines

↓

followed by second molars (with lower typically erupting before each upper).

During this time, infants will usually experience some pain, and the irritation causes excessive salivation, often causing excoriated skin around the mouth. Many parents also report a mild fever (the child feeling hot to touch but not above 37.9 °C) and diarrhoea associated with teething, although other infective causes do need to be eliminated.

These 'milk teeth' or deciduous teeth become loose between 6 and 13 years of age (Thibodeau and Patton, 2012). As the permanent teeth grow underneath the deciduous teeth, the pressure stimulates the roots to be reabsorbed, resulting in the deciduous teeth becoming loose. This needs consideration in children undergoing surgery, as loose teeth could be inadvertently dislodged and swallowed or aspirated.

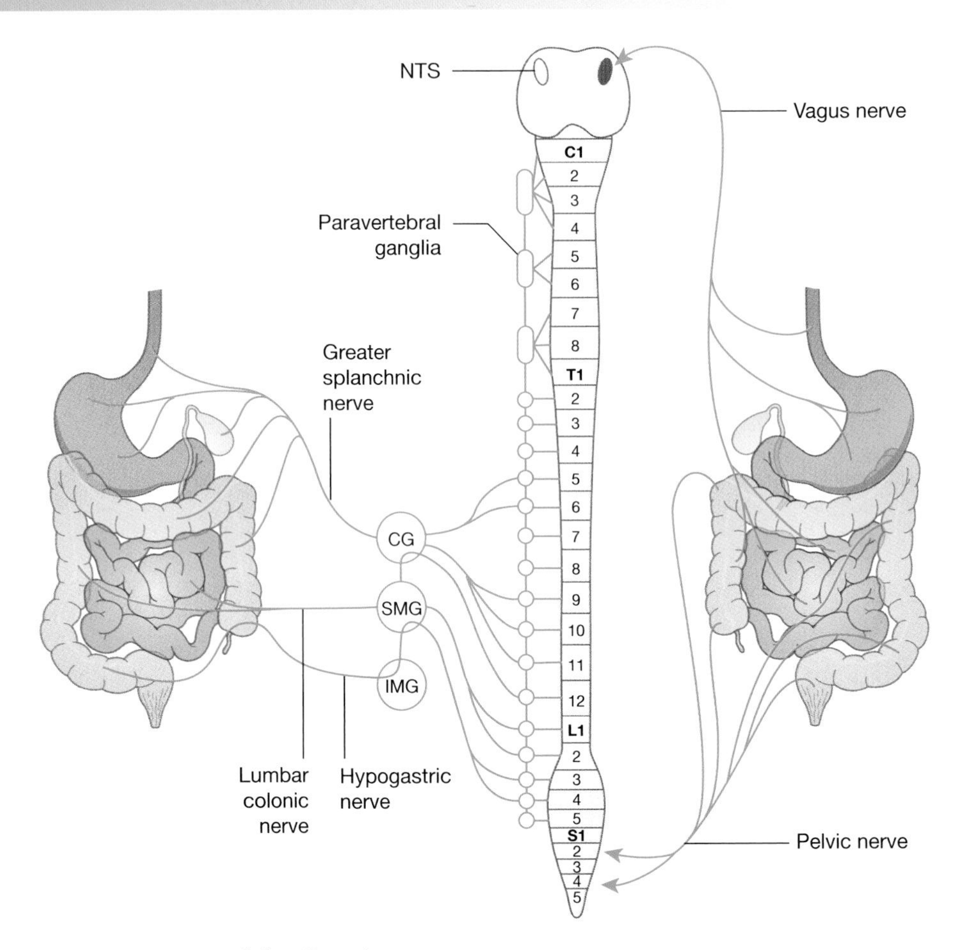


Figure 12.3 Innervation of the digestive tract.

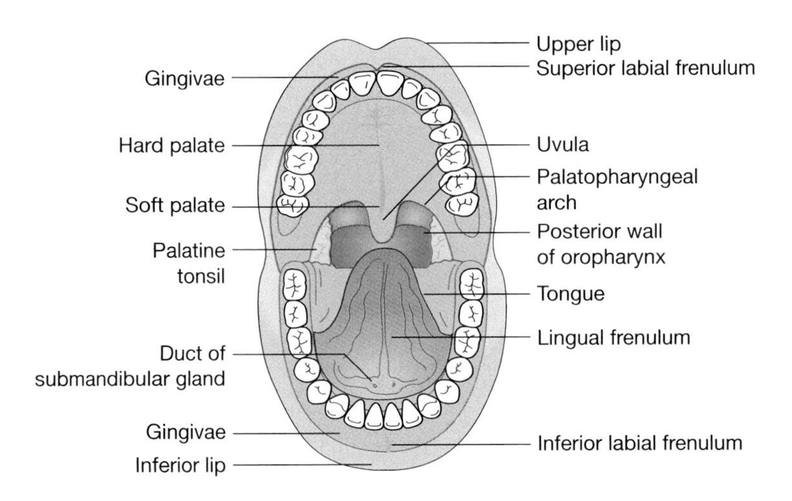


Figure 12.4 The oral cavity.

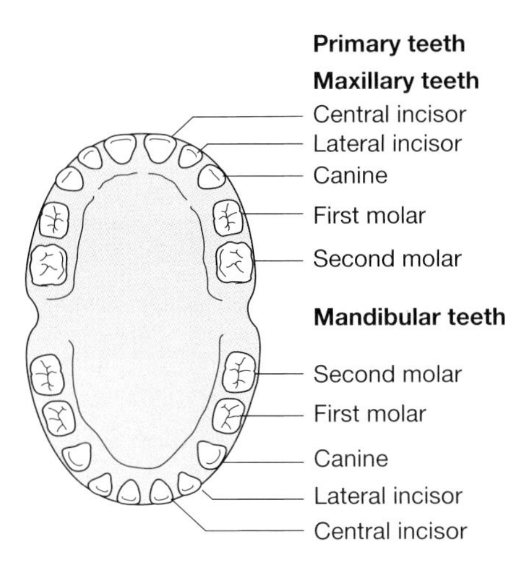




Figure 12.5 Primary dentition, or 'milk teeth'.

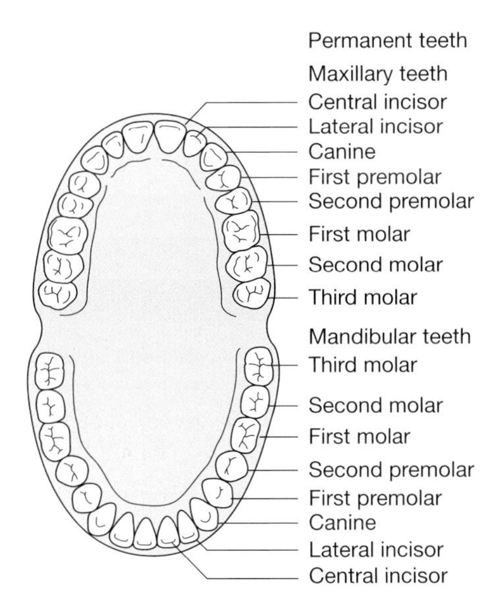


Figure 12.6 Secondary dentition, or permanent teeth.

The first and second primary molars are replaced by the first and second permanent premolars; a further three molars then erupt in each quadrant of the mouth. The third molar, or 'wisdom' tooth, will not always erupt in all adults, dependent on space along the gum line (Figure 12.6).

Mechanical digestion is aided by the teeth physically breaking the food into smaller particles and the movement of the tongue to form this into a small ball or 'bolus' which is then passed to the back of the throat, or pharynx, for swallowing. Chewing is controlled by cranial nerve V (the trigeminal nerve), the tongue is controlled by cranial nerve XII (the hypoglossal nerve) and swallowing requires control by both cranial nerves IX and XII (glossopharyngeal and

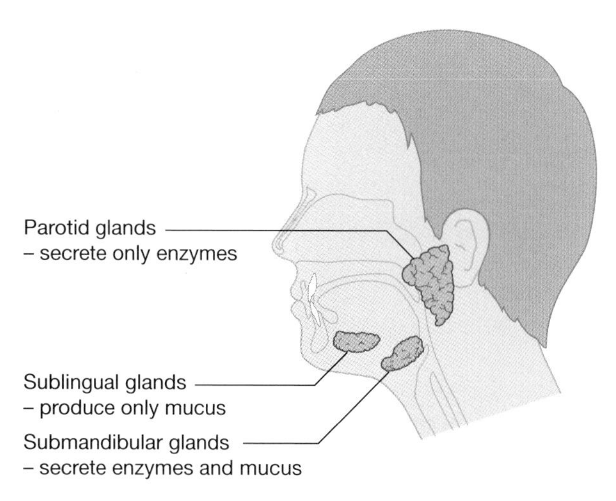




**Figure 12.7** Salivary gland positions in the mouth.

hypoglossal). Until swallowing happens, the movement of the food is under voluntary control. Once the food passes the pharynx, further movement through the digestive tract is under involuntary control until defecation, which is a learnt, voluntary action.

For information on the tongue and sense of taste, please see Chapter 18.

The chewing of food in the mouth mixes it with mucus and salivary amylase, starting the digestion of starches. Salivation is controlled by cranial nerves VII (facial) and IX (glossopharyngeal), and saliva is secreted from the three salivary glands (Figure 12.7).

## The oesophagus

This connects the back of the throat (or pharynx) to the stomach. It, again, has stratified epithelial cells, allowing the passage of solid food with minimal damage. The stratified epithelial cells are interspersed with mucus-secreting goblet cells, which help to lubricate the food during its passage. As food passes down the oesophagus, salivary amylase activity in digesting starches continues. Food passes by the mechanical process of peristalsis, which is possible due to the surrounding muscle structure, with alternate contraction and lengthening of circular and longitudinal muscles.

The wall of the gastrointestinal tract four has layers (Figure 12.8):

- The mucosa (mucus membrane) – the layer in contact with food. This has different epithelial cells dependent on its position (e.g. mouth, stratified to resist hard food; ileum, simple columnar for absorption). This membrane produces mucus to assist with lubrication of food.
- The submucosa – a layer of connective tissue below the mucosa, with blood vessels and nerves.
- The muscularis – this is two layers of muscle tissue at right angles to each other: one circular and one longitudinal. This assists in peristalsis (Figure 12.9) and further mechanical digestion.
- The serosa – this is the outermost layer in contact with the abdominal cavity connective tissue, and forming part of the visceral peritoneal membrane. This anchors the loops of the bowel with peritoneal tissue and mesentery.

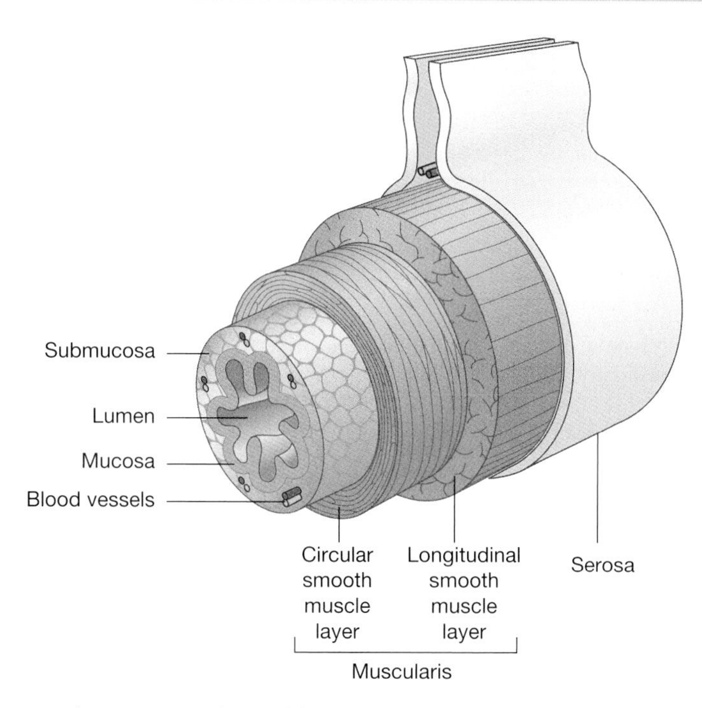


**Figure 12.8** Diagram showing circular and longitudinal muscles in the digestive tract.

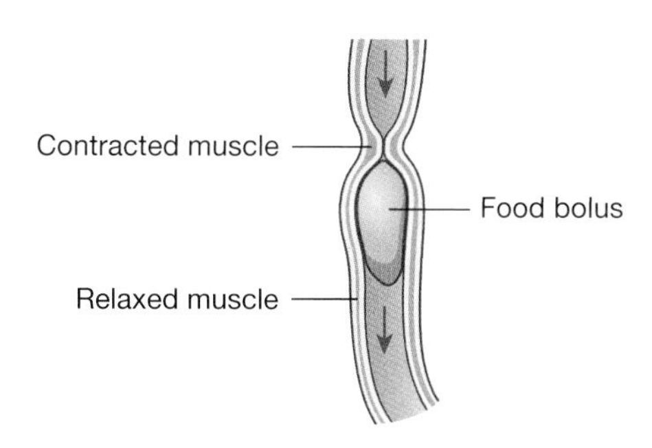


**Figure 12.9** Peristalsis.

## Clinical application

Some children have major difficulties ingesting food (for example, some children with special needs cannot coordinate swallowing, placing them at risk of aspiration). If this cannot be resolved through medication or the support of a speech and language therapist and dietician, then a decision may be made to insert a gastrostomy. This is a surgically created artificial opening directly into the stomach, through which specially prepared liquid feeds can be introduced, bypassing the mouth and oesophagus.

## The stomach

The distal end of the oesophagus joins the top of the stomach at the cardiac sphincter (or lower oesophageal sphincter). This is a round muscle that prevents food from travelling back up the oesophagus. The stomach is a j-shaped sac that performs both mechanical and chemical digestion, and then passes its partially digested contents into the duodenum through the pyloric sphincter. The stomach is only about 20–30 mL in size at birth, but by 1 year is 200 mL (Coad and Dunstall, 2005). The muscular outer layer of the stomach has layers running at right angles, allowing for the churning of food. After 2 years of age, the stomach is positioned in a vertical 'j' shape as in the adult; prior to this it lies more horizontally. This has implications for feeding a younger child with a nasogastric tube, as the cardiac sphincter is also unable to close completely due to the tube placement. Therefore, infants requiring nasogastric feeds should be fed in an upright position to prevent reflux of stomach contents and potential aspiration.

The inner mucous membrane is stratified columnar epithelium. This allows for expansion of the stomach when a large meal is ingested, and provides a relatively tough surface for hard food. There are goblet cells interspersed with the columnar epithelium, and these secrete mucus to assist in the breaking down and lubrication of solid food particles into semi-solid matter called *chyme*.



Parietal cells produce hydrochloric acid (HCl), which assists in the breakdown of the cell walls of the plant and animal matter ingested, allowing the nutrition from the cell contents to be released. It is also important in destroying microbes. It is this HCl that gives the stomach its characteristic pH of 3–5, which turns pH paper to an orange–red colour (Figure 12.10).

The gastric glands also secrete water, mineral salts and gastrin, which is a hormone causing increased acid production and gut motility in response to the presence of food. Chief cells secrete inactive enzyme precursors: pepsinogen, which is converted to pepsin in the presences of HCl, and prorennin, which is converted to rennin. Pepsin is responsible for digestion of proteins, about 10% of which occurs in the stomach. Rennin is responsible for converting the soluble proteins in milk into an insoluble form, so that pepsin can work.

## Clinical application

The National Patient Safety Agency has produced a flow chart for testing correct placement of nasogastric tubes (NGTs) in children (NPSA, 2011). This follows inadvertent placement of NGTs into the lungs, with devastating consequences if feed is administered.

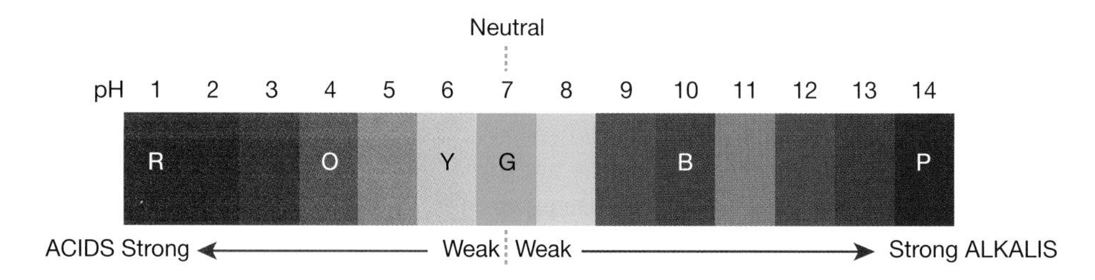


**Figure 12.10** pH paper universal indicator colours.

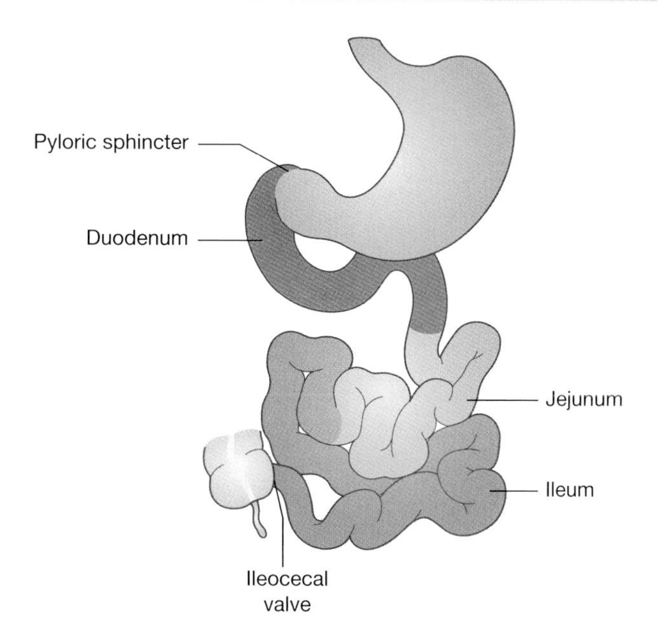


Figure 12.11 An overview of the small intestine.



## The small intestine (fig 12.11)

The small intestine (Figure 12.11) goes from 275 cm at birth to 575 cm by adulthood (Weaver *et al.*, 1991). Although it is much longer than the large intestine, its diameter is much smaller, hence the name. The small intestine is divided into three sections:

- the duodenum
- the jejunum
- the ileum.

### *Duodenum*

This is the proximal part of the small intestine, joining the stomach to the jejunum. It is where most of the chemical digestion takes place. The bile duct from the gallbladder empties into the duodenum, as does the pancreatic duct.

### *Jejunum*

The jejunum is the portion of the small intestine between the duodenum and the ileum. It is here that the pH of the digestive contents has changed from acid (pH 3–5) to alkali (pH 8–9). This provides a suitable environment that allows for different enzymes to work (see Chapter 2).

If the child has difficulty with digesting food in the stomach or duodenum, a specially prescribed, partially digested enteral formula can be introduced into the jejunum via a naso-jejunal tube or jejunostomy.

### *Ileum*

The ileum is the longest part of the small intestine, and it is here that the majority of nutrients are absorbed from the partially digested chyme passing from the jejunum, helped by membrane-bound enzymes on the epithelial lining of the gastrointestinal tract. The carbohydrates, proteins and some vitamins and minerals pass into the many capillaries surrounding the small intestines,

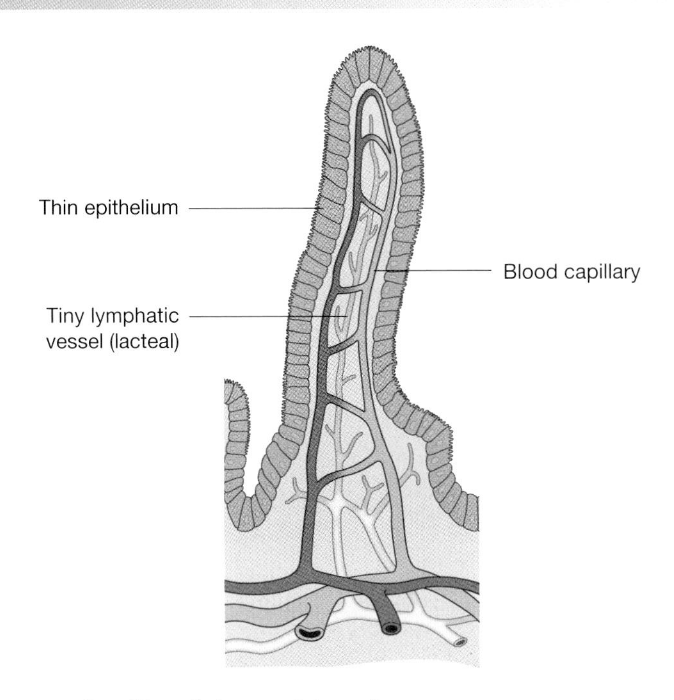




**Figure 12.12** Structure of a villus of the small intestine.

and then the blood goes to the hepatic portal vein and to the liver. The lipids and vitamins A, D, E and K pass into the lymphatic system. The ileum is highly specialized to perform these functions owing to the presence of villi (Figure 12.12). These greatly increase the surface area for absorption of nutrients into both the blood stream and lymphatic system.

The ileum is responsible for the absorption of nutrients and some absorption of water. It leads to the large intestine, the junction of which is called the ileo-caecal valve.

If the small intestine is unable to perform its function, then excessively watery stools will pass into the large intestine at a rate too great for water absorption to take place. This results in chronic diarrhoea.

## The liver and production of bile

The liver has two lobes: a large superior anterior lobe and a smaller posterior inferior lobe. It is positioned in the right upper quadrant of the abdomen (Figure 12.13).

The liver receives 20% of its blood flow from the hepatic artery and 80% of its blood flow from the hepatic portal vein (coming directly from the gastrointestinal tract) (Clancy and McVicar, 2009).

Boxes 12.2 and 12.3 outline the functions the liver is responsible for, which can be divided into nutrition, growth and repair, and elimination.

## The gallbladder

The liver continuously produces bile, but this is then stored in the gallbladder (Figure 12.14) and released when a meal is ingested.

A meal signals the gallbladder to contract and empty the stored bile into the duodenum. The role of bile is to emulsify fats. Think of it like washing-up liquid; it breaks a slick of fat down into

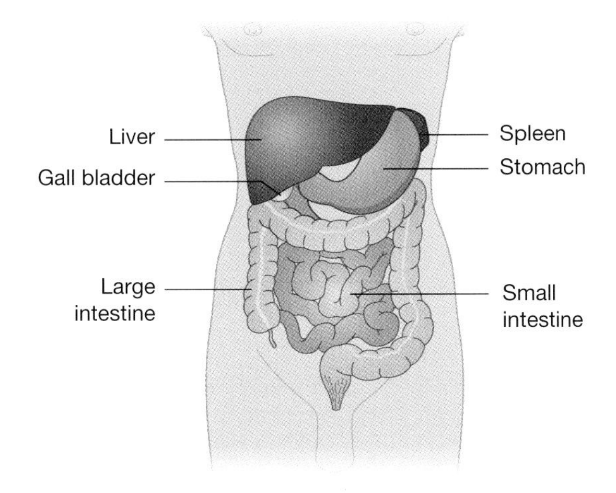




Figure 12.13 Position of the liver in relation to the gastrointestinal tract.

## Box 12.2 Functions of the liver: nutrition, growth, and repair.

- Carbohydrate metabolism
- Lipid metabolism
- Protein metabolism and storage
- Bile salts synthesis
- Bile production and secretion
- Mineral and vitamin storage
- Vitamin D activation
- Iron storage and reclamation from broken down red blood cells
- Synthesis of clotting factors
- Storage of blood (a 'reservoir').

## Box 12.3 Functions of the liver: elimination.

- Detoxification
- Urea formation
- Degradation of drugs
- Steroid catabolism
- Metabolizes hormones
- Breakdown and excretion of red blood cells.

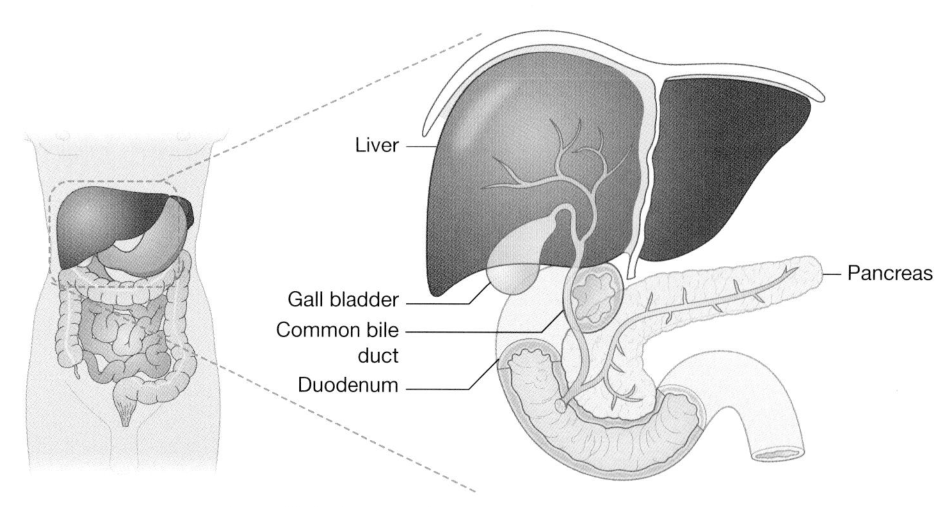




Figure 12.14 Position of the gallbladder and pancreas.

much smaller globules. These globules are then of a size where they can be surrounded by the digestive juices from the pancreas, and broken down into lipids which can be absorbed into the lymph. If bile is not released, then pancreatic enzymes (see section below on the pancreas) can only work on the outer part of the large fat molecules. This leads to inadequate amounts of lipids being absorbed. The child then has difficulty gaining weight and also difficulty absorbing the fat-soluble vitamins A, D, E and K, and the excess fat is excreted in the stools (steatorrhoea).

## The pancreas

The pancreas has two key functions: endocrine and exocrine. It is located below the liver and attaches to the duodenum at the same level as the gallbladder (see Figure 12.14). The endocrine function is the release of the insulin in response to a rise in blood glucose levels and the release of glucagon in response to a drop in blood glucose levels. This is explored in detail in Chapter 11.

The exocrine function is the production and secretion of bicarbonate ions and alkaline digestive enzymes by the acini cells, the functions of which are shown in Table 12.2.

## The large intestine

The large intestine (approximately 40 cm long at birth and growing to 1.5 m in adults) begins at the ileo-caecal junction, then the ascending colon, transverse colon, descending colon, sigmoid colon, the rectum, and ends at the anus (see Figure 12.15). It is responsible for the final absorption of nutrients, maintenance of fluid balance, and for compaction and defecation of waste products. The bacteria in the large intestine are also responsible for synthesis of vitamin $B_{12}$, thiamine ($B_1$), riboflavin ($B_2$) and vitamin K.

The caecum is the blind ending of the large intestine. Leading from this is a vestigial appendix (has no function in humans). It is possible that faecal matter may become stuck in the opening of the caecal end of the appendix, leading to occlusion of the lumen of the appendix beyond this. Bacteria from the faecal material then replicate and cause an infection in the appendix (appendicitis).

Table 12.2 Pancreatic enzymes and enzyme precursors.

| | |
|---|---|
| Trypsinogen is an inactive precursor, which is activated into active trypsin in the duodenum. | Continues protein metabolism, breaking down into amino acids |
| Chymotrypsinogen is an inactive precursor, which is activated into chymotrypsin to work with trypsin in the duodenum. | Continues protein metabolism, breaking down into amino acids |
| Carboxypeptidase | Continues protein metabolism, breaking down into amino acids |
| Pancreatic lipase | Breaks down fats into fatty acids and glycerol |
| Pancreatic amylase | Breaks down carbohydrates and sugars |
| Elastases | Break down the protein elastin |
| Nucleases | Break down nucleic acids (DNAse and RNAse) |



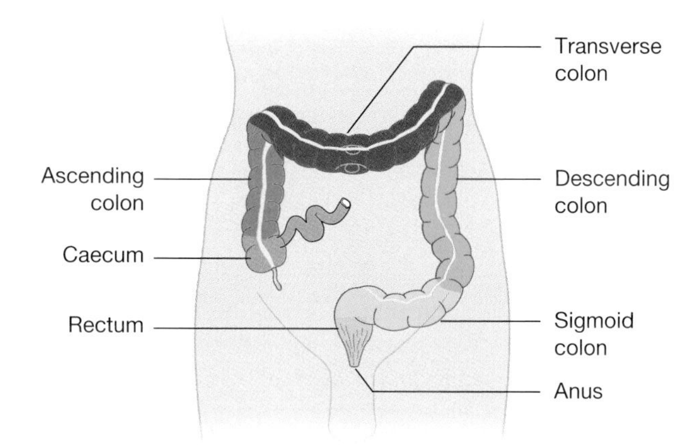


Figure 12.15 The large intestine.

As the lumen of the colon is larger than the ileum, it requires the bolus of faeces to be a reasonable size and consistency in order to be passed by peristalsis. If the stools are too watery, they will pass through too quickly with inadequate time for water to be absorbed, and diarrhoea will result. The large intestine has no villi; therefore, it has a smaller surface area for absorption than the small intestine. If the stools are too hard, then it takes longer for them to pass, resulting in more water absorption and more compaction of additional faecal material, resulting in a large, hard stool that is difficult to pass (constipation). Healthy diet advice includes the provision of fibre to allow a child to form the necessary size bolus to easily pass stools along the colon through peristalsis. However, high-fibre diets are not recommended for children, as these can make them feel full, and thus consume inadequate calories for growth (http://www.nhs.uk/Livewell/Goodfood/Pages/Underweightyoungchild.aspx). Water intake is also essential for preventing the stool from becoming too hard, making peristalsis and defecation difficult (Neill and Knowles, 2004) (see Chapter 13 – the renal system – for fluid balance).

## Clinical application

If a child has repeated (chronic) constipation, then they may suffer a temporary blockage of the colon. The large intestine then absorbs less fluid in an attempt to help it to pass. However, this results in the faeces behind the blockage becoming watery. This often leaks past the blockage, resulting in diarrhoea. Thus, constipation must be considered in a child who has intermittent diarrhoea (Slater-Smith, 2010).

Once the stool has passed through the ascending, transverse, descending and sigmoid colons it is stored in the rectum. When the rectum is full, stretch receptors are activated to signal involuntary smooth muscle of the internal anal sphincter to relax, allowing the faeces to descend to the external anal sphincter to be expelled (Slater-Smith, 2010). This is an involuntary response in the younger child. However, as children grow, they learn that they can control the constriction of the rectal muscles. Toilet training is about learning the feeling that faeces are stored in the rectum and can be expelled voluntarily.



The physical aspect of defecation is complicated by social and emotional aspects. Some people will only want to defecate in complete privacy, or in a known environment, and this can lead to the retention of faeces in the rectum, and resultant constipation.

## Clinical application

When younger children use a potty rather than a toilet, this places their bottoms below the level of their knees, in a squatting position. Because of the shape of their sigmoid colon, this can lead to part of the rectum extruding beyond the anal opening (rectal prolapse) if they strain on defecation. Parents can be shown how to safely push the rectum back through the anal sphincter, and the child should be encouraged to use a toilet rather than a potty to reduce the squatting position. This is in addition to dietary advice to ensure that constipation is not an underlying cause.

# Conclusion

This chapter has examined the structure and function of each part of the gastrointestinal tract, linking this to nutrition. Embryological development of the gastrointestinal system has been outlined and links made to abnormal development. The clinical application and discussion throughout has facilitated application of knowledge of the gastrointestinal system to the care of children and their families.

# Activities

**Now review your learning by completing the learning activities in this chapter. The answers to these appear at the end of the book. Further self-test activities can be found at www.wileyfundamentalseries.com/childrensA&P.**

## Exercises

1. Why does the mouth and stomach have stratified epithelium, and the intestines have simple columnar epithelium?
2. List the function of these structures within the stomach:
   (a) chief cells
   (b) parietal cells
   (c) gastric glands.
3. Why are vitamins A, D, E and K often considered separately to the other vitamins?
4. List some of the advantages to the infant of breastmilk compared with formula milk.
5. Where is bile (a) made (b) stored and (c) released to?
6. The pancreas has two key functions: endocrine and exocrine. What is the definition of each?
7. The following terms relate to the large intestine. Put in order from proximal to distal: sigmoid colon, caecum, anus, transverse colon, appendix, rectum, ileo-caecal valve, descending colon, ascending colon.
8. List five functions of the liver.



## Complete the sentence

The pH of the gastrointestinal lumen changes from _____ in the stomach to _____ in the duodenum.

## Putting your knowledge into clinical practice

The following three activities are related to disorders of the digestive system, allowing you to apply your knowledge to clinical practice.

1. Give a definition of the following conditions affecting the small intestine: (a) Crohn's disease; (b) short gut syndrome.

2. Case study. Jessica is 12 years old and has cystic fibrosis (CF). She takes pancrease sprinkled on her food. Explain how CF affects the digestive system, and why digestive enzymes, vitamin supplements and overnight feeds are often part of a child's care.

3. Impact on the family. One of the fundamental aspects of caring for any child is the provision of adequate nutrition in order to allow them to grow and develop.
   *Consider:* What is the impact on the family of having a child labelled 'underweight' or 'clinically obese' with no underlying medical disorder? What judgements do you think others may make about their parenting skills?

## Conditions

The following table contains a list of conditions. Take some time and write notes about each of the conditions. You may make the notes taken from text books or other resources; for example, people you work with in a clinical area or you may make the notes as a result of people you have cared for. If you are making notes about people you have cared for you must ensure that you adhere to the rules of confidentiality.

| Condition | Your notes |
|---|---|
| Appendicitis | |
| Constipation | |
| Parasitic infection | |
| Ulcerative colitis | |
| Gastroenteritis | |

## Glossary



**Constipation:** difficulty in passing faeces (straining or painful hard stools), lack of frequency (three times a week or less), or feeling that bowels are not emptied following defecation.

**Diarrhoea:** watery stools.

**Digestion:** the breaking down of food into their constituent particles which can then be absorbed into the circulation or lymphatic system.

**Enzymes:** protein molecules that speed up chemical reactions without being changed in any way themselves.

**Exomphalos:** the failure of the fetal gut to retract back into the abdominal cavity and close off the abdominal wall, resulting in the infant being born with intestines extruding from the abdomen.

**Haematemesis:** the presence of blood in the vomit.

**Ingestion:** the taking in of food and drink through the mouth.

**Meconium:** the black, tarry stools passed by a neonate in the first 48 h of life. This is made from bile pigments and waste skin cells from the developing intestine whilst *in utero*, and is passed following delivery.

**Melaena:** the presence of blood in the stools.

**Peristalsis:** the movement of food through the digestive tract in one direction from the mouth towards the anus, by the alternate contraction and relaxation of circular and longitudinal muscles.

**Steatorrhoea:** the presence of fat in the stools (appearance is white/grey colour).

**Transamination:** the making of new amino acids by the liver to compensate for low levels of ingestion.

**Villus/villi:** a projection from the epithelial lining of the small intestine, containing blood and lymph. This increases the surface area for absorption.

**Vomiting:** the forceful ejection of the stomach contents through the mouth.

**Weaning:** the expansion of the diet to include food and drinks other than breastmilk or formula.

## References

Chamley, C.A., Carson, P., Randall, D., Sandwell, M. (2005) *Developmental Anatomy and Physiology of Children: A Practical Approach*, Elsevier Churchill Livingstone, Edinburgh.

Clancy, J., McVicar, A.J. (2009) *Physiology and Anatomy for Nurses and Healthcare Practitioners: A Homeostatic Approach*, 3rd edn, Hodder Education, London.

Coad, J., Dunstall, M. (2005) *Anatomy and Physiology for Midwives*, 2nd edn, Churchill Livingstone Elsevier, Edinburgh.
Glasper, A., Aylott, M., Battrick, C. (eds) (2010) *Developing Practical Skills for Nursing Children and Young People*, Hodder Arnold, London.
Gross, L.S., Li, L., Ford, E.S., Liu, S. (2004) Increased consumption of refined carbohydrate and the epidemic of type 2 diabetes in the US: an ecologic assessment. *American Journal of Clinical Nutrition*, **79** (5), 774–779.
Neill, S., Knowles, H. (2004) *The Biology of Child Health*, Palgrave McMillan, Basingstoke.
NPSA (2011) *Alert 1253 – decision tree for nasogastric tube placement checks in children*, National Patient Safety Agency, www.nrlsnpsa.nhs.uk (accessed 30 October 2013).
Neu, J. (2007) Gastrointestinal development and meeting the nutritional needs of premature infants. *American Journal of Clinical Nutrition*, **85** (2), 629s–634s.
Sajjadian, N., Shajari, H., Jahadi, R. *et al.* (2010) Relationship between birth weight and time of first deciduous tooth eruption in 143 consecutively born infants. *Paediatrics and Neonatology*, **51** (4), 235–237.
Shaw, N.J., Mughal, M.Z. (2013) Vitamin D and child health Part 1 (skeletal aspects). *Archives of Disease in Childhood*, **98**, 363–376.
Slater-Smith, S. (2010) Promoting children's continence. In: Glasper, A., Aylott, M., Battrick, C. (eds), *Developing Practical Skills for Nursing Children and Young People*, Hodder Arnold, London, Chapter 14.
Thibodeau, G.A., Patton, K.T. (2012) *Structure and Function of the Body*, 14th edn, Elsevier Mosby, St. Louis, MO.
Tortora, G.J., Grabowski, S.R. (2000) *Principles of Anatomy and Physiology*, 9th edn, J. Wiley & Sons, Inc., New York, NY.
Weaver, L.T., Austin, S., Cole, T.J. (1991) Small intestinal length; a factor essential for gut adaptation. *Gut*, **32** (11), 1321–1323.
Wilcox, J. (2004) Feeding the body in childhood. In: Neill, S., Knowles, H. (eds), *The Biology of Child Health*, Palgrave McMillan, Basingstoke, Chapter 8.

# Chapter 13

# The renal system

## Elizabeth Gormley-Fleming

*Learning and Teaching Institute, University of Hertfordshire, Hatfield, UK*
*Children's Nursing, School of Health and Social Work, University of Hertfordshire, Hatfield, UK*

## Aim

The aim of this chapter is to help develop your knowledge and understanding of the anatomy and physiology of the child's renal system from birth until it becomes fully mature. The kidneys play an important role in homeostasis. This chapter should be read in consideration with Chapter 2 homeostasis.

## Learning outcomes

On completion of this chapter the reader will be able to:

- Describe the gross anatomy of the renal system.
- Describe the microscopic anatomy of the renal system.
- Understand and describe the function of the kidneys.
- Describe in detail the formation and composition of urine.
- Articulate how fluid balance and electrolyte balance is maintained.
- Describe the function of the bladder and micturition.

## Test your prior knowledge

- Name the organs of the renal system.
- List the functions of the kidneys.
- Identify the differences between the male's renal system and a female's renal system.
- List the functions of the nephrons.
- identify the vessels that supply blood to the kidney.
- Identify the three processes involved in the formation of urine.
- List the composition of urine.
- Describe the process of bladder control that leads to continence in early childhood.
- How much urine should an infant produce per hour?
- Describe the process of micturition.

*Fundamentals of Children's Anatomy and Physiology: A Textbook for Nursing and Healthcare Students*, First Edition. Edited by Ian Peate and Elizabeth Gormley-Fleming.

Companion website: www.wileyfundamentalseries.com/childrensA&P

# Introduction

The renal system plays a very important role in determining one's overall health status as it is essential in the maintenance of all the body systems. The renal system is one of the major excretory systems of the human body. It also assists in the maintenance of homeostasis. The kidneys continually filter waste products from the bloodstream that will later be excreted from the body via the bladder as urine. Substrates from this filtration process that are required for good health are returned to the blood. The kidneys also act as a regulator by maintaining the correct balance between water and sodium and acids and base. This enables the correct constituents of the blood. From birth to adulthood significant changes occur. The embryonic development of the kidney commences during the third week of gestation from the intermediate mesoderm (MacGregor, 2010). The developing baby's kidney will begin to produce urine between 9 and 12 weeks of gestation, and this is excreted into the amniotic fluid. Failure of any part of the renal system to develop correctly will have implications for the infant. As kidney development and genitourinary tract development are interdependent, if there is any abnormality of one system, then abnormalities in the other system should also be considered. This chapter will discuss the structure and function of the renal system. Consideration will be given to some common clinical conditions of childhood.



# The renal system

The renal system (Figure 13.1), or urinary system as it is also referred to as, consists of:

- two kidneys, which act as filters and produces urine;
- two ureters, which carry the urine to the bladder;

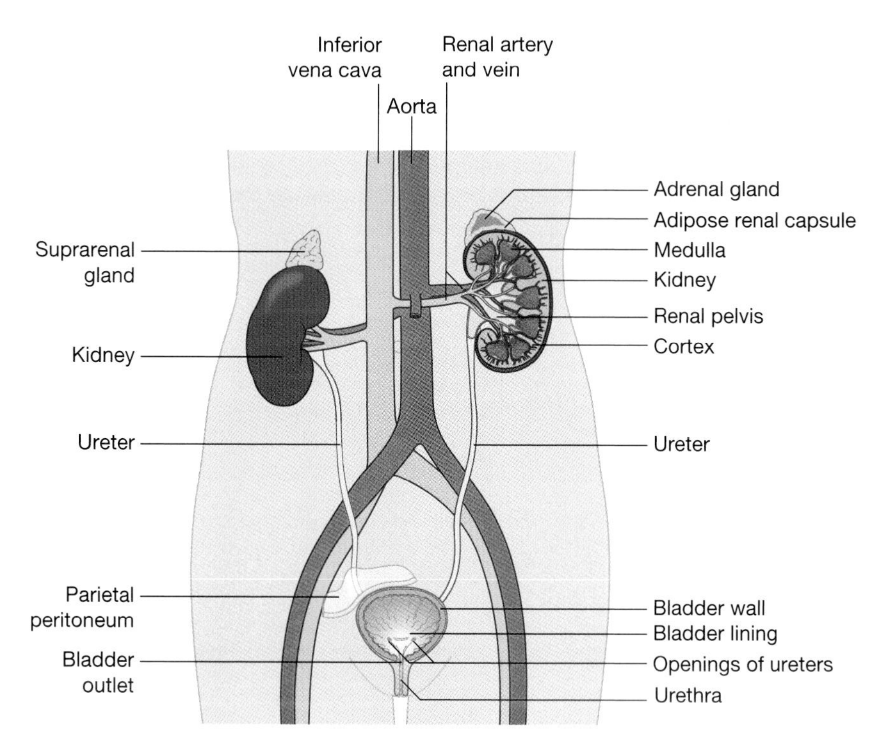


Figure 13.1 The renal system – anterior view.

- the urinary bladder, which acts a reservoir for urine;
- the urethra, which conveys urine out of the body.

The renal system has three major functions:

1. Excretion – removal of organics waste from body fluids.
2. Elimination – the releasing of those waste products from the body.
3. Homeostatic regulation – the balancing of the volume and solute concentration of the blood plasma.

# The kidney



Kidney development commences at the first week of gestation and continues until 36 weeks' gestation. Nephrogenesis is the term that refers to the development the kidney *in utero*. Significant development occurs in the first trimester of pregnancy. If born prematurely, nephrogenesis continues at the same rate *ex utero* as it does *in utero*. By this time, 36 weeks' gestation, nephrogenesis is complete as the total number of nephrons present reaches their maximum limit. The anatomical formation of the kidney occurs exclusively *in utero*; however, physiological function develops post birth. The differences between the newborn renal function and the mature kidney are outlined in Table 13.1. Post-birth renal growth is considered to involve the elongation of the proximal tubules and the loop of Henle (Ichikawa, 1990).

At birth, the term baby will have the adult complement of nephrons; that is, 800 000–1 000 000 nephrons (Bonilla-Felix *et al.*, 1998). The neonate's kidney is lobular (England, 1996), and this will remain so until the age of 4–5 years of age. By this age the renal system is mostly mature.

Table 13.1 Renal function in the newborn-key points. Adapted from Lissauer and Fanaroff (2011).

| Newborn infant | Renal function |
|---|---|
| Fetal kidneys have no function in homeostasis | Their function is to produce urine to add to the volume of amniotic fluid |
| Low glomerular filtration rate (GFR) in newborn | GFR in healthy newborn is 30 mL/min per 1.73 $m^2$ compared with an adult, whose GFR is 120 mL/min per 1.73 $m^2$ |
| Newborn kidney main aim is to optimize dietary solutes for growth and not excretion | Newborns undergo rapid growth and have little excess dietary solute that requires excretion; therefore, the newborn kidney is optimized for retention of essential substances such as sodium and other minerals<br>Also in the presence of periods where growth ceases due to ill health, the infant has insufficient renal reserve; thus, homeostasis can become unbalanced quickly |
| At term the infant's kidney conserves sodium, while the pre-term infant's kidney loses sodium | Term infant's urine is mostly salt free and can thrive on human milk that is low in sodium. Consequently, a pre-term infant will become sodium depleted and need added dietary sodium in order to prevent hyponatraemia |
| Low urine osmolarity | Inability to concentrate urine as a newborn |

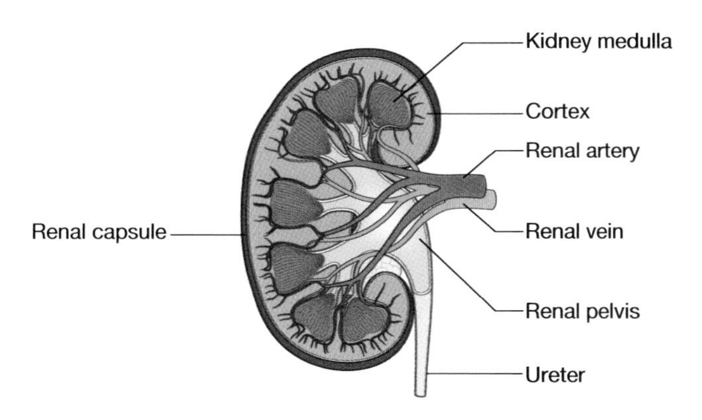


Figure 13.2 Gross anatomy of the kidney.



## Gross structure of the kidney

There are normally two kidneys, which are bean-shaped organs. The lateral surface of the kidney is convex and the medial surface is concave. It is on the medial surface that the hilum is located, and this leads into the renal sinus. It is here at the hilum that the renal veins, renal arteries, nerves and ureters enter and leave the kidneys (Figure 13.2).

There are three tissue layers surrounding each kidney:

1. The renal fascia.
2. Adipose tissue or perirenal fat capsule.
3. The renal capsule.

The renal fascia is a dense fibrous outer layer of connective tissue that anchors the kidney and the adrenal gland to the surrounding structures. The perirenal fat capsule's function is to protect the kidney from trauma, and this surrounds the renal capsule. The renal capsule surrounds the entire kidney. This is a transparent capsule composed of smooth connective tissue, and its function is to prevent infection from spreading to the kidneys from surrounding areas and also to protect the kidney from trauma and to maintain the shape of the kidney (Marieb and Hoehn, 2010).

The newborn's kidney is approximately 23 g in weight. By the age of 6 months this will have doubled, and by the age of 1 year this will have trebled (Sinclair, 1991). By adulthood the kidney will weigh approximately 150 g.

## Clinical application

While it is normal to be born with two kidneys, the absence of one kidney (renal agenesis) is relatively common. However, the growth of the kidneys depends on its work. If there is only one kidney then this will double in size. This would also happen if a kidney had to be removed.

There are several risk factors identified from *in utero* exposure that may lead to congenital renal disease. Some of these are environmental toxins, drugs and physical agents. It is thought that these may stop, retard or accelerate development, lead to abnormal development or alternate the pattern of maturation.

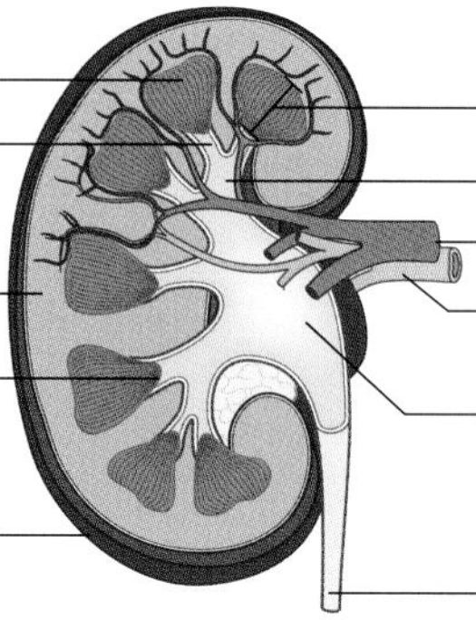

Figure 13.3 Internal anatomy of the kidney.



## Internal anatomy of the kidney

There are three distinct areas within the kidney (Figure 13.3):

- renal cortex
- renal medulla
- renal pelvis.

The renal cortex is a reddish brown layer, and this is underdeveloped at birth. This superficial layer has inward projections called the renal columns. These renal columns separate the renal pyramids.

The renal medulla, the innermost layer, consists of pale conical-shaped striations that are called the renal pyramids. This striped appearance of the renal pyramids is due to the fact that they are formed of parallel bundles of urine-collecting tubules and capillaries (Marieb and Hoehn, 2010). The base of the renal pyramids abuts the renal cortex. The tips of these pyramids, of which there are 8–18 in total, are referred to as renal papilla, and these project into the renal sinus. Each pyramid consists of a series of grooves that converge at the papilla. The renal pyramids are separated by the renal columns, which are inward projections of cortical tissue.

The renal pelvis is the funnel-shaped collecting chamber that collects the urine. Extensions of the renal pelvis called the major calyces, of which there are two or three, subdivide into the minor calyces. There are approximately 8–18 minor calyces. These minor calyces enclose the opening of the papillae. Urine drains continuously from the papillae as it is passed from the apex of the pyramids into a minor calyx, then into a major calyx. The major calyces unite to form the renal pelvis. Urine flows from the major calyces into the renal pelvis and then into the ureter. The wall of the renal pelvis is covered in smooth muscle and lined with transitional epithelium (Waugh and Grant, 2010). It is the peristaltic action of the smooth muscle in the walls of the calyces that propels the urine into the ureters towards the bladder.

After birth the kidneys increase in size, and this is due to enlargement of the existing structures such as the nephrons and the interstitial tissue (Matsumara and England, 1992).

## Location

The kidneys form in the pelvis, and by week 9 of gestation they have risen into the posterior abdominal wall to meet the supradrenal glands (England, 1996). As they ascend into the abdominal cavity they also rotate and locate themselves on either side of the vertebra between T12 and L3 when fully developed. During the migration into the abdominal cavity, changes to the blood supply to the kidneys occur. Initially blood is supplied from the common iliac artery, but as they migrate upwards the blood supply now comes from branches of the abdominal aorta, and this becomes the permanent renal artery eventually with all the accessory branches deleted (Moore and Persaud, 2003). When migration is complete the kidneys are retroperitoneal organs, as they are situated on the posterior aspect of the abdominal wall behind the peritoneum. The left kidney is located slightly more superior to the right kidney, with the right kidney slightly inferior due to the positioning of the right lobe of the liver. The adrenal glands are located on the superior aspects of each kidney.

## Clinical application

Failure of the kidney to migrate upwards into the abdominal cavity can lead to several conditions arising. These include:

- horseshoe kidneys (kidneys that fuse during ascent);
- pelvic kidney (kidney that fails to ascend);
- malrotation of the kidney;
- accessory branches of the renal artery.



## Microscopic structure

The kidney is composed of approximately 1 million function units called nephrons and a smaller number of collecting ducts. In the first 6 months of life the kidney will undergo a significant period of hyperplasia of its connective tissue, increased vascularization and hypertrophy (Ichikawa, 1990).

## The nephron

The role of the nephron is essentially to form urine. A nephron (Figure 13.4) consists of four main sections:

- Bowman's capsule;
- proximal convoluted tubule;
- loop of Henle;
- distal convoluted tubule.

There are two types of nephrons: cortical nephrons (85%), which lie almost entirely within the cortex, and juxtamedullary nephrons (15%), which have a long nephron loop and extend into the medulla. Cortical nephrons perform most of the reabsorptive and secretory function, while the juxtamedullary nephrons enable the kidney to produce concentrated urine. At birth, the juxtamedullary nephrons have a higher blood flow than the cortical nephrons. The implication of this for the neonate is their enhanced ability to conserve sodium rather than excrete it. The function of each section of the nephron is outline in Table 13.2.

### *Bowman's capsule*

This blind-end is cup shaped and called the glomerular capsule or Bowman's capsule. The glomerulus, a capillary network of approximately 50 capillaries, sits with Bowman's capsule. As a unit this is referred to as the renal corpuscle (Marieb and Hoehn, 2010). At birth, the diameter of the glomerulus is 100 μm, and this increases throughout childhood to reach a total of 300 μm by adulthood. The glomeruli become fully functioning by 6 weeks of age (England, 1996).

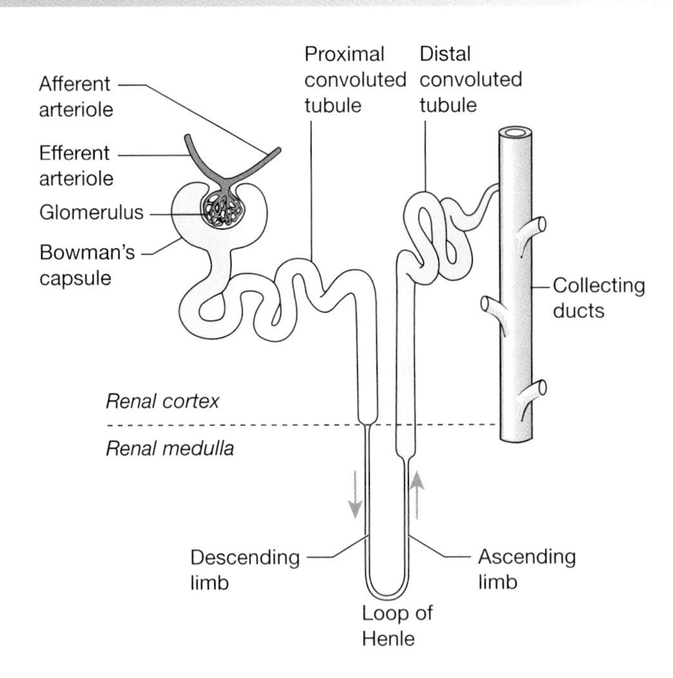


Figure 13.4 The nephron.



Table 13.2 Function of the structures within the nephron. Adapted from Marieb and Hoehn (2010).

| Structure | Specific function | Mechanism |
|---|---|---|
| Renal corpuscle | Filtration of water, solutes from plasma. Retention of plasma protein and blood cells | Hydrostatic pressure and filtration under pressure |
| Proximal convoluted tubule | Active reabsorption of ions, sugars, amino acids and vitamins. Passive reabsorption: urea, water, soluble lipids and chloride ions Secretion: hydrogen, ammonia, creatinine, drugs and toxins | Carrier mediated transport. Diffusion |
| Loop of Henle | Reabsorption of sodium and chloride ions and water | Active transport by ion transporters. Osmosis |
| Distal convoluted tubule | Reabsorption: sodium, chloride and calcium ions. Water. Secretion of hydrogen, ammonia, creatinine, drugs and toxins | Counter transport. Osmosis. ADH and aldosterone regulated |
| Collecting ducts | Reabsorption sodium, water, bicarbonate, urea | Counter transport: aldosterone regulated; osmosis: ADH regulated. Diffusion |

The walls of the glomerulus and Bowman's capsule are formed with cuboid epithelium at birth. By the end of the first year of life they transform to become a single layer of flattened epithelial cells. The walls of the glomerulus are more permeable that those of other capillaries. Blood enters the glomerulus via the afferent arteriole and leaves via the efferent arteriole; the blood then flows through a network of capillaries called the peritubular capillaries, and these surround the renal tubule. The afferent arteriole has a larger diameter than the efferent arteriole, and it is this that increases the pressure within the glomerulus that enables filtration to take place across the glomerular capillary wall. The glomeruli of an infant have less fenestrae, a lower hydrostatic pressure and a reduced renal plasma flow (Neill and Knowles, 2004). The process of filtration takes place in the renal corpuscle. The blood is separated into two different components: filtrated blood and a filtrate. The filtrate is collected in the capsule at a rate known as the glomerular filtration rate (GFR). This filtrated blood is dissolved out of the glomerular capsules into the capsular space. The filtrate then enters the proximal convoluted tubule.

## Clinical application



### Glomerulonephritis

Glomerulonephritis is inflammation of the glomeruli, and it affects the filtration ability of the kidney. It is thought to be an autoimmune complex disorder and may be primary or a manifestation of a system disorder that can range from mild to severe. Most cases are post infectious and occur following a streptococcus bacterial infection.

The glomeruli become oedematous and infiltrated with polymorphonuclear leucocytes that occlude the lumen of the capillaries. There is decreased filtration, resulting in an excessive accumulation of water and retention of sodium, which expands the plasma and interstitial fluid volume. This leads to circulatory congestion and oedema occurs. Hypertension occurs, which may also be as a result of increased renin production.

The condition peaks at 6–7 years of age, with males more commonly affected than females by 2:1.

Common features include oliguria, haematuria and proteinuria.

The child may present with:

- oedema – periorbital and facial oedema are more prominent, and this may extend to the extremities and abdomen during the day;
- decrease in appetite;
- decreased urine output, and urine is brown in colour and cloudy;
- pallor;
- headache;
- abdominal discomfort;
- vomiting;
- lethargy;
- mild to moderate hypertension.

## *Proximal convoluted tubule*

The proximal convoluted tubule leads from Bowman's capsule into the loop of Henle. Situated in the renal cortex, the proximal convolute tubule consists of epithelial cells that are densely packed with microvilli. Reabsorption is the prime function of the proximal convoluted tubule.

The microvilli increase the total surface area, thus enhancing reabsorption of salt, water, plasma proteins and glucose from the filtrate. These are then released into the peritubular fluid, which is the interstitial fluid that surrounds the renal tubule. Sodium pumps are also present. Other substances (e.g. uric acid) are secreted into the renal tubules for excretion at the same time as this reabsorption is occurring.

The mean length of the proximal convoluted tubule is approximately 2 mm in the neonate, compared with 20 mm in an adult's kidney.

### *Loop of Henle*

As the proximal convoluted tubule descends it curves to form the loop of Henle or nephron loop. This then divides into a descending limb and an ascending limb. Each limb has both thick and thin segments, and this refers to the cellular structure. The thick segment is composed of cuboidal epithelium and the thin segment is lined with squamous epithelium. Fluid in the descending limb flows towards the renal pelvis, and fluid in the ascending limb flows towards the renal cortex. The descending limb pumps sodium and chloride ions out of the tubular fluid. The long ascending limb in the medulla forms an unusually high solute concentration in the peritubular fluid. The thin segment is permeable to water but not to the solute, so water moves out of these segments, thus concentrating the tubular fluid. The loop of Henle is short in infants.



### *The distal convoluted tubule*

The distal convoluted tubule commences at the point where the ascending limb of the loop of Henle forms a sharp turn near the renal corpuscle. Three essential processes occur within the distal convoluted tubules. These are:

1. The active secretion of ions, acids and drugs into the tubule
2. The selective reabsorption of sodium and calcium ions from the tubular fluid.
3. The selective reabsorption of water, and this enables the further concentration of the tubular fluid.

The distal convoluted tubule in the infant is relatively resistant to aldosterone. This results in a limited concentrating ability.

## Juxtaglomerular complex

The renal corpuscle and the epithelial cells of the distal convoluted tubule are in close proximity and their nuclei are clustered. This area is called the macula densa. The walls of the afferent arteriole contain smooth muscle fibres known as juxtaglomerular cells. The fibres and the cells of the macula densa form the juxtaglomerular complex, which secretes the hormone erythropoietin and the enzyme renin.

## The collecting ducts

The distal convoluted tubule opens into a collecting duct. Several collecting ducts merge to form papillary ducts, which in turn empty into the minor calyx. The urine in the minor calyces will enter the major calyx and then the renal pelvis. The composition of the urine is altered here as sodium and water are reabsorbed. This will determine the volume of urine formed and its osmolality. The amount of water reabsorbed here is influenced by antidiuretic hormone (ADH), while the amount of sodium reabsorbed is controlled by aldosterone. There is a resistance to aldosterone in the distal convoluted tubules, and this continues until the nephron matures.

The nephron and glomeruli are immature at birth. This immaturity results in a reduced GFR and a limiting concentrating ability. The concentrating capacity of a neonate's kidney is approximately half that of a fully mature adult kidney (Sharma *et al.*, 2010). At birth, renal blood flow is low and peripheral vascular resistance is high. The GFR is gestational age related. The more premature the infant, the more reduced the GFR is. At 1 week of age the GFR is 1.5 mL/(kg min), 20–40 mL/min per 1.73 $m^2$, and this increases to adult parameters by 2 years of age: 2.0 mL/(kg min), 120 mL/min per 1.73 $m^2$ (Rennie, 2005). GFR doubles in the first 2 weeks of life, and this is a result of increased blood flow to the kidney and a lengthening of the cortical nephrons (Gomez and Norwood, 1999).

The low GFR in a newborn is due to:

- low renal blood flow secondary to high renal vascular resistance;
- low arterial perfusion, resulting in low intraglomerular capillary hydrostatic pressure;
- small glomerular capillary surface area;
- low water permeability of glomerular capillaries;
- high red blood cell volume (Ichikawa, 1990).

It is these two factors, the limited ability to concentrate urine and the reduced GFR, that contribute to a neonate's susceptibility to dehydration and fluid overload.



# Functions of the kidney

At birth there is a period of rapid growth and development of the physiological function of the kidney that is essential to enable adaptation to extra-uterine life (Solhaug *et al.*, 2004). At birth the function of the kidneys alters significantly: they are no longer required to contribute to the volume of amniotic fluid but are now vital to sustain life by maintaining homeostasis.

The main functions of the kidneys are fluid balance, electrolyte balance and acid–base balance (Table 13.3).

# Blood supply to the kidneys

The kidneys continually function to cleanse the blood and have a large supply of blood vessels to assist in this process (Figure 13.5). Blood is supplied to the kidney by the:

- renal artery
- segmental arteries
- interlobar arteries
- arcuate arteries
- cortical radiate arteries.

The renal artery arises from the abdominal aorta and enters the kidney at the renal hilum where it subdivides into five segmental arteries. This segmental artery further subdivides again into several interlobar arteries. At the junction of the medulla and the renal cortex the interlobar arteries subdivide into the arcuate arteries, and these arch over the base of the medullary pyramids. Cortical radiate arteries project outwards from the arcuate arteries, and these branch out to become the afferent arterioles. One of these is located in each nephron. Further division of the afferent arterioles occurs and becomes the glomerulus. As the glomerulus capillaries leave Bowman's capsule they unite and form the efferent arteriole. These then form to become the peritubular capillaries and then these become the interlobar veins. The vein pathway traces the arterial flow in reverse before finally reaching the inferior vena cava.

Table 13.3 Functions of the kidney.

| Function of the kidney | Notes |
|---|---|
| Fluid balance | At birth 75% of the infant's weight is water. There is an excess of extracellular fluid. As the percentage of body water decreases, a corresponding decrease in extracellular fluid occurs, so by adulthood this is approximately 20%. The high proportion of extracellular fluid, which consists of plasma, interstitial fluid and lymph, predisposes the infant to a rapid loss of body fluid, and thus dehydration. The kidneys maintains the balance between the amount of fluid entering the body and the amount of fluid leaving the body in the mature kidney, but in the immature kidney the reduced GFR makes the infant vulnerable to fluid overload, particularly when receiving intravenous fluids |
| Removal of waste | A newborn baby will produce a total volume of 200–300 mL of urine per 24 h by the end of the first week of life. The infant will void up to 20 times a day and the urine will be odourless, colourless and have a specify gravity of 1.020. The infant is particularly vulnerable to sodium overload. The pre-term infant has difficulty in reabsorbing sodium due to immaturity of the proximal and distal convoluted tubules |
| Renin, angiotensin and erythropoietin secretion | These hormones are secreted by the kidneys. Renin and angiotensin play a role in the regulation of sodium, and therefore water reabsorption, thus resulting in peripheral vasoconstriction and blood pressure control |
| | Erythropoietin is transported in the blood to the bone marrow where it stimulates the production of red blood cells. This is essential to maintain the transportation of oxygen around the body via the red blood cells |
| Vitamin D synthesis | Renal immaturity affects vitamin D formation and calcium homeostasis. The kidney produces the hormone calcitrol, which is essential for the maintenance of calcium and phosphate levels in the blood and bones. The developing fetus and neonate have a high calcium and phosphate requirement for bone formation and growth. Active calcium transport *in utero* provides higher fetal calcium level than the maternal level. At birth this source is removed, leading to a rapid alteration in the calcium homeostasis mechanism and calcium levels fall initially to adult values. As parathyroid hormone and vitamin D control matures, the levels rise again |
| Acid–base maintenance | At birth the regulation of the acid–base balance is established, but there is a limited ability to excrete hydrogen ions, so infants will have a limited response to a metabolic acidosis. This is in part due to the immaturity of the renal cortex and the juxtamedullary nephrons having a higher blood flow than the cortical nephrons. As a result, the neonate is able to conserve sodium but has difficulty in excreting ammonium compounds, which affects the kidneys' ability to correct acidosis. Subsequently, the serum pH is slight lower than the norm at 7.3–7.35 |



# Formation of urine

Urine is formed when the kidneys begin to function *in utero* at around 10 weeks' gestation. This urine is not like the urine that is formed after birth; its function is not to excrete waste, as this is the function of the placenta, but to supplement the amniotic fluid volume (Larsen, 1998). The fetus may produce up to 200 mL per day. It is expected that all newborn infants will produce urine within the first 24 h of birth; 90% usually do. Failure to do so may be associated with abnormalities of perfusion–filtration or obstruction. It may also be due to renal failure; 0.1% of infants may require peritoneal dialysis to manage their acute renal failure.

GFR is low in the newborn (20 mL/min per 1.73 $m^2$) compared with the adult rate of 80 mL/min. The GFR continues to rise after birth due to increase in systemic blood pressure, the fall in

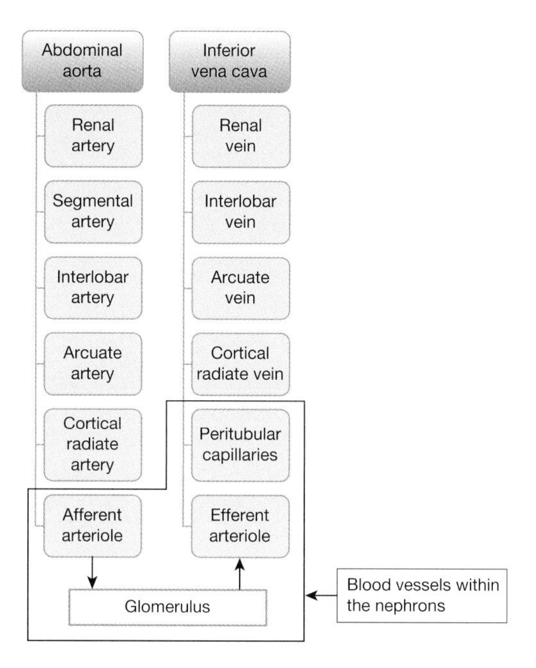


Figure 13.5 Blood supply to and from the kidneys.



renal vascular resistance and the increase in renal blood flow. The minimum acceptable amount of urine produced by a term infant is 1 mL/(kg h). Solute accumulation occurs below this level, and this may be detrimental to the infant if not recognized and managed.

Urine is formed in three phases:

- filtration
- selective reabsorption
- secretion.

## Filtration

The initial stage in the formation of urine is filtration. This continuous process takes place through the semipermeable walls of the glomerulus that is situated in Bowman's capsule. Blood enters the kidney via the renal artery; this is further divided into smaller arterioles that eventually become the afferent arterioles that carry blood into Bowman's capsule. This then becomes a network of capillaries called the glomerulus. Water and small molecules that pass through these capillaries are removed, but larger molecules such as blood and plasma remain in the capillaries. This process occurs as a result of osmosis and diffusion. It is the hydrostatic pressure in the capillaries that forces water and solute through the capillary wall into the collecting unit (Chamley *et al.*, 2005). The water and small molecules such as salt, glucose and other waste products are referred to as the glomerular filtrate. The fluid from the filtered blood contains electrolytes and waste products of metabolism (Table 13.4). The filtered blood is then returned back into circulation via the efferent arterioles and eventually the renal vein. The immature kidney has limited filtration ability, so results in a limited concentrating ability, reduced GFR and a maximum osmolarity of approximately 800–1200 osmol/L until the age of 2 years (Davenport, 1996).

Table 13.4 Outcome of simple filtration.

| Constituents of glomerular filtrate | Constituents remaining in glomerulus |
|---|---|
| Water<br>Amino acids<br>Mineral salts<br>Glucose<br>Uric acid<br>Urea<br>Creatinine<br>Toxins<br>Hormones<br>Drugs | Erythrocytes<br>Leucocytes<br>Platelets<br>Blood proteins |



## Selective reabsorption

The volume and the composition of the glomerular filtrate are altered during the process of selective reabsorption, with any substance that is essential for the maintained of homeostasis and fluid and electrolyte balance reabsorbed. These include sodium, calcium, chloride and potassium. Some substances are never identifiable in the urine because they are completely reabsorbed unless present in excessive qualities. The membranes develop as the child grows and they become less permeable, so there is an increase in the renal tubular glucose reabsorption (Avner *et al.*, 1990). Once the kidney reaches maturity, blood glucose is entirely reabsorbed into the bloodstream from the proximal tubules as it a valuable nutrient. The neonate has reduced reabsorption of glucose, so glycosuria may be evident.

This process of reabsorption takes place by osmosis, diffusion or active transportation. Sodium, amino acids, potassium, calcium, phosphate and chloride ions may all be absorbed by active transport, whereas sodium and chloride ions may also be reabsorbed by active and passive methods (Wilson and Waugh, 1996). In the neonate there is reduced reabsorption and an increased excretion of amino acids, which makes them more prone to the effect of malnutrition.

There is a progressive increase in the tubular reabsorption of sodium in the neonate. At birth the ability of the infant of handle the sodium load is approximately 70%, and this increases to 85% at the end of week 2 due to the immaturity of the distal tubule. High sodium excretion in the infant decreases during the first 12 months of life. The implication of this is the infant can maintain a positive sodium balance and achieve optimal growth and development on a relatively low sodium intake when breast fed (Neill and Knowles, 2004). However, a high solute feed or concentrated formula feeding could cause sodium overload.

## Secretion

The final part of the process of urine formation is secretion. This involves the secreting of substances that have not been removed through filtration in the glomerulus being secreted directly into the convoluted tubules and then excreted from the body in urine. This may take place by active transport. Substances that are secreted into the tubular fluid include potassium, hydrogen, ammonia, creatinine, urea and hormones. Tubular secretion is important for removing excess potassium, controlling the pH of the blood, removal of the end products that have been reabsorbed, such as uric acid and urea, and removal of drug metabolites that are bound to plasma proteins.

## Hormonal control

There are four hormones that play a role in the regulation of fluid balance in the body. Secretion of these is regulated through a negative-feedback system. These are:

- angiotensin II
- aldosterone
- ADH
- atrial natriuretic peptide (ANP).

### *Angiotensin II*

The renin–angiotensin–aldosterone system results in peripheral vasoconstriction and sodium and water reabsorption.

Renin is produced by the juxtaglomerular cells in response to a decrease in blood volume and blood pressure. Renin acts on angiotensin, which is a plasma protein, and this is converted to angiotensin I. Angiotensinogen is produced by the liver and transported to the lungs via the blood. In the lungs, angiotensin-converting enzyme (ACE) converts angiotensin to angiotensin II. In the proximal convoluted tubule, angiotensin II reabsorbs sodium, chloride and water.



### *Aldosterone*

Secreted by the adrenal cortex, which has been stimulated by the presence of angiotensin II, aldosterone increases the reabsorption of sodium and water and the excretion of potassium. Aldosterone levels are higher at birth (Martinerie, Viengchareun, Meduri, *et al*., 2011), and this impacts on the ability of the distal tubules to handle a sodium load.

### *Antidiuretic hormone*

Produced by the hypothalamus and stored in the posterior pituitary gland, ADH increases the permeability of the cells in the distal convoluted tubules and collecting ducts to reabsorb water. If ADH is reduced or absent then less water is reabsorbed and more urine is voided.

ADH levels rise gradually during the first 3 months of life. The distal convoluted tubules are relatively insensitive to ADH initially also. The impact of this is a greater fluid intake is required and a larger production of urine is required to excrete the solute load (Ichikawa, 1990). Once the level of ADH rises, the concentrating ability is increased.

### *Atrial natriuretic peptide*

Secreted by the atria of the heart due to stretching of the wall, this hormone increases the reabsorption of water and sodium in the proximal convoluted tubule and collecting ducts. ANP levels have been found to be increased in children with end-stage renal failure. It is thought that ANP plays a significant role in volume homeostasis (Rascher *et al*., 1985).

# Composition of urine

Assessment of renal function is important in children as they are more vulnerable to renal disorders. Examination of the constituents of urine is an important aspect of the assessment of the child as it may reveal disorders of the renal system.

Urine is normally amber. The amount of urine secreted is dependent on the fluid intake of the child. Pale urine is usually dilute urine, and the neonate's urine is dilute at birth. The characteristics of urine are outlined in Table 13.5.

**Table 13.5** Composition of urine.

| Characteristics | Normal findings | Abnormal findings |
|---|---|---|
| Colour | Amber in colour due to the presence of bile pigment containing urobilin called urochrome<br>Pale urine is usually dilute and neonates tend to have pale urine | Very pale – dilute urine may be due to excessive fluid intake or renal disease where there is an inability to concentrate urine.<br>Dark orange – concentrated urine-fluid deficit as kidneys are conserving water |
| Transparency | Clear when first voided. Will become cloudy if allowed to stand | Very cloudy urine may indicate the presence of white blood cells |
| Odour | Ammonia or mildly aromatic odour. Ammonia odour becomes more apparent if the urine is left to stand | Fishy smell – putrefaction occurs due to the decomposition of protein by bacteria |
| Protein | Small amount of protein usually albumin may be present due to immaturity of nephrons | Proteinuria >1000 mg/day is indicative of renal impairment |
| Glucose | Negative | Glycosuria indicates that renal threshold has been reached and the plasma glucose is >11 mmol/L. This is indicative of:<br>• diabetes type 1<br>• gluconeogenesis from steroid therapy |
| Ketones | Negative | Ketonuria occurs when more fat is metabolized than necessary so excess is excreted in the urine |
| Blood | Negative, but menstrual blood may contaminate urine. Some foods and food dye may colour the urine | Haematuria indicates either active bleeding in the renal system or renal disease |
| Bilirubin | Breaks down quickly in the presence of light | May indicate jaundice secondary to liver disease |
| Specific gravity | Neonate: 1.002–1.008<br>Child: 1.002–1.030 | Static specific gravity – inability of kidneys to concentrate urine due to renal failure.<br>Low specific gravity – excessive fluid intake.<br>High specific gravity – dehydration, glycosuria |
| pH | Neonate 5–7<br>Child 4.5–8 | ↓ pH occurs as hydrogen ions are excreted in urine and metabolic acidosis is present |



# Clinical application

## Urinary tract infection

Urinary tract infections (UTIs) are relatively common in infants and young children. Symptoms of UTIs in the neonatal period may be overlooked as the signs and symptoms are not always obvious, so this should always be considered as a potential diagnosis and excluded. During the first year of life, 1.2% of male infants and 1.1% of female infants will have a UTI. After this period, UTIs are more common in girls (3%) than in boys (0.1%). Renal scarring will occur in 5–10% of these children, and 1% of this group will need management for hypertension.

UTIs may be due to bacterial infection, vesico-ureteral reflux and congenital obstructions. They may involve the urethra, bladder, ureters, renal pelvis, calyces and renal parenchyma.

The infant may present with sepsis, fever, vomiting and prolonged jaundice (Jones and Smith, 1999).

The child may present with fever, vomiting, dysuria, frequency, haematuria, abdominal or back pain, dark, cloudy, malodourous urine and a burning sensation when passing urine.

Treatment with antibiotics and prophylactic antibiotics is usually successful. Surgery is indicated if the primary cause is due to obstruction.

# The ureters

The ureters are a pair of muscular tubes that extend from the renal pelvis of the kidney to the posteriolateral surface of the bladder. They enter the bladder at an oblique angle at the trigone, and this helps prevent the reflux of urine from the bladder back up into the ureters (Matsumara and England, 1992). Their final position is achieved as the kidneys ascend into their position in the posterior abdominal wall. They pass through the psoas major muscle.



# The bladder

The bladder is a hollow muscular organ (Figure 13.6). Its function is to act as a reservoir for urine. The bladder forms from the cranial part of the urogenital sinus. As it enlarges, the ureters become incorporated into the dorsal wall separately. In the newborn the bladder lies at a higher level than the adult bladder does. It is an abdominal organ with the internal urethral orifice at the level of the upper border of the symphysis pubis, so approximately one-half of the bladder lies above the superior aperture of the pelvis. When fully distended the infant's bladder is

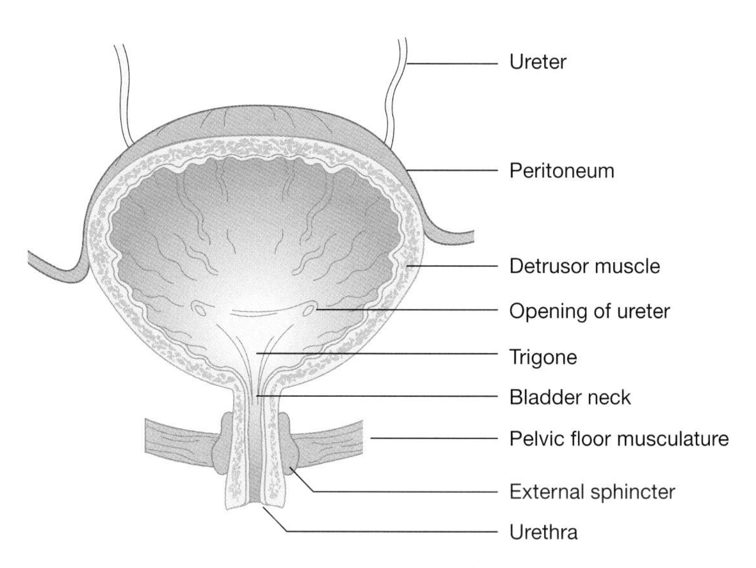


Figure 13.6 The bladder – cross-section.

entirely abdominal and may extend up to the umbilicus (England, 1996). The internal urethral orifice descends rapidly during the first year of life and continues slowly until the ninth year of life. This descent then continues until it finally reaches its adult position as a pelvic organ. The implication of this is that the infant's ureter is shorter than that of an older child's.

The bladder is cylindrical shaped in early childhood and becomes more pyramidal by the age of 6 years. There will be a small amount of urine in the bladder at birth. At puberty the bladder becomes a true pelvic organ as a result of growth of the pelvis and maturation of the pelvic bones.

The bladder is formed of four layers. The innermost layer is mucosa. This mucosal lining is in folds called rugae and these disappear as the bladder fills with urine. This layer is connected to the muscle layer by connective tissue. This muscle, the detrusor, is a circular muscle surrounded by a layer of longitudinal muscle on each side. The area of the bladder where the ureters and urethra enter is called the trigone; here, the mucosa is thicker. Small flaps of mucosa cover these openings, and these act as valves preventing the back flow of urine up into the ureters. There are a large number of stretch receptors in this area which are sensitive to bladder filling. The trigone acts as a funnel that channels the urine into the urethra when the bladder contracts.



At the base of the bladder, the bladder neck joins the urethra and the urinary sphincters (Figure 13.6). The bladder neck has long muscle fibres that open and close to facilitate the voiding of urine. There is a ring of smooth muscle at the neck of the bladder; the male bladder will have a ring of circular muscle, while the female will have longitudinal muscle in the internal sphincter (Colborn, 1994). The external sphincter is composed of smooth and striated muscles. Both sphincters are normally closed due to the action of the pelvic floor striated muscle.

Bladder capacity is estimated at 30 mL per year of age plus 30 mL. The sensor receptors become activated when 15 mL of urine is present in the bladder in the child's bladder; an adult's bladder has capacity of 300–400 mL of urine. Table 13.6 outlines the normal urine volume for children.

## The urethra

This is the muscular tube that conveys urine out of the body from the bladder. It is composed of three layers: muscular, erectile and mucosa. The urethra is encompassed by both the internal and external sphincter muscles. The internal sphincter is created by the detrusor muscle and is located at the urethra junction (Peate and Nair, 2012). The internal sphincter is under involuntary

Table 13.6 Urine output in 24 h for infants and children.

| Age of child | Amount of urine | Number of voids |
|---|---|---|
| Pre-term baby–32 weeks' gestation | 12 ml/kg every 24 h | 20 in 24 h |
| Term baby | 1–3 mL/(kg h) | 10–20 in 24 h |
| 6 months–2 years | 540–600 mL/24 h | 10 in 24 h |
| 5–8 years | 600–1200 mL/24 h | 6–8 times a day |
| 8–14 years | 100–1200 mL/24 h | 6–8 times a day |
| >14 years | 1500 mL/day | 4–8 times a day |

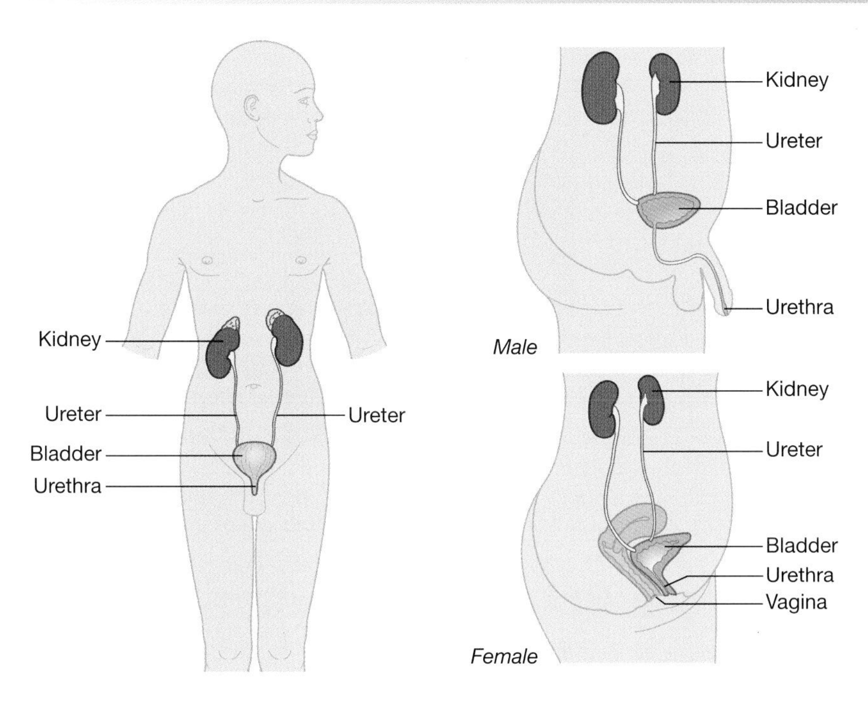


Figure 13.7 Female and male urethra.



control and the external sphincter is under voluntary control once detrusor stability is achieved. There are differences between the male and female urethra (Figure 13.7).

## The female urethra

The epithelium layer in the female urethra is derived from the ectodermal urogenital sinus, and the connective tissue and smooth muscle is developed from the splanchnic mesoderm. The urogenital fold fuses during weeks 9–12 of fetal development, and the urethral groove forms a tube that becomes the whole of the female urethra. It is adhered to the anterior wall of the vagina, and the female urethra exits the body via the urethral orifice, which is located in the labia minora. This external opening is anterior to the vagina and posterior to the clitoris. The female urethra is approximately 4 cm in length at maturity and its only function is to transport urine from the body.

## The male urethra

Like the female urethra, the male urethra develops from the urogenital sinus and the splanchnic mesoderm, with the exception of the penile urethra. This develops from an ectodermal ingrowth that splits at the tip of the penis to form the urethral groove, which is lined with endodermal cells and develops into the urethral plate. The distal aspect of the male urethra is derived from the granular plate. This granular plate grows from the tip of the glans penis into the developing gland penis. This then connects to the urethra, which has developed from the phallic part of the urogenital sinus (Moore and Persaud, 2003). The granular plate develops a canal allows the urethra to open at the urinary meatus, which is located on the glans penis. Ectodermal cells of the glans penis penetrate inwards towards the midline to form the chordee, which leads to the external urethral meatus. The prostate gland develops as a result of multiple budding of the mesenchyme. The male urethra is approximately 20 cm long when fully developed. The male urethra has a dual function: to transport urine from the body and to facilitate ejaculation of semen.

## Clinical application

Hypospadias in the male is a result due to failure or incomplete fusion of the urogenital fold. This results in an incomplete penile urethra opening. It may be located behind the glans penis or anywhere along the ventral surface of the penile shaft. Surgical correction is required, and the aims of surgery are to enable the child to void urine in the standing position and have a direct stream of urine, to improve the physical appearance and to produce a sexually adequate organ.



# Function of the bladder and micturition

Micturition is the voiding of urine and is a simple spinal reflex that is under both unconscious and conscious control (Figure 13.8). The ability to control the bladder is learnt in early childhood, usually between the ages of 18 months to 4 years.

In the infant bladder, control is impossible and bladder emptying is dependent on the actions of the reflex arc. The bladder distends and the stretch receptors in the area of the trigone send

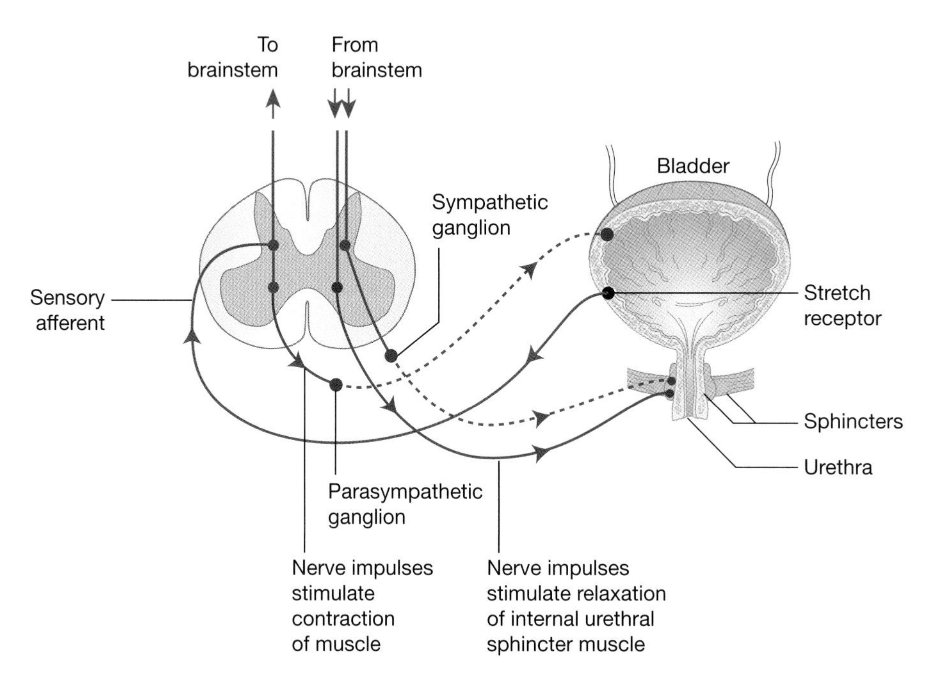


Figure 13.8 Bladder control sensory pathway.

impulses to the sacral area of the spinal cord via the autonomic nervous system. This initiates a response from the internal sphincter on the bladder, which relaxes, the detrusor muscle contracts and urine is expelled from the bladder. Voluntary emptying occurs when the bladder reaches a capacity of 15 mL of urine.

Around the age of 18 months the child develops an awareness of wanting to pass urine. They can retain their urine for a short length of time only. Maturation of the diaphragm, abdominal and perineal muscles all assist in the child's ability to increase the volume of urine in their bladder. Development of the frontal and parietal lobes of the cerebral cortex is also required for full bladder control, as the conscious desire to void urine is expressed from here.

Girls achieve bladder control, detrusor stability, earlier than boys. This is an important milestone of childhood development.

# Conclusion

The renal system is composed of the kidneys, ureters, bladder and urethra. This system is essential for homeostasis by the maintenance of fluid balance, removal of waste products and the secretion of hormones.



The development of the renal system is complex, and growth of the renal system continues post birth. This growth enables the structures to achieve maturity and the functionality required for healthy childhood growth and development.

The functions of the kidneys and the formation of urine have been discussed, along with the process of micturition. Relevant clinical considerations have been identified in this chapter.

# Activities

**Now review your learning by completing the learning activities in this chapter. The answers to these appear at the end of the book. Further self-test activities can be found at www.wileyfundamentalseries.com/childrensA&P.**

## True or false?

1. The urinary system plays a vital role in maintaining homeostasis of water and electrolytes in the body.
2. Renin is a hormone that is secreted by the kidney which is important in the control of blood pressure.
3. Erythropoietin is produced and secreted by the kidney.
4. Urine is stored in the bladder and excreted by the process known as menstruation.
5. The kidney contains approximately 1–2 million nephrons.
6. The glomerular capsule is also known as Bowman's capsule.
7. Antidiuretic hormone decreases the permeability of the distal convoluted tubules and collecting tubules.
8. The transport maximum for glucose is 9 mmol/L.
9. Glomerular filtration rate is the volume of filtrate formed by both kidneys each hour.

## Wordsearch

There are several words linked to this chapter hidden in the following square. Can you find them? A tip – the words can go from up to down, down to up, left to right, right to left, or diagonally.

| | | | | | | | | | | | | | | |
|---|---|---|---|---|---|---|---|---|---|---|---|---|---|---|
| P | F | O | X | I | C | R | A | N | S | S | T | S | D | E |
| U | O | I | M | O | I | H | L | U | L | O | S | A | L | L |
| C | R | T | L | M | T | N | L | S | O | D | N | C | S | E |
| E | A | E | A | T | P | U | E | P | E | I | S | S | U | T |
| P | O | P | T | S | R | D | E | T | R | U | S | O | R | E |
| I | E | L | S | E | S | A | T | T | M | M | M | W | N | I |
| P | T | L | M | U | R | I | T | P | B | I | H | O | R | S |
| S | S | O | V | R | L | S | U | I | E | S | R | S | N | R |
| L | L | W | S | I | O | E | M | M | O | E | E | A | E | M |
| G | T | R | A | N | S | P | O | R | T | N | M | D | P | A |
| N | T | G | T | E | E | M | T | S | O | W | D | R | H | X |
| M | E | D | U | L | L | A | O | G | O | A | N | S | R | I |
| R | E | N | I | N | S | D | I | B | L | L | O | O | O | M |
| D | I | O | E | E | L | R | A | B | S | P | P | B | N | U |
| U | T | M | I | A | T | E | T | E | G | F | I | A | I | M |

## Conditions

The following table contains a list of conditions. Take some time and write notes about each of the conditions. You may make the notes taken from text books or other resources; for example, people you work with in a clinical area or you may make the notes as a result of people you have cared for. If you are making notes about people you have cared for you must ensure that you adhere to the rules of confidentiality.

| Condition | Your notes |
|---|---|
| Urinary tract infection | |
| Renal colic | |
| Chronic kidney disease | |
| Acute kidney disease | |
| Nephrotic syndrome | |

# Glossary

**Angiotensin:** a plasma protein that leads to vasoconstriction which raises blood pressure.

**Aldosterone:** a steroid hormone that is secreted by the outer cortex of the adrenal gland.

**Anuria:** absence of urine.

**Bladder:** a sac that acts as a reservoir for urine.

**Calyces:** branch of the renal pelvis.

**Cotex:** the reddish brown layer of tissue immediately below the capsule.

**Diuresis:** excess production of urine.

**Dysuria:** pain or discomfort on passing urine.

**Erythropoietin:** hormone produced by the kidney whose function is to regulate red blood cell production.



**Fibrous capsule:** this surrounds the kidney.

**Glomerulus:** the network of capillaries located in the Bowman's capsule.

**Haematuria:** presence of blood in the urine.

**Hilus:** the concave border of the kidney where the blood vessel, lymph vessels and nerves enter and leave.

**Medulla:** the innermost layer of the kidney.

**Micturition:** the voiding of urine.

**Nephrogenesis:** refers to the development the kidney in utero.

**Nephrons:** the functional unit of the kidney.

**Oliguria:** urine of output of $<1$ ml/kg/hour.

**Renal cortex:** the outermost aspect of the kidney.

**Renal medulla:** the middle layer of the kidney.

**Renal pelvis:** the funnel shaped structure that acts as a receptacle for the urine formed by the kidney.

**Renin:** a renal hormone that assists in the maintenance of blood pressure.

**Secretion:** the secretion of non-threshold substances into the convoluted tubules.

**Selective reabsorption:** the reabsorption of essential products of filtrate necessary to maintain electrolyte and fluid balance.

**Simple filtration:** a process that occurs between the semipermeable walls of the glomerulus and the glomerular capsule as the first process in the formation of urine.

**Trigone:** the three orifices in the bladder wall that form a triangle.

**Ureter:** a tube that conveys urine to the bladder.

**Urethra:** a canal that extends from the neck of the bladder to the exterior.

# References

Avner, E.D., Ellis, D., Ichikawa, I. (1990) Normal neonates and the maturational development of homeostatic mechanism. In: Ichikaw, I. (ed.), *Pediatric Textbook of Fluids and Electrolytes*, Williams & Wilkins, Baltimore, MD.

Bonilla-Felix, M., Brannan, P., Portman, P. (1998) Neonatal nephrology. In: Merenstein, G.B., Gardner, S.L. (eds), *A Handbook of Neonatal Intensive Care*, 4th edition. Mosby, St. Louis, MO, pp. 535–570.

Chamley, C., Carson, P., Randall, D., Sandwell, W.M. (2005) *Developmental Anatomy and Physiology. A Practical Approach*, Churchill Livingstone, Edinburgh.

Colborn, D. (1994) *The Promotion of Continence in Adult Nursing*, Chapman and Hall, London.

Davenport, M. (1996) Paediatric fluid balance. *Care of the Critically Ill*, **12** (1), 26–31.

England, M. (1996) *Life before Birth*, 2nd edn, Mosby, London.

Gomez, R.A., Norwood, V.F. (1999) Recent advances in renal development. *Current Opinion in Pediatrics*, **11** (2), 135–140.

Ichikawa, I. (1990) *Pediatric Textbook of Fluids and Electrolytes*, Williams & Wilkins, Baltimore, MD.

Jones, K.V., Smith, C.G. (1999) Urinary tract infections in childhood. *Medicine*, **27** (7), 71–75.

Larsen, W.J. (1998) *Essential of Human Embryology*, Churchill Livingstone, Edinburgh.

Lissauer, T., Fanaroff, A.A. (2011) *Neonatology at a Glance*, 2nd edn, Wiley–Blackwell, Oxford.



Marieb, E., Hoehn, K. (2010) *Human Anatomy & Physiology*, 8th edn, Pearson Benjamin Cummings, San Francisco, CA.

Martinerie, L., Viengchareun, S., Meduri, G. *et al.* (2011) Aldosterone postnatally, but not at birth, is required for optimal induction of renal mineralocorticoid receptor expression and sodium reabsorption. *Endocrinology*, **152** (6), 2483–2491.

Matsumara, G., England, M.A. (1992) *Embryology Colouring Book*, Mosby, St Louis, MO.

MacGregor, A. (2010) *Introduction to the Anatomy and Physiology of Children*, 2nd edn, Routledge, London.

Moore, K.L., Persaud, T.V.N. (2003) *The Developing Human: Clinically Orientated Embryology*, 6th edn, WB Saunders, Philadelphia, PA.

Neill, S., Knowles, H. (2004) *The Biology of Child Health. A Reader in Development and Assessment*, Palgrave Macmillian, Basingstoke.

Peate, I., Nair, M. (2012) *Fundamentals of Anatomy and Physiology for Student Nurses*, 2nd edn, Wiley–Blackwell, Oxford.

Rascher, W., Tulassay, T., Lang, R.E. (1985) Atrial natriuretic peptide in plasma of volume overloaded children with chronic renal failure. *Lancet*, **326**, 303–305.

Rennie, J.M. (ed.) (2005) *Roberton's Textbook of Neonatology*, 4th edn, Elsevier, Oxford.

Sharma, A., Ford, S., Calvert, J. (2010) Adaptation for life: a review of neonatal physiology. *Anaesthesia and Intensive Care Medicine*, **12** (3), 85–90.

Sinclair, D. (1991) *Human Growth after Birth*, 5th edn, Oxford Medical Publication, Oxford.

Solhaug, M.J., Bolger, P.M., Jose, P.A. (2004) The developing kidney and environmental toxins. *Pediatrics*, **113** (4 Suppl.), 1084–1091.

Waugh, A., Grant, A. (2010) *Ross and Wilson. Anatomy and Physiology in Health and Illness*, 11th edn, Churchill Livingstone Elsevier, Edinburgh.

Wilson, K.J.W., Waugh, A. (1996) *Ross and Wilson. Anatomy and Physiology in Health and Illness*, 8th edn, Churchill Livingstone, Edinburgh.

# Chapter 14

# The reproductive systems

Ann L. Bevan

*School of Health and Social Care, Bournemouth University, Bournemouth, UK*

## Aim

The aim of this chapter is to help you to develop insight and understanding of the male and female reproductive systems. By gaining further understanding and developing your insight you will be able to provide high-quality, safer and effective informed care.

## Learning outcomes

On completion of this chapter the reader will be able to:

- Describe the process of sexual differentiation and development in the fetus.
- Name and locate the male reproductive organs.
- Name and locate the female reproductive organs.
- Describe and understand the role and functions of the male and female reproductive systems.
- Understand the process of puberty in males and females.
- Discuss the phases of the menstrual cycle.

## Test your prior knowledge

- What is the process of differentiating the sexes in the fetus?
- What happens at puberty in males and females?
- Where are female eggs stored and released from in the female reproductive system?
- What structures would you expect to find in the female reproductive system?
- What is the function of the prostate gland in the male reproductive system?
- What structures would you expect to find in the female reproductive system?
- What is the function of the hormone oestrogen?
- What is the function of the hormone testosterone?
- What is the function of the female breast?
- What conditions may occur in boys' and girls' reproductive systems?

*Fundamentals of Children's Anatomy and Physiology: A Textbook for Nursing and Healthcare Students*, First Edition. Edited by Ian Peate and Elizabeth Gormley-Fleming.

Companion website: www.wileyfundamentalseries.com/childrensA&P

# Introduction

Regardless of the fact that the reproductive system is not essential to physical survival, all living things reproduce. They reproduce for survival of the species and to pass on genetic material from generation to generation.

In humans, as in most animals, reproduction is sexual; that is, there needs to be one male and one female, each with specialized cells designed to reproduce another human being on their coming together. Males produce sperm and females produce eggs or ova in order to reproduce. Once the male sperm has been delivered to the female ova and fertilization has taken place the female reproductive system then has the responsibility of growing and developing the embryo into a fetus over a period of 9 months.

Children are not necessarily aware of the differences between the sexes until after the age of 3 years, at which time they notice that boys and girls have different external genital appearances (MacGregor, 2008). It is when both girls and boys reach puberty that under the influence of hormones the reproductive system comes out of its apparent sleep (Marieb and Hoehn, 2013) and fully begins to develop.

This chapter provides an overview of the anatomy and physiology of the human reproductive system from the developing embryo to the young person. The reproductive systems include in the male the testes, penis, ducts and glands, and in the female the ovaries and fallopian tubes, uterus, vagina and vulva.



# Fetal embryology: sexual differentiation

Sexual reproduction involves the union of two cells from different organisms of the same species. Males produce sperm and females eggs for this purpose, and these special cells are known as gametes. These gametes are produced by cell division called meiosis. In humans the total number of chromosomes is 46. Each gamete produces 23 chromosomes in order to fuse together with the opposite gamete, making up the total number of required chromosomes in the new fertilized cell.

The determination of the sex of an embryo occurs at fertilization by the addition of an X (female) or Y (male) chromosome to the X chromosome already present in the ovum. Until around 7 weeks' development the embryo follows the same path regardless of sex as the gonads appear identical; this is known as the indifferent stage. The indifferent gonad differentiates in embryos with an XX chromosome complex into an ovary and with an XY complex into a testis (Moore and Persaud, 2008). The type of sexual differentiation is then governed by the type of gonad present. The male testes produce testosterone, which determines male sexual differentiation. However, hormones produced by the ovaries do not play a part in female differentiation in the fetal period as this process still continues even in the absence of ovaries (Moore and Persaud, 2008).

In the male, under the influence of anti-Müllerian hormone secreted by pre-Sertoli cells, Müllerian ducts develop, causing the embryonic female tract to regress. Sertoli cells are situated in the testis and they act upon Leydig cells to produce androgens that act upon the embryonic genital organs producing maleness (MacGregor, 2008). At 8–10 weeks, testosterone stimulates the Wolffian system, which triggers growth and development of the structures of the male reproductive system. In the absence of a Y chromosome the female Müllerian system is fully developed by age 3 months. The superior part of the vagina, uterus and fallopian tubes all develop from the paramesonephric duct or Müllerian ducts (Figures 14.1 and 14.2).

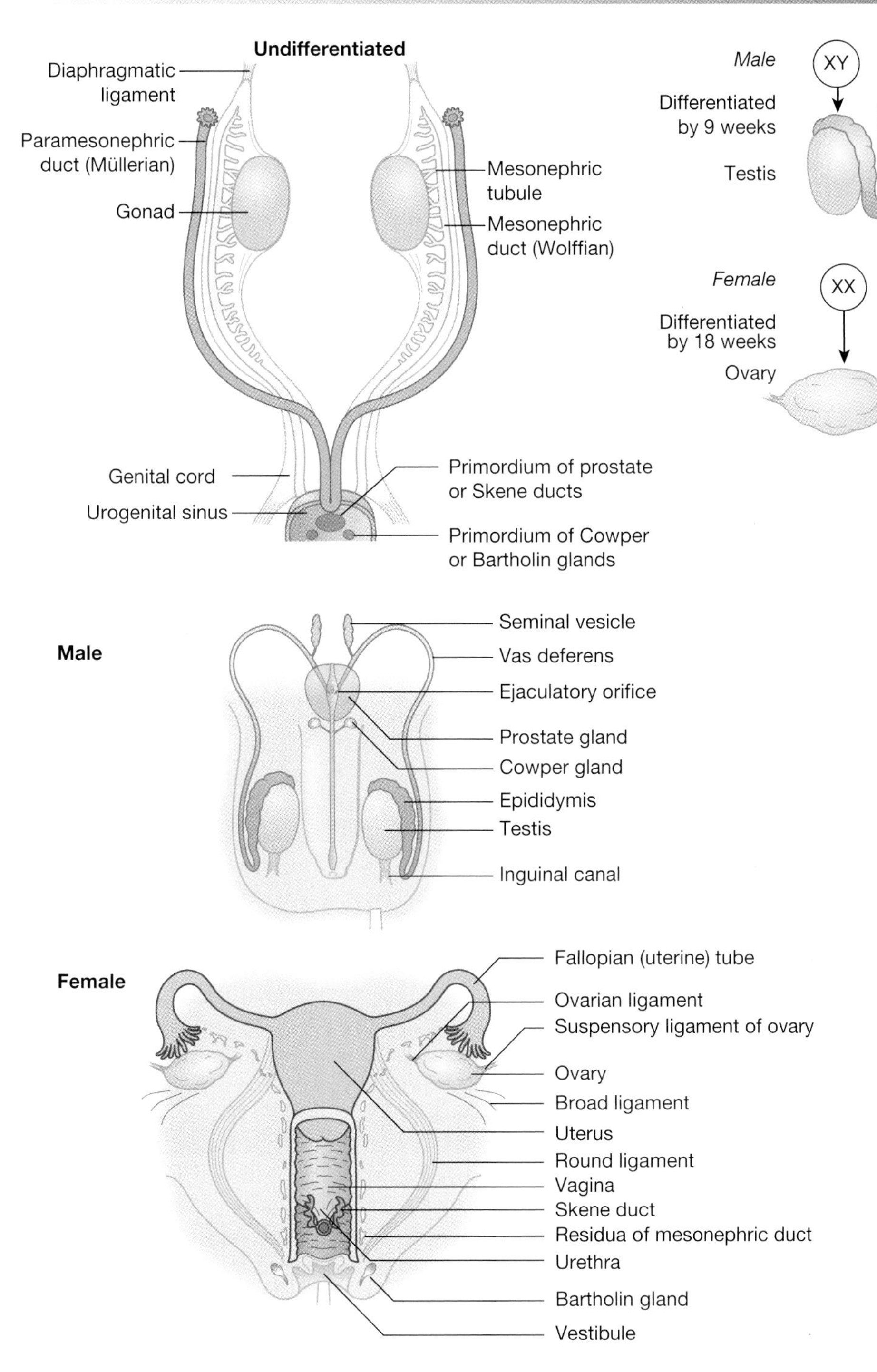




**Figure 14.1** Development of internal genitalia.

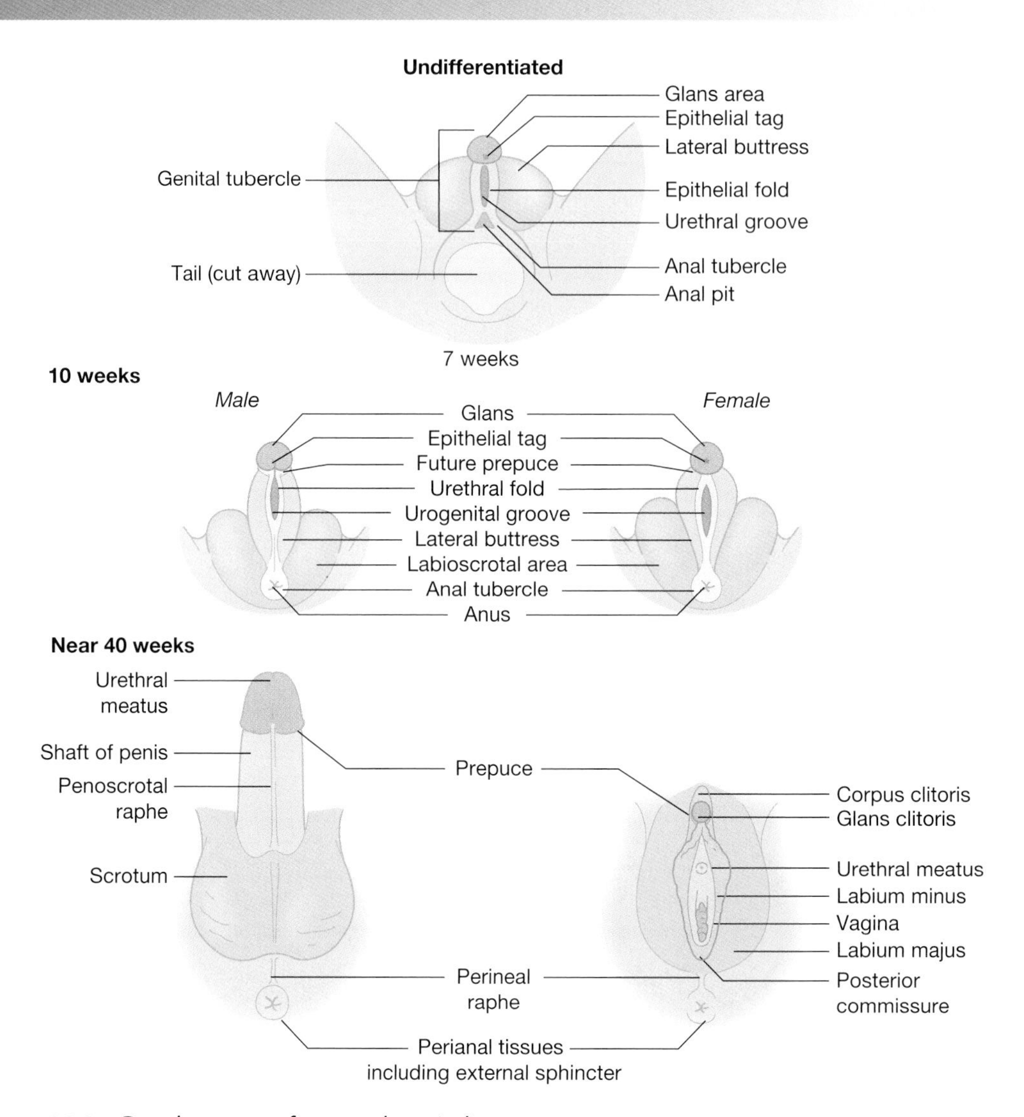




**Figure 14.2** Development of external genitalia.

For a few months before birth the male infant has plasma gonadotropin and testosterone levels nearly two-thirds those of an adult male. Following birth these hormones drop precipitously, likely due to the withdrawal of placental steroid hormones; similarly, hormonal levels drop in females. The negative-feedback system on the hypothalamus and pituitary gland is removed and the gonadotropins leuteinizing hormone (LH) and follicle-stimulating hormone (FSH) are released. After the first year of life the gonadotropins are suppressed until puberty, after which the adult pattern of hormone interaction is achieved.

# The male reproductive system

The male reproductive system has the majority of its organs outside of the body, unlike the female system where most of the reproductive organs are out of sight (see Figure 14.3 for an

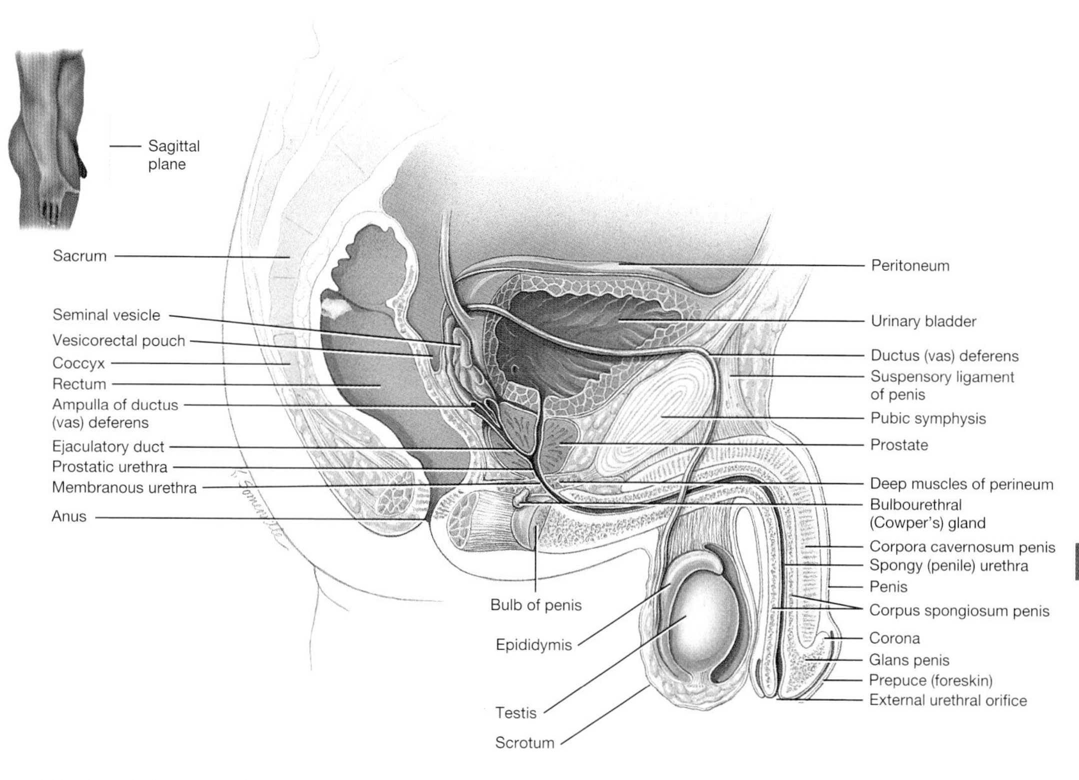


**Figure 14.3** Male reproductive system.
*Source:* Tortora and Derrickson (2009), Figure 28.1, p. 1082. Reproduced with permission of John Wiley and Sons, Inc.



illustration of the male reproductive and genitorunial system). The reproductive system works alongside other body systems such as the neuroendocrine system to produce hormones important to development and sexual maturation. In males the reproductive system is integral with the urinary system and essential for its functioning.

The functions of the male reproductive system include producing, maintaining and transporting sperm, producing the transport fluid called semen, the discharge of sperm from the penis, and the production and secretion of male sex hormones.

## The testes and scrotum

In the male fetus, about 1 month before birth, the testes usually descend from their development position on the posterior abdominal wall down the inguinal canal to the scrotal sac. Each testis is attached to an inferior ligament or gubernaculum, and these extend to enable the testes to descend (see Figure 14.4 for the descent of the male testes). This movement is triggered by the growth of abdominal viscera and an increase in testosterone action. Without this descent sperm will not develop and the tubules for sperm storage and maturation will not function correctly as they then become fibrous (MacGregor, 2008).

The male gonads in the form of a pair of sperm-producing testes are situated within the scrotal sac, which hang outside the body at the root of the penis. The reason they are in this relatively precarious position is because in order to produce viable sperm they need to be produced and stored in a temperature approximately 3 °C lower than the core body temperature of 37 °C. External temperature changes cause the scrotal sac to either pull the testes closer to the pelvic floor in cold weather, or become loose to move the testes away from the pelvic floor in warm weather. This phenomenon is called the cremasteric reflex. The cremaster muscle, which is part of the spermatic cord, contracts and the spermatic cord shortens; this moves the testicles in closer to the body. When the cremaster muscle loosens, the testicle is moved away in order to cool it down. Without this reflex temperature control mechanism the sperm could become sterile and unable to reproduce.

## Clinical application

Cryptorchidism is the failure of one or both testes to descend from the abdominal cavity into the scrotal sac prior to birth. Although more common in pre-term infants (i.e. those newborns less than 37 weeks' gestation), it does occur in about 3–4% of term males. If the testes have not descended by the age of 1 year they usually do not do so spontaneously and will need an orchidopexy before the age of 2 years to move the testes down and fix them in the scrotal sac (Sandwell, 2005).



The testes are divided into about 250 wedge-shaped lobules. Each lobule consists of four tightly coiled seminiferous tubules. Within these tubules is thick stratified epithelium consisting of spheroid spermatogenic cells; these are situated within sustentocytes, which are supporting cells that also contribute to the formation of sperm. Spaces exist between the tubules, and these are called Leydig cells, which synthesize and secrete testosterone, among other androgens. The seminiferous tubules of each lobule connect to a straight tubule through which sperm enters called the rete testis, which then connects to the efferent ductules and epididymis. There are three to four layers of smooth muscle – like myoid cells, which contract rhythmically and help squeeze sperm and fluids through the tubules and out of the testes.

The newborn testes are small and grow slowly over time until they reach their adult size in their teen years, of approximately 5 cm long, 3 cm wide and 2.5 cm thick, and weighing 10–15 g; this is approximately 40 times that of the newborn. One way to measure the development of the external genitalia throughout childhood and into adolescence is by comparing it with the Tanner scale. This scale can be used by nurses and other health professionals to inform of any arrested development or as an indicator of congenital problems in development.

## Clinical application

### Male genital growth and development

The Tanner Scale (Tanner, 1962) can be used as an aid to determine physical development in children, young people and adults by nurses and other health professionals. Comparing the physical characteristics of external genitalia with the scale can help determine normal and/or abnormal development. The scale consists of five stages on the development of the penis, testicular growth and pubic hair (Figure 14.4).

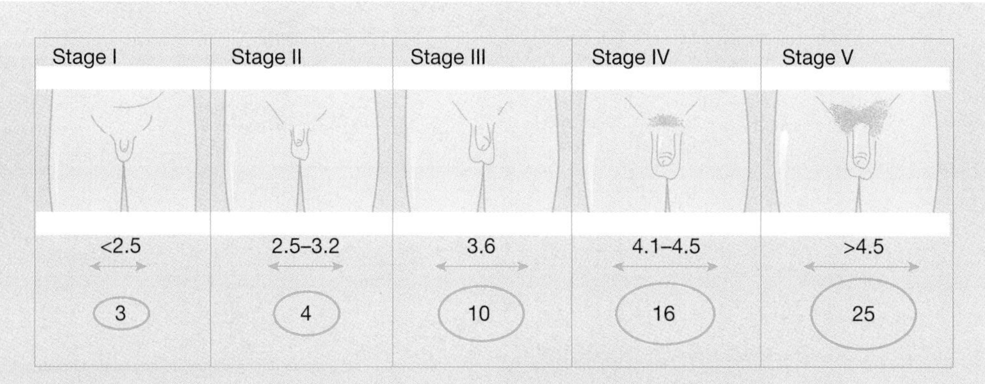


**Figure 14.4** The Tanner scale for male genital growth.

Stage I. Pre-adolescent. Testes, scrotum and penis are of about the same size and proportion in early childhood.

Stage II. Enlargement of scrotum and testes. The skin of the scrotum reddens and changes in texture. Little or no enlargement of penis at this stage. Pubic hair begins to appear.

Stage III. Enlargement of penis, which occurs at first mainly in length. Further growth of testes and scrotum.

Stage IV. Increased size of penis with growth in breadth and development of glands. Hair becomes coarser and continues to spread.

Stage V. Genitalia adult in size and shape. Pubic hair adult in type and quantity.

The Tanner scale is only one method of assessing physical development.



## Hormonal influences

The hypothalamic–pituitary–gonadal axis regulates the production of gametes and sex hormones through hormonal interactions between the hypothalamus, anterior pituitary gland and the gonads. Hormones play a part in the function of the male reproductive system. These male sex hormones are known as androgens, and the majority are produced by the testes. A small number are also produced by the adrenal cortex. Testosterone is the main androgen produced by the testes.

With the onset of puberty the hypothalamus increases its secretion of gonadotropin-releasing hormone (GnRH). This stimulates the anterior pituitary gland to increase the release of LH. These hormones both work on the negative-feedback system that controls the secretion of testosterone in the blood and sperm production (spermatogenesis) (Figure 14.5).

## Male duct system

The epididymis is a comma-shaped duct about 3.8 cm long (Marieb and Hoehn, 2013). The tube consists of the head, which contains the efferent ductules, situated on the superior aspect of the testes, the body and a tail situated on the posterolateral aspect of the testes. Immature sperm pass through the head and body and are then stored in the tail of the epididymis until ejaculation. The epididymis is coiled and if unravelled would be 6 m in length. It is made of psuedostratified cilia, epithelium and smooth muscle. Within the epididymis the sperm continue to mature and gain motility in order that they can fertilize the ova. This maturation takes about 14 days to occur (Jenkins *et al.*, 2010). The sperm then move into the vas deferens through

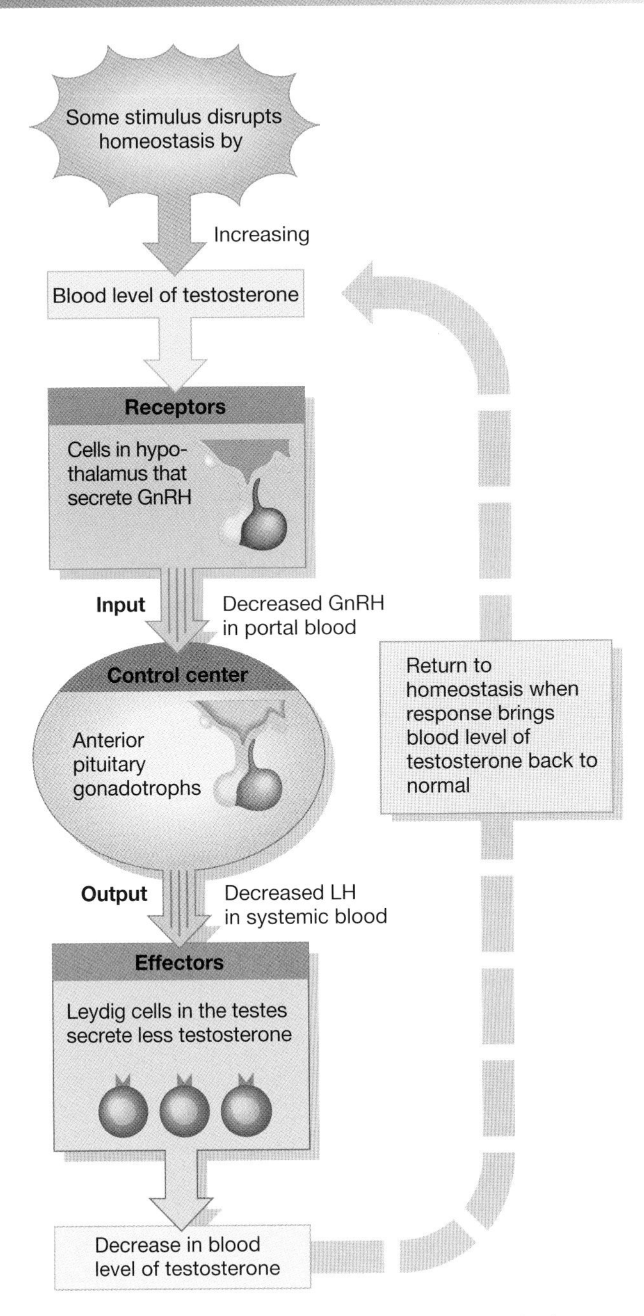




**Figure 14.5** Negative-feedback system associated with the control of testosterone in the blood.
*Source:* Tortora and Derrickson (2009), Figure 28.8, p. 1089. Reproduced with permission of John Wiley and Sons, Inc.

the action of peristaltic action generated by smooth muscle contractions during sexual activity.

The vas deferens, or ductus deferens, is about 45 cm long and is situated from the epidiymis through the inguinal canal into the pelvic cavity. This tube contains ciliated epithelium and is surrounded by a thick muscular layer. The vas deferens runs through the spermatic cord, which is a tube connecting the scrotal sac and inguinal canal; this cord contains blood vessels and nerves. The terminus of each vas deferens then joins with the ejaculatory ducts. The ejaculatory ducts enter the prostate gland, discharging fluid into the urethra.

## The prostate gland

The prostate is a single doughnut-shaped gland wrapped about the urethra just inferior to the bladder (see Figure 14.3 for position of the prostate gland). It grows slowly until puberty, at which point over a short period of time it doubles in size and then continues to grow. It is composed of three zones: the central zone, the peripheral zone and the transition zone (Peate, 2011). It is enclosed by a thick connective tissue capsule and is made up of 20–30 tubuloalveolar glands embedded in smooth muscle and connective tissue. Prostatic smooth muscle contracts during ejaculation, forcing prostatic fluid into the prostatic urethra via a number of ducts. The fluid makes up one-third of the semen volume and plays a role in activating sperm (Marieb and Hoehn, 2013). It is a slightly acidic (pH 6.5) milky substance that contains several enzymes, prostate-specific antigen and citrate. Although the prostate gland is most active during and after puberty, at birth it does release a glycogen-rich secretion into the ducts (Sandwell, 2005).

## The penis



The penis is a copulatory organ that is also part of the genitourinary tract. It hangs alongside the scrotum suspended from the perineum (Figure 14.6). It is a highly vascular organ and encloses the urethra. The penis consists of an attached root and a shaft that ends in a tip called the glans penis. The glans is covered by a prepuce known as the foreskin. The foreskin is sometimes removed surgically following birth in a procedure called circumcision. Some cultures or religions require this procedure be undertaken. Proponents assert that circumcision results in less risk of acquiring HIV, as well as reduces the risks for reproductive system infections in both males and females (Marieb and Hoehn, 2013). However, circumcisions remain a controversial practice owing to the risks of bleeding, infection and other complications (Cohen, 2009; Martini *et al.*, 2012).

The penis usually hangs flaccid until it is sexually aroused. Internally, the penis contains erectile tissue, which consists of a spongy network of connective tissue and smooth muscle with numerous vascular spaces. The vascular spaces fill with blood during sexual arousal, causing the penis to enlarge and become rigid. This reaction is in response to an impulse that stimulates the parasympathetic nervous system. This then allows the penis to serve as a penetrating organ for the female vagina to allow sperm to be delivered in close proximity to the female ova.

The seminiferous tubules that transport the sperm to the penis are solid at birth. The rectum and the prostate gland are the only two major organs in the pelvis of the newborn male. The penis is relatively large in the newborn and consists of spongy material that fills with blood, as a newborn can experience an erection.

## Clinical application

Hypospadias refers to the external urethral orifice being located on either the glans or the body of the ventral surface of the penis. It is the most common anomaly of the penis, seen in 1 in every 300 male infants. In addition to the abnormal position of the meatus, the penis is curved ventrally, called *chordee*, and is underdeveloped (Moore and Persaud, 2008) The only recourse is surgery to reconstruct the penis before the age of 2 years in order to allow urine to pass in a steady straight stream (Sandwell, 2005).

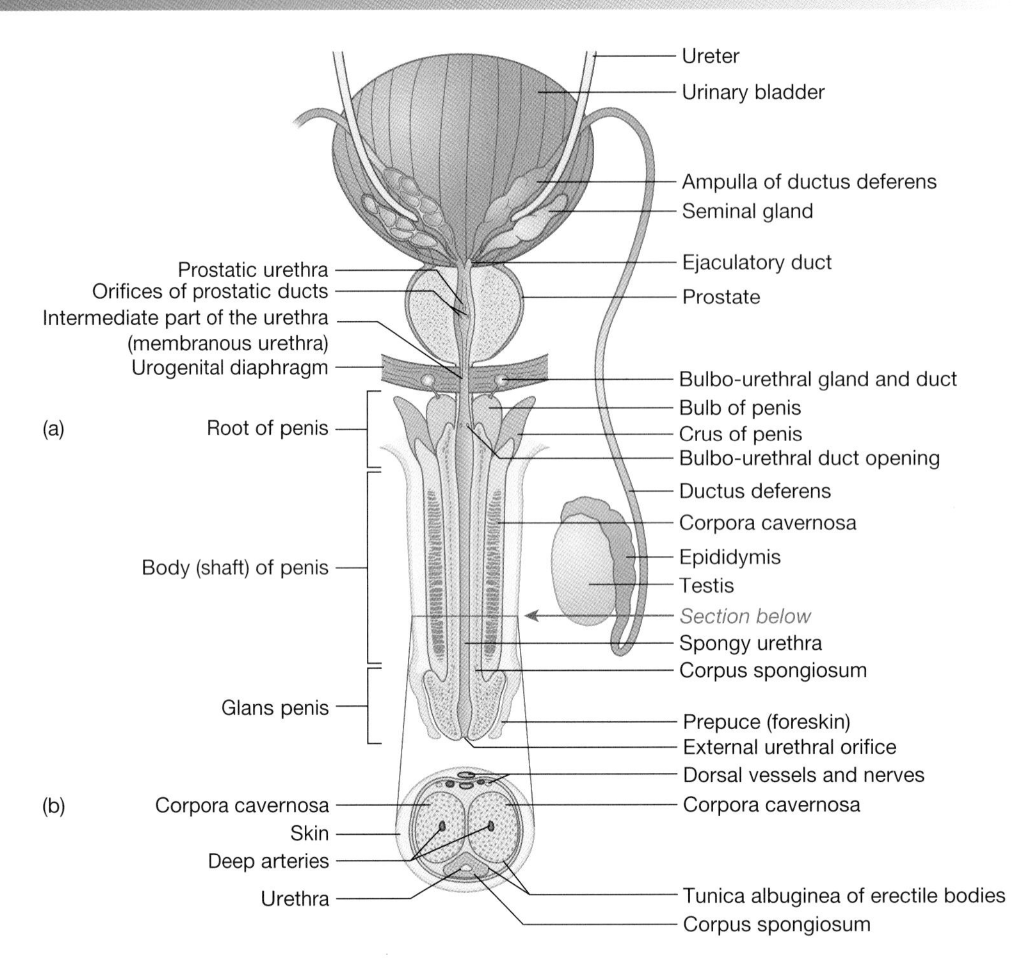


**Figure 14.6** Male reproductive organs. (a) Posterior aspect of the longitudinal (frontal) section of the penis. (b) Transverse view of the penis with the dorsal aspect at the top.



## Male puberty

Puberty in the male commences between ages 9 and 15 years (Sandwell, 2005). The timing of puberty is determined by genetic inheritance and usually commences with a growth spurt. Interstitial cells of the testes are stimulated by luteinizing hormone from the anterior pituitary gland to increase production of testosterone. This in turn stimulates the reproductive organs and secondary male sex characteristics to develop. This stimulation occurs within a cascade system called the hypothalamic–pituitary–gonad axis. Other hormones, such as thyroxine and cortisol, also combine with testosterone to activate bone and muscle growth (Marieb and Hoehn, 2013).

One of the first signs of puberty is testicular growth, which can be measured using an orchidometer to measure the increasing volume of the testes. Adult volume is approximately 20 mL; puberty commences when a volume of 6 mL is attained. The growth of axilla and pubic hair, skeletal growth, libido and changes in sweat and sebaceous glands are stimulated by androgens from the adrenal cortex. The changes in sweat and sebaceous glands in particular can be embarrassing to young people as it can lead to body odour and acne. In addition, as vocal cords enlarge in the expanding larynx the male voice 'breaks', taking on a deeper tone.

Secondary sexual characteristics, which include outward signs of changes, are seen in boys initially with the growth of testes and scrotum. Penis growth, pubic hair and facial hair occur simultaneously with increasing size and growth of the skeleton and muscles. Seminiferous tubules also mature at this time and produce spermatozoa. Seminal discharge or 'wet dreams' (nocturnal emissions) may occur at night due to canalization of the seminal vesicles (MacGregor, 2008).

## Spermatogenesis

Sperm production is a continuous process commencing at puberty and usually lasts for life. A young healthy man can produce as many as a hundred million sperm a day. Sperm production takes place in the testes and is termed spermatogenesis.

The spermatogonia, or sperm stem cells, undergo mitosis to form primary spermatocytes. Cell division continues with these primary spermatocytes forming two secondary spermatocytes which then go on to complete meiosis to form spermatids. These spermatids develop to form immature spermatozoa or sperm. This process takes place inside the seminiferous tubules in the testes and takes approximately 65–75 days to complete (Peate, 2011) (Figures 14.7 and 14.8).



Primary spermatocytes have 46 chromosomes; secondary spermatocytes have 23 chromosomes due to the process of meiosis. The sperm cells formed following the various stages of cell division also have 23 chromosomes each. These 23 chromosomes then join with the other 23 chromosomes provided by the female ova, resulting in the required 46 chromosomes necessary for human development.

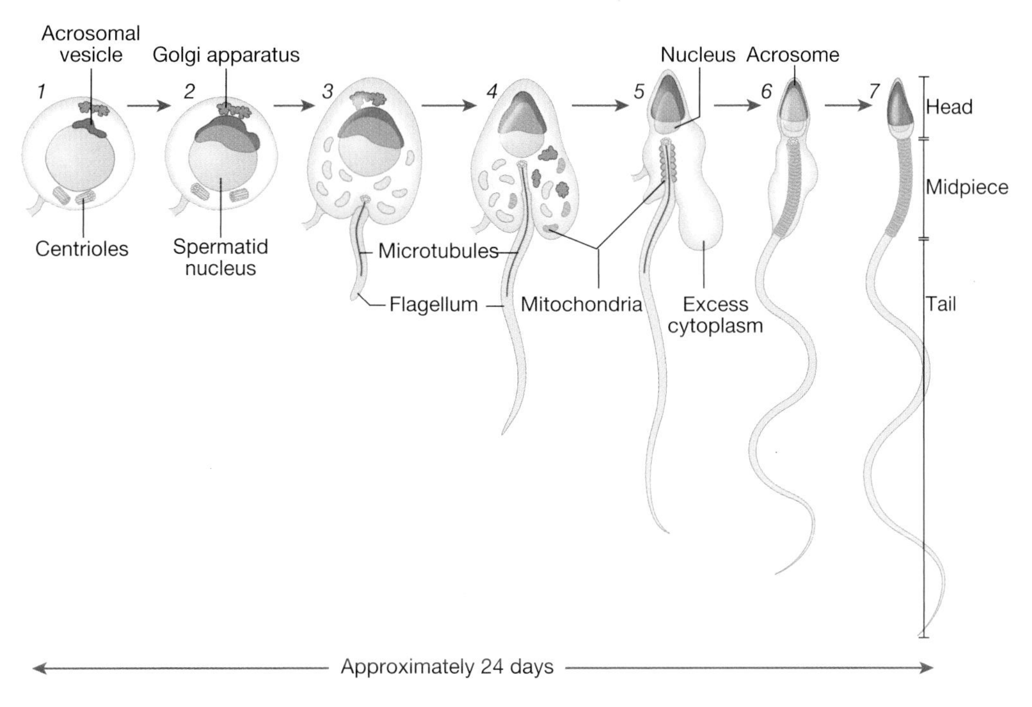


**Figure 14.7** Events in the transformation of a spermatid into a sperm cell.

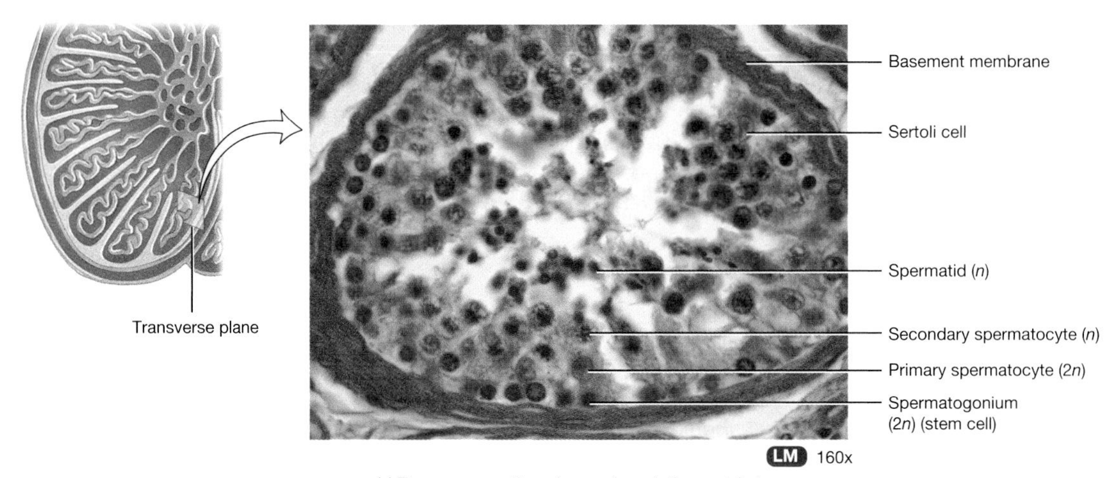


(a) Transverse section of several seminiferous tubules



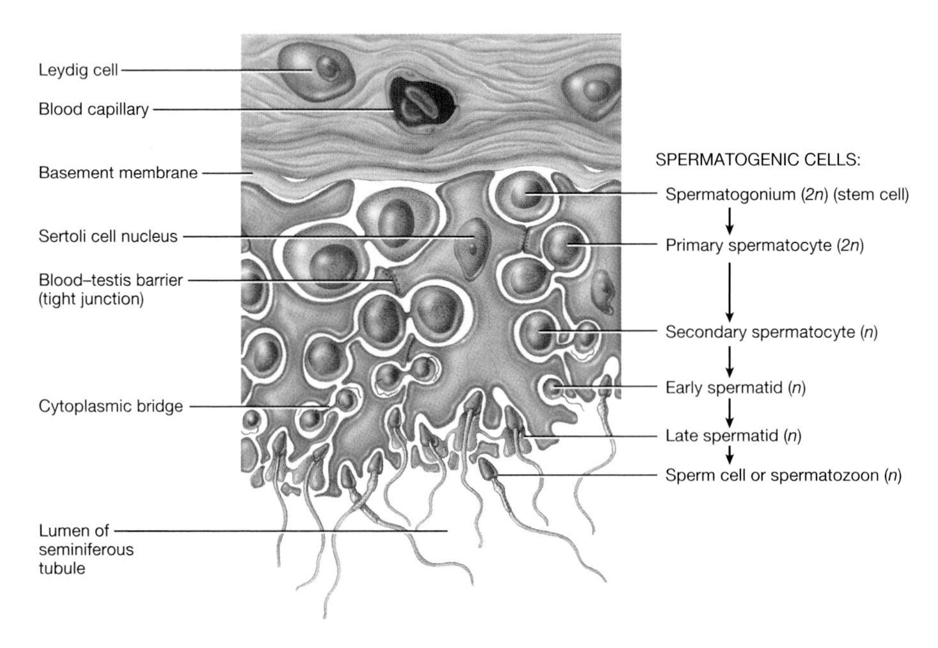


(b) Transverse section of a portion of a seminiferous tubule

**Figure 14.8** Stages of spermatogenesis.
*Source:* Tortora and Derrickson (2009), Figure 28.6, p. 1088. Reproduced with permission of John Wiley and Sons, Inc.

## Sperm

Each spermatozoon has four distinct components: head, neck, middle piece and tail. The head contains a nucleus with densely packed chromosomes and a compartment called the acrosome which contains enzymes that assist with penetration essential to fertilization. The neck and middle contain many mitochondria, which provide energy for movement. The tail is the only flagellum in the human body. It is a whip-like organelle that moves the sperm cell through a

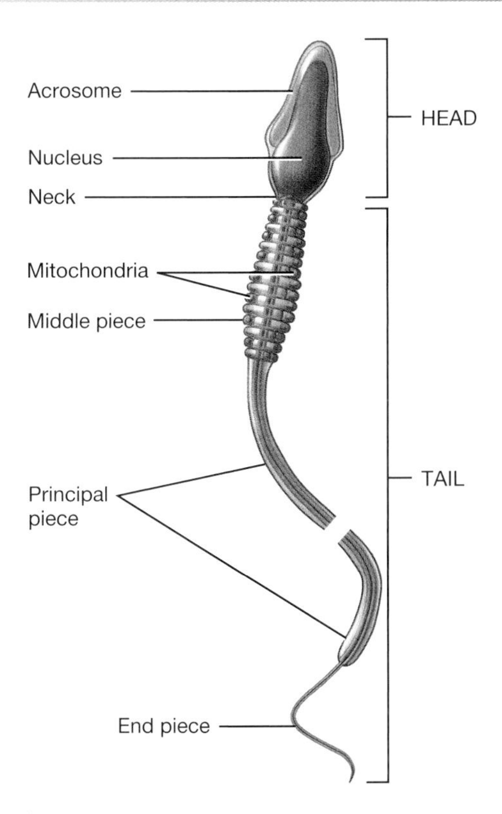


**Figure 14.9** Components of a sperm.
*Source:* Tortora and Derrickson (2009), Figure 28.6, p. 1088. Reproduced with permission of John Wiley and Sons, Inc.



whip-like, corkscrew movement through the female reproductive system to the ovum (see Figure 14.9).

Formed sperm travel into the epidydimis via small ducts called the rete testes. The crescent-shaped coiled epididymis acts as a holding place where sperm mature and take on nutrients to grow for several weeks prior to continuing their travels. As sperm continue to mature they develop motility. Approximately 300 million sperm mature each day (Tortora and Derrickson, 2010). Arrival at the two vas deferens is the final stage for the sperm, as from there they travel into the seminal vesicles .The fluids are then released into the ejaculatory ducts, within the prostate gland.

Fluid secreted by the prostate gland is of a milky alkaline consistency and acts as a friendly protective environment in which sperm can survive when they are ejaculated into the acidic environment of the vagina. The sperm-laden fluid then leave the ejaculatory ducts and move into the urethra from where they are ejaculated during sexual intercourse or masturbation. Sperm generally last only 48 h once deposited in the female reproductive system.

# The female reproductive system

The female reproductive system and reproductive role is more complex than that of the male as it also has to grow a fetus and provide nourishment after birth (Marieb and Hoehn, 2013). Similar to the male, alongside the endocrine system the female reproductive system produces

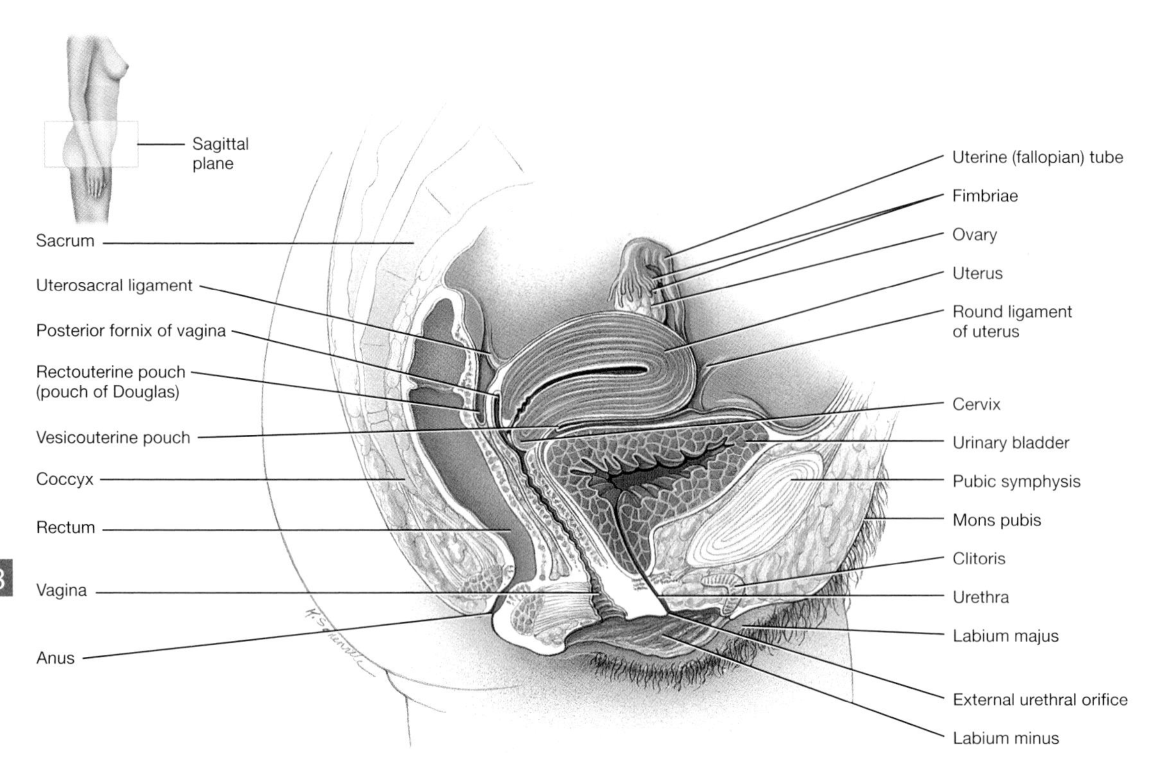


**Figure 14.10** The female reproductive system (sagittal section) and surrounding structures.
*Source:* Tortora and Derrickson (2009), Figure 28.11(a), p. 1096. Reproduced with permission of John Wiley and Sons, Inc.

hormones essential in the biological development and sexual arousal and activity around puberty.

Unlike the male reproductive system, the majority of the female reproductive system is hidden inside the body. The system consists of the ovaries, fallopian tubes, uterus, vagina and external genitalia (Figure 14.10). The breasts are also part of the reproductive system. The urinary system is separate to the reproductive system in females, although very close in proximity; therefore, health problems in one can often affect the other (Peate, 2011).

The function of the female reproductive system is to produce ova and contain and develop the fetus once fertilization has taken place through vaginal intercourse with the male penis or artificial insemination. The breasts are designed to produce milk to feed and nourish the newborn. The uterus and ovaries function on a monthly cycle to prepare the uterus to receive a fertilized ovum through hormonal influence; and if pregnancy does not occur, menstruation occurs and the cycle continues.

## The ovaries

Female ovaries are small at birth, although larger than the testes. They lie in the abdominal cavity in the newborn, not descending into the pelvis until about 6 years of age. There is minimal growth of the ovaries until puberty, when they increase 20-fold (Sandwell, 2005). However, female newborn ovaries already contain the full complement of potential ova. These lay largely dormant until the onset of puberty, at which time ovulation occurs each month.

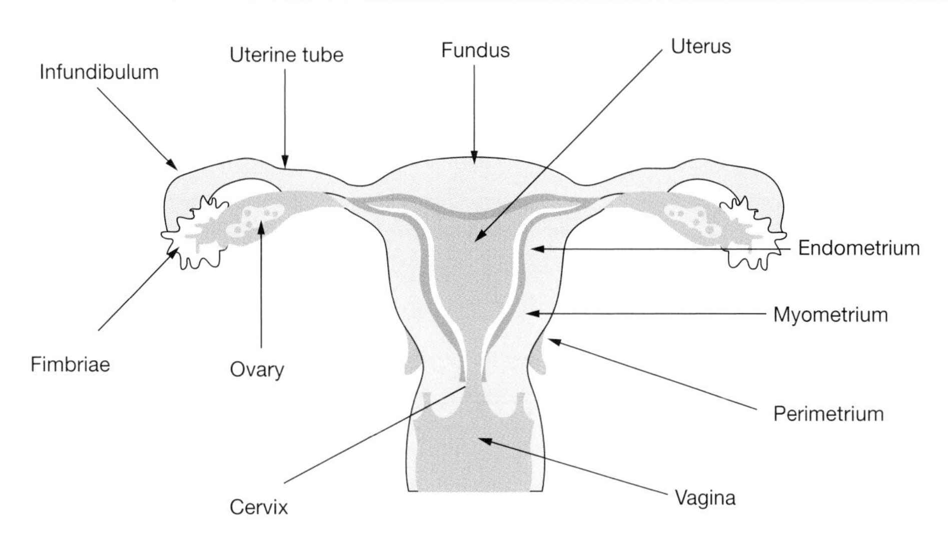


**Figure 14.11** The uterus and associated structures.
*Source:* Nair and Peate (2013), Figure 14.4, p. 404. Reproduced with permission of John Wiley and Sons, Ltd.



There are two ovaries, shaped like almonds, one on either side of the uterus. They are attached to the uterus and body wall by the ovarian ligament, broad ligament and others (see Figure 14.11). The ovary has an outer single layer of epithelium beneath which the female gametes or ova are produced. The ovaries also produce the female hormones of oestrogen and progesterone.

Maturation of the ovum takes place in a fluid-filled cluster of cells called ovarian follicles. Oestrogen hormone is secreted from cells in the follicle wall as the follicle develops, which stimulates growth of the uterine lining. Follicles are stimulated to mature each month by FSH and LH. Mature follicles are called Graafian follicles. Ovulation occurs when the follicle ruptures, allowing an ovum to escape into the pelvic cavity where it is swept into the nearest fallopian tube. Developing ova not released after maturing just degenerate.

After the ovum is expelled, the remaining follicle becomes a solid glandular mass named the corpus luteum. The corpus luteum then secretes oestrogen and progesterone. If the ovum is not fertilized then the corpus luteum shrinks and is replaced by scar tissue (Figure 14.12), however, if pregnancy occurs then the structure remains active for a while. In some cases following normal ovulation, the corpus luteum persists and forms a small ovarian cyst that usually resolves itself.

## Oogenesis

Specialized nuclear division called meiosis that occurs in the testes to produce sperm also occurs in the ovaries. In the fetal period, oogonia multiply rapidly by mitosis. A female fetus has a fixed number of oogonia, between 2 million and 4 million, whereas spermatogonia are continuously regenerated at puberty (Peate, 2011). As the oogonia transform into primary oocytes, primordial follicles appear. The primary oocytes begin the first meiotic division, which then halts late in this initial stage (Figure 14.13).

At birth, a female has approximately 1 million oocytes already in place in the cortical region of the immature ovary. By puberty, around 300 000 oocytes remain (Marieb and Hoehn, 2013).

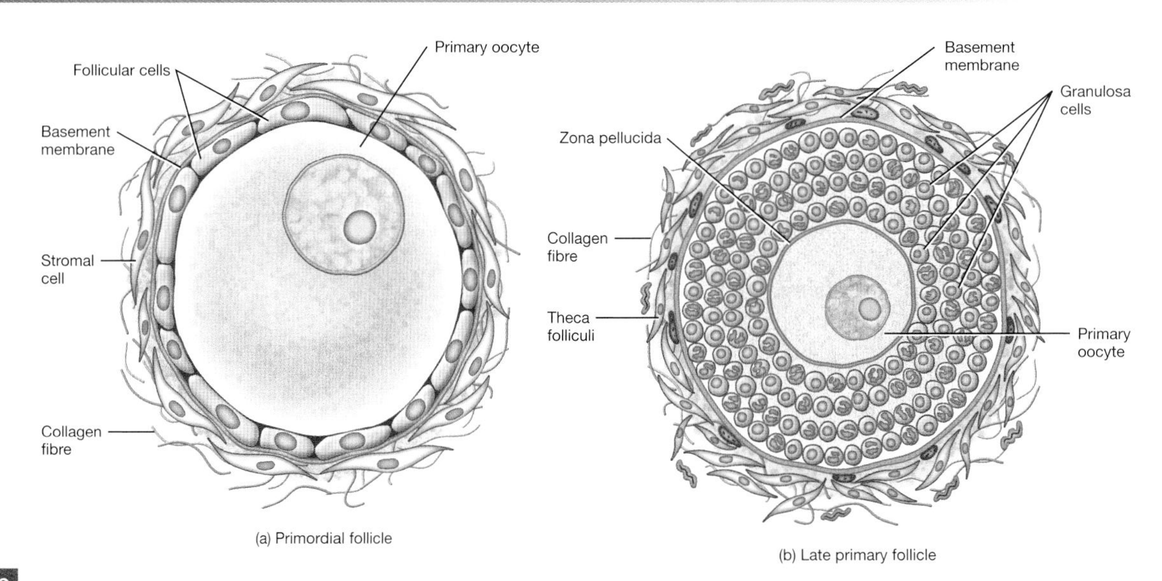




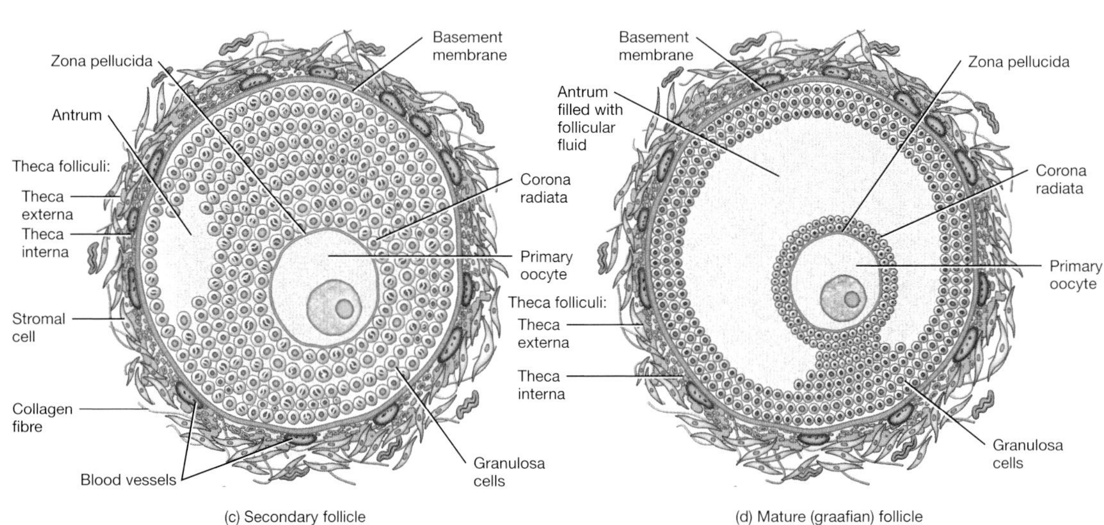


**Figure 14.12** The developmental sequences associated with maturation of an ovum.
*Source:* Tortora and Derrickson (2009), Figure 28.14(a)–(d), pp. 1099–1100. Reproduced with permission of John Wiley and Sons, Inc.

The primordial follicles are stimulated by the FSH and LH that are released by the anterior aspect of the pituitary gland; usually, only one reaches the maturity required for ovulation. The pool of primary follicles continues throughout life until the supply is depleted, at a time called menopause.

## Female puberty

Female puberty commences approximately 2 years earlier in girls than in boys, commencing with a growth spurt as early as 8 years or as late as 14 years. It has been suggested that a critical weight of 47 kg for girls in the UK needs to be reached to change metabolic rate and trigger

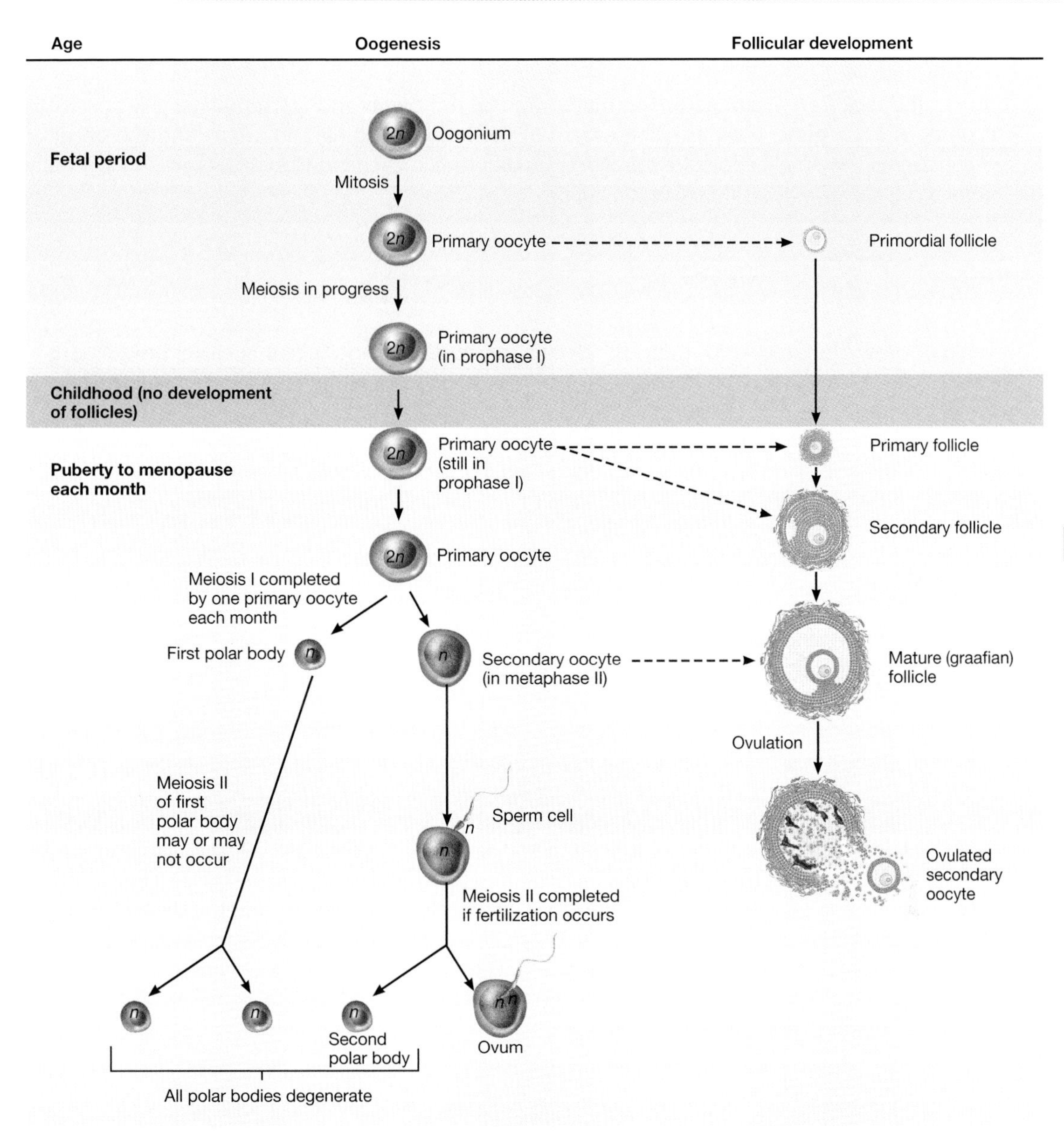


**Figure 14.13** Oogenesis and follicular development.
*Source:* Tortora and Derrickson (2009), Table 28.1, p. 1102. Reproduced with permission of John Wiley and Sons, Inc.



hormonal changes (MacGregor, 2008). This has ramifications for the growing overweight and obesity trends in children and young people, as girls will be reaching this weight earlier and, therefore, experiencing earlier puberty.

The production of oestrogen in the female ovary is stimulated within the cascade system of hypothalamic–pituitary–gonad axis similar to the male. Androgens from the adrenal cortex also stimulate the skeletal growth and a widening of the pelvis, growth of axilla and pubic hair, libido and changes in sweat and sebaceous glands, which can result in increased body odour and acne in girls also.

In the ovary, FSH stimulates the maturation of the ovum and LH stimulates theca cells to produce androgens. There is enlargement of the breasts, vagina and uterus and the onset of menarche. The timeline for these processes is different in each individual, but puberty is usually completed in 3–5 years (MacGregor, 2008).

## Clinical application

Delayed puberty and precocious puberty are when pubertal changes occur in either boys or girls outside of the normally expected periods. Delayed puberty is when there is a failure to menstruate or develop beyond the age of 16 years in girls or genital growth in boys that takes over 5 years to complete. Precocious puberty is when girls younger than 8 years have breast development or pubic hair. In boys, precocious puberty is when penile and testicular enlargement, pubic hair and textured scrotum occur before 9.5 years. In both cases, at a time when young people are conscious of how they look, this can be a difficult time for them psychologically.



### Hormonal influences

Ovaries grow and continuously secrete small amounts of oestrogens during childhood. This inhibits the release of GnRH from the hypothalamus. Prior to puberty the hypothalamus begins to release GnRH in a rhythmic manner, stimulating the anterior pituitary to release FSH and LH, which prompts the ovaries to secrete hormones. This continues during early puberty until the onset of the menarche, at which point the cycle becomes more regular.

Oestrogens, progesterone and androgens are the hormones produced by the ovaries. Oestrogens are needed in combination with other hormones to stimulate the uterus to prepare for the growth of a fetus. They play a part in regulating the ovarian and uterine cycles (Figure 14.14). Oestrogens also play a part in supporting the growth spurt at puberty and the development of other secondary sex characteristics, such as breast development, increasing depositions of subcutaneous fat and the development of the wider pelvis necessary for childbirth. In addition, oestrogens help to reduce the rate of bone reabsorption, decrease cholesterol levels and increase blood clotting (Peate, 2011).

### The menstrual cycle

The menstrual cycle is a series of cyclic changes that the uterine endometrium undergoes each month as it responds to the changes in oestrogen and progesterone levels. These changes occur in order to prepare for the implantation of a fertilized embryo. These endometrial changes coordinate with the phases of the ovarian cycle, dictated by gonadotropins released by the anterior pituitary gland. The menstrual cycle can last anything from 22 to 45 days, but 28 days is the average, and the first day of the menstrual flow is considered as the first day of the cycle.

The first phase of the menstrual cycle is the menstrual phase that lasts for the first 5 days. During this phase, menstruation or menses occurs, where the uterus sheds all but the deepest layer of endometrium. This thick layer of the endometrium detaches from the uterine wall; this detached tissue and blood then passes out through the vagina as menstrual flow. The average duration of menstruation is 3–5 days, although this can vary. By day 5, however, the growing ovarian follicles start to produce more oestrogen.

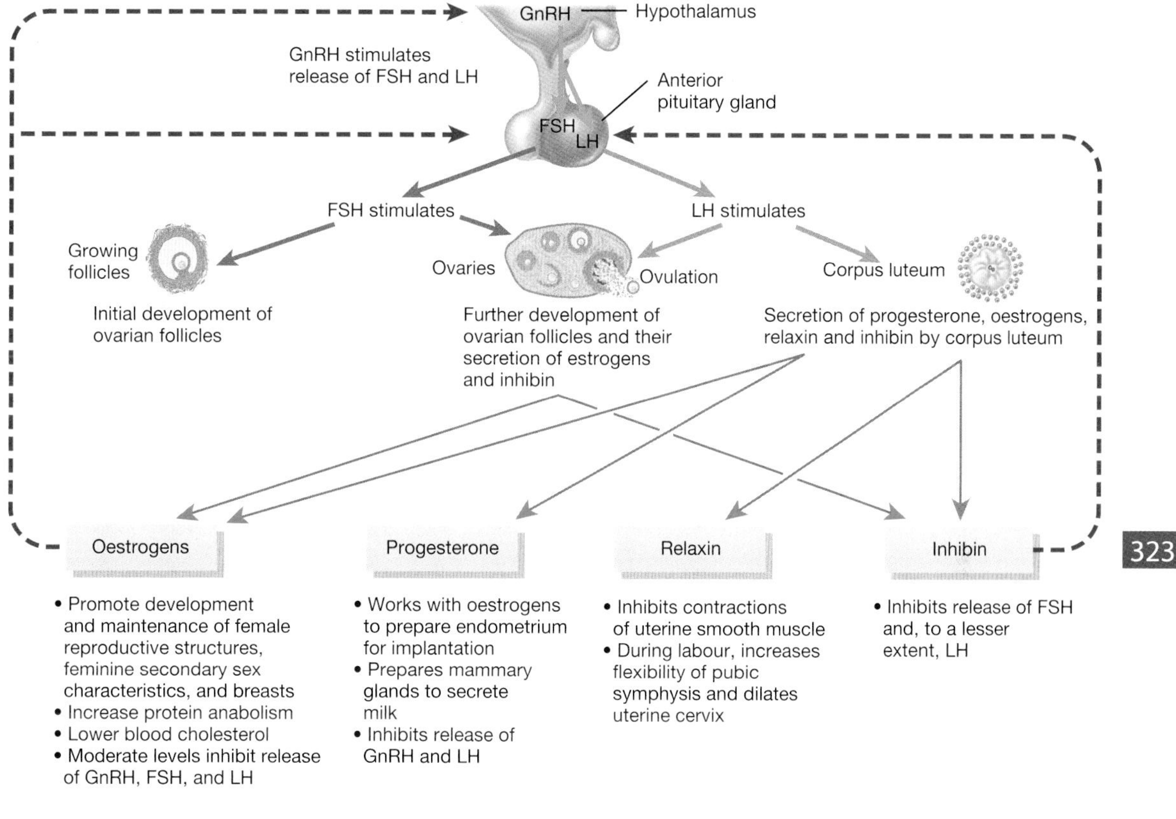


**Figure 14.14** The ovarian and uterine cycles.
*Source:* Tortora and Derrickson (2009), Figure 28.23, p. 1113. Reproduced with permission of John Wiley and Sons, Inc.



The proliferative phase from days 6 to 14 is where the endometrium rebuilds itself under the influence of rising oestrogen levels. The basal layer of the endometrium generates a new functional layer, during which time the glands enlarge and the spiral arteries multiply. Cervical mucous is normally thick and sticky, but during this phase it becomes thin and forms channels to facilitate sperm travel into the uterus. Ovulation occurs at the end of the proliferative stage in response to the release of LH from the anterior pituitary gland on day 14.

The secretory phase from days 15 to 28 is the phase when the endometrium prepares for embryonic implantation. The rising levels of progesterone produced by the corpus luteum cause increased vascularity in the endometrium, changing the inner layer to secretory mucosa. This activity stimulates the secretion of nutrients into the uterine cavity necessary to sustain the embryo until it has finished implanting in the blood-rich endometrial lining. In response to rising progesterone levels, the cervical mucosa again becomes thick and viscous, forming a cervical plug in order to block the passage of more sperm, pathogens or other foreign materials. If fertilization does not occur, the corpus luteum degenerates as LH blood levels fall. Progesterone levels fall and deprive the endometrium of hormonal support, and the spiral arteries kink and spasm; this, in turn, causes hypoxia (lack of oxygen) of the endometrial cells, which then die and glands regress. Menstruation then begins, commencing the cycle over again (Figure 14.15).

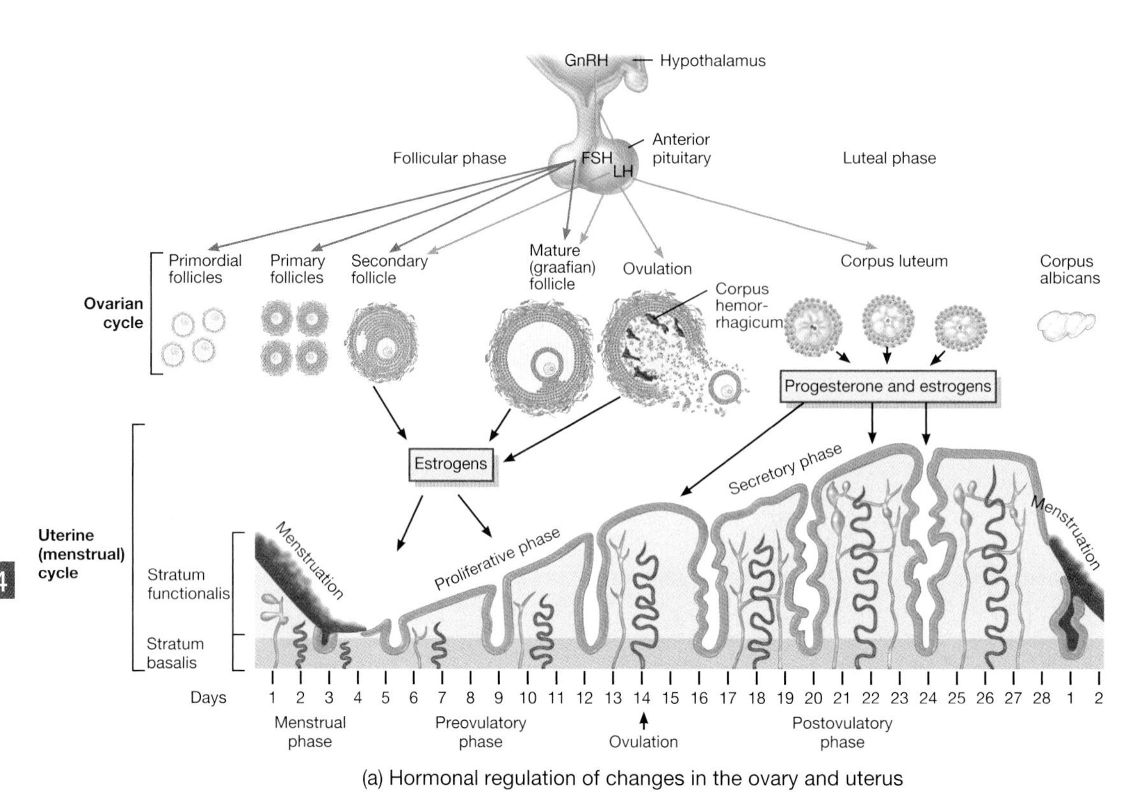


(a) Hormonal regulation of changes in the ovary and uterus

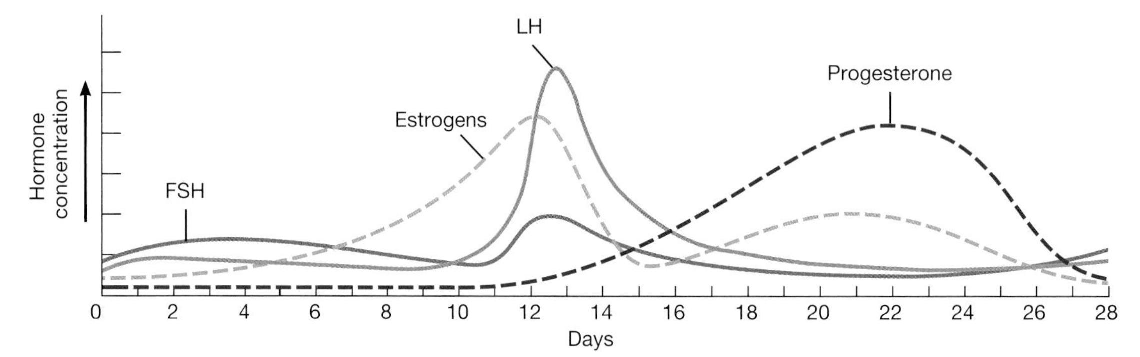


(b) Changes in concentration of anterior pituitary and ovarian hormones

**Figure 14.15** The female reproductive cycle: changes in anterior pituitary and ovarian hormones and their effect on the menstrual cycle.
*Source:* Tortora and Derrickson (2009), Figure 28.24, p. 1114. Reproduced with permission of John Wiley and Sons, Inc.

## The fallopian tubes

The fallopian tubes receive the ovulated oocyte as it is expelled from the ovary into the pelvic cavity. They are long, delicate tubes that are supported by the broad ligaments. Each fallopian tube is about 10 cm long (Marieb and Hoehn, 2013); they enter the uterus at one end and the other open end drapes fimbriae over the ovary.

The fallopian tubes consist of the isthumus, which is the constricted region next to the uterus, the ampulla at the other end, which curves around the ovary, and the infundibulum, which ends in the fimbriae that drape over the ovary. Cilia on the fimbriae create currents that carry an oocyte into the uterine fallopian tube. Within the fallopian tubes are sheets of smooth muscle and thick mucosa that contain ciliated and nonciliated cells. The oocyte is moved towards the uterus in nourishing secretions, produced by the nonciliated cells, by the movement of the cilia and muscular peristaltic action.

## The uterus

The uterus (Latin word for womb) is a hollow muscular organ located in the pelvis anterior to the rectum and posterosuperior to the bladder. The uterus is relatively large at birth, due to the influence of maternal oestrogens that circulate through the placenta. It adopts the normal adult position of anteversian and anteflexion at about 6 years of age after the bladder descends into its place in the pelvis. By the end of adolescent growth the uterus is approximately 7.5 cm long (Peate, 2011). The function of the uterus is to grow and nurture the implanted embryo into a viable fetus over a 40-week period of gestation.



The uterus consists of the body, which is the major portion of the uterus. The upper rounded region superior to the entrance of the fallopian tubes is the fundus. The cervix is the narrow neck that projects into the vagina. Within the cervix is the cervical canal, which includes the external os that connects with the vagina and the internal os that communicates with the uterine body.

The uterus has three layers. The outer serous layer that merges with the peritoneum is called the perimetrium. The myometrium is the muscular wall of the uterus, which consists of a number of muscle layers going in different directions to allow contractions to occur during menstruation and childbirth. The inner lining of the uterus is called the endometrium and is a specialized epithelial layer that changes during the menstrual cycle.

## The vagina

The vagina is a thin-walled fibromuscular structure approximately 8–10 cm in length (Marieb and Hoehn, 2013). It lies between the bladder and rectum and extends from the cervix to the external genitalia. It functions as the canal for menstrual flow and the passageway for the delivery of an infant. It is also the organ of female sexual response, as it receives the penis and semen during sexual intercourse (Figure 14.16).

The vaginal wall distends and consists of three coats: a smooth muscle muscularis and outer fibroelastic adventitia layers and an inner mucosa consisting of transverse ridges or rugae that stimulate the penis during intercourse. The vaginal mucosa is lubricated by the cervical mucous glands and mucosal transudate from the vaginal walls. The pH of the vagina is normally acidic, ranging from 3.8 to 4.2 (Peate, 2011). This acidic environment helps keep the vagina infection free and healthy; it is also hostile to sperm. However, conversely, in adolescence the vaginal fluid tends to be alkaline, which puts teenagers who are sexually active at higher risk of contracting sexually transmitted infections.

In young girls prior to sexual penetration the mucosa near the distal vaginal orifice forms a partition called the hymen. The durability of this differs in different females; in some it may rupture during sports or inserting tampons, whereas in extreme cases it has to be surgically opened. This hymen is very vascular, so when it is breached or stretched it bleeds.

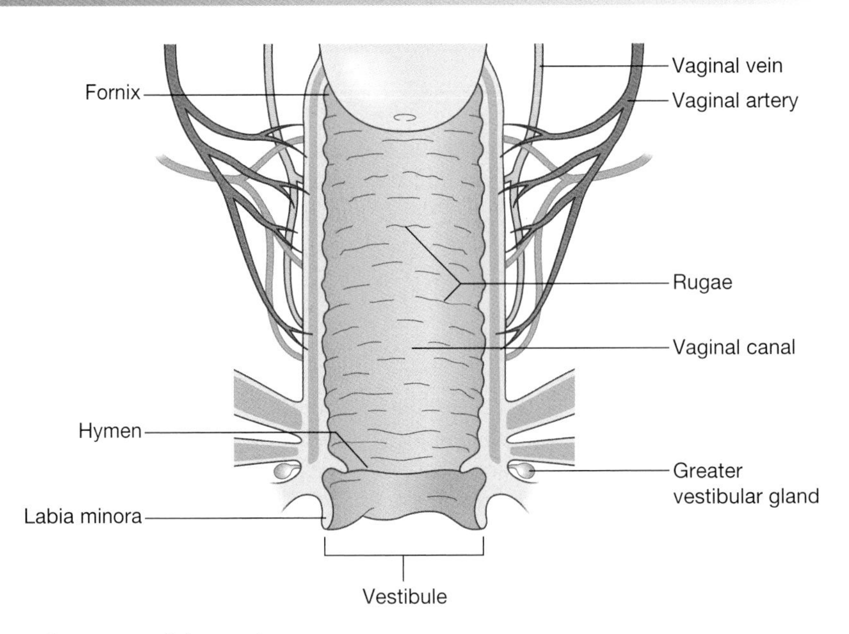




**Figure 14.16** Anatomy of the vagina.

## Clinical application

### Vaginitis

This is an inflammatory condition of the vagina characterized by redness and swelling of vaginal tissues, vaginal discharge, burning and itching. Pain may occur with sexual intercourse and urination. The causes of vaginitis may include infectious agents, foreign bodies or chemical irritants. In pre-pubertal girls the most common causes include poor hygiene, intestinal parasites or the presence of foreign bodies. During and after puberty, *Candida albicans* or bacterial vaginosis are the most common causes, and these can be transmitted sexually. Treatment is directed at the cause of the disorder.

## The external genitalia

The external genitalia, also called the vulva, includes the mons pubis, labia majora and labia minora, greater vestibular glands, clitoris, and vaginal and urethral openings (Marieb and Hoehn, 2013) (Figure 14.17).

The mons pubis is a rounded fatty area overlying the pubic symphysis. During childhood this area is bare, but it starts to grow hair in early puberty, around 11–12 years of age. Hair cover slowly increases to all external areas outside of the labia minora during puberty.

Running posteriorly from the mons pubis are two fatty skin folds which become covered in hair at the same time as the mons pubis called the labia majora; these are the counterpart of the male scrotum as they derive from the same embryonic tissue. The labia minora sit inside the labia majora and are two, thin, hair-free skin folds. The clitoris located to the anterior end of the labia minora has many sensory nerve endings and becomes swollen with blood and erect during tactile stimulation contributing to sexual arousal. This is likened to the penis as it has dorsal erectile columns (corpora cavernosa).

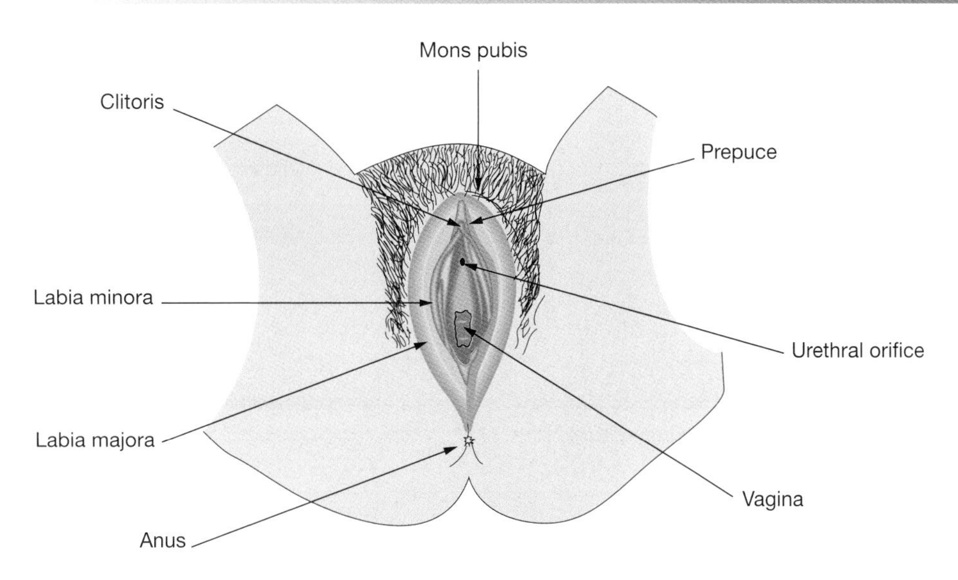


**Figure 14.17** The female external genitalia (vulva).
*Source:* Nair and Peate (2013), Figure 14.2, p. 402. Reproduced with permission of John Wiley and Sons, Ltd.



## Clinical application

### Sexually transmitted infections

A rise in sexually transmitted infections, such as chlamydia, gonorrhoea, genital warts and genital herpes infections, is thought to be caused when young people have unsafe sex for the first time at an early age. This may be associated with a lack of knowledge, having consumed alcohol when having sex and unable to correctly use contraception (Tripp and Viner, 2005). Long-term pelvic inflammatory disease, infertility, neonatal morbidity and genital cancers are consequences of untreated sexually transmitted infections during this teenage period, particularly in young girls due to an immature cervix and immunity (Porth, 2012).

A quadrivalent human papillomavirus vaccine is now offered to females around the age of 11–12 years as this infection can cause cervical dysplasia and cervical cancer. An estimated 50% of women are infected during adolescence and the symptoms are often silent (McCance and Huether, 2010).

## The breasts

The breasts, also called mammary glands, are present in both sexes but are usually only functional in females. The role of the breasts is to nourish a newborn baby through the production of milk, which is controlled by the hormone prolactin. The breasts are rounded skin-covered domes located anterior to the pectoral muscles of the thorax at the level between the third and seventh ribs. Slightly below the centre of each breast is the areola, a ring-shaped area of pigmented skin that contains glands that secrete sebum. Sebum is a fatty substance that reduces drying and cracking of the skin around the nipple; the nipple becomes erect when cold or during sexual arousal.

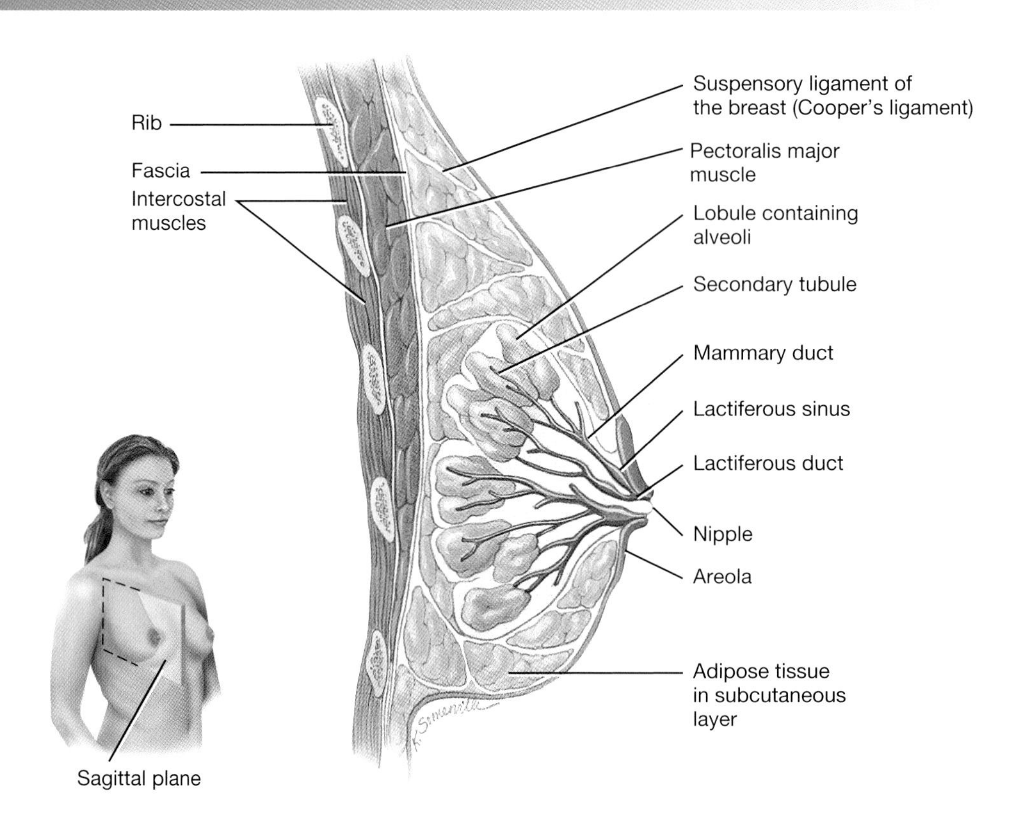


**Figure 14.18** The breast.
*Source:* Tortora and Derrickson (2009), Figure 28.22, p. 1111. Reproduced with permission of John Wiley and Sons, Inc.



Internally, each breast has 15–25 lobes radiating around the nipple. The lobes are surrounded by fat and connective tissue (Figure 14.18). The connective tissue provides support as it forms suspensory ligaments that attach the breast to underlying muscle fascia. Each lobe contains lobules which in turn contain alveoli that produce milk during the last part of pregnancy and postpartum. Milk travels from the alveoli down the lactiferous ducts to open outside of the nipple.

## Clinical application

### Breast growth

The Tanner scale (Tanner, 1962) can be used as an aid to determine physical development in children, young people and adults by nurses and other health professionals. Comparing the physical characteristics of the female breast with the scale can help determine normal and/or abnormal development. The scale consists of five stages on the development of the breast (Figure 14.19).

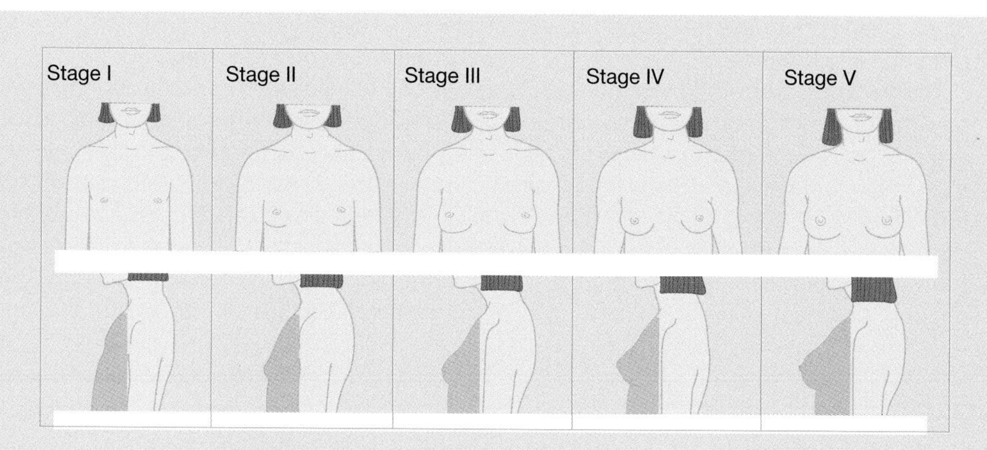


**Figure 14.19** The Tanner scale for female breast growth.

Stage I. Pre-adolescent: elevation of papilla only.

Stage II. Breast bud stage: elevation of breast and papilla as small mound. Enlargement of areolar diameter.

Stage III. Further enlargement and elevation of breast and areola, with no separation of their contours.

Stage IV. Projection of areola and papilla to form a secondary mound above the level of the breast.

Stage V. Mature stage: projection of papilla only, due to recession of the areola to the general contour of the breast.

The Tanner scale is only one method of assessing physical development.



# Conclusion

There are a number of factors that play a role in coordinating the correct functioning of the reproductive systems. Hormones play a major role in regulating both male and female reproductive systems, the major hormones being testosterone, oestrogen and progesterone. In males, the sperm counts and pH levels have to be correct, there needs to be the right balance of nutrients, and erection and ejaculation must occur in the proper sequence. In the female, the ovarian and uterine cycles must be coordinated, ovulation and transport of the oocyte must occur normally, and the reproductive tract needs to provide a hospitable environment for the survival and movement of the sperm for fertilization to take place. In addition, in order for the reproductive system to be able to fulfil its function, the digestive, endocrine, nervous, cardiovascular and urinary systems must all be functioning normally.

The reproductive system is unique, in that it is nonfunctional during the first 10–15 years (Marieb and Hoehn, 2013) and in order to fulfil its biological function of pregnancy and birth it is able to interact with the complementary system of another person. This is a complex system that is essential for life, as without it human life would eventually end. The reproductive system's primary function of producing offspring does not play a role in maintaining homeostasis. However, reproduction depends on a number of physical, physiological and psychological factors, many of which require intersystem cooperation. In addition, the sex hormones have direct effects on other organs and tissues; for example, testosterone and oestrogen affect both muscular and bone development in the growing child and young person.

The transition from child to adult occurs during the teenage years and through the process of puberty. It is a period of rapid and extreme change for young people, and its successful completion is dependent not only on all of the physical factors described in this chapter but also on psychological and social factors and cultural expectations. Adolescent sexual health is an important aspect of health promotion with young people as it can affect their future reproductive and general health due to the risk of contracting sexually transmitted diseases during the early years of sexual maturation. It is also an important time for educating young people about the consequences of sexual activity and their responsibilities in producing offspring.

# Activities

**Now review your learning by completing the learning activities in this chapter. The answers to these appear at the end of the book. Further self-test activities can be found at www.wileyfundamentalseries.com/childrensA&P.**

## Methods of birth control

Complete the following table with the advantages and disadvantages for young people of each method of contraception.

| Method | Advantages | Disadvantages |
|---|---|---|
| *Surgical* | | |
| Vasectomy/tubal ligation | | |
| *Hormonal* | | |
| Birth control pills | | |
| Birth control injection | | |
| Birth control patch | | |
| Birth control ring | | |
| Barrier | | |
| Male condom | | |
| Diaphragm (with spermicide) | | |
| Contraceptive sponge (with spermicide) | | |
| *Other* | | |
| Spermicide | | |
| Fertility awareness | | |

## Conditions

The following tables contain a list of conditions. Take some time and write notes about each of the conditions. You may make the notes taken from text books or other resources; for example, people you work with in a clinical area or you may make the notes as a result of people you have cared for. If you are making notes about people you have cared for you must ensure that you adhere to the rules of confidentiality.

Reproductive system complaints in boys: complete the table using a variety of sources.

| Complaint | Assessments | Condition/cause |
|---|---|---|
| Dysuria | | |
| Deflected urinary stream | | |
| Penile pain, lesions or discharge | | |
| Testicular swelling or mass | | |
| Testicular pain | | |
| Scrotal bulging or swelling | | |
| Scrotum feels empty | | |



Reproductive system complaints in girls: complete the table using a variety of sources.

| Complaint | Assessments | Condition/cause |
|---|---|---|
| Vulvar itching, pain or rash | | |
| Vaginal discharge | | |
| Pelvic pain | | |
| Breast tenderness or pain | | |
| Dysuria | | |
| Haematuria | | |

## Glossary

**Adrenal cortex:** the outer portion of an adrenal gland.

**Androgens:** masculinizing male sex hormones produced by the testes in the male and the adrenal cortex in both sexes.

**Anterior:** near to the front.

**Bilateral:** related to both sides of the body.

**Broad ligament:** a double fold of parietal peritoneum attaching the uterus to the side of the pelvic cavity.

**Canal:** a channel or passageway, a narrow tube.

**Connective tissue:** the most prominent type of tissue in the body; this tissue provides support.

**Corpus luteum:** a yellowish body found in the ovary when a follicle has discharged its secondary oocyte.

**Distal:** further away from the attachment of a limb to the trunk of the body.

**Endometrium:** the mucous membrane lining the uterus.

**Fetus:** the developing organism *in utero*.

**Fimbriae:** finger-like structures found at the end of the fallopian tubes.

**Follicle:** a secretory sac or cavity containing a group of cells that contains a developing oocyte in the ovary.

**Follicle-stimulating hormone:** secreted by the anterior pituitary gland initiates the development of an ovum.

**Gamete:** a male or female sex cell.



**Glans penis:** the enlarged region at the end of the penis.

**Gonad:** a gland that produces hormones and gametes – the testes in males and the ovaries in females.

**Hormone:** a secretion of endocrine cells that alters the physiological activity of target cells.

**Inferior:** away from the head or towards the lower part of a structure.

**Inguinal canal:** passage in the lower abdominal wall of the male.

***In utero*:** within the uterus.

**Lateral:** furthest from the midline of the body.

**Leydig cell:** a type of cell that secretes testosterone.

**Ligament:** dense, regular connective tissue.

**Luteinizing hormone:** a hormone secreted by the anterior pituitary that stimulates ovulation and readies glands in the breast to produce milk; stimulates testosterone secretion in the testes.

**Meatus:** a passage or opening.

**Meiosis:** a kind of cell division occurring during the production of gametes.

**Menopause:** the termination of the menstrual cycle.

**Myometrium:** the smooth muscle layer of the uterus.

**Oestrogens:** feminizing sex hormones produced by the uterus.

**Oocyte:** an immature egg cell.

**Oogenesis:** formation and development of the female gametes.

**Ovarian cycle:** a series of events in the ovaries that occur during and after the maturation of the oocyte.

**Ovarian follicle:** a general name for immature oocytes.

**Ovary:** the female gonad.

**Ovulation:** the rupture of a mature Graafian follicle with discharge of a secondary oocyte after penetration by a sperm.

**Ovum:** the female egg cell.

**Penis:** the male organ of urination and copulation.

**Peristalsis:** consecutive muscular contractions along the walls of a hollow muscular organ.

**pH:** a measure of acidity and alkalinity.

**Placenta:** an organ attached to the lining of the uterus during pregnancy.

**Progesterone:** a female sex hormone produced by the ovaries.

**Prolactin:** a hormone secreted by the anterior pituitary that initiates and maintains milk production.

**Rete:** the network of ducts in the testes.

**Scrotum:** the skin-covered pouch containing the testes.

**Semen:** fluid discharged by ejaculation.

**Spermatogenesis:** the maturation of spermatids into sperm.



**Testes:** the male gonads.

**Testosterone:** male sex hormone.

**Urethra:** the tube from the urinary bladder to the exterior of the body conveys urine in females and urine and semen in males.

**Uterus:** hollow muscular organ in the female; also called the womb.

**Vagina:** a muscular tubular organ in the female leading from the uterus to the vestibule.

**Vas deferens:** the main secretory duct of the testicle, through which semen is carried from the epididymis to the prostatic urethra, where it ends as the ejaculatory tract.

**Vulva:** the female external genitalia.

# References

Cohen, B.J. (2009) *The Human Body in Health and Disease*, 11th edn, Lippincott Williams & Wilkins, Philadelphia, PA.

Jenkins, G.W., Kemnitz, C.P., Tortora, G.J. (2010) *Anatomy and Physiology: From Science to Life*, 2nd edn, John Wiley & Sons, Inc., Hoboken, NJ.

MacGregor, J. (2008) *Introduction to the Anatomy and Physiology of Children: A Guide for Students of Nursing, Child Care and Health*, 2nd edn. Routledge, Abingdon.

Marieb, E.N., Hoehn, K. (2013) *Human Anatomy & Physiology*, 9th edn, Pearson, London.

Martini, F.H., Nath, J.L., Bartholomew, E.F. (2012) *Fundamentals of Anatomy & Physiology*, 9th edn, Pearson, San Francisco, CA.

McCance, K.L., Huether, S.E. (2010) *Pathophysiology: The Biologic Basis for Disease in Adults and Children*, 6th edn, Mosby Elsevier, St Louis, MO.

Moore, K.L., Persaud, T.V.N. (2008) *The Developing Human: Clinically Oriented Embryology*, 8th edn, Saunders Elsevier, Philadelphia, PA.

Nair, M., Peate, I. (2013) *Fundamentals of Applied Pathophysiology*, Wiley–Blackwell, Oxford.

Peate, I. (2011) The reproductive system. In: Peate, I., Nair, M. (eds), *Fundamentals of Anatomy and Physiology for Student Nurses*, Wiley-Blackwell, Chichester.

Porth, C.M. (2012) *Essentials of Pathophysiology*, 3rd edn, Lippincott Williams & Wilkins, Philadelphia, PA.

Sandwell, M. (2005) The reproductive tract. In: Chamley, C.A., Carson, P., Randall, D., Sandwell, M. (eds), *Developmental Anatomy and Physiology of Children: A Practical Approach*. London: Elsevier Churchill Livingston.

Tanner, J.M. (1962) *Growth at Adolescence*, 2nd edn, Blackwell, Oxford.

Tortora, G.J., Derrickson, B.H. (2009) *Principles of Anatomy and Physiology*, 12th edn, John Wiley & Sons, Inc., Hoboken, NJ.

Tortora, G.J., Derrickson, B. (2010) *Essentials of Anatomy and Physiology*, 8th edn, John Wiley & Sons, Inc., New York, NY.

Tripp, J., Viner, R. (2005) Young people's health: the need for action. *British Medical Journal*, **330**, (7496), 901–903.

# Chapter 15

# The nervous system

Petra Brown

*School of Health and Social Care, Bournemouth University, Bournemouth, UK*

## Aim

The aim of this chapter is to provide the reader with an understanding of the anatomy and physiology of the nervous system, both in terms of structure and function. Childhood development from fetus to adulthood is discussed and illustrated with clinical considerations.

## Learning outcomes

On completion of this chapter the reader will be able to:

- Outline the organization of the nervous system.
- List the functions of neurone and neuroglia cells and describe the transmission of nerve impulses.
- Discuss the development of fetal and childhood maturation of the nervous system.
- Discuss the anatomy and physiology of the central and peripheral nervous systems.
- Understand the function of specific brain areas and describe reflex and motor development maturation of the nervous system.
- Identify the role of the autonomic nervous system in maintaining homeostasis.

## Test your prior knowledge

- Describe the functions of the nervous system.
- List the main four anatomical areas of the brain.
- List the two anatomical areas of the central nervous system.
- What does the abbreviation CSF mean?
- Describe the functions of the spinal cord.
- How many pairs of spinal nerves are there?
- Name three common neurotransmitters that occur in the central nervous system.
- Which ions are involved in the transmission of action potentials?
- The nervous system works closely with another system to maintain homeostasis in the body. Can you name it?
- List and describe the function of the four main types of peripheral sensory neurones.

*Fundamentals of Children's Anatomy and Physiology: A Textbook for Nursing and Healthcare Students*, First Edition. Edited by Ian Peate and Elizabeth Gormley-Fleming.

Companion website: www.wileyfundamentalseries.com/childrensA&P

# Introduction

The nervous system is a complex system that interacts with all body systems to maintain homeostasis in conjunction with the endocrine system. It controls many of these systems through a complex communication system, and the major functions can be summarized as follows:

1. Orientates us to internal and external environment, such as pain (internal) and seeing danger (external).
2. Coordinates and maintains homeostasis of body activities.
3. Assimilates experiences and information, through memory, learning, intelligence and dreaming.
4. Development of instinctual information at birth.

# Organization of the nervous system

It is necessary to understand the complexity and organization of the nervous system by deconstructing its component parts (Figure 15.1). Each component part is intimately connected through a network of neurones (nerves) and supportive neuroglia (nerve fibres). The nervous system can be divided into the central nervous system (CNS) and peripheral nervous system (PNS).



Anatomically, the CNS incorporates the:

- brain
- spinal cord.

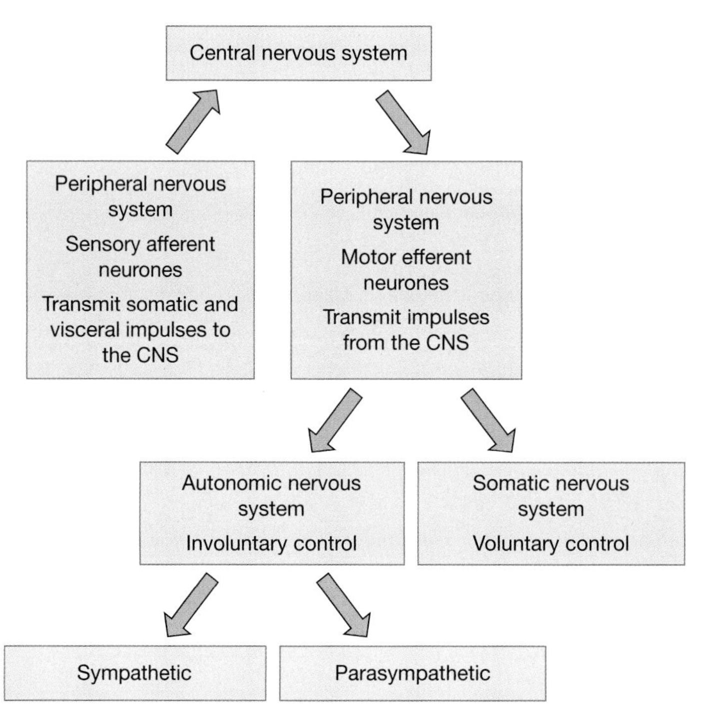


**Figure 15.1** Organization of the nervous system.

The PNS includes the

- cranial nerves
- spinal nerves
- sensory (afferent) neurones
- motor (efferent) neurones
- somatic nervous system (voluntary)
- autonomic nervous system (involuntary)
  - sympathetic
  - parasympathetic.

# Cellular structure of the nervous system

The cellular building blocks of the nervous system are the highly specialized neurones and neuroglia. Neurones are the primary transmission and communication cells, whilst neuroglia cells perform a supportive role.

## Neurones



There are an estimated 100 billion neurones in the adult human brain. Neurones consist of an axon, dendrites and a cell body (Figure 15.2). Their main function is the transmission of electrical impulses when stimulated.

- **Axon.** A collection of axons forms a nerve. Axon length varies in the nervous system from microscopic to 1 m in an adult. Peripherally, axons can branch off to form an axon collateral (Figure 15.2). Axons carry electrical impulses away from the cell body towards axon terminals and synapses.
- **Dendrites.** These are the numerous branched projections of a neurone that carry electrical impulses towards the cell body. Peripheral sensory neurone dendrites form large branching networks to transmit somatic and visceral information. Motor neurone dendrites are involved in transmission across synapses in multiple neurone networks in the somatic and autonomic nervous systems.
- **Cell body.** This contains the nucleus of the neurone and other organelles crucial for cellular life. Nuclei are a collection of cell bodies in the CNS, usually located in the grey matter. In the PNS they are called ganglia.

## Neuroglia

Neuroglial cells support neurones to function in several important ways, as indicated in the following (Figure 15.3):

- Structural support
  - Satellite cells give structural support and regulate the micro-environment of cranial nerve ganglions, which are a collection of neurone cell bodies in the CNS.
- Phagocytosis
  - Microglia cells are phagocytic cells in the CNS. They are involved in cleaning up cells that have undergone aptosis (programmed cell death). They play an important part in removing unwanted 'dead' and 'redundant' cells during fetal and childhood development.
- Myelination
  - Myelin forms a protective insulating sheath around the neurone axon. The sheath contains 80% lipids and 20% protein and is produced in the PNS by Schwann cells. These wrap

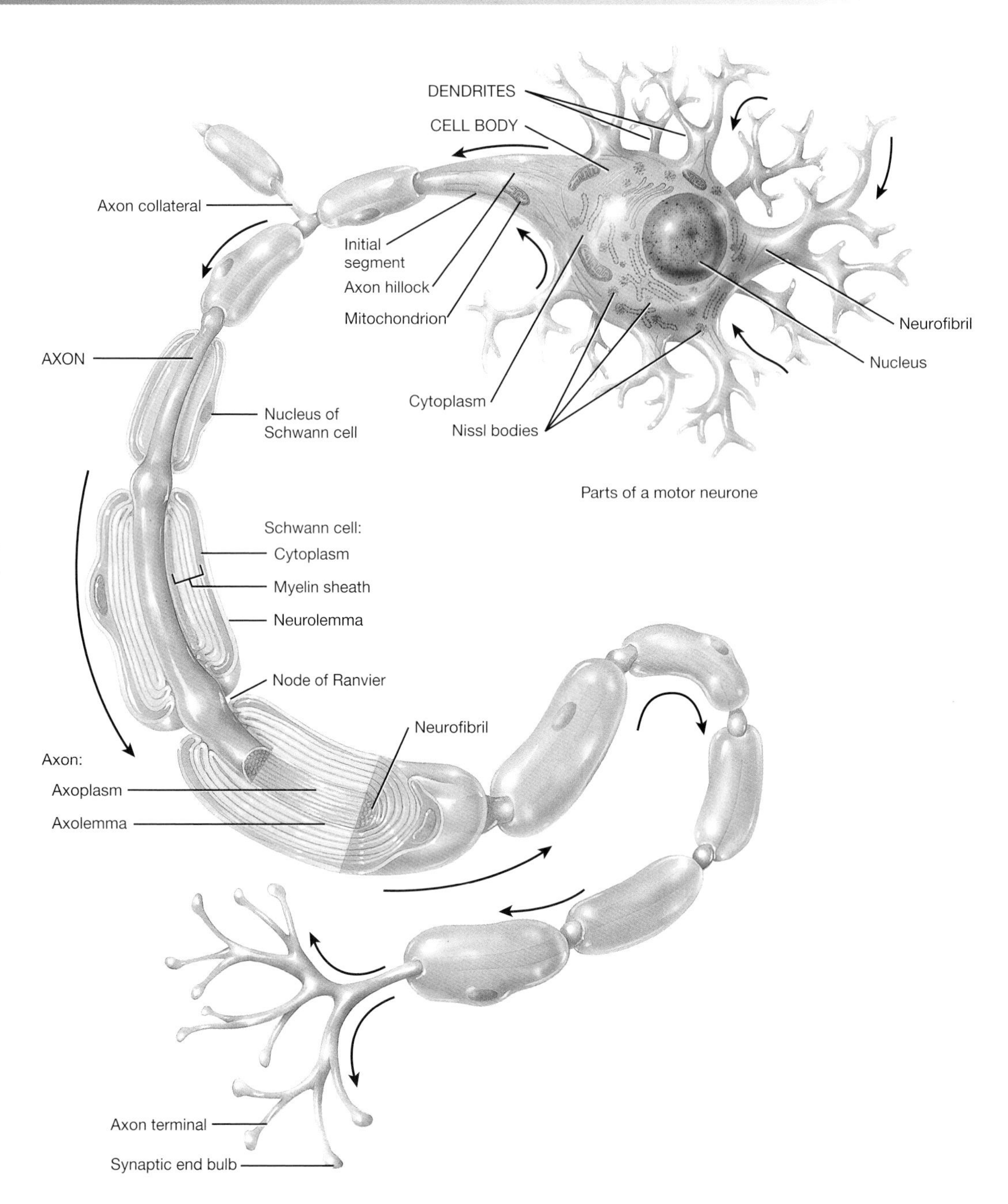


**Figure 15.2** Motor neurone cell.
*Source:* Tortora and Derrickson (2009), Figure 12.2, p. 418. Reproduced with permission of John Wiley and Sons, Inc.

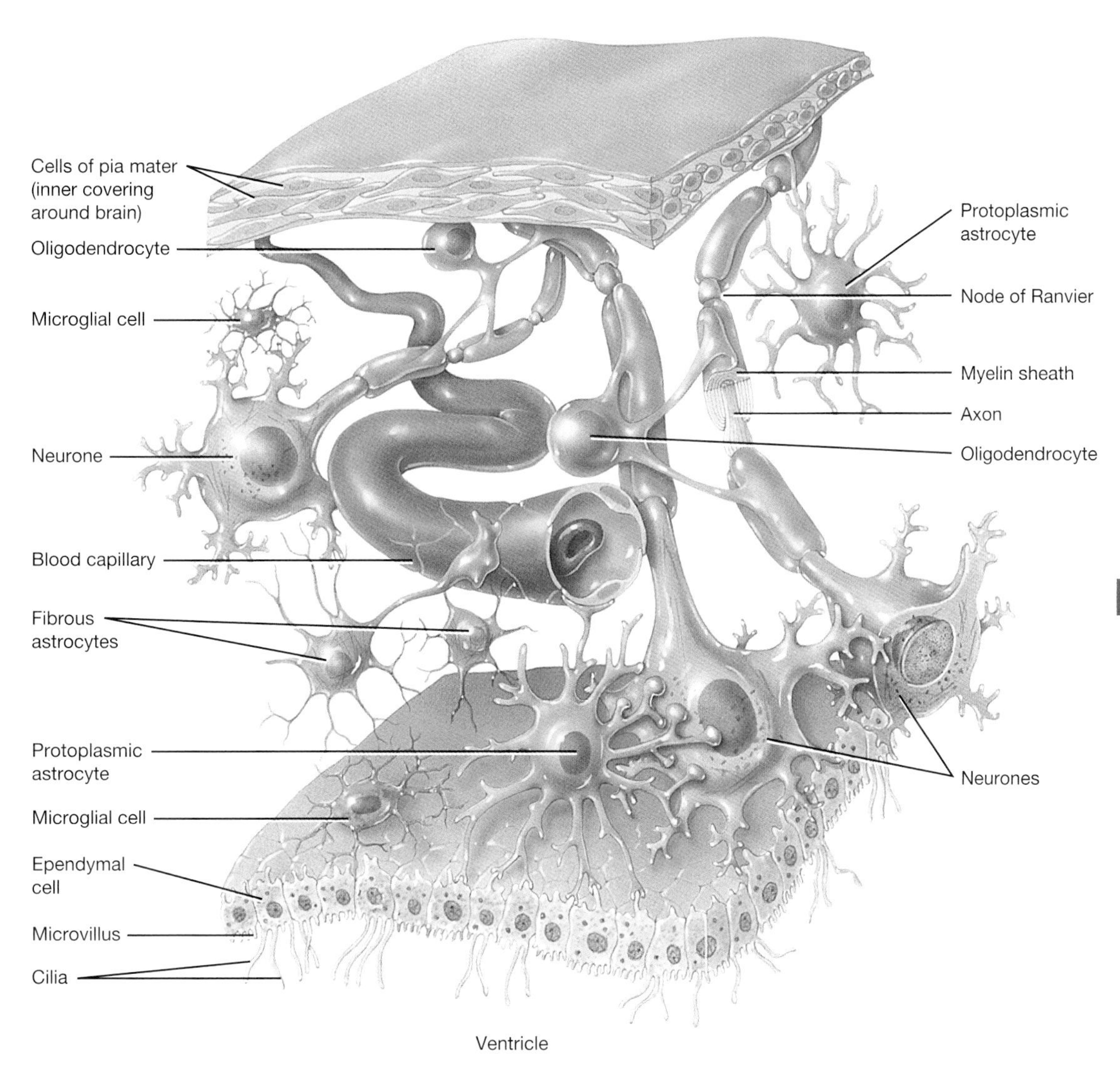


**Figure 15.3** Neuroglia cell of the CNS.
*Source:* Tortora and Derrickson (2009), Figure 12.6, p. 422. Reproduced with permission of John Wiley and Sons, Inc.



around axons in regularly spaced intervals called the nodes of Ranvier (Figure 15.2). These unmyelinated intervals allow an electrical charge to jump from one node to another in a rapid-fire sequence termed salutatory conduction. This allows for faster transmission of nerve impulses, compared with simple propagation of nerve impulses via non-myelinated fibres. Myelination of peripheral nerves continues after birth, and the speed of transmission increases as the child grows older, allowing the child to develop fine motor control. Schwann cells aid in the regeneration of axons after injury.

- Oligodendrocytes are responsible for myelination of axons in the CNS. They demonstrate negligible regrowth after injury.

- Blood–brain barrier
  - Astrocytes wrap around synaptic endings of neurones and blood capillaries. They mediate the permeability of endothelial cells in the blood capillaries which form the blood–brain barrier. This prevents potentially harmful substances such as bacteria and toxic agents entering the CNS by limiting the free diffusion of substances. They provide nutrients to neurones, maintain extracellular ion balance and provide structural support in both the grey and white matter of the brain.
  - Ependymal cells line the fluid-filled spaces in the brain and spinal cord. They assist with cerebrospinal fluid (CSF) circulation and are part of the blood–brain barrier.
  - In the fetus and newborn the blood–brain barrier is indiscriminately permeable, allowing passage of protein and other large and small molecules to pass freely between the cerebral vessels and the brain. Conditions such as hypertension, hypercapnia, hypoxia and acidosis cause cerebral vasodilation and disrupt the blood–brain barrier.

## Clinical considerations



### Newborn jaundice

Bilirubin is measured in newborns with jaundice to detect elevated unconjugated bilirubin levels, using a heel prick blood sample (Labtests Online, 2014). A common cause is immaturity of the liver at birth (physiologic jaundice), which cannot process bilirubin quickly enough, leading to a build-up. It usually resolves within a few days after birth as the liver matures. Other causes of raised bilirubin levels may include genetic disorders, hypoxia, hepatitis and haemolytic disease of the newborn.

In the newborn, the blood–brain barrier is not fully developed for 2–4 weeks (MacGregor, 2008). This allows water and unconjugated bilirubin to enter the interstitial spaces in the brain. Excessive amounts of bilirubin damage the developing brain cells and may cause developmental disabilities, and possibly sight and hearing deficits also. In the older child, elevated bilirubin levels will be less toxic to the brain, but investigations should be carried out to determine the cause.

Prompt treatment is required in the newborn and includes phototherapy. Exposure to blue light converts bilirubin in the skin into a water-soluble form that can be excreted in the bile. In severe cases, blood-exchange transfusions may be used, or surgery if the jaundice is a result of obstruction (NICE, 2010).

# Transmission of nerve impulses

Neurones facilitate communication between the CNS and PNS via afferent and efferent neurones. Transmission can be via electrical action potentials along the length of the neurones and by chemical neurotransmitters at synaptic endings.

In simple linear transmission, action potentials are generated and transmitted by neurones through the exchange of sodium and potassium ions across the cell membrane (see Figure 15.4). The process of positive and negative ion exchange occurs in a wave of polarization and depolarization along the neurone. In salutatory conduction, action potentials travel much faster along insulated myelinated axons where the action potential jumps between the nodes of Ranvier.

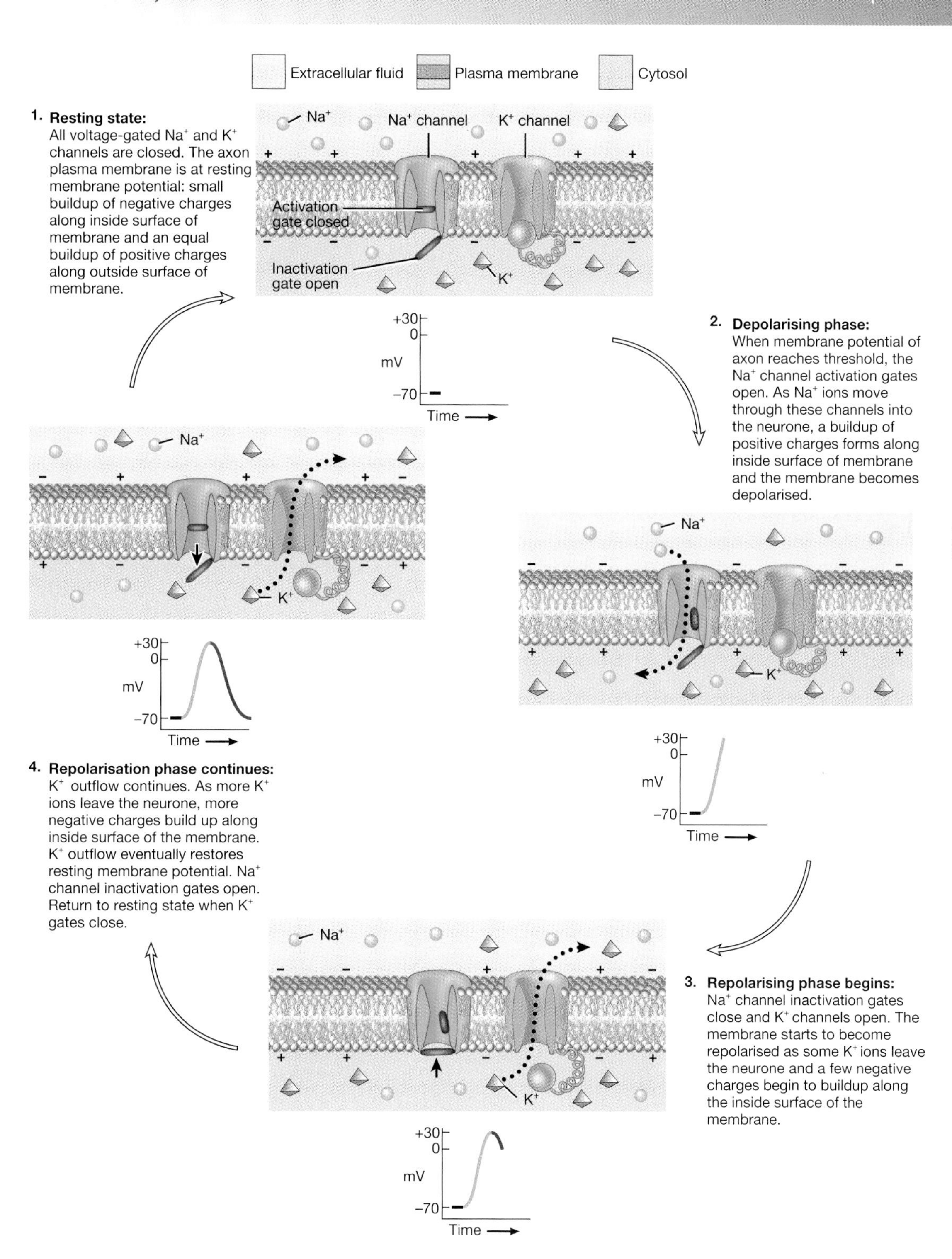


**Figure 15.4** Action potential.
*Source:* Tortora and Derrickson (2009), Figure 12.21, p. 437. Reproduced with permission of John Wiley and Sons, Inc.

Neurotransmitters facilitate communication with adjacent neurones, glands and muscle effector cells. Neurones are not physically connected to each other; a microscopic gap called a synapse exists between them. When an action potential impulse arrives at an axon terminal, it triggers the release of a neurotransmitter into this synaptic space. Depending on the type of neurotransmitter excreted, the adjacent neurone or effector cell's action will be stimulated or inhibited. Acetylcholine is released within the CNS and at neuromuscular junctions. Norepinephrine (noradrenaline) and dopamine are released within the CNS and at autonomic nervous system synapses.

# Fetal development

The nervous system develops early in the third week of embryonic development after the formation of the three primary germ layers: the endoderm, mesoderm and ectoderm. All nervous tissue and ophthalmic and auditory systems develop from a thickened area of the ectoderm. This grows into a neural plate from which two neural folds advance towards each other. By the end of the third week the neural folds fuse to form a neural tube in a process called neurulation (Price *et al.*, 2011).



During the fourth week the neural tube cells differentiate to develop into three enlarged areas called the primary brain vesicles (Figure 15.6). This include the fore-, mid- and hindbrain. Between the fifth week and 11th weeks, secondary brain vesicles develop from these to form distinct parts of the brain (Figures 15.5 and 15.6) (Price *et al.*, 2011).

These areas continue to grow exponentially until maximum brain volume and growth reach their peak at 30 weeks' gestation. Growth continues at a slower rate until birth (MacGregor, 2008).

The PNS develops from the neural crest, which forms alongside the neural tube at around 4 weeks. Tissues in the neural crest differentiate to form the root ganglia of spinal nerves and the ganglia of cranial nerves, the autonomic nervous system and the meninges (Price *et al.*, 2011).

# Childhood development

At birth all the major structures are present in the nervous system. The brain weighs approximately 12–20% of the newborn's body weight in comparison with an adult, where the brain

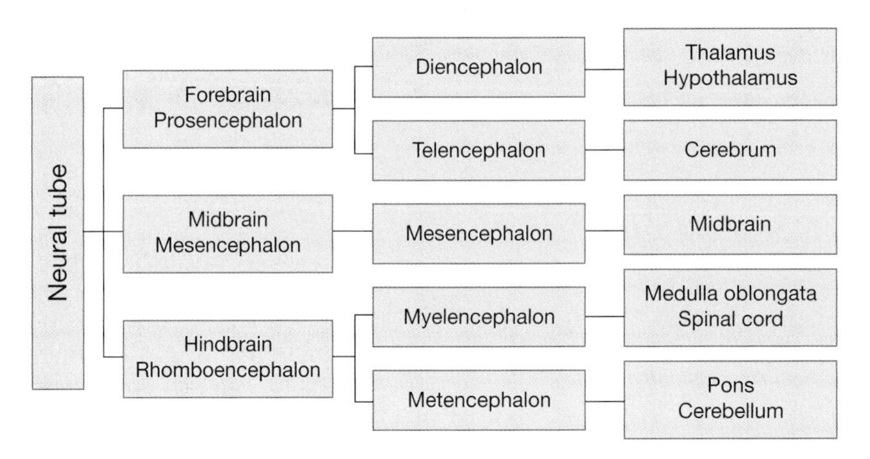


**Figure 15.5** Embryonic brain development.

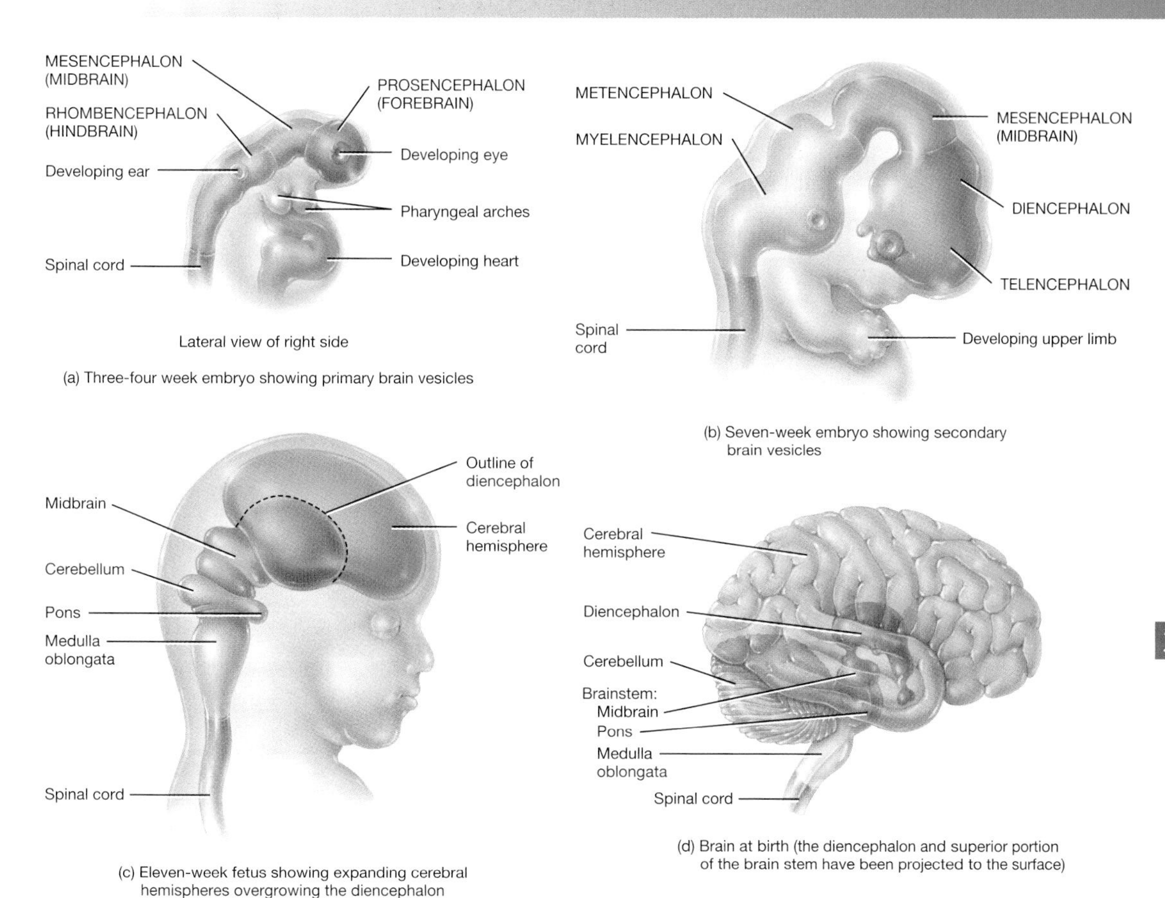


**Figure 15.6** Development of the brain and spinal cord.
*Source:* Tortora and Derrickson (2009), Figure 14.28, p. 538. Reproduced with permission of John Wiley and Sons, Inc.



comprises only 2% of the total body weight. Neurogenesis, the formation of new neurones, continues after birth, doubling the size of the brain in the first year of life. To accommodate this initial growth within the rigid skull, the anterior fontanelle closes slowly over a period of 18 months. A 2-year-old child's brain is 75% of its future adult brain weight, increasing to 90% by the age of 6, reaching adult size by the age of 10 (MacGregor, 2008).

During this rapid developmental change, nerve connections in the brain and spinal cord are constantly developing in response to a range of stimuli. Stable permanent connections develop as a result of repetitive exposure to stimuli and movement. Permanent nerve pathways develop, and the range of stimuli and interaction with a baby and young child directly affects the development of the brain and spinal cord. Some brain and neural apoptosis (pre-programmed cell death) continues until the age of 2 to remove redundant pathways (Price *et al.*, 2011).

This process of neurogenesis, formation of neural connections and apoptosis continues throughout the lifespan at a much slower rate and is referred to as the plasticity of the nervous system (Pascual-Leone *et al.*, 2011). Plasticity is an intrinsic function of the nervous system that enables modification of its function and anatomy in response to environmental changes. These changes can include injury, learning a new motor skill or cognitive skill.

# Central nervous system

The CNS consists of the brain and spinal cord. It acts as the control centre of body activities by assimilating internal and external information. It uses this information to maintain homeostasis by mediating motor and endocrine responses.

## Brain

The brain is a remarkable structure that defines who we are as individuals and how we experience the world. Recent advances in neuroimaging have allowed researchers to look inside the brain, providing vivid pictures of its subcomponents and their associated functions.

At birth the brain is fully formed but not fully mature or developed. It continues to develop for 20 years, when it is considered to have reached full maturation.

### *Cerebral blood supply*

Blood is supplied to the brain from the aorta via the internal carotid and vertebral arteries, which further subdivide to supply blood to the anterior and posterior cerebral circulation. The brain is very susceptible to interruption in blood supply, which leads rapidly to oxygen and glucose deprivation of the delicate neural tissues. Postural changes, haemorrhage and occlusion of an artery can cause sudden pressure changes in the circulating blood volume. A system of autoregulation is present to prevent sudden pressure changes that could cause rapid cell death. The circle of Willis connects cerebral arteries, allowing for a rapid redistribution of nutrient- and oxygen-rich arterial blood to be diverted to dependent areas, should the need arise (Figure 15.7).



Venous blood returns to the heart via the jugular veins and the superior vena cava.

### *The ventricles and cerebrospinal fluid*

The four interconnected cerebral ventricles in Figure 15.8 are filled with CSF, which is produced in the choroid plexus of the ventricle walls. It continuously circulates through the ventricles and in the subarachnoid space around the brain and the spinal cord (Figure 15.8).

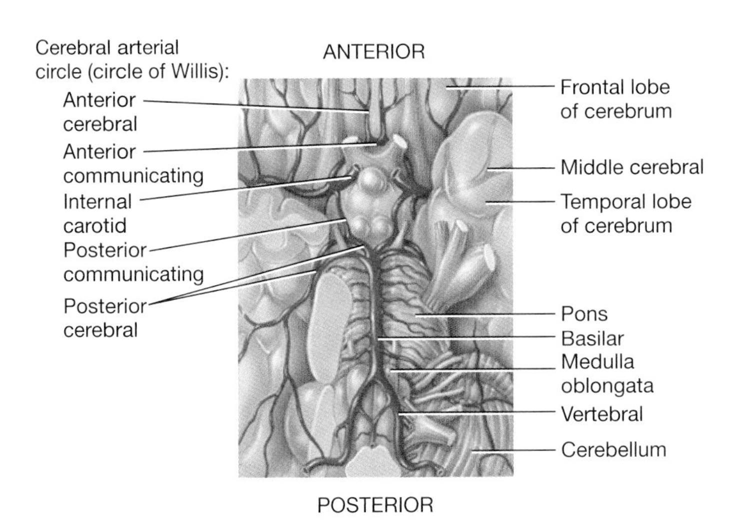


**Figure 15.7** Circle of Willis.
*Source:* Tortora and Derrickson (2009), Figure 21.19(c), p. 791. Reproduced with permission of John Wiley and Sons, Inc.

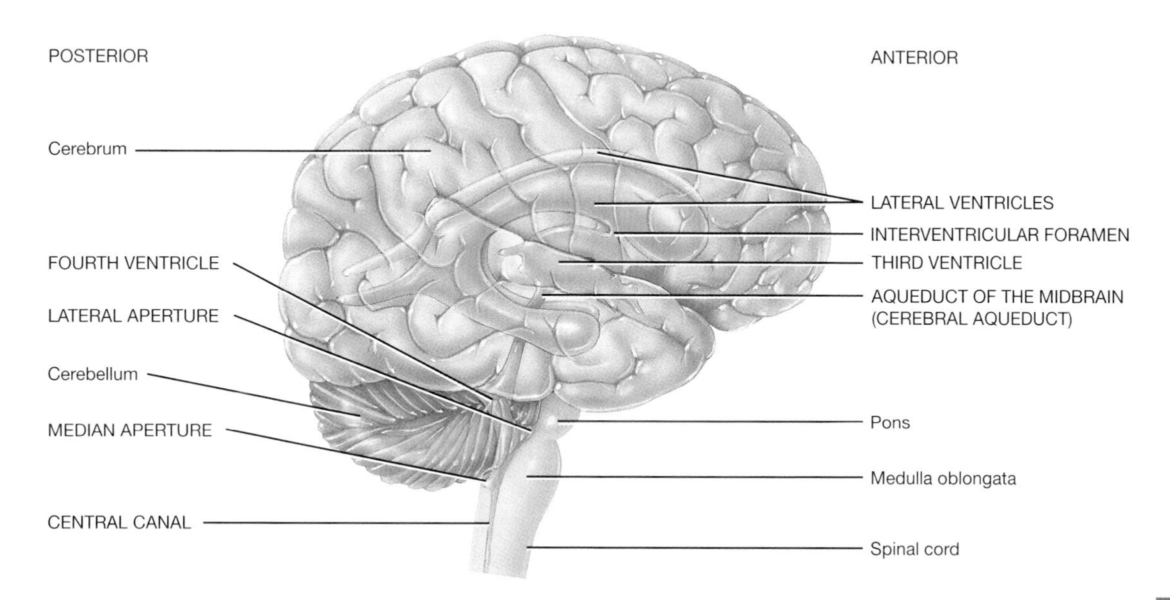




**Figure 15.8** Cerebral ventricles (right lateral view).
*Source:* Tortora and Derrickson (2009), Figure 14.3, p. 499. Reproduced with permission of John Wiley and Sons, Inc.

The CSF has four main functions:

- Mechanical protection of the delicate brain tissue against sudden jolts that may cause it to come into contact with the hard bones of the skull.
- Buoyancy, by supporting the mass of the brain and preventing ischaemia in the lower parts of the brain.
- Chemical protection against fluctuations in pH and ionic composition.
- Circulation of a limited amount of nutrients, such as oxygen and glucose. The removal of waste products from cerebral metabolic processes.

## *The meninges*

The meninges consist of three protective layers surrounding the brain, and they are continuous with the spinal cord meninges (Figure 15.9). The three cranial meninges are the dura mater, arachnoid mater and pia mater.

- The dura mater lies close to the bone of the skull and consists of two mainly fused layers. In the spinal meninges, only the internal layer is present.
- The arachnoid mater lies underneath the dura mater and the space between them is called the subdural space. The space between the arachnoid mater and the pia mater is called the subarachnoid space. CSF circulates within this space. It also contains some of the larger blood vessels in the brain.
- The pia mater covers the brain and contains many small blood vessels.

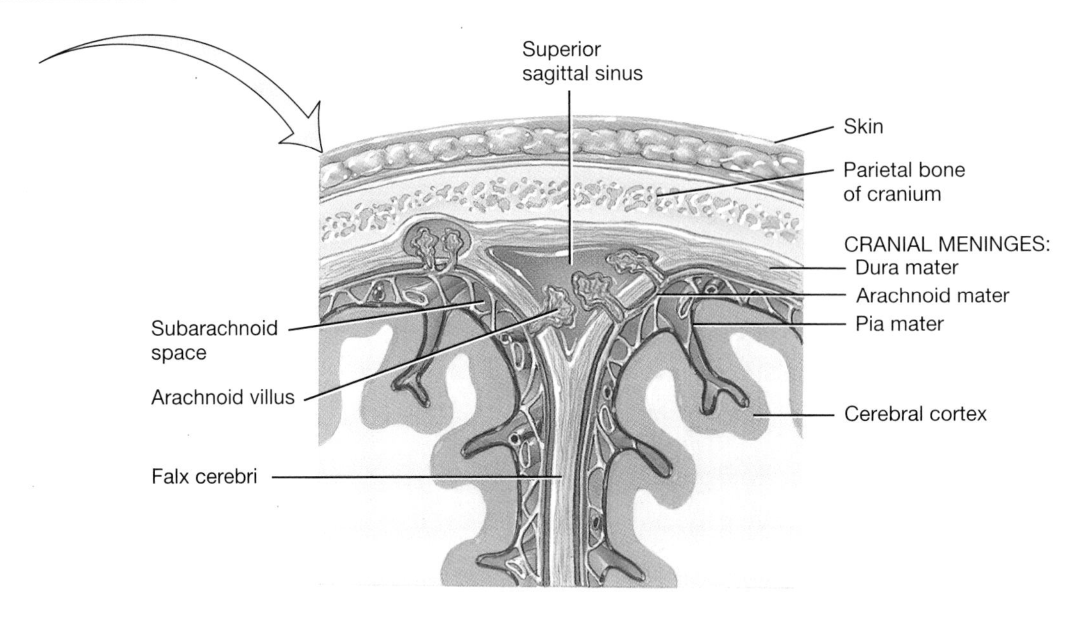




**Figure 15.9** Anterior view of frontal section through the skull showing the cranial meninges.
*Source:* Tortora and Derrickson (2009), Figure 14.2(a), p. 498. Reproduced with permission of John Wiley and Sons, Inc.

## Clinical considerations

### Hydrocephalus

In the healthy hydrated baby, the fontanelles are flat and soft to touch. CSF protects the brain and lies underneath the arachnoid membrane. The absence of these small areas of skull means that it normal to feel the heartbeat or a slight temporary rise if the baby coughs, due to increased intercranial pressure in the CSF.

In hydrocephalus the CSF pressure becomes elevated. This could be due to abnormalities in the brain such as tumours, inflammation or developmental irregularities. These prevent CSF draining from the ventricles, into the subarachnoid space, leading to an increase in intercranial pressure. If this increased pressure persists, the fluid build-up compresses and damages brain tissue. Clinical signs will include (Corns and Martin, 2012):

- bulging anterior fontanelle;
- head enlargement at birth (in children with congenital hydrocephalus);
- dilated scalp veins and separation of skull sutures (later sign);
- downward eye gaze (also known as the setting-sun sign);
- convulsions.

Hydrocephalus is an acute life-threatening emergency, which must be treated promptly. Early diagnosis is critical in the older child whose skull bones have fused, leaving little room for swelling and which will rapidly lead to raised intercranial pressure. Cushing's reflex, also known as the vasopressor response, will result in the clinical signs know as Cushing's triad (Corns and Martin, 2012):

- raised systolic blood pressure with widening pulse pressures
- bradycardia, reduction in the heart rate
- irregular or slowed respiration.

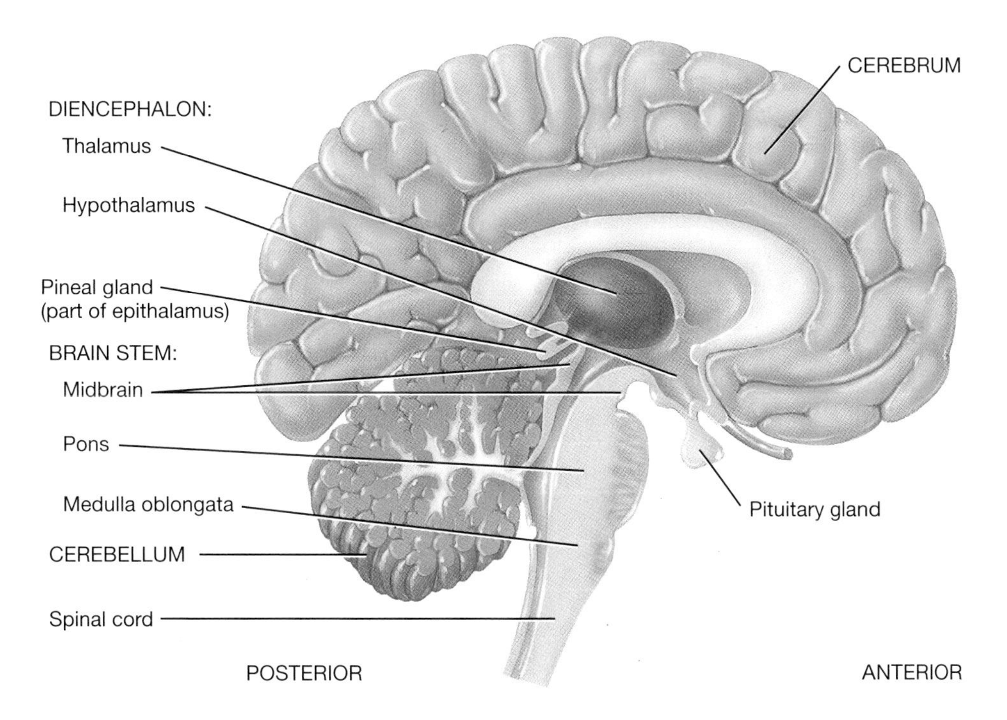


**Figure 15.10** Structure of the brain.
*Source:* Tortora and Derrickson (2009), Figure 14.1, p. 497. Reproduced with permission of John Wiley and Sons, Inc.



As can be seen from Figure 15.10, there are many anatomical subdivisions in the brain and these are subdivided into the

- cerebrum
- diencephalon
- cerebellum
- brainstem.

Within each area are further anatomical and physiological regions.

## *Cerebrum*

The cerebrum is the largest part of the brain and consists of the cerebral cortex, cerebral hemispheres and basal ganglia. It processes motor, sensory and association information. The cerebrum controls all voluntary motor action in the body and is responsible for consciousness, thought and learning.

Anatomically, the outer layer of cerebrum is called the cerebral cortex or cortex. It contains folds, called gyri, and fissures, termed sulci, which give it a distinctive 'wrinkled' appearance. The outer cortex consists of unmyelinated neurones known as grey matter. This is in contrast to the white matter, in the layers below, which consist of predominately myelinated axons.

There are two hemispheres in the cerebrum, the left and right, and each consists of four lobes, namely the frontal, parietal, temporal and occipital. Each of the lobes is named after the bone that covers it. Figure 15.11 demonstrates a functional map of this region of the brain, first published in 1906 by Brodmann. Each area is numbered for ease of identification, and these are still

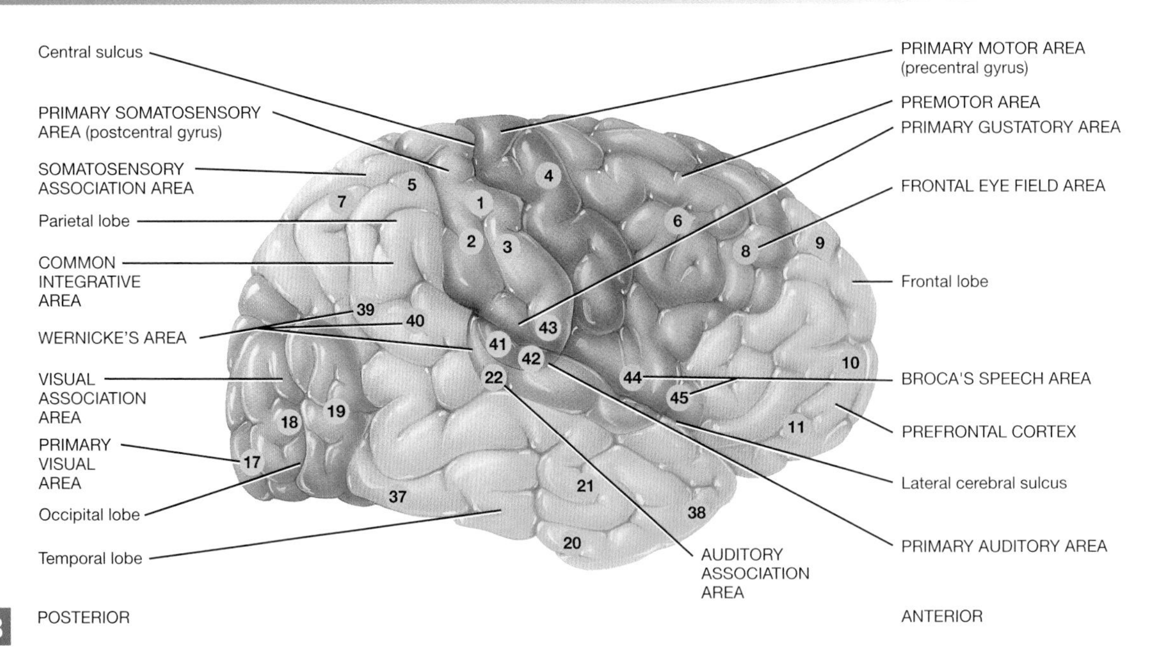




**Figure 15.11** Lateral view of the right cerebral hemisphere showing Brodmann's functional areas.
*Source:* Tortora and Derrickson (2009), Figure 14.15, p. 519. Reproduced with permission of John Wiley and Sons, Inc.

in current use. In the majority of people Wernicke's area and Brocca's speech area are usually in the left hemisphere and have been added to Figure 15.11 to indicate their relative location.

- **The frontal lobes** contain a number of important areas involved in cognitive function, voluntary motor movement, speech and sight:
  - The *prefrontal cortex*, which is the main site involved in cognitive functions and voluntary behaviours such as planning, problem solving, thinking, attention and intelligence. It is also involved in the acquisition of appropriate social skills and behaviours.
  - The *premotor area*, which is involved in planning and executing motor movements. It uses sensory information from other cortical regions to select appropriate muscle movements. It is also important for learning through imitation.
  - The *primary motor area*, which initiates and coordinates voluntary motor movements. Areas of the motor area correspond precisely to specific body parts. Not all body parts are equally represented in terms of proportionate size. The hand motor area occupies a larger space than the ankle because it has to be able to do a far more complex range of movements and, therefore, requires a larger amount of neural innervation. This can be visually demonstrated on a homunculus map, which is pictorial representation of how much the respective motor area innervates certain body parts. Practice or training of a particular muscle group can increase the representative space in the primary motor area.
  - *Broca's area*, which is only present in the left hemisphere in the majority of people. It is involved mainly in the motor production of spoken language and language-associated gestures. Broca's area also plays a significant role in language comprehension and the processing of complex sentences.
- **The parietal lobes** perceive and integrate sensory information to build a coherent somatosensory and visual picture of our surroundings. They integrate information from the ventral and dorsal visual pathways, which process what and where things are. Spatial mapping and

spatial awareness allow us to coordinate our movements in response to the objects in our environment. They are also involved in manipulating objects, number representation and comparison with previous experiences, thereby allowing for recognition of familiar objects.
  - The *somatosensory association area* receives tactile information from the body, such as touch, pressure, temperature and pain. Sensory information is carried to the brain by neural pathways to the spinal cord, brainstem, and thalamus, which project to the somatosensory association area. It integrates sensory information, producing a homunculus map, similar to that of the primary motor area. Sensory information about the feet, for example, map to the medial somatosensory association area.
- **The temporal lobes** contain a large number of substructures, whose functions include perception of hearing, vision and smell, face and object recognition, memory acquisition, language comprehension, autobiographical information, memory, word retrieval and learning:
  - The *primary auditory area*, which is responsible for processing sounds and their comprehension. Specific sound frequencies map precisely onto the primary auditory area. Auditory memory is associated with this area.
  - *Wernicke's area*, which is associated with language comprehension. In a similar way to Broca's area, in the majority of people it is situated in the left hemisphere.
- **The occipital lobes** contain the primary visual area of the brain.
  - The *primary visual area* receives visual information from the retina and integrates different visual information, such as colour, orientation and motion.



The **basal ganglia** are connected to other motor areas and link the thalamus with the primary motor area. They regulate the initiation of voluntary movements, balance, eye movement and posture. They are also associated with reward and reinforcement, addictive behaviours and habit formation. The basal ganglia are linked to the limbic system.

The **corpus callosum** consists of a large bundle of fibres connecting the right and left hemispheres of the brain, thus allowing information to move between hemispheres. Each hemisphere controls movement in the opposite side of the body; therefore, this is an important integrative structure.

## Diencephalon

The diencephalon extends upwards from the brainstem towards the cerebrum and is almost completely surrounded by it (Figure 15.10). It includes:

- *The thalamus*, which is a large, two-lobed structure that acts as a relay station for sensory and motor impulses. It receives sensory information, via the brainstem, which it processes and relays to the appropriate areas in the cerebral cortex. It relays motor impulses to and from the cerebral cortex to the brainstem. The thalamus contributes to many processes in the brain, including perception, attention, timing and movement.
- *The epithalamus* is part of the forebrain and comprises pineal body and surrounding structures. The pineal gland secretes melatonin and, therefore, plays a central role in alertness, awareness and sleep cycles.
- *The hypothalamus*, which controls many autonomic nervous system functions and behavioural activities. It is part of the limbic system and integrates information from many different parts of the brain. It is closely associated with the pituitary gland and is involved in stimulating the release of oxytocin, antidiuretic hormone and epinephrine (adrenalin).

## Cerebellum

The cerebellum is the second largest part of the brain, and it contains more neurones than the rest of the brain combined. Whilst the cerebrum plans and executes voluntary motor movement, the cerebellum monitors and regulates motor behaviour. It constantly calibrates and corrects any deficits, ensuring the movement is precise, timely and coordinated.

### Brainstem

The brainstem lies deep within the base of the brain above the spinal cord. It consists of three areas: the midbrain, pons and medulla oblongata. While the brainstem can organize motor movements such as reflexes, it coordinates with the motor cortex and associated areas to contribute to fine movements of limbs and the face. The brainstem plays an important part in maintaining homeostasis by controlling autonomic functions.

The *midbrain* contains the major motor nuclei controlling eye movement. It contains descending neural pathways that carry signals down from the cerebral hemispheres to the lower brain structures and spinal cord. The midbrain also contains the ascending sensory pathways from the spinal cord to the higher brain centres.

The *pons* contains cranial nerve nuclei associated with sensory input from and motor outflow to the face. Eleven of the 12 cranial nerves enter or leave the brainstem here, carrying motor and sensory information for the head and neck. The pons is the region in the brain most closely associated with breathing and respiratory rhythm. It forms a bridge between the cerebrum and cerebellum and is involved in motor control, posture and balance. It is also involved in sensory analysis and is the site at which auditory information enters the brain.



The *medulla*, also known as the medulla oblongata, is a continuation of the spinal cord and contains axons, which are a continuation of those in the spinal cord, as well as motor and sensory nerves for the throat, neck and mouth. The medulla plays an important part in the reflex control of the respiratory and cardiovascular systems.

The *reticular formation* is a functional neural network extending from the spinal cord, through the brainstem into the diencephalon. It has both ascending and descending sensory and motor roles. Other functions include pain modulation, cardiovascular control, sleep and alertness. The reticular activating system filters out repetitive meaningless stimuli in a process called habituation.

### The limbic system

The limbic system is a group of functional brain structures including the amygdala, hippocampus and hypothalamus that are involved in processing and regulating emotions, memory, olfactory stimuli and sexual arousal. The limbic system has an important role in the body's response to stress and is highly connected to the endocrine and autonomic nervous systems.

- The *amygdala* is a complex structure adjacent to the hippocampus. The amygdala is involved in processing emotions, including fear, and coordinates physiological responses based on cognitive information. It links areas of the cortex that process 'higher' cognitive information with hypothalamic and brainstem systems that control 'lower' metabolic responses, such as touch, pain and respiration.
- The *hippocampus* is the area of the brain most closely aligned to memory formation. It is important as an early storage place for long-term memory, and the transition of long-term memory to permanent memory. The hippocampus also plays an important role in spatial navigation. The subiculum, which is part of the hippocampus, plays a role in learning, information processing and regulation of the body's response to stress via the hypothalamus and pituitary gland.

## Spinal cord

The spinal cord extends from the medulla oblongata to the second lumbar vertebrae. The spinal cord has three main functions:

- transmission of efferent motor information from the brain to skeletal muscles and other muscles, primarily in the white matter of the spinal cord;
- transmission of afferent sensory information to the brain, which is also primarily via the white matter;
- coordination of autonomic and somatic reflex arcs, mediated by the central grey matter.

The spinal column protects the spinal cord externally, and the spinal meninges form internal protective layers. The three spinal meninges are:

- The *dura mater*, the outermost single layer. The epidural space lies between the vertebral bone and the dura mater. It is filled with adipose tissue and blood vessels. This space does not exist in the brain.
- The *arachnoid mater* is the middle protective layer. Similar to the structure in the brain, the space between the arachnoid mater and the pia mater is called the subarachnoid space and contains CSF.
- The *pia mater* is closely adhered to the spinal cord and forms the final protective layer.

Blood supply to the spinal cord consists of three main arteries that travel down the subarachnoid space. This is supplemented by arteries originating in the aorta, which enter alongside the spinal nerves into the spinal column (Figure 15.12).



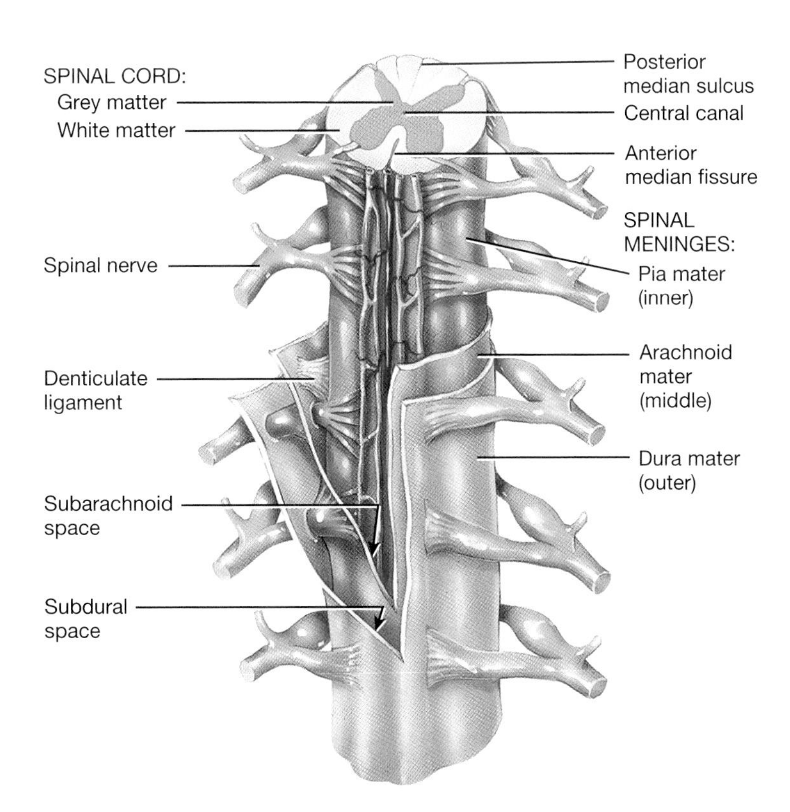


**Figure 15.12** Anterior view of the spinal cord.
*Source:* Tortora and Derrickson (2009), Figure 13.1, p. 462. Reproduced with permission of John Wiley and Sons, Inc.

# Peripheral nervous system

The PNS includes:

- 12 pairs of cranial nerves
- 31 spinal nerves
- sensory (afferent) neurones
- motor (efferent) neurones
- somatic nervous system (voluntary)
- autonomic nervous system (involuntary)
  - sympathetic
  - parasympathetic.

Afferent neurones transmit sensory impulses to the CNS and efferent neurones transmit motor impulses away from the CNS to the muscles, organs and glands of the body. Efferent neurones can be further subdivided into the autonomic involuntary nervous system and the somatic voluntary nervous system. The autonomic nervous system is further subdivided into the sympathetic and parasympathetic division, which maintains homeostasis.



## Cranial nerves

There are 12 pairs of cranial nerves, which leave the brainstem to supply the majority of the head and neck region. The exception is the vagus nerve, which innervates the thorax and abdominal region. Some cranial nerves are purely sensory or motor, whilst some have a mixed function. In Figure 15.13, sensory pathways are coloured blue and motor pathways are red.

In clinical practice, each cranial nerve is tested as part of a full neurological assessment. This allows clinicians to ascertain which part of the brain has been affected in an injury or disease process (Rawles *et al.*, 2010). Table 15.1 provides a list of cranial nerves, innervation and function.

## Spinal nerves

There are 31 pairs of spinal nerves that leave the spinal cord and column to transmit information to and from the CNS (Figure 15.14). Each spinal nerve carries both sensory and motor information to the skin, muscles and glands in a specific area of the body.

These specific areas are referred to as dermatomes (sensory innervation) and myotomes (motor innervation). Figure 15.15 indicates the areas of innervation for each spinal nerve. They are numbered according to the exit region and level of the spinal cord.

Typical spinal nerve fibres have both a posterior and anterior root which connects to the spinal cord (Figure 15.16). When they exit the spinal cord, they initially fuse to become a mixed afferent and efferent nerve fibre.

Sensory receptors generate an impulse in response to a trigger. This travels down the sensory afferent neurone to a connecting interneurone in the grey matter of the spinal cord. This lies in the integration centre from where impulses can be passed upwards to the brain via sensory neurones in the spinal cord. Motor impulses travelling down the spinal cord exit this area via the efferent motor neurone to the muscle of endocrine gland.

## Reflex and motor development

A reflex is a fast, involuntary sequence of actions that occurs in response to a particular stimulus. A reflex arc pathway controls either an autonomic or a somatic reflex action. Autonomic action reflexes control organs and smooth muscle, and somatic reflexes innervate skeletal muscles.

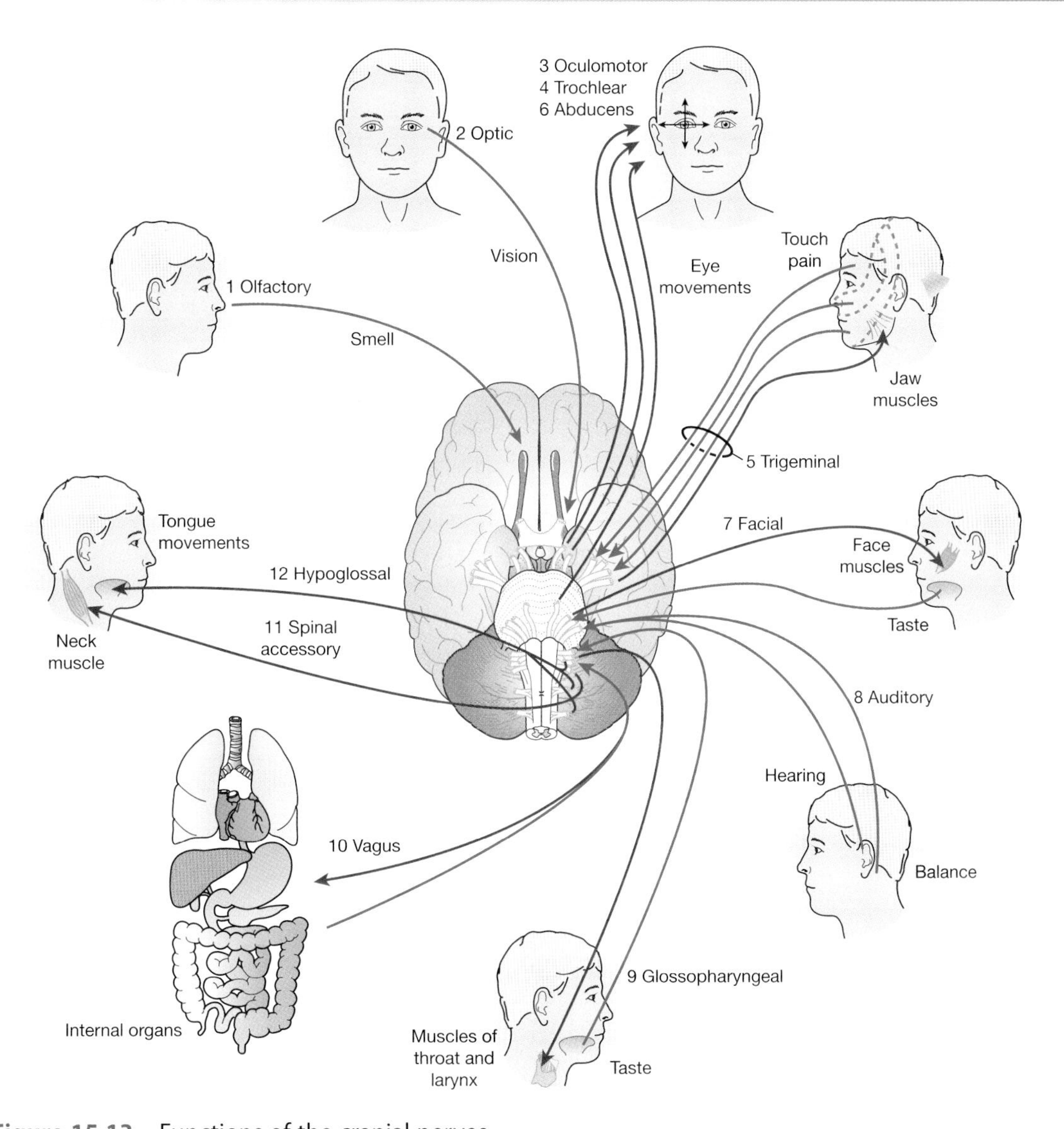


**Figure 15.13** Functions of the cranial nerves.
*Source:* Peate and Nair (2011), Figure 6.10, p. 171. Reproduced with permission of John Wiley and Sons, Ltd.



An impulse from a sensory neurone enters the spinal cord and synapses in the integration centre of the grey matter of the spinal cord. The impulse leaves the spinal cord via a motor neurone to a skeletal muscle to initiate a muscular movement (Figure 15.16). For example, a child touching a hot surface (sensory receptors in the skin) will remove their hand very quickly (motor response) because the reflex arc bypasses the brain, allowing for a quicker response. Sensory information is passed to the brain that the reflex action has occurred, but the motor response is involuntary.

Primitive reflexes develop in the fetus alongside spinal cord growth from 6 weeks onwards.

At birth these include grasping with hands and feet (palmar and plantar reflex), a sucking reflex in response to oral stimuli, corneal and blinking reflexes, and the Moro and startle reflex,

Table 15.1 Cranial nerves.

| Cranial nerve | No. | Innervation | Location and function |
|---|---|---|---|
| Olfactory | 1 | Sensory | Olfactory receptors and smell |
| Optic | 2 | Sensory | Retina and vision |
| Oculomotor | 3 | Motor | Eye muscles, including eye lid, eyeball, pupil and lens |
| Trochlear | 4 | Motor | Eye muscles, downward and lateral eye movement |
| Trigeminal | 5 | Sensory and motor | Eye and jaw. Chewing movement and sensation of touch, pain and temperature in face, eyes, teeth, tongue and mouth |
| Abducens | 6 | Motor | Eye muscles, lateral movement |
| Facial | 7 | Sensory and motor | Most facial expressions, secretion of tears and saliva, taste, ear sensation |
| Auditory | 8 | Sensory | Hearing and balance |
| Glossopharyngeal | 9 | Sensory and motor | Sensation from tongue, tonsil and pharynx. Taste. Assist in swallowing. Monitoring of blood pressure and $O_2$ and $CO_2$ in the bloodstream, via carotid sinus receptors |
| Vagus | 10 | Sensory and motor | Sensory, motor and autonomic functions of viscera – glands, digestion, heart rate, breathing rate, aortic blood pressure |
| Spinal accessory | 11 | Motor | Head and shoulder muscle movement |
| Hypoglossal | 12 | Motor | Tongue muscles movement |



which are triggered by sudden movement or sound. These assist newborns to survive outside the womb. As the neurological system matures, these reflexes usually start to disappear at around 2 months of age, although some, such as the gag and swallowing reflexes, persist throughout the lifespan.

A secondary set of motor reflexes that are not under voluntary control is also present. These include stepping, standing and swimming movements.

A third set of reflexes in the newborn is the postural reflexes that develop gradually, usually from 3 months onwards. These postural reflexes include the tonic neck, righting and labyrinthine reflexes. These help the baby develop and maintain its balance against gravity when disturbed.

Motor and postural reflexes gradually disappear during the first 2 years of life. Voluntary muscle control improves as the infant develops motor skills, muscle strength and finer prehensile control. Between the ages of 15 months and 2 years most children will be able to progress from sitting unaided to crawling and ultimately walking without support. A mature walking pattern is achieved by the age of 4, and 60% of 6- to 7-year-olds will have achieved the fundamental motor skills of climbing, jumping, throwing and catching (MacGregor, 2008).

There is variation in the rate of this development owing to the plasticity of the nervous system and the influence of genetic and environmental factors. Gender also seems to play a part. Boys are able to throw and kick a ball earlier than girls are, who are more proficient at hopping and skipping at an early age (MacGregor, 2008).

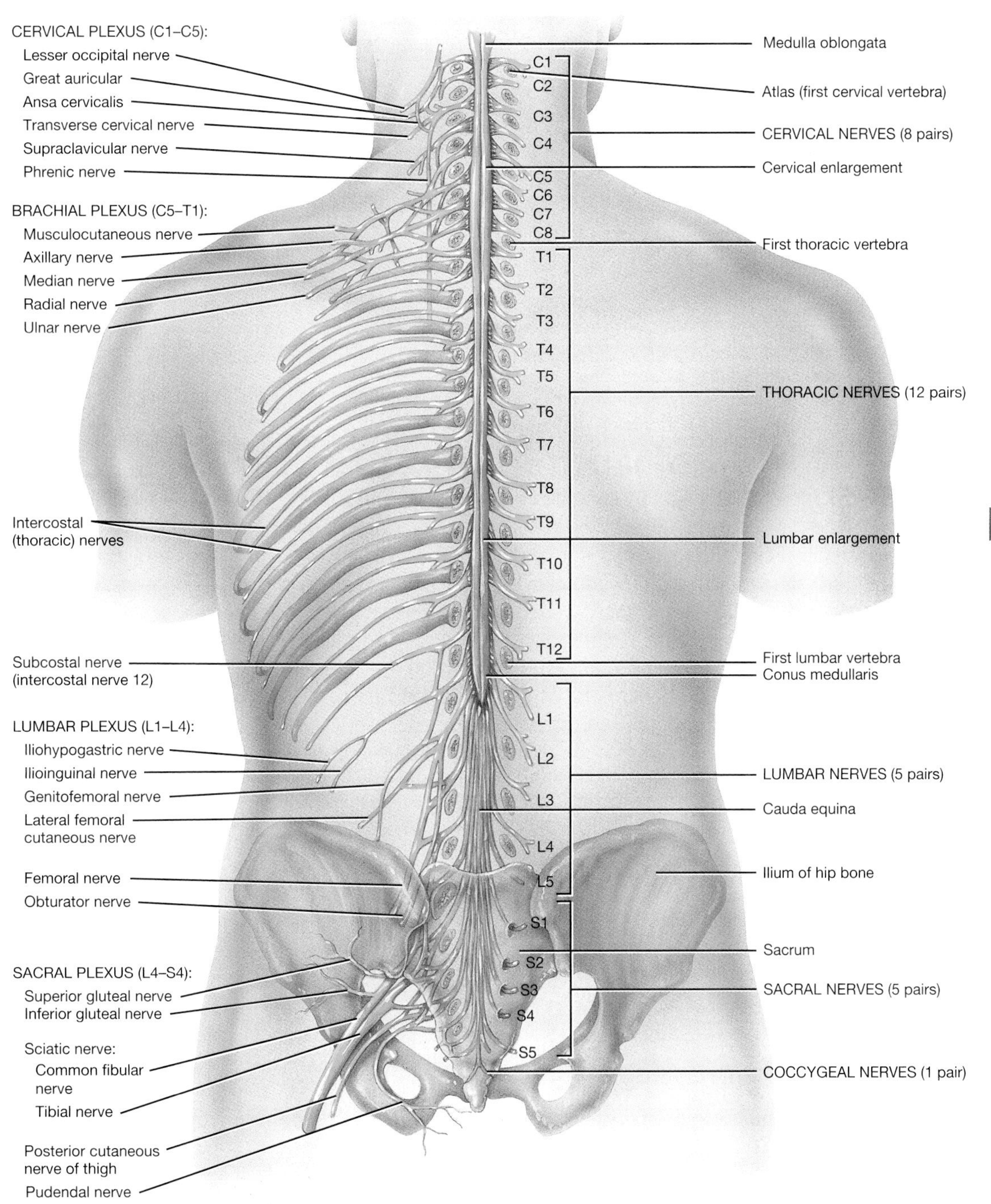


**Figure 15.14** The spinal cord and spinal nerves.
*Source:* Tortora and Derrickson (2009), Figure 13.2, p. 463. Reproduced with permission of John Wiley and Sons, Inc.

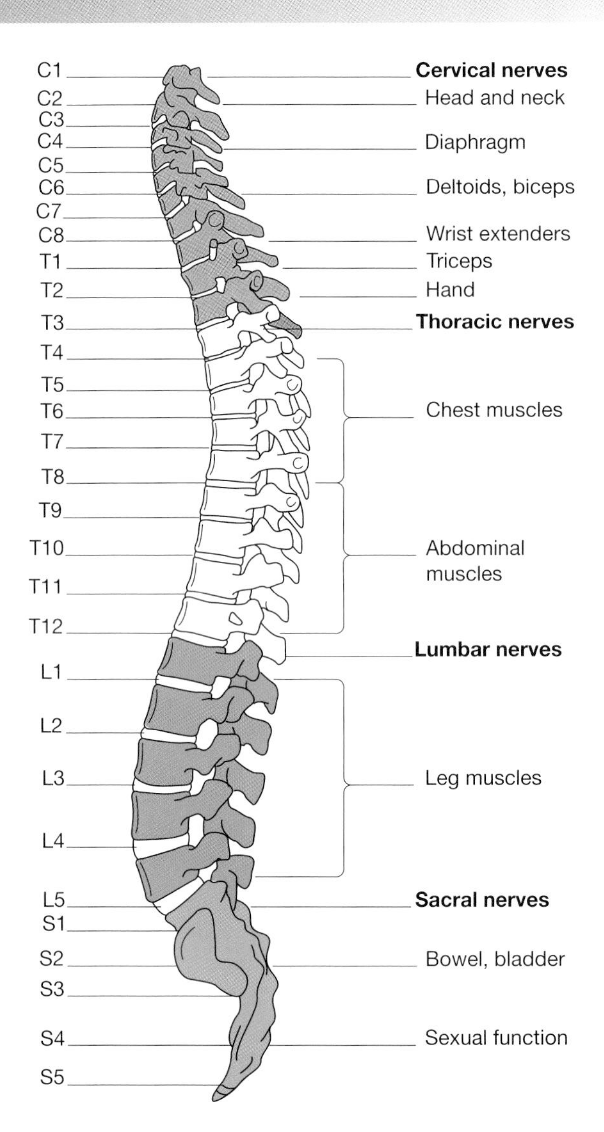


**Figure 15.15** The spinal nerves and their areas of innervation.
*Source:* Peate and Nair (2011), Figure 6.13, p. 175. Reproduced with permission of John Wiley and Sons, Ltd.



## Peripheral sensory (afferent) neurones

The dendrites of sensory neurones are often also sensory detectors and are highly specialized throughout the body. They send impulses towards the CNS and are therefore afferent neurones (Figure 15.16). They can be classified into the following:

- Somatic sensory neurones, which are situated in the skin and are responsible for relaying information about touch, temperature, pain, limb position, vibration and pressure.
- Visceral sensory neurones, which are situated in smooth muscles, visceral organs, cardiac muscles and include baro receptors and chemo receptors. They are the sensory part of the autonomic nervous system, conveying unconscious sensory information to the CNS. Examples include heart rate, blood pressure, blood gas composition and visceral pain perception.

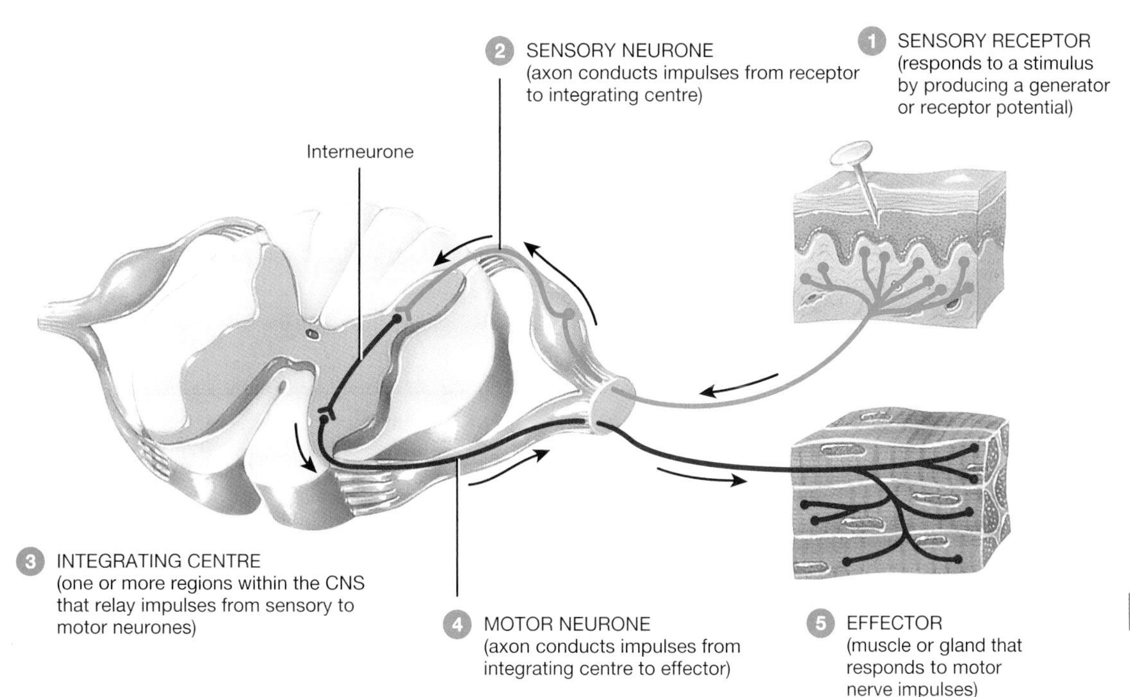


**Figure 15.16** A typical spinal nerve.
*Source:* Tortora and Derrickson (2009), Figure 13.14, p. 482. Reproduced with permission of John Wiley and Sons, Inc.



- Mechanoreceptors, which are mainly situated in muscles and joints. They monitor movement, stretch and pain giving us a sense of proprioception.
- Special senses neurones, which include vision, taste, smell and auditory information.

## Peripheral (efferent) motor neurones

The motor neurones send impulses from the CNS to the PNS and are therefore efferent neurones. The somatic motor neurones are part of the somatic nervous system and cause a voluntary skeletal muscle action. The autonomic neurones are part of the autonomic nervous system and they cause an involuntary action in smooth muscle, organs or glands.

## Somatic nervous system

The somatic nervous system coordinates all voluntary motor systems, except those innervated by reflex arcs. Impulses from the primary motor cortex travel through the CNS to the peripheral motor neurones, causing voluntary muscle contraction.

## Autonomic nervous system

The autonomic nervous system coordinates all involuntary motor responses to maintain homeostasis in organ systems. The autonomic nervous system maintains this balance through the opposing actions of the sympathetic and parasympathetic nervous system. The sympathetic nervous system causes excitation, whilst the parasympathetic has an inhibitory effect (Figure 15.17).

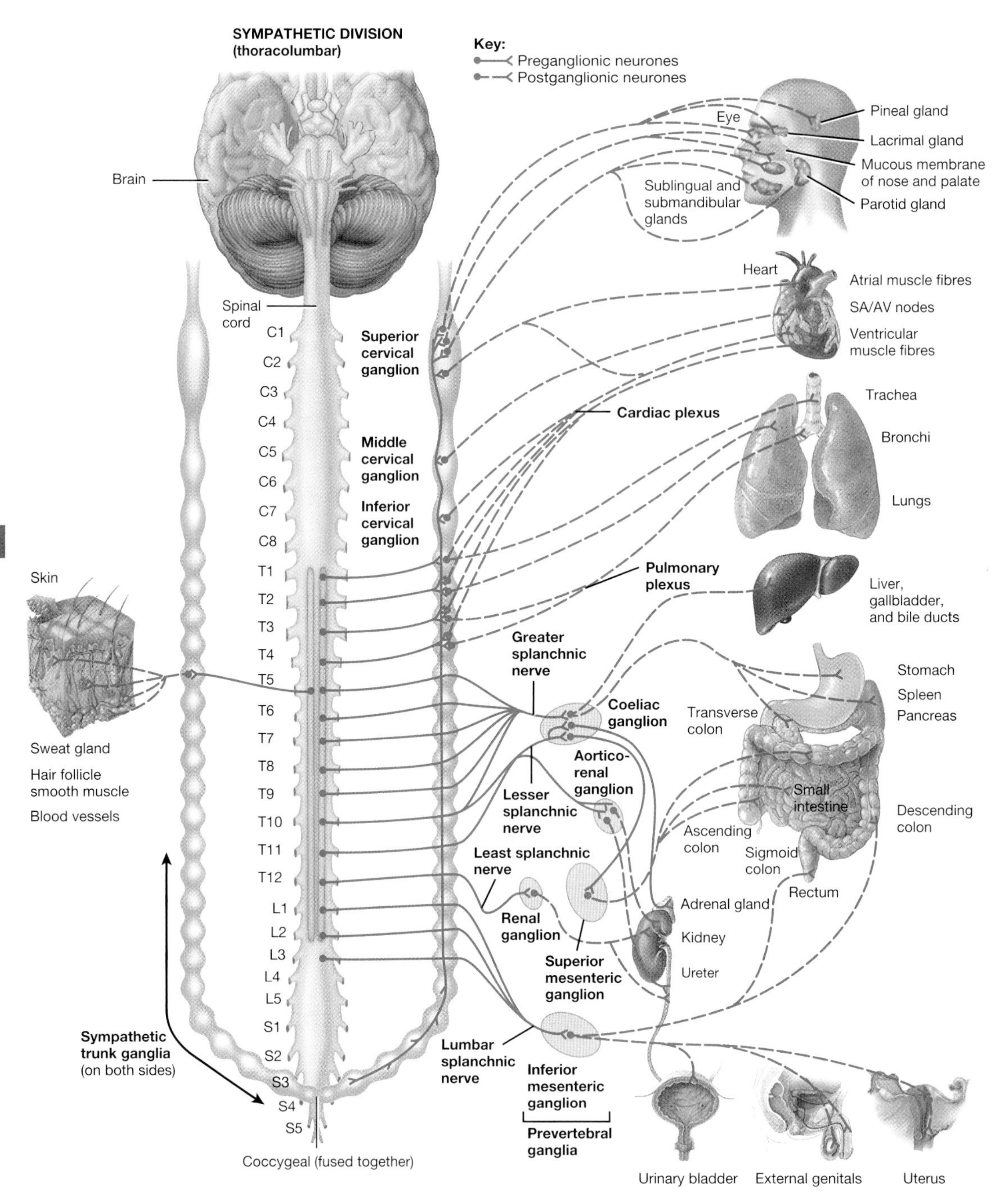


**Figure 15.17** Sympathetic nervous system.
*Source:* Tortora and Derrickson (2009), Figure 15.2, p. 550. Reproduced with permission of John Wiley and Sons, Inc.

**Table 15.2** Effects of the sympathetic and parasympathetic divisions of the autonomic nervous system. Source: Peate and Nair (2011), Table 6.2, p. 178. Reproduced with permission of John Wiley and Sons, Ltd.

| Organ/system | Sympathetic | Parasympathetic |
|---|---|---|
| Cell metabolism | ↑ Metabolic rate and blood sugar levels. Stimulates fat breakdown | None |
| Blood vessels | Constricts visceral and skin blood vessels<br>Dilates heart and skeletal muscle blood vessels | None |
| Eye | Dilates pupils | Constricts pupils |
| Heart | ↑ Heart rate and contractility | ↓ Heart rate |
| Lungs | Dilates bronchioles | Constricts bronchioles |
| Kidneys | ↓ Urine output | None |
| Liver | Glucose released | None |
| Digestive system | ↓ Peristalsis and constricts digestive system sphincters | ↑ Peristalsis and dilates digestive system sphincters |
| Adrenal medulla | Stimulates release of epinephrine and norepinephrine | None |
| Lacriminal glands | Inhibits production of tears | ↑ Production of tears |
| Salivary glands | Inhibits production of saliva | ↑ Production of saliva |
| Sweat glands | Stimulates production of perspiration | None |



## *Sympathetic nervous system*

The sympathetic nervous system (thoracolumbar) innervates body systems and glands during activity or emotional stress. Efferent motor neurones transmit impulses to internal organs and endocrine and exocrine glands (Figure 15.17). During times of extreme stress it triggers the release of norepinephrine (noradrenaline), to activate the 'fight or flight' response. Together with epinephrine (adrenaline), this causes glucose to be released from energy stores, increases blood flow to skeletal muscles and increases heart rate and contractility (Table 15.2). It also has a potent effect on the amygdala to increase alertness and our emotional responses.

## *Parasympathetic nervous system*

The parasympathetic nervous system (cransiosacral) innervates body systems and glands during rest and sleep periods (Figure 15.18). It operates in opposition to the sympathetic nervous system (Table 15.2) through the release of acetylcholine.

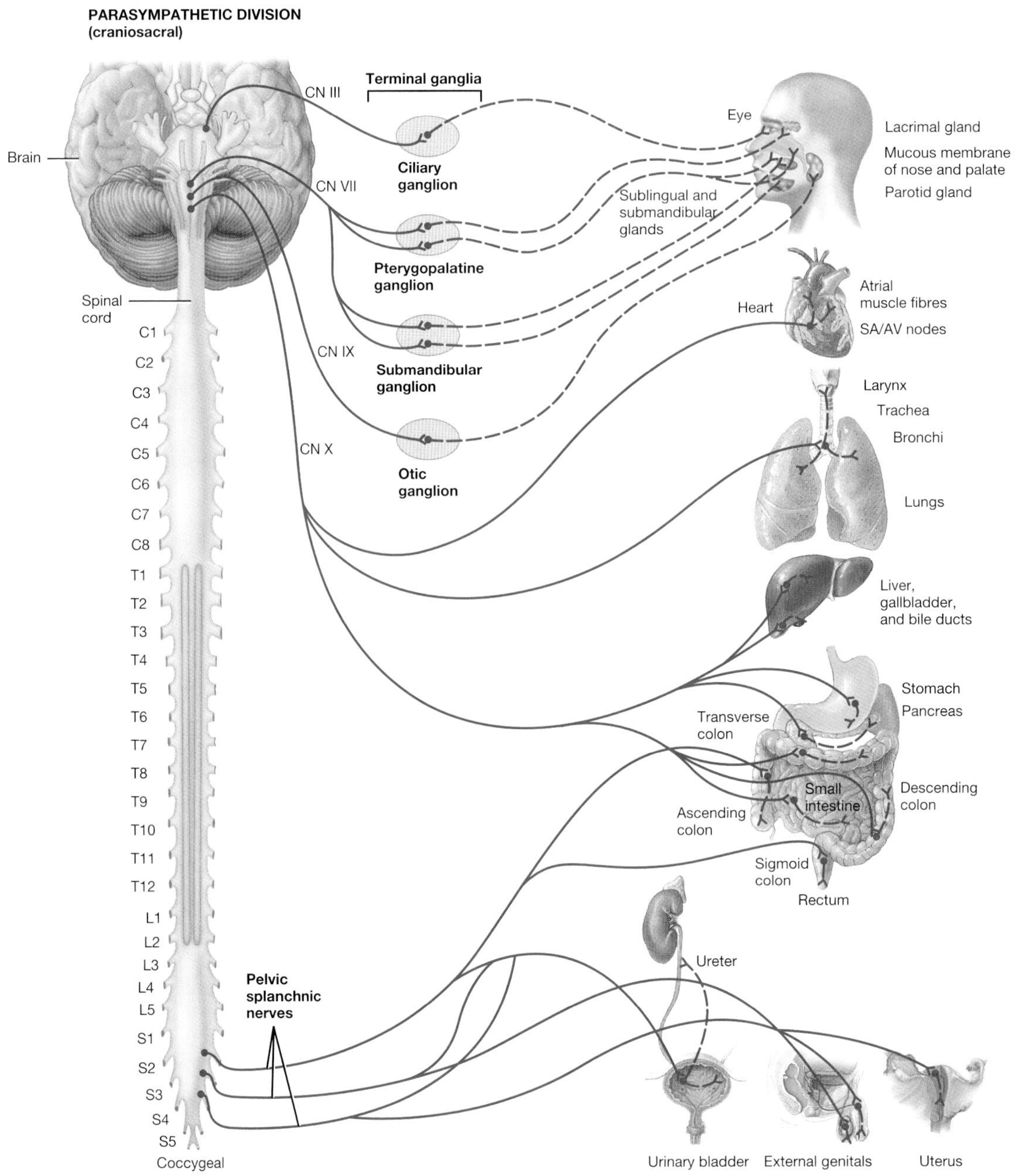

PARASYMPATHETIC DIVISION
(craniosacral)

Brain
Spinal cord
C1 C2 C3 C4 C5 C6 C7 C8
T1 T2 T3 T4 T5 T6 T7 T8 T9 T10 T11 T12
L1 L2 L3 L4 L5
S1 S2 S3 S4 S5
Coccygeal
CN III
CN VII
CN IX
CN X
Terminal ganglia
Ciliary ganglion
Pterygopalatine ganglion
Submandibular ganglion
Otic ganglion
Pelvic splanchnic nerves
Eye
Lacrimal gland
Mucous membrane of nose and palate
Parotid gland
Sublingual and submandibular glands
Heart
Atrial muscle fibres
SA/AV nodes
Larynx
Trachea
Bronchi
Lungs
Liver, gallbladder, and bile ducts
Stomach
Pancreas
Transverse colon
Small intestine
Descending colon
Ascending colon
Sigmoid colon
Rectum
Ureter
Urinary bladder
External genitals
Uterus



**Figure 15.18** Parasympathetic nervous system.
*Source:* Tortora and Derrickson (2009), Figure 15.3, p. 551. Reproduced with permission of John Wiley and Sons, Inc.

# Conclusion

Most of the structural development of the nervous system occurs before birth. Primary reflexes protect us and keep us safe in the initial few months of life, whilst our nervous system undergoes rapid growth. Throughout childhood this continues, and we develop fine motor control and cognitive ability as nerve fibres continue to myelinate and neural pathways are made permanent within the brain. The nervous system is complex in form, structure and function. It enables us to make sense of the world around us and interact with it accordingly. It helps maintain homeostasis in body systems without our awareness, and philosophically it is what makes us human.

# 'Did you know' information

1. There are no pain receptors in the brain, so the brain can feel no pain.
2. Neurons develop at the rate of 250 000 neurons per minute during early pregnancy.
3. You can't tickle yourself because your brain distinguishes between unexpected external touch and your own touch.
4. The brain uses 20% of the total oxygen in your body at any one time.
5. Reading aloud and talking often to a young child promotes brain development.



# Activities

**Now review your learning by completing the learning activities in this chapter. The answers to these appear at the end of the book. Further self-test activities can be found at www.wileyfundamentalseries.com/childrensA&P.**

## True or false?

1. Motor and postural reflexes disappear by the age of 2.
2. The brain is considered to have reached full maturity at the age of 12.
3. There are 10 pairs of cranial nerves
4. Axons carry impulses away from the cell body
5. Peripheral efferent neurones transmit impulses to the central nervous system.

## Exercise

- Draw a diagram of the two main divisions and their subdivisions of the nervous system.
- Identify the functions of each brain region.
- Locate the areas of innervation for each spinal nerve.
- Differentiate between the functions of the sympathetic and parasympathetic nervous system.
- Identify the stages of reflex development from newborn to adulthood.

## Wordsearch

There are 20 words linked to this chapter hidden in the following square. Can you find them? A tip – the words can go from up to down, down to up, left to right, right to left, or diagonally.



| T | C | Y | N | F | M | B | U | E | N | L | O | R | A | Y | M | P | H | N | Q | U | O | E | D | P |
|---|---|---|---|---|---|---|---|---|---|---|---|---|---|---|---|---|---|---|---|---|---|---|---|---|
| M | J | B | R | A | I | N | S | T | E | M | H | E | E | N | F | C | R | I | U | S | M | N | P | A |
| H | O | G | C | B | W | F | B | R | T | X | S | W | Z | H | E | M | O | O | C | H | I | S | O | G |
| E | X | E | T | C | A | N | N | E | U | R | O | G | L | I | A | V | T | J | Y | E | C | M | W | O |
| M | S | P | I | M | F | J | R | Q | B | R | R | O | H | E | D | S | C | H | Z | B | R | Q | E | G |
| I | V | O | P | I | L | G | A | F | Y | T | G | T | Q | A | P | Z | E | H | K | C | O | U | B | A |
| S | H | N | E | X | S | N | W | X | L | Y | C | W | K | U | D | S | F | S | A | Y | G | X | K | N |
| P | K | S | G | R | D | O | B | K | O | V | U | P | E | W | F | N | F | M | E | S | L | E | O | G |
| H | C | J | D | O | T | A | N | E | Y | N | O | I | L | D | A | L | E | U | Y | R | I | I | N | L |
| E | L | R | B | K | M | U | P | T | D | D | Y | C | D | P | L | T | E | T | M | S | A | T | C | I |
| R | A | F | F | A | C | N | Y | T | O | I | M | I | D | B | R | A | I | N | E | X | C | S | Y | A |
| E | G | C | U | L | A | R | S | P | U | Q | U | T | L | O | G | O | E | S | P | A | N | Y | S | L |
| S | U | O | E | Z | H | G | M | O | D | C | I | E | C | B | J | T | U | E | B | C | P | U | Q | I |
| Q | C | I | X | R | A | R | O | Y | O | V | C | S | M | N | Q | U | L | B | T | M | M | I | E | S |
| S | A | E | W | P | P | O | D | V | E | X | L | Y | M | A | W | O | A | I | N | A | R | Z | J | T |
| C | T | E | R | V | E | G | Y | W | I | L | S | Z | P | A | B | S | H | H | L | P | I | R | P | F |
| U | I | N | J | E | X | T | A | N | F | S | I | W | D | E | T | J | N | A | D | T | L | A | M | V |
| N | O | D | B | R | B | P | L | M | E | E | P | N | M | H | L | L | H | M | V | H | E | S | E | E |
| P | N | I | F | C | T | R | Q | H | T | N | C | Y | S | I | B | T | D | E | N | D | R | I | T | E |
| K | I | J | E | O | V | D | U | U | M | E | S | F | N | H | C | W | M | R | X | T | I | W | G | V |
| A | R | T | U | D | K | G | I | M | B | X | C | L | J | S | E | M | G | C | E | A | H | T | R | R |
| M | H | O | D | L | O | X | A | V | T | I | Y | H | A | C | R | A | Y | I | H | G | J | D | O | E |
| D | I | E | N | C | E | P | H | A | L | O | N | I | G | H | W | N | T | O | K | F | E | N | Y | N |
| G | J | V | A | F | O | F | T | S | U | P | M | A | C | O | P | P | I | H | A | Z | X | L | A | L |
| W | A | H | E | U | B | R | O | G | I | E | K | O | N | P | N | T | S | C | L | E | R | B | S | I |

## Conditions

The following is a list of conditions. Take some time and write notes about each of the conditions. You may make the notes taken from text books or other resources, for example, people you work with in a clinical area or you may make the notes as a result of people you have cared for. If you are making notes about people you have cared for you must ensure that you adhere to the rules of confidentiality.

| Condition | Your notes |
|---|---|
| Epilepsy | |
| Multiple sclerosis | |
| Muscular dystrophy | |
| Cerebro-vascular accident | |
| Autism | |
| Spina bifida | |
| Poliomyelitis | |
| Cerebral palsy | |
| Dyspraxia | |



## Glossary

**Action potential:** conduction along a nerve or muscle cell membrane caused by a large, transient depolarization.

**Antidiuretic hormone (ADH):** hormone that acts on the kidneys to reabsorb more water, thus reducing urine output.

**Afferent fibres:** carry nerve impulses towards the central nervous system.

**Arachnoid mater:** middle layer of the meninges.

**Astrocyte:** neuroglial cell that helps for the blood–brain barrier.

**Autonomic nervous system:** involuntary motor division of the motor nervous system.

**Axon:** process of a neurone that carries impulses away from the cell body.

**Brainstem:** collective name given to the pons, medulla and midbrain.

**Cation:** an ion with a positive charge.

**Central nervous system:** brain and spinal cord.

**Cerebellum:** anatomical region of the brain responsible for coordinated and smooth skeletal muscle movements.

**Cerebral hemispheres:** division of the cerebrum.

**Cerebrospinal fluid:** fluid that surrounds the central nervous system.

**Cerebrum:** large anatomical region of the brain thst is divided into the cerebral hemispheres.

**Circle of Willis:** part of arterial blood supply to the brain.

**Cranial nerves:** 12 pairs of nerves that leave the brain and supply sensory and motor neurones to the head, neck, part of the trunk and the viscera of the thorax and abdomen.

**Dendrite:** part of neurone that transmits impulses towards the cell body.

**Diencephalon:** anatomical region of the brain consisting of the thalamus, hypothalamus and epithalamus.

**Dura mater:** tough outer layer of the meninges.

**Effector:** muscle, gland or organ stimulated by the nervous system.

**Efferent fibres:** carry nerve impulses away from the central nervous system.

**Ependymal cells:** neuroglial cells that line the cavities of the central nervous system.

**Epinephrine:** hormone produced by the adrenal medulla that is also a neurotransmitter.

**Epithalamus:** part of the brain that forms the diencephalon.

**Ganglia:** a group of neuronal cell bodies lying outside the central nervous system.

**Hypothalamus:** part of the diencephalon with many functions.

**Limbic system:** part of the brain involved in emotional responses.

**Lobe:** a clear anatomical division or boundary within a structure.

**Medulla oblongata:** part of the brainstem.



**Meninges:** three layers of tissue that cover and protect the central nervous system (dura, arachnoid and pia maters).

**Midbrain:** part of the brainstem that links the brainstem to the diencephalon.

**Microglia:** neuroglia that has the ability to phagocytose material.

**Motor area:** area located in the cerebral cortex that controls voluntary motor function.

**Motor nerves:** neurones that conduct impulses to effectors that may be either muscle or glands.

**Myelin sheath:** fatty insulating layer that surrounds nerve fibres responsible for speeding up impulse conduction.

**Neuroglia:** cells of the nervous system that protect and support the functional unit – the neurone.

**Neuromuscular junction:** region where skeletal muscle comes into contact with a neurone.

**Neurone:** functional unit of the nervous system responsible for generating and conducting nerve impulses.

**Nuclei:** cluster of cell bodies within the central nervous system.

**Oligodendrocytes:** glial cells that help produce the myelin sheath.

**Peripheral nervous system:** all nerves located outside of the brain and spinal cord (the central nervous system).

**Pia mater:** innermost layer of the meninges.

**Pineal gland:** part of the diencephalon that has an endocrine function.

**Pituitary gland:** an endocrine gland located next to the hypothalamus that produces many hormones.

**Receptor:** sensory nerve ending or cell that responds to stimuli.

**Refractory period:** the period immediately after a neurone has fired when it cannot receive another impulse.

**Reticular formation:** area located throughout the brainstem that is responsible for arousal, regulation of sensory input to the cerebrum and control of motor output.

**Saltatory conduction:** transmission of an impulse down a myelinated nerve fibre where the impulse moves from node of Ranvier to node.

**Sensory area:** area of the cerebrum responsible for sensation.

**Sensory nerves:** neurones that carry sensory information from cranial and spinal nerves into the brain and spinal cord.

**Somatic nervous system:** voluntary motor division of the peripheral nervous system.

**Spinal nerves:** 31 pairs of nerves that originate on the spinal cord.

**Synapse:** junction between two neurones or neurones and effector site.

**Thalamus:** part of the diencephalon.

**Ventricle:** cavity in the brain.

**White matter:** myelinated nerve fibres.

## References




Corns, R., Martin, A. (2012) Neurosurgery: hydrocephalus. *Surgery (Oxford)*, **30**, 142–148.

Labtests Online (2013) Bilirubin: the test. http://labtestsonline.org/understanding/analytes/bilirubin/tab/test (accessed 9 September 2014).

MacGregor, J. (2008) *Introduction to the Anatomy and Physiology of Children [Electronic Resource]: A Guide for Students of Nursing, Child Care, and Health*, 2nd edn, Routledge, Abingdon.

NICE (2010) NICE Guidelines (CG98). Neonatal Jaundice. http://www.nice.org.uk/guidance/cg98 (accessed 9 September 2014).

Pascual-Leone, A., Freitas, C., Oberman, L. *et al.* (2011) Characterizing brain cortical plasticity and network dynamics across the age-span in health and disease with TMS-EEG and TMS-fMRI. *Brain Topography*, **24** (3–4): 302–315.

Peate, I., Nair, M. (eds) (2011) *Fundamentals of Anatomy and Physiology for Student Nurses*, Wiley–Blackwell, Chichester.

Price, D.J., Jarman, A.P., Mason, J.O., Kind, P.C. (2011) *Building Brains [Electronic Resource]: An Introduction to Neural Development*, Wiley–Blackwell, Chichester.

Rawles, Z., Griffiths, B., Alexander, T. (2010) *Physical Examination Procedures for Advanced Nurses and Independent Prescribers*, Hodder Arnold, London.

Tortora, G.J., Derrickson, B.H. (2009) *Principles of Anatomy and Physiology*, 12th edn, John Wiley & Sons, Inc., Hoboken, NJ.

# Chapter 16

# The muscular system

Elizabeth Gormley-Fleming

*Learning and Teaching Institute, University of Hertfordshire, Hatfield, UK*
*Children's Nursing, School of Health and Social Work, University of Hertfordshire, Hatfield, UK*

## Aim

The aim of this chapter is to enable the reader to develop their understanding and knowledge of the musculature system of the body. This will include understanding the function of the different types of muscles, the location of the muscles and how muscles contract and relax.

## Learning outcomes

On completion of this chapter the reader will be able to:

- Describe the structure and function of the muscular system.
- Identify the main characteristics of the various muscle types.
- Describe the nature of muscle tone and how a muscle contracts.
- Name and locate the major muscles of the body.
- Identify the energy sources that muscles use.

## Test your prior knowledge

- What is the function of the muscular system?
- What are the energy sources required for muscle contraction?
- List the stages of muscle contraction.
- List the different types of muscle tissue in the human body.
- Identify three characteristics of a muscle.
- List the different types of anatomical movements in the human body.
- What types of contractions occur when skeletal muscle contracts?
- Identify four different muscles in the lower limbs.
- What are the difference between the muscles of male and females?
- What are the three types of muscles fibres in skeletal muscle?

*Fundamentals of Children's Anatomy and Physiology: A Textbook for Nursing and Healthcare Students*, First Edition. Edited by Ian Peate and Elizabeth Gormley-Fleming.

Companion website: www.wileyfundamentalseries.com/childrensA&P

# Introduction

All physical movement and functioning of the human body involves the action of muscles, be it walking, the contraction of ventricles of the heart or the voiding of urine. All movement that alters the position of the body occurs through the joints; hence, the musculature system cannot be considered in isolation but must be considered in conjunction with the skeletal system, which is discussed in Chapter 17. This chapter will identify the structure and functions of the musculature system.

# Muscle development in early life

Muscle development occurs very early in embryonic life. Muscles form from the myoblasts that have differentiated from the mesoderm, with the exceptions of the iris and the arrector pili muscle. These develop from the neuroectoderm. The mesoderm develops and is arranged in columns beside the developing nervous system. Following a process of segmentation, these columns then form blocks called somites. The lower body at the front of the somites contributes to the development of the cartilage, bone of the ribs and vertebral column. The posterior aspect of the somite contributes to the skeletal muscle development of the body and the limbs with the exception of the skeletal muscle of the head, which develops from the general mesoderm (Chamley *et al.*, 2005).



Cardiac muscle development occurs during the 3–4 weeks of fetal development and the first heart beat can be heard at this stage. At week 7, the neck and trunk muscles contract spontaneously and some arm and leg movement occurs at this stage. At 11 weeks of fetal development the fetus will swallow and may suck its thumb as its muscles have developed sufficiently to enable this (England, 1996). By 16 weeks of fetal development the mother is able to feel her baby move *in utero*; this is called quickening.

At birth the musculature system is not yet fully developed. The composition of muscles alters with maturity, and *in utero* the muscle fibres contain more water and intracellular matrix. Post birth, both of these structures reduce and a cytoplasm appears. The muscles increase in size due to their diameter increasing along with an increase in the length and width. Muscle growth occurs in tandem with bone growth. Maximum muscle strength is achieved at approximately 25 years of age and it declines after this age.

# Types of muscle tissue

There are three types of muscle tissue in the human body, classified according to location, structure and nerve supply. They are:

- smooth muscle
- cardiac muscle
- skeletal muscle.

## Smooth muscle

Smooth muscle contains small, thin spindle-shaped cells of variable size that have one centrally located nucleus and is arranged in parallel lines. It is found in sheets in the blood vessels and hollow internal organs, such as the oesophagus, urinary bladder, reproductive system and respiratory tract. It is non-striated, so does not have a striped appearance (Figure 16.1).

It is controlled by the medulla oblongata, so movement is involuntarily. This muscle type has the intrinsic ability to contract and relax. When located in the longitudinal layer, the muscle

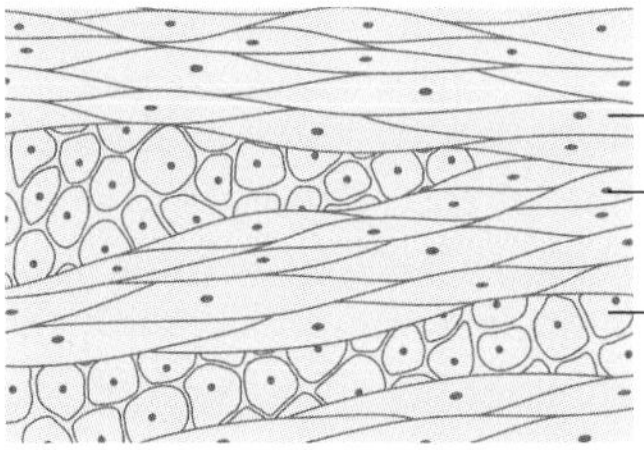

Muscle cell
Nucleus
Smooth muscle cells in cross-section

**Figure 16.1** Smooth muscle – cross-section.

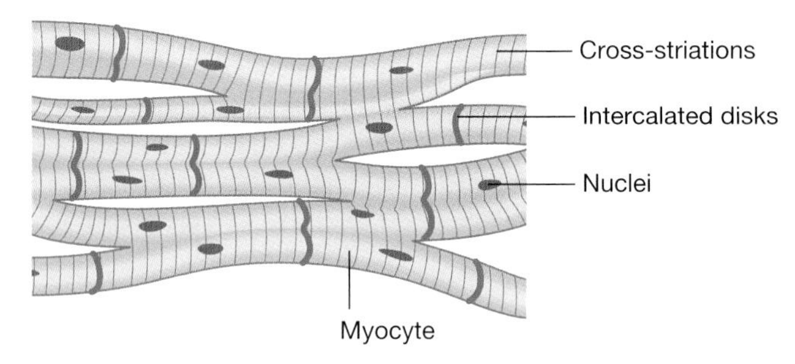




**Figure 16.2** Cardiac muscle.

fibres run parallel with the long axis of the organ, causing the organ to dilate and shorten when the muscle contracts. When located in the circular layer, the muscle fibres run around the circumference of an organ causing the organ to elongate when the muscle contracts as the lumen of the organ or vessel constricts (Marieb and Hoehn, 2010).

Contractions are stimulated by the autonomic nerve impulses, some hormones and local metabolites, although a degree of muscle tone is always present. This allows the smooth muscles to relax for only short periods of time.

## Cardiac muscle

This is found exclusively in the wall of the heart, the myocardium. It is thicker to allow contraction of the heart moving the blood around the body. Cardiac muscle has a single nucleus (Figure 16.2). It is an involuntary responsive muscle, is quadrangular in shape and striated. It develops from the splanchnic mesenchyme which surrounds the heart tube and is recognizable by week 3–4 of embryonic development, when it commences pumping blood.

As this muscle continues to develop, specialized fibre bundles develop with fewer myofibrils in the final stages of embryonic development. These are the Purkinje fibres, which are an essential component of the conducting circuit of the heart. Intercalated discs are present in cardiac muscle fibres only. The muscle cells lie end to end, and it is these discs that lie at the junction of the cells. Their function is to allow the spread of electrical activity through the cardiac muscle. Cardiac muscle has some ability to regenerate and also has the ability to thicken and grow as the child develops. Cardiac muscle is discussed in more detail in Chapter 9.

## Skeletal muscle

Composed of over 600 muscles, the skeletal muscles are the only voluntary muscles of the body. Skeletal muscle is mostly developed before birth and completely developed by the end of the first year of life. A discrete organ, skeletal muscle is made of various different tissue types. Unlike the other muscle types, skeletal muscle cannot contract on its own, so each skeletal muscle fibre

**Table 16.1** Summary of the muscle tissue. Adapted from Peate and Nair (2011).

| Muscle type | Cell types | When is it found? | Type of control |
|---|---|---|---|
| Smooth muscle | Non-striated, single nucleus, rod | Hollow organs, e.g. bladder, oesophagus | Involuntary |
| Skeletal muscle | Striated. single long cylindrical cells | Attached to and covering bones | Voluntary |
| Cardiac muscle | Striated, single nucleus | Wall of the heart | Involuntary |

**Table 16.2** Principal movement functions of muscles. Adapted from Kingston (2005).

| Functional name | Definition |
|---|---|
| Flexor | Moves anterior surfaces closer |
| Extensor | Moves posterior surfaces closer |
| Abductor | Moves body part away from midline |
| Adductor | Moves body part towards midline |
| Rotator | Rotates the body part around longitudinal axis |
| Supinator | Turns palm of hand anteriorly |
| Pronator | Turns palm of hand posteriorly |
| Sphincter | Allows orifice to open |
| Levator | Upward movement |
| Depressor | Downwards movement |
| Tensor | Produces a degree of tension |



is supplied with a nerve ending that controls its activity (Marieb and Hoehn, 2010). These are the muscles involved in moving bones and generating external movement. Skeletal muscles are cylindrically shaped striated fibres that lie parallel to each other, and it is this that gives them their striated appearance.

Skeletal muscle is underdeveloped in the extremely premature neonate, hence why its posture is hypotonic. At 30 weeks' gestation the neonate will have flexion of its feet and knees, and by 35 weeks gestation flexion has extended to the upper limbs as well as the hips and thighs. At term the infant has the ability to fully flex all of its limbs with immediate recoil (Crawford and Hickson, 2002).

A summary of the various muscle tissues is given in Table 16.1.

# Function of musculature system

There are four main functions of the musculature system. These are to:

1. Maintain posture and tone – by adjustment to the skeletal muscles.
2. Allow movement, when contraction of the muscles pulls on the tendons of the bones (Table 16.2).

3. Joint stabilization – muscle tendons reinforce the joints to allow free movement and frequently cross over where major joints are concerned to provide stronger re-enforcement of the joint (i.e. knee/shoulder).
4. Heat generation – muscles generate heat as they contract. The cells produce adenosine triphosphate (ATP), giving the muscles the energy to contract. This is more common in skeletal muscles.

In addition to the key functions listed above there are other important functions, and these include:

1. Protection of internal organs – especially the abdomen, where layers of muscle fibres protect the visceral organs.
2. Cardiac movement, and elevation of blood pressure in stress response.
3. Aiding digestion by peristalsis and the movement of waste from the body.
4. Regulating the passage of fluids and substances through internal body openings and from the body.
5. Acting as a shock absorber, thus protecting the internal organs.
6. Enabling facial expression.



## Clinical application

### Muscular dystrophy

Muscular dystrophies consist of the largest group of muscle diseases in children. It is genetic in origin. There is a gradual degeneration of the muscle fibres. This leads to weakness and wasting of the skeletal muscle, with increasing deformities and disability. The most common form is Duchenne muscular dystrophy (DMD). DMD is an X-linked recessive trait located on the short arm of the X chromosome. There is a mutation of the gene that encodes dystrophin, which is a protein found in skeletal muscle. Males are almost exclusively affected. The incidences are 1 per 3600 male births.

DMD is characterized by:

- early onset – usually 3–5 years of age;
- progressive muscle weakness and wasting;
- contractures of the joints;
- calf muscle hypertrophy;
- loss of independent ambulation by 9–11 years of age;
- generalized weakness by early teenage years.

Learning difficulty is present in 25–30% of patients with DMD. This disease progresses relentlessly until death, which usually occurs from respiratory or cardiac failure. Diagnosis is possible antenatally. DNA analysis establishes diagnosis from either blood or muscle sample obtained from muscle biopsy. Electromyography studies may also be used as part of the evaluative process.

No effective treatment currently exists. Maintenance of optimal function in all muscles for as long as possible is the primary goal of therapy. This includes stretching and strengthening exercises, breathing exercises and a range of motion exercises. Surgery may be required to release contractures. As the disease progresses, non-invasive intermittent positive-pressure ventilation may be required in the home setting.

Genetic counselling is recommended for the parents and female siblings.

Myoblast transfer from the unaffected father has occurred (Sanat, 2004). Treatment is still being researched.

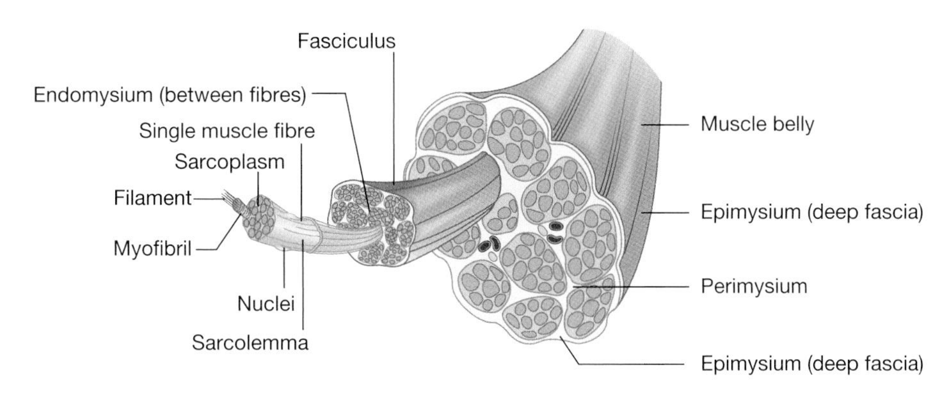


**Figure 16.3** The gross anatomy of striated skeletal muscle.

# Gross anatomy of skeletal muscle

Each skeletal muscle is a discrete organ that is composed of several different types of tissue. Blood vessels, nerve fibres and connective tissue are also present along with the skeletal muscle fibres. The basic structure of each muscle consists of a bundle of muscle fibres that are bound closely together by connective tissue, and the resulting bundle is known as the muscle belly (Kingston, 2005). The architecture of a muscle may change, but the basic structure remains constant (Figure 16.3).



The individual muscle fibres are held together by many layers of connective tissue sheaths. These sheaths support each cell, and this serves to reinforce the muscle as a whole, especially during strong contraction of the muscle, as otherwise it may burst. These connective sheaths are continuous with one another and also with the tendons and bones.

These connective tissue sheaths are the:

- epimysium – outermost sheath;
- perimysium and fascicles;
- endomysium.

The epimysium is composed of dense irregular tissue that surrounds the whole of the muscle. It may intermingle with the fascia of adjoining muscles or to the superficial fascia.

The muscle fibres of each skeletal muscle are grouped into fascicles. These resemble bundles of twigs. Each fascicle is surrounded by a layer of connective tissue called the perimysium.

The endomysium is a thin sheath of fine areolar connective tissue that surrounds each individual muscle fibre. The collagen fibres of the perimysium and the endomysium are interwoven and blend into one another. At the end of the muscle the endomysium, perimysium and the epimysium come together to form a bundle. This is called a tendon, or if it is a broad sheet bundle it is called an aponeurosis. These attach skeletal muscle to bone (Martini *et al.*, 2012).

As the muscle fibres contract, they pull on the sheath, in turn transmitting the force to move the bone.

## Blood and nerve supply

Generally, each muscle is served by one artery, one or more veins and one nerve. The endomysium and the perimysium contain the blood vessels and nerves that supply the muscle fibres. The blood vessels and nerves enter the muscle together and follow the same branching course through to the perimysium.

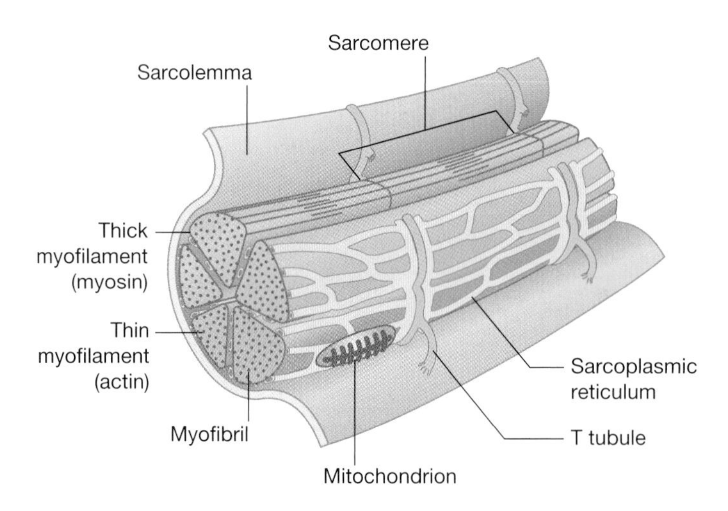


**Figure 16.4** Micro-anatomy of skeletal muscle fibre.



Skeletal muscle has a rich blood supply as they have large energy requirements, thus demanding a continual oxygen supply. As a result of this there is a corresponding volume of waste that requires transportation in the venous system away from the muscle in order to maintain healthy and efficient muscle contractions.

Skeletal muscle is under mostly voluntary control, but some are controlled at a subconscious level, such as the skeletal muscles involved in breathing. Nerve fibres enter the epimysium and then branch through the perimysium and then enter the endomysium to innervate each muscle fibre.

# The micro-anatomy of the muscle

Microscopically, each skeletal muscle fibre is a long cylindrical cell and is different from the typical cells described in Chapter 4. These are extremely large cells and have multiple oval nuclei (Figure 16.4).

During myogenesis it is the embryonic muscle cells, myoblasts, that fuse to form these multinucleated muscle cells or muscle fibres. Myofibrils then appear in the cytoplasm. Most of the skeletal muscle is developed by birth. It is the composition of the muscle fibres that alters during development, as during fetal development the muscle fibres contains mostly water and intracellular matrix. At birth, the cells grow in size and the water and intracelluar matrix are reduced. The diameter of the muscle fibrils remains constant; it is their length that increases.

It is the number of muscle fibres that varies between girls and boys from birth to maturity. Boys will have a 14-fold increase, whereas girls will have a 10-fold increase.

The muscle fibres of a female child will achieve their maximum diameter at the age of 10 years, whereas a male will not achieve this until 14 years of age. A summary of the functional components and organization of the skeletal muscle fibres is presented in Table 16.3.

## The sarcolemma and transverse tubes

The plasma membrane of the muscle fibre is the sarcolemma, and this surrounds the sarcoplasm. The surface of the sarcolemma has multiple openings, and these form a network of

**Table 16.3** The organization and function of the components of skeletal muscle fibres.

| Component | Function |
|---|---|
| Sarcolemma | Plasma membrane of muscle fibre |
| Transverse tubules (T-tubules) | Narrow tubes that are continuous with the sarcolemma. Filled with extracellular fluid, these T-tubules conduct electrical impulses into the cell interior |
| Sarcoplasm | The cytoplasm of the muscle fibre. Contains myofibrils |
| Myofibrils | Key role in muscle contraction |
| Myoglobin | A red pigment that stores oxygen |
| Glycosomes | Granules of stored glycogen that provide glucose during muscle cell activity |
| Myofilament | Bundles of myofibrils – two types: thick and thin, containing proteins that give the striated appearance to the muscle tissue. Role in muscle contraction |
| Sarcoplasmic reticulum | Stores calcium ions |
| Sarcomere | Smallest functioning unit of the muscle fibre and is responsible for muscle contraction |



tubules called transverse tubules or T-tubules. These are continuous with the sarcolemma and extend into the sarcoplasm. Filled with extracellular fluid, the T-tubules form a network of passages through the muscle fibre. Electrical impulses conducted by the sarcolemma travel to the T-tubule into the interior aspect of the cell, thus triggering muscle fibres to contract.

## The sarcoplasm

This sarcoplasm consists of large number of mitochondria. From here, ATP is produced during muscle contraction (Peate and Nair, 2011). The membrane complex in the muscle fibre is called the sarcoplasmic reticulum (SR). This forms a tubular network around each myofibril. The T-tubule surrounding the myofibril is tightly bound to the membranes of the SR. The other aspect of the T-tubule, the tubules of the SR, enlarge, fuse together and form the chambers called the terminal cisternae. Two terminal cisternae and the T-tubule form a triad.

Muscle fibres pump calcium ions from the cell via its plasma membrane. Calcium ions are also removed by actively transporting them from the sarcoplasm into the terminal cisternae of the SR. Stored calcium ions are released into the sarcoplasm when a muscle contraction commences. Glycosomes (stored glycogen granules) and myoglobin (a red pigment that stores oxygen) are contained within the sarcoplasm.

## Myofibrils

Myofibrils are rod-like structures that run parallel to each muscle fibre. The contracting element of the muscle fibre, the myofibrils appear in the sarcoplasm of the muscle cell; the cross-striations develop from these, thus forming striated muscle (Chamley *et al.*, 2005). These myofibrils are densely packed together,and the mitochondria and other organelles have to squeeze past them. The myofibrils play a substantial role in muscle contraction. The types of protein filaments that are present in the myofibrils are myosin (which is a thick filament) and actin (which is the thin filament). Two other proteins, troponin and tropomyosin, are also present.

## The sarcomeres

Myofibrils are bundles of thick and thin myofilaments, and these myofilaments are organized into functional units called sarcomeres, and there are approximately 10 000 sarcomeres in a myofibril (Tortora and Derrickson, 2010). Z-like discs separate the sarcomeres. The sarcomeres are the smallest functioning unit of a muscle fibre, and it is the interaction between the thick and thin filaments that is responsible for muscle contraction (Martini *et al.*, 2012). A sarcomere contains a thick filament, a thin filament and proteins. The protein stabilizes the positions of the filaments and regulates the interactions between the filaments.

It is the distribution of the thick and thin filaments that gives each myofibril its banded appearance. Each sarcomere has A bands (which are dark) and I bands (which are light).

# Types of muscle fibres

There are three types of muscle fibres in skeletal muscle:



1. fast oxidative–glycolytic (FOG) fibres;
2. slow oxidative (SO) fibres;
3. intermediate fibres.

## Fast oxidative–glycolytic fibres

Most of the skeletal muscle fibres in the body are FOG fibres. These fibres are red to pink in colour and have a very fast speed of contraction. The myoglobin content is high and the glycogen stores are of intermediate level. These FOG fibres are moderately fatigue resistant. They have an abundance of mitochondria and capillaries. They are able to generate ATP by aerobic respiration, and ATP is also generated by anaerobic glycolysis due to the high glycogen content. These fibres are also referred to as type 2 fibres.

## Slow oxidative fibres

These small red fibres have many mitochondria and capillaries and have a high myoglobin content. They are also referred to as type 1 fibres. Their primary pathway for ATP synthesis is aerobic respiration. They have a low glycogen store. These are fatigue-resistant fibres and produce slow, prolonged contractions.

In the young child, particularly with regard to the muscles of breathing – the intercostal and the diaphragm – they have fewer type 1 muscle fibres. So any factors that contribute to the work of breathing will impact significant on the infant's ability to sustain effective respirations (Goldsmith, 2003). The infant has approximately 25% type 1 fibres compared with the 50% type 1 fibres of an adult. They do not achieve the adult configuration of type 1 fibres until approximately 2 years of age.

## Fast glycolytic or intermediate fibres

These are large, pale fibres due to low levels of myoglobin. They have an intermediate capillary network and they generate APT mainly by anaerobic glycolysis. As they have few mitochondria, these fibres fatigue quickly and are mostly used for short-term powerful actions.

## Clinical application

Children under the age of 2 react more acutely to respiratory tract infections than older children do. This is primarily due to the anatomical difference of this age group. The tongue is large, the trachea has incomplete rings of cartilage and is shorter, the epiglottis lies at the level of C3–4 and is omega shaped, the larynx is funnel shaped, the narrowest point is the sub-glottic area, the ribs are horizontal and the chest wall is compliant and small. They also have a small force residual capacity. There is a large amount of type 1 fibres. The accessory muscle contributes less to the work of breathing in young children. They have functional diaphragmatic breathing. Because of their high metabolic rate, the work of breathing can account for 40% of their cardiac output. In the presence of respiratory disease, airway resistance is increased significantly in children. This increases the work of breathing.

The signs of increased work of breathing are visible by close inspection of the child's face, neck and chest wall.

These signs include:

- nasal flaring
- tracheal tug
- subcostal and intercostal recession
- sub-sternal recession
- head bobbing
- pursing of the lips.

Early recognition, accurate and rapid assessment of the airway and breathing and prompt treatment are essential if positive outcomes are to be achieved in the child who has ineffective breathing.



# Skeletal muscle relaxation and contraction

The ability of skeletal muscle to contract and to relax is under the control of the pyramidal and extra-pyramidal tract systems.

## Contraction

The nerve cells that activate skeletal muscle are called somatic motor neurones. Located in the brain or spinal cord, theses motor neurones have long thread-like extensions called axons that travel to the muscle cell they serve (Marieb and Hoehn, 2010). These neurones divide as they enter the muscle and form several curly branches. These then collectively form a motor end plate. The name given to this area where the synapse occurs is the neuromuscular junction, and this is located approximately halfway along the muscle fibre (Figure 16.5). Each muscle fibre has one neuromuscular junction. The axon terminal and the muscle fibre are separated by a gap called the synaptic cleft. There are small sac-like structures within the axon terminals called synaptic vessels that contain acetylcholine (ACh). There are millions of ACh receptors located in the sarcolemma of the muscle fibre. This is a gap where the neurotransmitter ACh is active and passes the impulse from the neurone to the muscle cell.

The nerve impulse reaches the end of the axon and the axon terminal releases ACh into the synaptic cleft. This comes in contact with the ACh receptors of the sarcolemma of the muscle fibre. The conduction action is spread down to the T-tubules into the cisternae which encircle

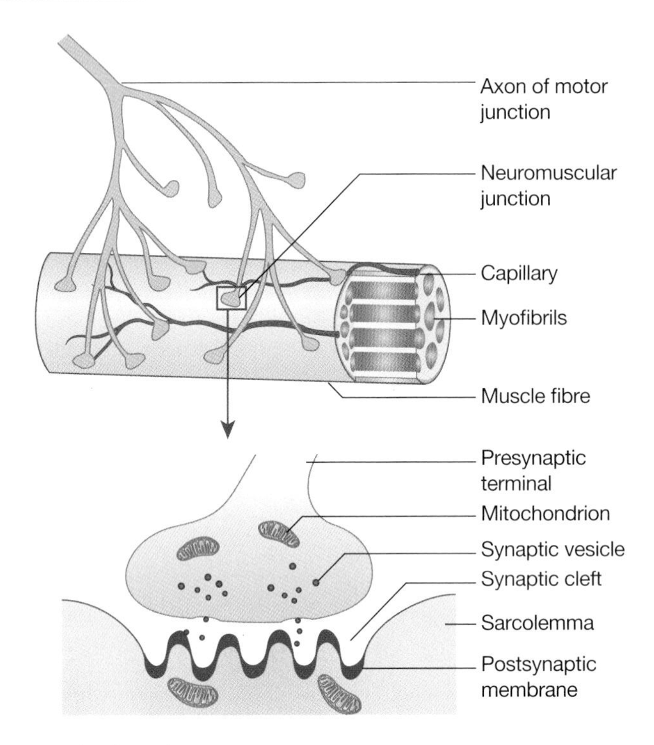


**Figure 16.5** Neuromuscular junction.



the sarcomeres of the muscle fibres. Calcium ions are then released. This triggers an electrical event that causes the muscle to contract. The length of the contraction is dependent on three factors:

- the period of stimulation of the neuromuscular junction;
- the presence of calcium ions in the sarcoplasm;
- the availability of ATP.

## Relaxation

Once the ACh binds with the ACh receptors, the actions of acetyl cholinesterase (AchE) – an enzyme located in the synaptic cleft – breaks down and terminates the action of ACh. The concentration of calcium ions in the sarcoplasm reduces and returns to normal resting levels.

The muscle contraction now ceases and it returns to a relaxed state.

## Clinical application

Myasthenia gravis (MG) is an autoimmune disease. It is a disorder of the neuromuscular junction. ACh is unable to bind with the ACh receptors properly and the muscle cells are unable to sustain repeated contractions during any sustained period of exercise so subsequently tire very quickly.

The receptors are blocked by antibodies, and there is also a reduced amount of ACh available.

There are three types of MG that may affect children:

- Neonatal – this is transient due to the passage of maternal antibodies from the mother with MG to the unborn baby. The symptoms commence usually within 2 days of birth and last for a few weeks.
- Congenital childhood MG – this is not due to an autoimmune defect but hereditary factors. There is a genetic mutation that causes synaptic malformation.
- Juvenile MG – may be diagnosed any stage after the first month of life.

MG is often difficult to diagnose, but a hallmark sign is increasing muscle fatigue during periods of exercise but a return to normal after periods of rest. Involvement of the eye muscles, muscles of chewing and facial expression are often involved. Other muscles, such as those of breathing, may also be involved, necessitating ventilator support. Diagnosis is confirmed by physical examination, blood test for antibodies against ACh receptors and electromyography.

### Treatment

The thymus gland maybe responsible for the production of the antibodies, so surgical removal of this is a potential option as part of the treatment. AchE inhibitors and immunosuppressant medication are commonly used.

Education about the undertaking of physical activity is essential to maintain health.



# Energy requirements for muscle contraction

In order for muscle fibres to contract, energy is required. Initially, this is provided as ATP and is stored in the muscle fibres. The primary function of ATP is to transfer energy from one area to another. The demand for ATP is high when the muscle fibre is contracting, and only a small amount of this is stored in the cell. During the muscle contraction, ATP is generated at the same rate as it is utilized, so it becomes depleted quickly when the muscle is in use. Subsequently, additional pathways are required to produce energy. These are the ATP and creatine phosphate (CP) pathway and anaerobic respiration.

## ATP and creatine phosphate pathway

During rest, the skeletal muscle fibre produces more ATP than it requires. ATP transfers energy to creatine, which is a small molecule that the muscle cells assemble from amino acids. The energy transfer creates CP. When the muscle contracts, myosin breaks down ATP producing adenosine diphosphate (ADP) and phosphate. The energy stored in CP is then used to recharge ADP, converting it back to ATP. The creatine that is not used is excreted via the kidneys.

## Anaerobic respiration

There are reserves of glycogen in the sarcoplasm, and typically skeletal muscle contains large amounts of glycogen. During glycolysis, glucose is broken down into pyruvate. When the muscle fibres begin to run out of ATP and CP, the glycogen molecules are split by enzymes, thus releasing glucose, and this is then used to generate more ATP.

When activity is intense, ATP demands are excessive and ATP production is maximized from the mitochondria, but they can only produce a third of the required amount of ATP. The remainder comes from glycolysis; subsequently, pyruvate builds up in the sarcoplasm. This is converted to lactic acid. This process enables the cell to generate sufficient ATP during periods of intense activity.

## Aerobic respiration

Aerobic respiration accounts for 95% of the ATP demands of a resting cell. This occurs in the mitochondria. The mitochondria absorb oxygen, pyruvate, ADP and phosphate ions from the surrounding cytoplasm. These then enter the Krebs cycle (Chapter 2). A large amount of energy is released, and this is used to make ATP. Glycolysis occurs during aerobic respiration, and from the reactions occurring in the mitochondria glucose is broken down to yield water, carbon dioxide and large amounts of ATP (Marieb and Hoehn, 2010):

glucose + oxygen → carbon dioxide + water + ATP

Carbon dioxide diffuses from the muscle tissues into the circulation where it is then excreted from the body via the lungs.

## Oxygen deficit

Oxygen deficit is defined as the extra amount of oxygen required by the body in order to restore the muscle chemistry to its normalized state. For a muscle to return to its resting state it must replenish its oxygen reserves, remove accumulated lactic acid by reconverting it to pyruvic acid, replace the glycogen stores, and resynthesize ATP and CP reserves (Marieb and Hoehn, 2010). The liver will convert lactic acid to glucose or glycogen. It is during muscle contraction that all of the activities requiring oxygen occur more slowly or are reduced until sufficient oxygen is available again; hence the term oxygen deficit.



## Muscle fatigue

This is defined as when an active skeletal muscle can no longer perform at the required level despite neural stimulation. It is a cumulative process. Normal muscle function requires four conditions: substantial intracellular energy reserves, normal circulation, normal oxygen levels and normal blood pH. An interference with any of these factors will promote muscle fatigue.

# Organization of skeletal muscle

Skeletal muscle is attached to bone and connective tissue at a minimum of two points: a fixed end called the origin and the site where the movable end attaches to another structure called the insertion (Figure 16.6). The origin is proximal to the insertion.

Skeletal muscle may be divided in to four areas:

- head and neck muscles;
- muscles of the upper limb – shoulder, arm and forearm;

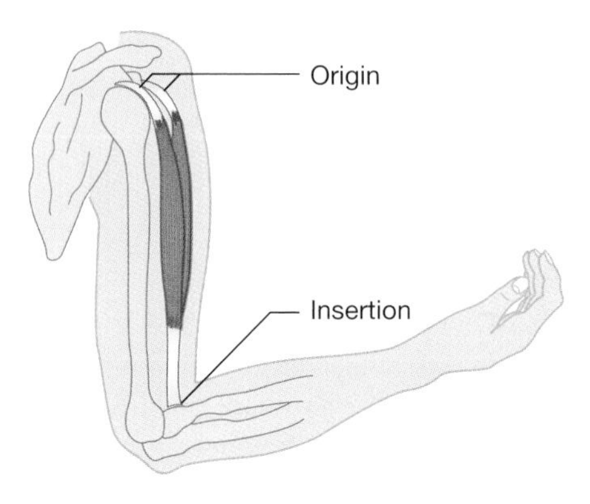


**Figure 16.6** Origin and insertion.

- trunk – abdomen and thorax;
- muscles of the lower limbs – pelvis, hip, leg and foot.

At birth, a newborn's movements are uncoordinated and mostly reflexive. Development is head to toe, and this reflects muscle development, which in turn reflects the level of neuromuscular coordination. Head lifting will occur before walking, for example. There are considerations for clinical practice; for example, intramuscular injection administration. These will be administered in the vastus lateralis as opposed to the gluteal maximus or deltoid; this is because the gluteal maximus and deltoid lack bulk in the young child.

During childhood, skeletal muscle control becomes more sophisticated, and this peaks during adolescence. Hormonal changes influence muscle growth during adolescence, with boys and girls taking a different growth pathway (Neu *et al.*, 2001). Muscle mass and bone mass are closely associated, and during growth spurts there is a correlation between bone mass and muscle mass development as the skeleton needs to continually adapt its strength to the increased load – muscle development drives bone development (Rauch *et al.*, 2004).

## The muscles of the head and neck

The muscles of the head and neck (Figures 16.7, 16.8 and 16.9) can be subdivided into different functioning groups:



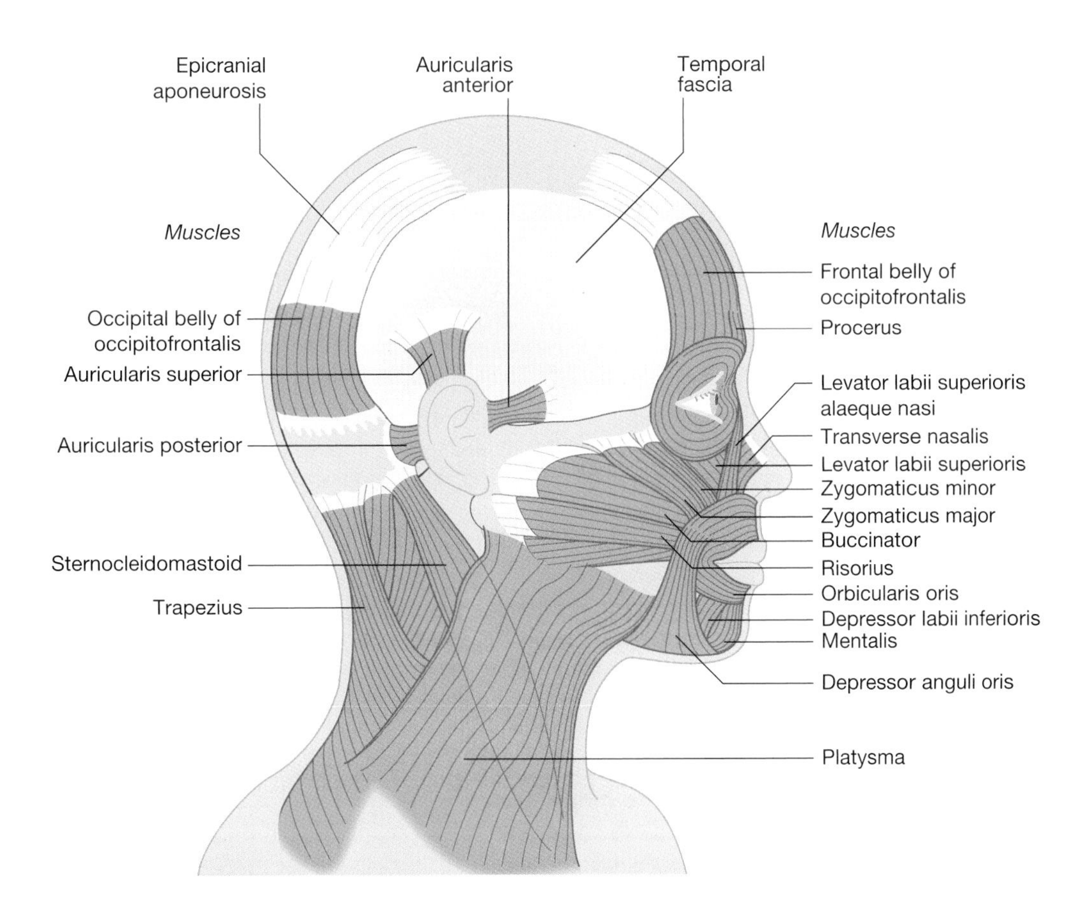


**Figure 16.7** Lateral view of the muscles of facial expression.

- the muscles of expressions (Table 16.4);
- the muscles mastication – chewing (Table 16.5);
- the muscles of the tongue (Table 16.5);
- the muscles of the neck (Table 16.6);
- intrinsic eye muscles (Chapter 19).

## The muscles of facial expression

The muscles of facial expression are shown in Figures 16.7 and 16.8, and their origin, insertion and action are listed in Table 16.4.



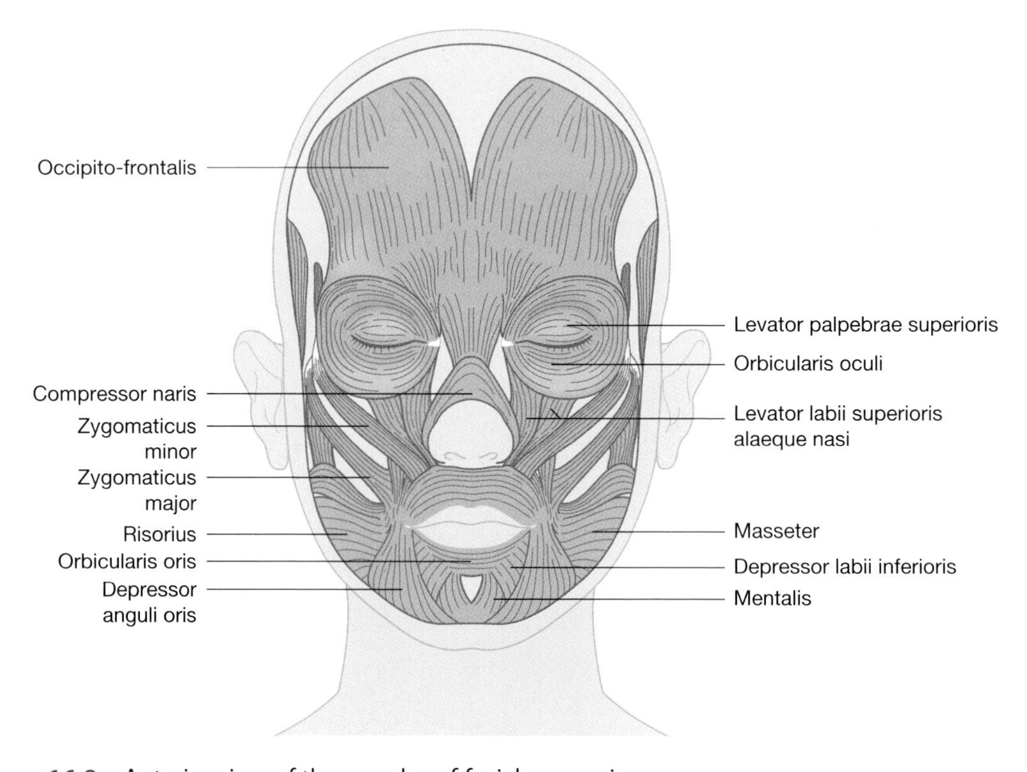


**Figure 16.8** Anterior view of the muscles of facial expression.

**Table 16.4** Muscles of facial expression. Adapted from Martini *et al.*, (2012).

| Muscle | Origin | Insertion | Action |
|---|---|---|---|
| *Mouth* | | | |
| Buccinator | Mandible and maxilla | Orbicularis oris | Compresses cheek |
| Depressor labii inferioris | Mandible | Lower lip | Depresses lower lip |
| Levator labii superioris | Infra orbital foramen | Obicularis oris | Elevates upper lip |

Table 16.4 (*Continued*)

| Muscle | Origin | Insertion | Action |
|---|---|---|---|
| Levator anguli oris | Maxilla | Corner of the mouth | Moves corner of mouth upwards |
| Mentalis | Mandible | Skin of the chin | Elevation and protrusion of lower lip |
| Obicularis oris | Maxilla and mandible | Lips | Purses lips |
| Risorius | Fascia of parotid gland | Angle of mouth | Brings corner of mouth toward the side |
| Depressor anguli oris | Anterior surface of mandibular body | Skin at angle of mouth | Depresses corner of mouth |
| Zygomaticus major | Zygomatic bone | Angle of mouth | Elevates and retracts corner of mouth |
| Zygomaticus minor | Zygomatic bone | Upper lip | Elevates and retracts upper lip |
| *Eye* | | | |
| Corrugator supercilli | Orbital rim | Eyebrow | Wrinkles brow |
| Levator palpebrae superioris | Tendinous band around optic foramen | Upper eyelid | Elevates upper eyelid |
| Orbicularis oculi | Margin of orbit | Skin around eyelid | Closes eye |
| *Nose* | | | |
| Procerus | Nasal bone | Aponeurosis of bridge of nose and skin of forehead | Moves nose and alters shape of nostrils |
| Nasalis | Maxilla and nasal cartilage | Bridge of nose | Elevates corner of nostrils, depresses tip of nose |
| *Ear* | | | |
| Temporoparetalis | Fascia around external ear | Epicranial aponeurosis | Moves auricle of ear |
| *Scalp* | | | |
| Occiptofrontalis frontal belly | Epicranial aponeurosis | Skin of eyebrow and bridge of nose | Raises eyebrow |
| Occipital belly | Occipital bone and mastoid bone | Epicranial aponeurosis | Tenses scalp |
| *Neck* | | | |
| Platysma | Acromion of scapula and second rib | Mandible | Tenses skin of neck |

**Table 16.5** The muscles of chewing and of the tongue. Adapted from Martini *et al.*, (2012).

| Muscle | Origin | Insertion | Action |
|---|---|---|---|
| Masseter | Zygomatic arch | Mandibular ramus | Elevates mandible and closes jaw |
| Temporalis | Temporal line of skull | Coronoid process of mandible | Elevates mandible |
| Pterygoids | Pterygoid plate | Mandibular ramus | Closes jaw, moves mandible from side to side |
| *Tongue* | | | |
| Genioglossus | Mandible | Hyoid bone | Depresses and protracts tongue |
| Hypoglossus | Hyoid bone | Side of tongue | Depresses and retracts tongue |
| Palatoglossus | Soft palate | Side of tongue | Elevates tongue, depresses soft palate |
| Styloglossus | Styloid process of temporal bone | Side, tip and base of tongue | Retracts tongue |



## Muscle of chewing and of the tongue

At birth the tongue is large if it is to be compared with that of an adult's tongue. As the infant develops and grows, the muscles of the tongue (Table 16.5) play a complex role in the development of speech and in the preparation of food for swallowing. Infants can use their tongues to imitate adult behaviour from a very early age, such as poking it out in response to imitating their carer.

## The muscles of the pharynx and neck

The muscles of the pharynx are responsible for initiating the swallowing mechanism. They move food into the oesophagus. The muscles of the neck include the muscles that control the position of the larynx, muscles that give foundation to the floor of the mouth and muscles that provide stability for the muscles of the tongue and pharynx (Figure 16.9, Table 16.6). In the young child some of these muscles may be used to enhance the work of breathing in times of respiratory distress, particularly the sternocleidomastoid.

## The muscles of the shoulder

The muscles of the shoulder function to provide movement, protection and support, as this is a vulnerable joint owing to the shallow gleno-humeral joint and its exposure to impact injuries. The shoulder joint consists of more than one articulation: the main joint is the ball-and-socket joint, the clavicle attaches medially to the sternum and the scapula further increases the movement of the shoulder by sliding over the rib cage. The individual function, origin and insertion are identified in Table 16.7.

The muscles are the

- superficial muscles (Figure 16.10);
- deeper musculo-tendinous rotator cuff (Figure 16.11).

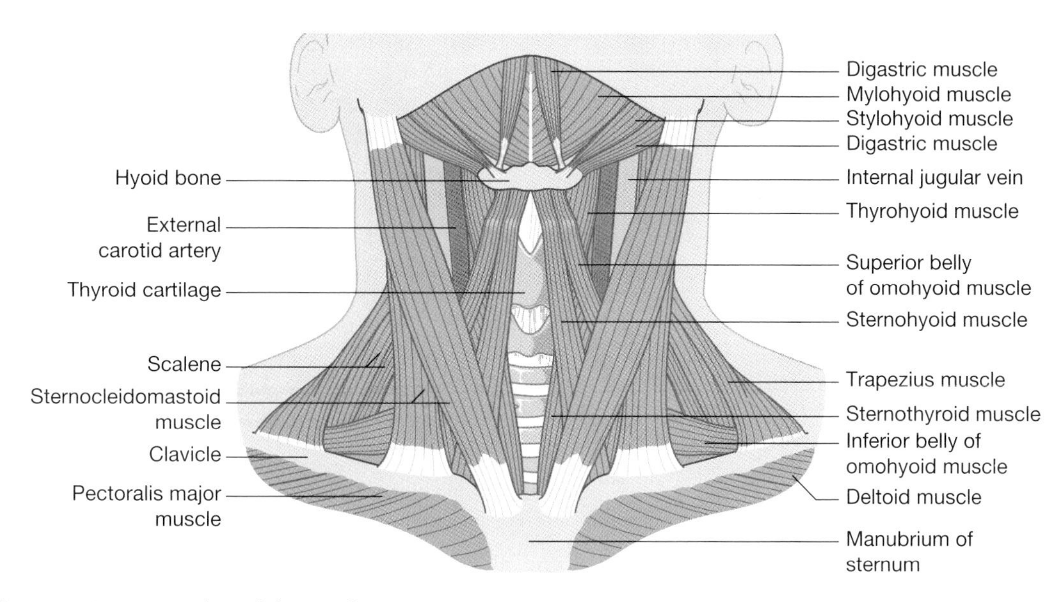


**Figure 16.9** Muscles of the neck.



**Table 16.6** Muscles of the neck – function, origin and insertion. Adapted from Kingston (2005) and Jarman (2003).

| Muscle | Origin | Insertion | Action |
|---|---|---|---|
| Digastric | Mandible and mastoid | Hyoid bone | Depresses mandible and elevates larynx |
| Geniohyoid | Mandible | Hyoid bone | Depresses mandible and elevates larynx, anterior movement of hyoid bone |
| Mylohyoid | Mandible | Hyoid bone | Elevates floor of mouth and hyoid bone, depresses mandible |
| Omohyoid | Scapula notch | Hyoid bone | Depresses hyoid bone and larynx |
| Sternohyoid | Clavicle and manubrium | Hyoid bone | Depresses hyoid bone and larynx |
| Sternothyroid | Manubrium | Thyroid cartilage of larynx | Depresses hyoid bone and larynx |
| Stylohyoid | Temporal bone | Hyoid bone | Elevates larynx |
| Thyrohyoid | Thyroid cartilage of larynx | Hyoid bone | Elevates thyroid |
| Sternocleidomastoid | Sternal end of clavicle and manubrium | Mastoid, superior nuchal line | Flex neck towards shoulder |

**Table 16.7** Muscles of the shoulder – function, origin and insertion. Adapted from Jarman (2003) and Kingston (2005).

| Muscle | Origin | Insertion | Action |
|---|---|---|---|
| Pectoralis major | Clavicle, sternum, costal cartilage and humerus | Greater tubercle and groove of humerus | Flexion and extension of shoulder. Respiratory function |
| Deltoid | Clavicle, scapula | Deltoid tuberosity | Abducts, flexion, medial and lateral rotation and extension of shoulder |
| Latissimus dorsi | Spinous process of inferior thoracic and all lumbar vertebrae, ribs 8–12 | Groove of humerus | Extension, adduction and medial rotation of shoulder |
| Serratus anterior | Anterior seven superior margins of ribs 1–9 | Scapula | Protracts shoulder |
| Biceps brachii | Scapulas | Radial tuberosity | Flexion and supination of shoulder. Flexion of elbow |
| Bicipital aponeurosis | | | |
| Brachialis | Humerus and ulna | Ulna | Flexion of forearm |
| Triceps brachii | Scapula | Ulna | Extension of forearm |
| Rhomboideus major and minor | Superior thoracic verterbrae C7–T1 | Scapula | Adducts scapula and downward rotation |
| *Rotator cuff muscles* | | | |
| Subscapularis | Anterior surface of scapula | Lesser tuberosity of humerus | Medial rotation of shoulder. |
| Supraspinatus | Scapula | Greater tuberosity of humerus | Abduction of deltoid. |
| Infraspinatus | Scapula | Greater tuberosity of humerus | Lateral rotation of shoulder |
| Teres minor | Lateral border of scapula | Bicipital groove of humerus | Extension, medial rotation and abduction of shoulder |



## The muscles of the upper arm, forearm and hand

The muscles of the upper arm forearm and hand are identified in Figures 16.12 and 16.13. Their function, insertion and origin are identified in Table 16.8. These muscles are immature in the young child and develop through use.

## The muscles of the thorax and abdomen

The primary function of the muscles of the thorax is to assist with the work of breathing (Figure 16.14). Breathing is cyclic (inspiration, expiration, pause) and becomes regular after the age of 2 years approximately. Prior to this it is irregular. The diaphragm is the major muscle of breathing in the young child. This also forms the divide between the thorax and abdominal cavities. The

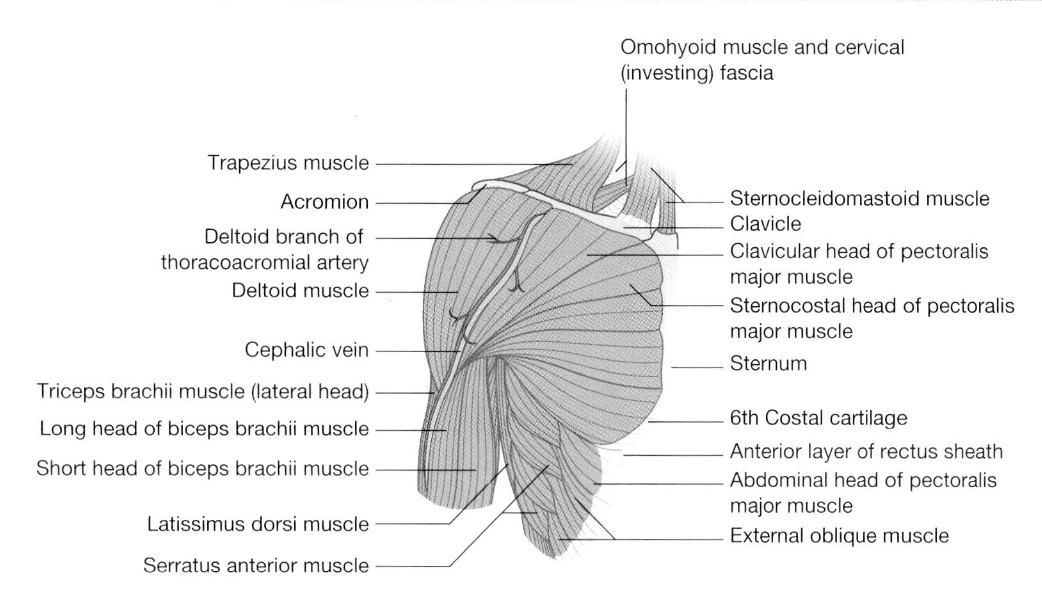


**Figure 16.10** The superficial muscles of the shoulder.



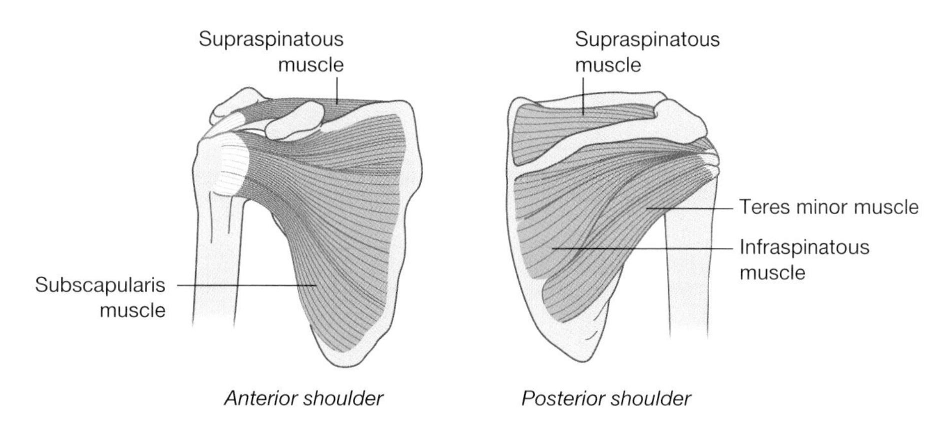


**Figure 16.11** Rotator cuff muscles.

chest wall is very compliant and the ribs lie horizontally. All of this, along with the presence of type 1 muscle fibres, impacts on respiratory effort. The muscles of the thorax are identified in Table 16.9.

The wall of the abdomen is composed of four pairs of muscles (Figure 16.15), and unlike other body compartments it only has bony protection on its posterior aspect. The anterior and lateral areas rely on its musculature to provide protection for the abdominal organs. The muscles of the abdomen are identified in Table 16.9.

## The muscles of the hips and pelvis

The main function of the muscles of the pelvis (Figure 16.16) is to provide a floor and support to the organs of the abdominal cavity. There are differences between the male and female

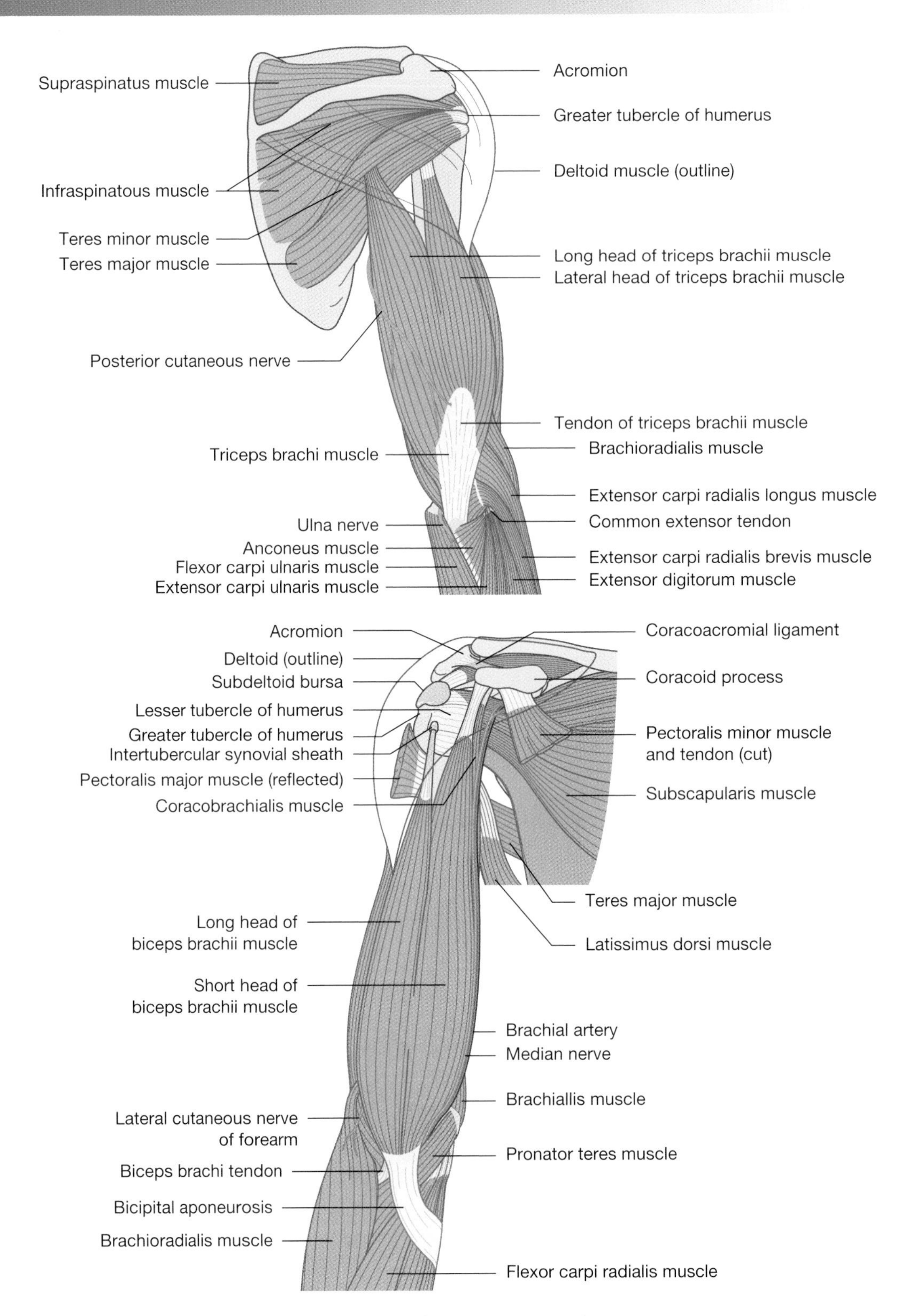


**Figure 16.12** Anterior and posterior views of upper arm muscles.

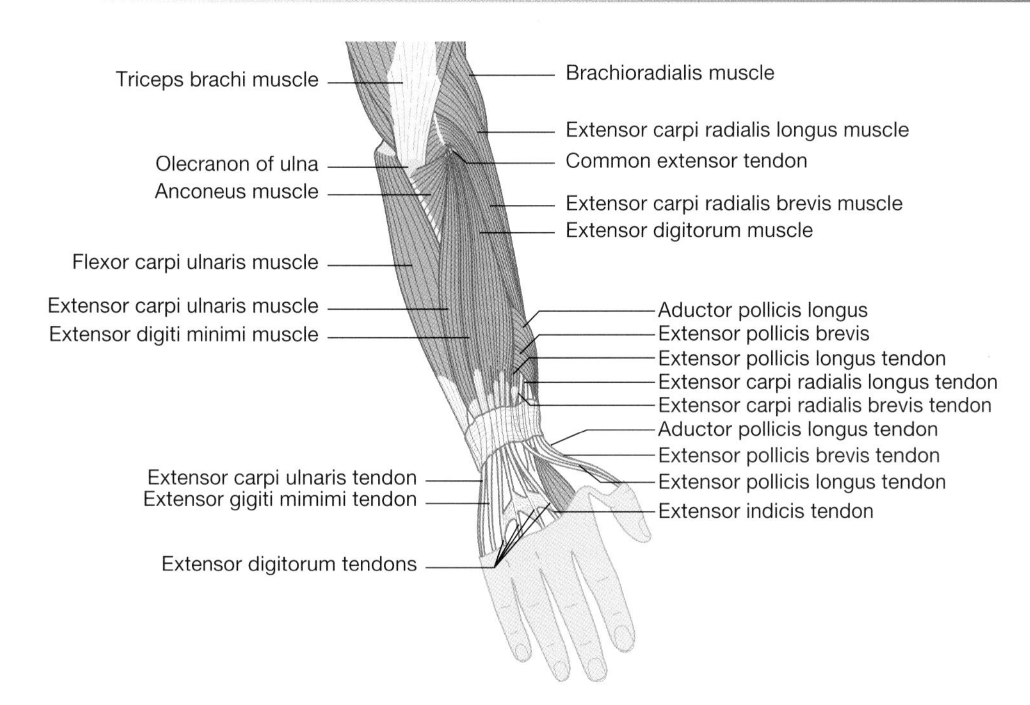


**Figure 16.13** Muscles of the forearm.



**Table 16.8** Muscles of the arm and hand – function, origin and insertion. Adapted from Kingston (2005) and Jarman (2003).

| Muscle | Origin | Insertion | Action |
|---|---|---|---|
| *Arm* | | | |
| Brachialis | Humerus | Coronoid process of ulna | Elbow flexion |
| Biceps brachii | Scapula | Radial tuberosity | Flexion of elbow and shoulder, supination of forearm |
| Brachioradialis | Supracondylar ridge of humerus | Styloid process | Flexion of elbow |
| Triceps brachii | Scapula and humerus | Medial head of humerus | Extension of elbow |
| Anconeus | Epicondyle of humerus | Olecranon process | Extension of elbow |
| *Elbow* | | | |
| Supinator | Epicondyle of humerus | Crest of ulna and lateral upper radius | Supinates forearm |

*(Continued)*

**Table 16.8** (*Continued*)

| Muscle | Origin | Insertion | Action |
|---|---|---|---|
| Pronator quadratus | Distal radius | Antero-medial surface of radius | Pronation of forearm |
| Pronator teres | Humeral head | Lateral radius | Pronation of forearm |
| *Wrist* | | | |
| Flexor carpi radialis | Common flexor tendon | Base of second metacarpal | Flexion and abduction of wrist |
| Flexor carpi ulnaris | Ulna | Pisiform bone | Flexion and adduction of wrist |
| Palmaris longus | Flexor tendon | Palmar aponeurosis | Flexion of wrist and tightening of palmar fascia |
| Extensor carpi radialis longus | Humerus | Base of second metacarpal | Extension and abduction of wrist |
| Extensor carpi radialis brevis | Extensor tendon | Base of third metacarpal | Flexion and abduction of wrist |
| Extensor carpi ulnaris | Posterior ulna | Base of fifth metacarpal | Extension and adduction of wrist |
| *Hand* | | | |
| Flexor pollicis brevis | Trapezium, trapezoid and capitate | Proximal phalanx | Flexion of thumb |
| Opponens pollicis | Trapezium | First metacarpal | Flexion and opposition of thumb |
| Flexor pollicis longus | Anterior radius | Distal phalanx | Flexion of thumb and wrist |
| Extensor pollicis longus | Posterior ulna | Distal phalanx and tuberosity of radius | Extension of thumb and wrist |
| Extensor pollicis brevis | Posterior radius | Proximal phalanx | Extension of thumb |
| Abductor pollicis longus | Posterior ulna | Trapezium and first metacarpal | Extension and abduction of thumb |
| Abductor pollicis brevis | Scaphoid, trapezium | Proximal phalanx | Abduction of thumb |
| Abductor pollicis | Third metacarpal | Proximal phalanx | Adduction of thumb |
| Flexor digitorum superficialis | Humero-ulnar head, radial head | Palmar surface of each phalanx | Flexion of interphalangeal and metacarpophalangeal joints |
| Flexor digitorum profundus | Anterior ulna | Second–fifth distal phalanges | Flexion of all fingers and wrist |

**Table 16.8** (*Continued*)

| Muscle | Origin | Insertion | Action |
|---|---|---|---|
| Flexor digiti minimi brevis | Hamate | Proximal phalanx of little finger | Flexion of little finger |
| The lumbricals | Tendon of flexor digitorum | Extensor aponeurosis of each finger | Flexion of metacarpophalangeal and extension of interphalangeal joints |
| The interossei | Metacarpals | Proximal phalanges | Flexion of second–fourth metacarpophalangeal joints and extension of interphalangeal joints |
| Extensors; digitorum, indicis, digiti minimi | Extensor tendon, lower ulna | Distal phalanges and proximal, middle and distal phalanges. | Extension of second–fifth fingers |



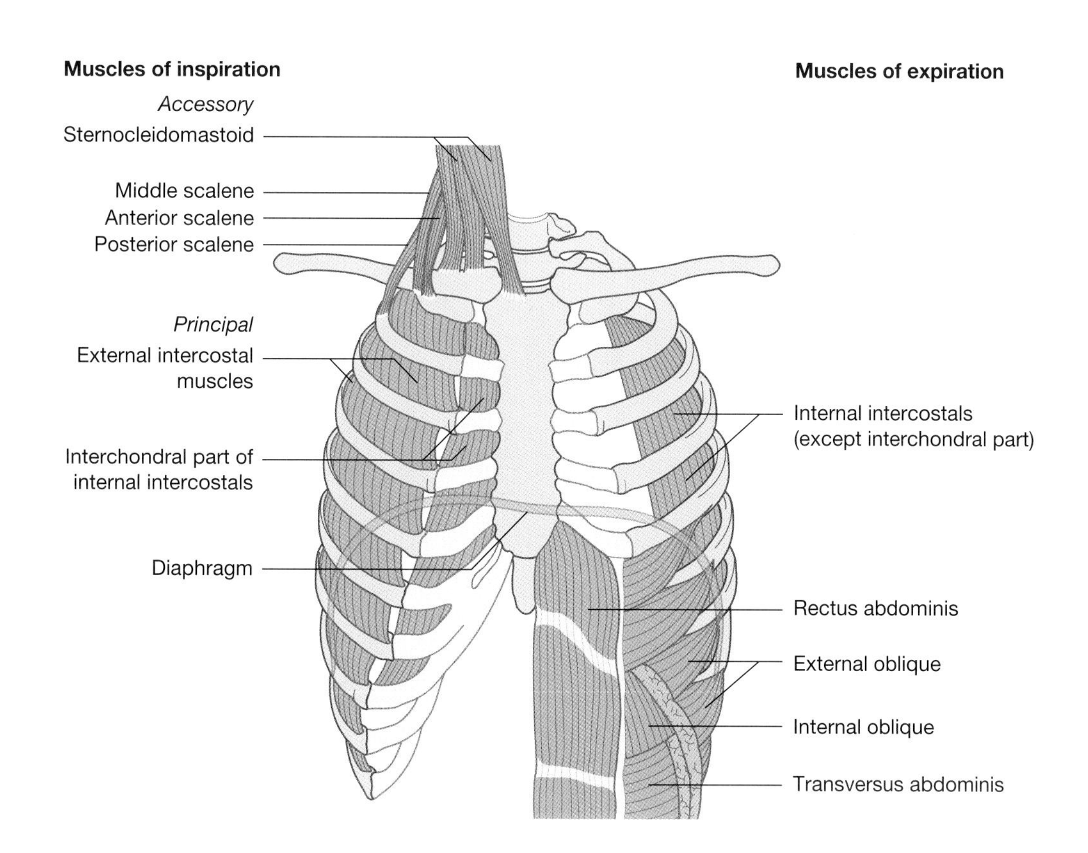


**Figure 16.14** Muscles of breathing.

**Table 16.9** Muscles of the thorax and abdomen – function, origin and insertion. Adapted from Kingston (2005).

| Muscle | Origin | Insertion | Action |
|---|---|---|---|
| *Thorax* | | | |
| External intercostals | Lower surface of rib | Upper surface of rib | Elevates each ribs in inspiration |
| Internal intercostals | Superior surface of rib | Superior border of rib below | Muscles of expiration – draw ribs closer together after inspiration |
| Pectoralis major | Sternum and costal cartilage | Humerus | Accessory muscle of deep inspiration |
| Pectoralis minor | Ribs 3, 4 and 5 | Scapula | An accessory muscle of deep inspiration |
| Scalenes | | | |
| anterior | Cervical vertebrae C3–6 | First rib | Elevates first rib in inspiration |
| medius | C2–7 | Upper surface of first rib | Elevates first rib in inspiration |
| posterior | C4–6 | Second rib | Elevation of second rib |
| Sternocleidomastoid | Manubrium, mastoid process | Clavicle | Elevation of thorax in deep inspiration |
| Diaphragm | Xiphisternum, lower six ribs | Central tendon | Main muscle of inspiration pulls central tendon down. Assists with 'Valsava manoeuvre'. Protects aorta and oesophagus. Enhances venous return to the heart |
| *Abdomen* | | | |
| Rectus abdominis | Pubic crest and symphysis pubis | Cost cartilage of fifth–seventh ribs. | Stabilizes pelvis. Increases abdominal pressure. Flexes and rotates lumbar region |
| Oblique externus | Ribs 5–8 anteriorly and 9–12 laterally | Iliac crest and linea alba | Flexes vertebral column and compresses abdominal wall |
| Oblique internus | Inguinal ligament, iliac crest | Linea alba, crest of pubis, inferior surface of ribs | Flexes vertebral column and compresses abdominal wall |
| Transversus abdominis | Inguinal ligament, costal cartilage, iliac crest | Linea alba, aponeurosis passing to rectus abdominus | Compression of abdominal content |
| Pyramidalis | Anterior pubis and symphysis pubis | Linea alba | Contraction of linea alba |

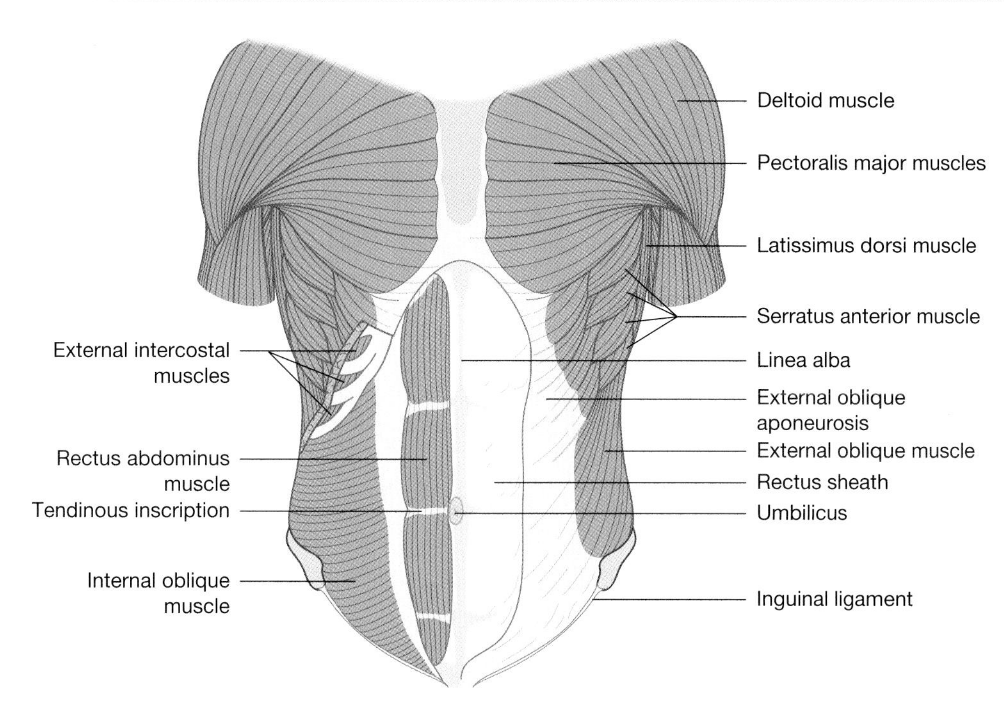


**Figure 16.15** The abdominal muscles.



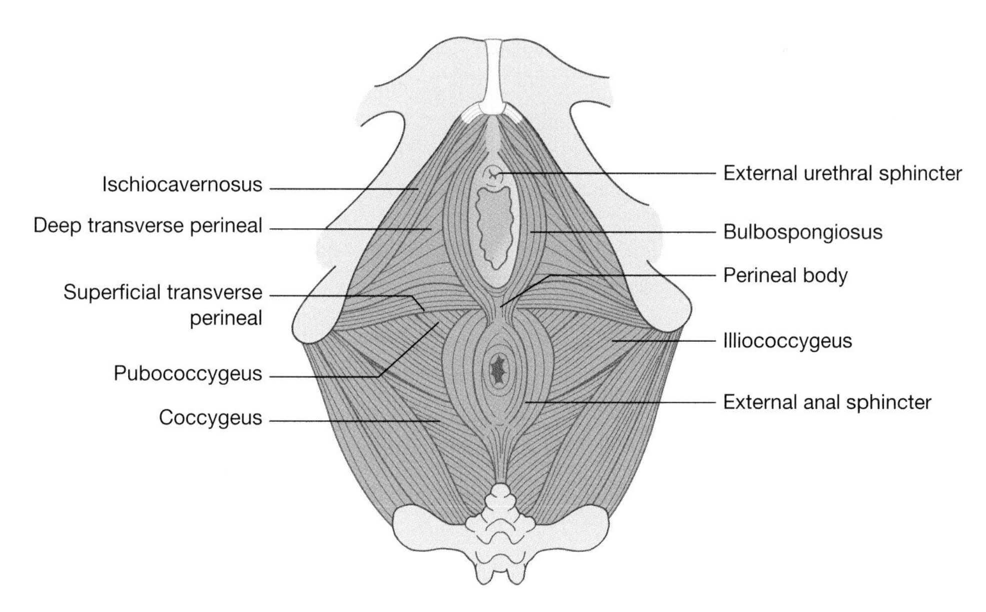


**Figure 16.16** Muscles of the pelvis.

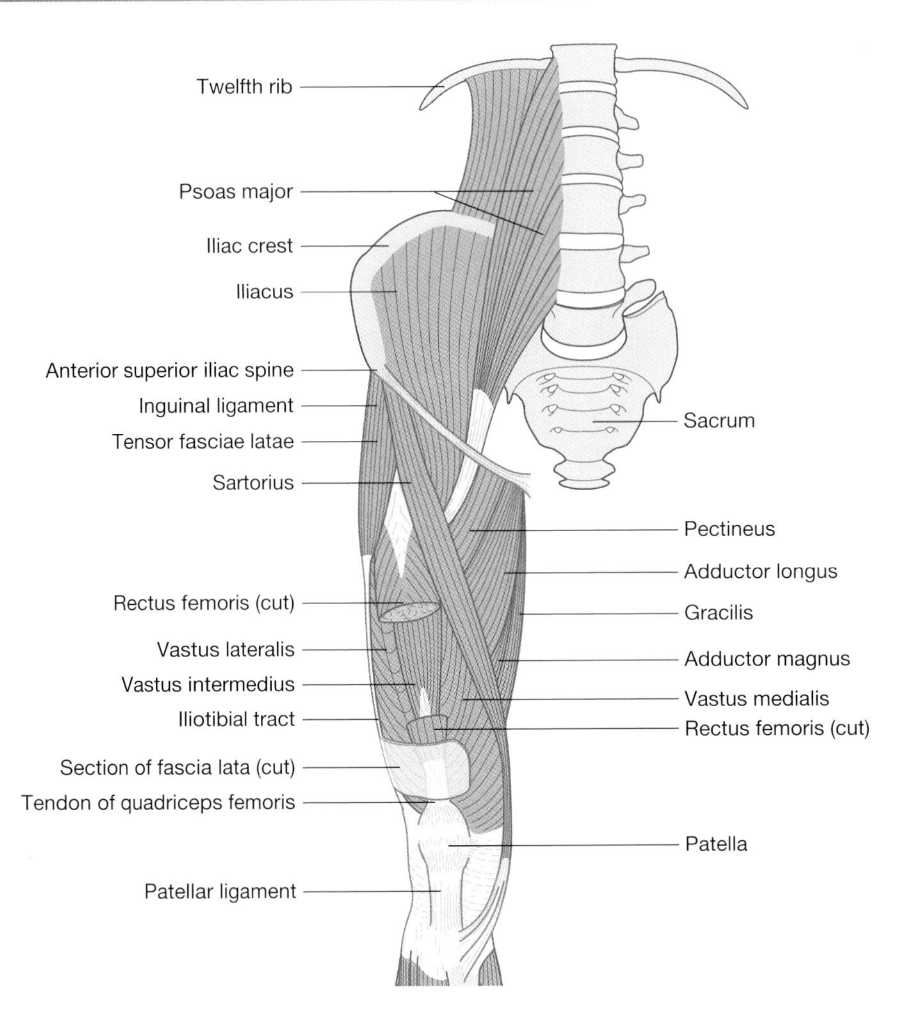




**Figure 16.17** Muscles of the hip.

musculature when considering the pelvis. The muscles of the hip (Figure 16.17) have multiple purposes: they contribute to movement of the hip, movement of the knee and some muscles contribute to movement of both the hip and knee (Table 16.10).

## The muscles of the leg and foot

The muscles of the leg and foot can be divided into three distinct areas: the muscles of the thigh and the lower leg (Figure 16.18) and the muscles of the foot (Figure 16.19). The function, origin and insertion of these muscles are described in Table 16.11.

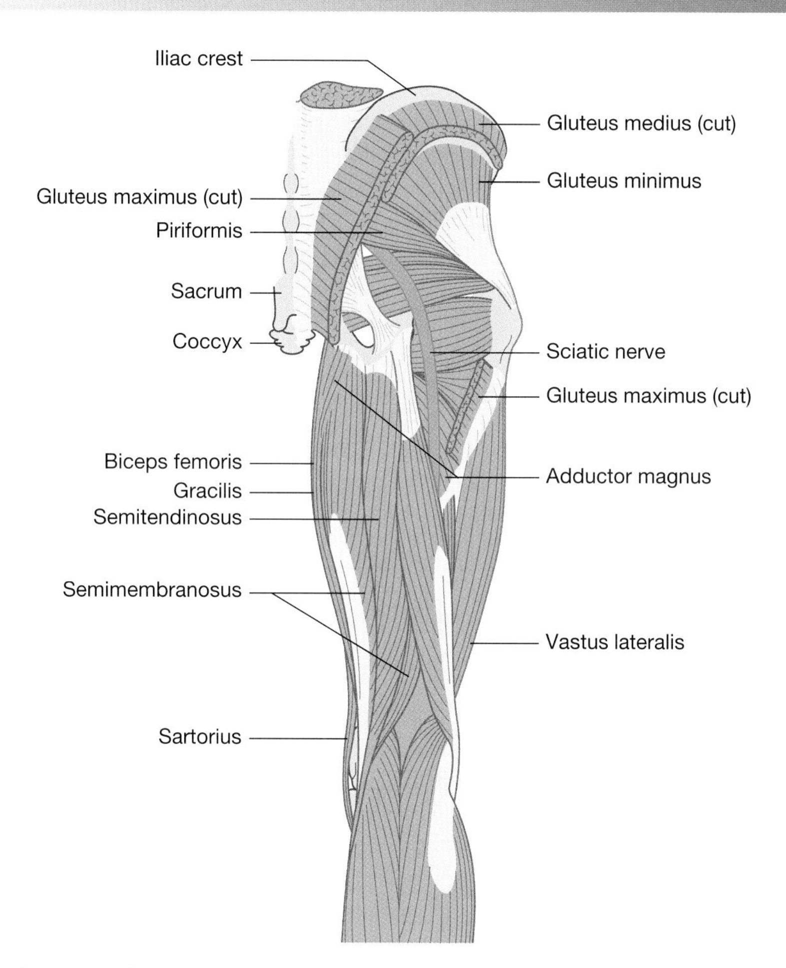




**Figure 16.17** (*Continued*)

# Conclusion

This chapter has identified the three types of muscle tissue. Skeletal muscle has been considered from a micro-anatomical perspective to its location and function within the human body. Differences between male and female musculature have been identified, as has the growth and development of muscle from embryology to maturity. To function efficiently the musculature system is supported by many other systems, and these must be considered.

**Table 16.10** The function of the hip and pelvis muscles. Adapted from Kingston 2005 and Jarmey 2003.

| Muscle | Origin | Insertion | Action |
|---|---|---|---|
| Psoas major | T12–L5 anterior processes of vertebrae | Lesser trochanter | Flexion of hip |
| Iliacus | Iliac fossa of ilium | Distal to lesser trochanter | Flexion of hip |
| Pectineus | Superior ramus of pubis | Inferior to lesser trochanter of femur | Flexion, adduction and medial rotation |
| Rectus femoris | Iliac spine | Patella | Flexion of hip. extension of knee |
| Sartorius | Superior iliac spine | Upper tibia | Flexion of hip and knee |
| Gluteal maximus | Iliac crest, sacrum, coccyx | Tuberosity of femur | Extension and lateral rotation at hip |
| Biceps femoris | Ischial tuberosity | Head of fibula | Extension of hip |
| Semitendinosus | Ischial tuberosity | Superior surface of tibia | Extension of hip, flexion of knee |
| Semimembranosus | Ischial tuberosity | Tibial condyle | Extension of hip, flexion of knee |
| Gluteal medius | Iliac crest of ilium | Greater trochanter of femur | Abduction and medial rotation of hip |
| Gluteal minimus | Lateral surface of ilium | Greater trochanter of femur | Abduction and medial rotation of hip |
| Tensor fasciae latae | Iliac crest | Iliotibial tract | Flexion and medial rotation of hip |
| Adductor longus | Inferior ramus of pubis | Linea aspera of femur | Adduction, flexion and medial rotation of hip |
| Adductor brevis | Inferior ramus of pubis | Linea aspera of femur | Adduction, flexion and medial rotation of hip |
| Adductor magnus | Inferior ramus of pubis posterior | Linea aspera and adductor tubercle of femur | Adduction, flexion, lateral and medial rotation of hip, extension of hip |
| Gracillis | Inferior ramus of pubis | Medial surface of tibia | Flexion at knee, adduction and medial rotation |
| Obturators internus and externus | Obturator foramen | Greater trochanter | Lateral rotation of hip |
| Gemelli | Ischial spine | Greater trochanter | Lateral rotation of hip |
| Quadratus femoris | Lateral body of tuberosity | Intertrochanteric crest of femur | Lateral rotation of hip |
| Piriformis | Sacrum | Greater trochanter | Lateral rotation of hip |
| Levator ani | Ischial spine | Coccyx | Tenses floor of pelvis, elevates and retracts anus |
| Coccygenus | Ischial spine | Sacrum and coccyx | Supports pelvic floor |

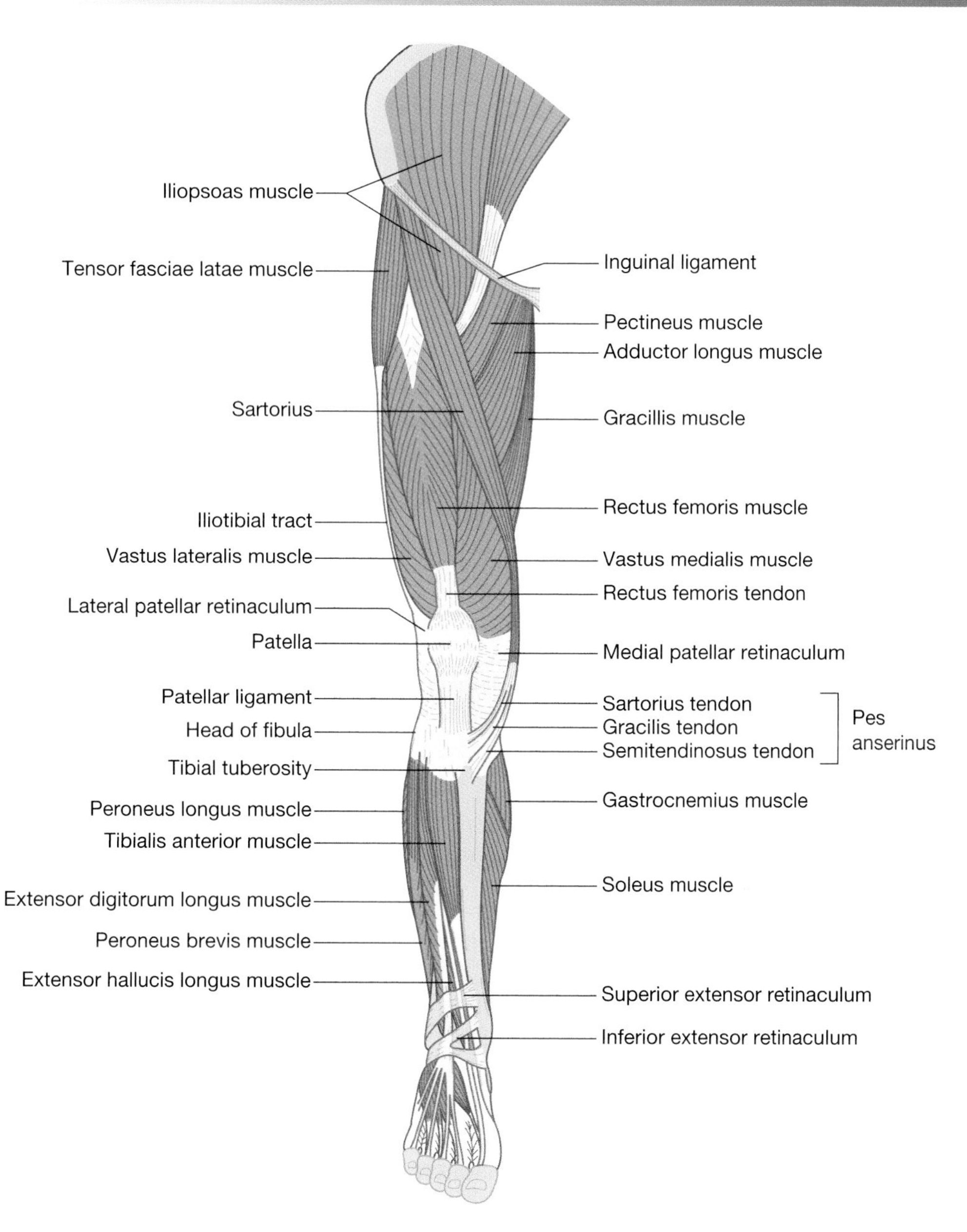




**Figure 16.18** Anterior and posterior views of the muscles of the leg.

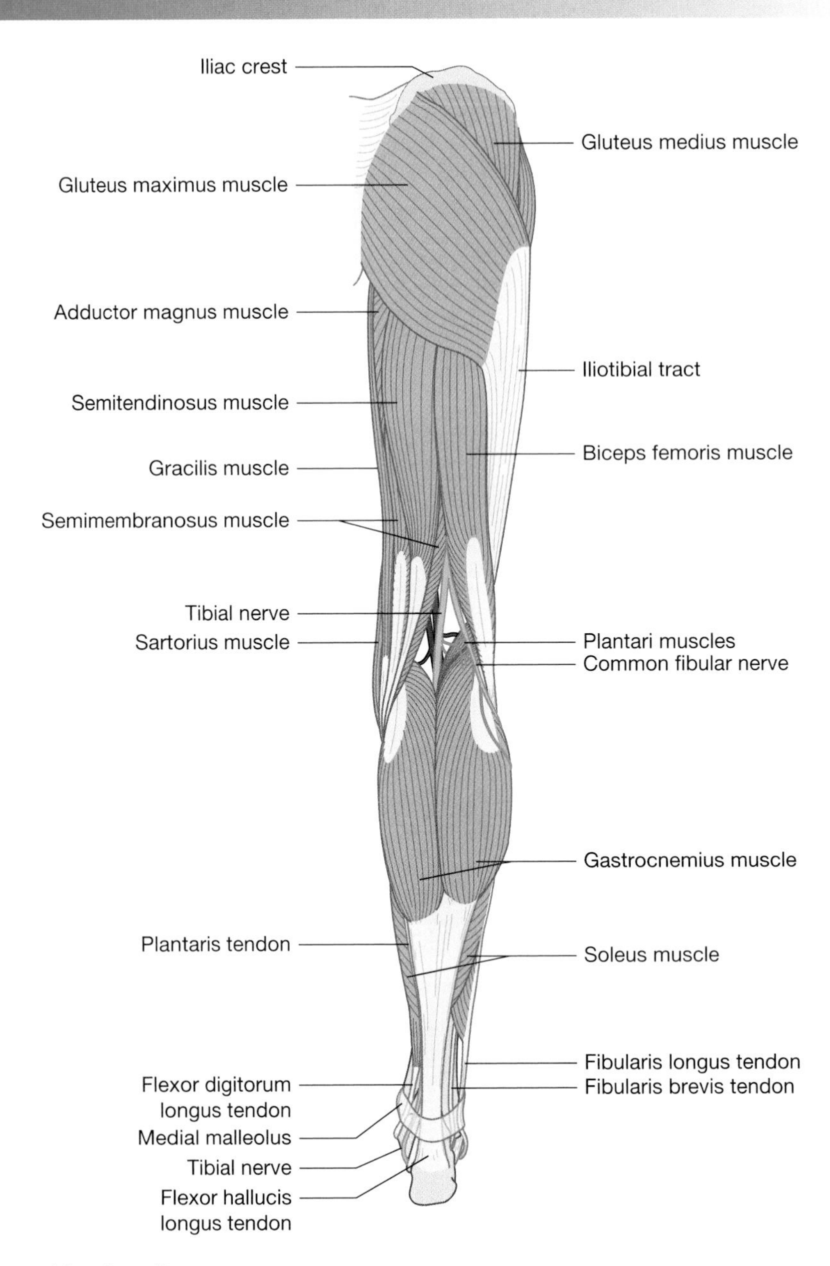




**Figure 16.18** (*Continued*)

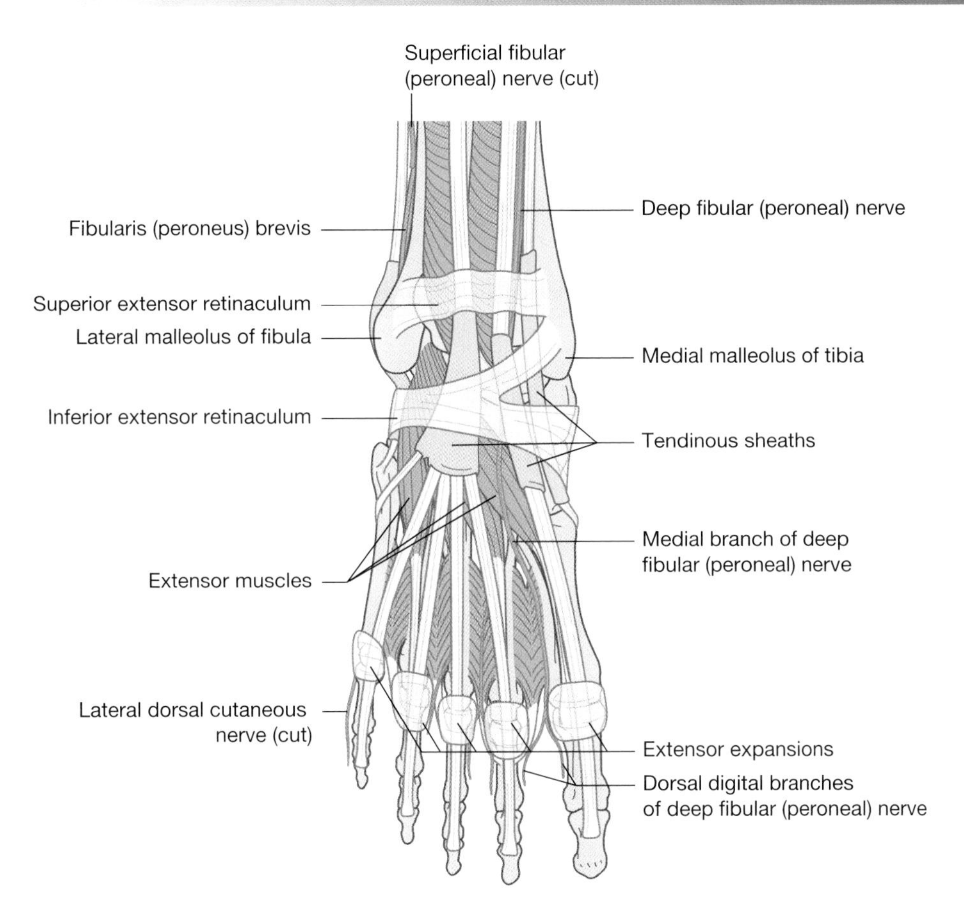


**Figure 16.19** The muscles of the foot.



**Table 16.11** Function, origin and insertion of the muscles of the thigh, lower leg and foot. Adapted from Kingston (2005).

| Muscle | Origin | Insertion | Action |
|---|---|---|---|
| *Thigh* | | | |
| Gluteal maximus | Outer surface of ilium | Posterior area of femur | Extends and lateral rotation of hip |
| Gluteal medius | Ilium | Greater trochanter of femur | Abducts hip joint |
| Gluteal minimus | Outer surface of ilium | Greater trochanter of femur | Abducts and medially rotates hip joint |
| Obturators externus and internus | Ischium, pubis and ilium | Greater trochanter of femur | Lateral rotation of hip. Keeps head of femur in acetabulum |
| Piriformis | Front of sacrum | Greater trochanter of femur | Lateral rotation of hip |

(*Continued*)

**Table 16.11** *(Continued)*

| Muscle | Origin | Insertion | Action |
|---|---|---|---|
| Gemelli | Ischial spine | Greater trochanter of femur | Lateral rotation of hip |
| Quadratus femoris | Ischial tuberosity | Intertrochanteric crest of femur | Lateral rotation of hip |
| Adductors | | | |
| brevis | Inferior rami of pubis | Linea aspera of femur | Adduction and lateral rotation of hip joint |
| longus | Inferior rami of pubis | Linea aspera of femur | As above, and also flexion of hip |
| magnus | Ischial tuberosity | Linea aspera of femur and tubercle of femur | Adduction and lateral rotation of hip joint |
| Gracilis | Pubic bone | Upper surface of tibia | Adducts hip, flexes knee, medial rotation of knee |
| Pectineus | Superior ramus of pubis | Medical shaft of femur | Adducts hip, flexes knee |
| Iliacus | Iliac fossa of ilium | Distal to lesser trochanter | Flexion at knee |
| Psosa major | Anterior surface of T12–L5 | Lesser trochanter | Flexion at hip |
| *The knee* | | | |
| Biceps femoris | Ischial tuberosity | Head of fibula and lateral condyle of tibia | Flexion of knee, extension and lateral rotation of hip |
| Semimembraneous | Ischial tuberosity | Medial condyle | Flexion of knee, extension and rotation of hip |
| Semitendinous | Ischial tuberosity | Medial surface of condyle | Flexion of knee, extension and rotation of hip |
| Satorius | Iliac spine | Medial surface of tibia | Flexes hip brining knee forward. Flexes knee joint. |
| Popiteus | Lateral condyle of femur | Tibial shaft | Medial rotation of tibia, flexion of knee |
| Rectus femoris | Inferior iliac spine | Tibial tuberosity | Extension at knee, flexion at hip |
| Vatus intermedius | Antrolateral surface of femur and linea aspera | Tibial tuberosity | Extension at knee |
| Vatus lateralis | Greater trochanter and linea aspera | Tibial tuberosity | Extension at knee |
| Vatus medialis | Entire linea aspera | Tibial tuberosity | Extension at knee |
| *Lower leg* | | | |
| Tibialis anterior | Lateral condyle of tibia | Cuneiform bone | Dorsiflexes foot |

**Table 16.11** (*Continued*)

| Muscle | Origin | Insertion | Action |
|---|---|---|---|
| Gastrocnemius | Medial condyle of femur | Calcaneus | Plantar flexion of foot |
| Fibularis | | | |
| longus | Lateral surface of fibula – upper two-thirds | First metatarsal | Everts foot |
| brevis | Lateral surface of fibula – lower two-thirds | Fifth metatarsal | Everts foot |
| Plantaris | Supracondylar ridge | Calcaneus | Extension at ankle, flexion of knee |
| Soleus | Head of fibula, shaft of tibia | Calcaneus | Extension at ankle |
| Tibialis posterior | Posterior surface of tibia and fibula | Tarsal bones | Inverts foot |
| *Toes* | | | |
| Flexor digitorum longus | Posterior tibia | Distal phalanges of second–fifth toes | Flexes all joints of lateral four toes. Plantar flexion of ankle |
| Flexor hallucis longus | Lower two-thirds of fibula | Distal phalanx of great toe | Flexes great toe, plantar flexion and inversion of foot |
| Extensor digitorum longus | Lateral condyle of tibia and fibula | Superior surface of phalanges of second–fifth toes | Extension of joints of second–fifth toes |
| Extensor hallucis longus | Anterior surface of fibula | Distal phalanx of great toe | Extension of joint of great toe |
| Lumbricals | Tendons of flexor digitorum longus | Insertions of extensor digitorum longus | Flexion at metatarsophalangeal joints of second–fifth toes |
| Plantar interosseus | Metatarsal bones | Medial side of toe 2, lateral side of toes 3 and 4 | Abduction of metatarsophalangeal joint of toes 3 and 4 |



# Activities

**Now review your learning by completing the learning activities in this chapter. The answers to these appear at the end of the book. Further self-test activities can be found at www.wileyfundamentalseries.com/childrensA&P.**

## True or false?

1. Cardiac muscle is non-striated and is uninucleated.
2. The body's skeletal muscles can be dived into five areas.
3. The buccinators muscle is found in the abdomen.
4. Each muscle begins at an origin and ends at an insertion.
5. Type 2 muscle fibres are fast contracting fibres.
6. Skeletal muscle is under involuntary control.
7. The diaphragm is the muscle that contributes most to the work of breathing in a young child.
8. Cardiac muscle does not regenerate.
9. The function of myoglobin in muscle tissue is to break down glycogen.
10. A fascicle is a bundle of muscle fibres.

## Wordsearch

There are several words linked to this chapter hidden in the following square. Can you find them? A tip – the words can go from up to down, down to up, left to right, right to left, or diagonally.

| t | s | m | o | o | t | h | m | u | s | c | l | e | r | n |
|---|---|---|---|---|---|---|---|---|---|---|---|---|---|---|
| o | r | h | s | e | j | n | i | k | r | u | p | n | o | o |
| h | l | i | p | l | d | e | l | t | o | i | d | e | s | i |
| t | f | m | c | p | i | r | a | m | u | s | n | u | n | t |
| o | n | o | n | e | h | r | a | h | e | e | i | r | e | c |
| t | p | c | w | y | p | t | b | c | a | m | t | o | t | u |
| n | o | i | t | r | e | s | n | i | e | a | t | x | x | d |
| n | j | n | e | f | d | r | b | n | f | n | h | e | e | b |
| m | u | s | c | l | e | a | c | r | a | o | r | l | l | a |
| e | r | b | i | f | t | o | e | l | a | o | y | f | f | n |
| d | i | a | p | h | r | a | g | m | g | c | y | m | a | i |
| t | n | r | o | t | c | u | d | d | a | o | h | l | s | g |
| l | e | l | s | k | e | l | e | t | a | l | n | i | c | i |
| c | n | a | m | r | o | f | i | s | u | f | t | t | i | r |
| o | g | m | y | o | f | i | l | a | m | e | n | t | a | o |

## Complete the sentence

1. Smooth muscle contains small, thin _____-shaped cells of variable size that have ___ centrally located nucleus and is arranged in _____ lines.
2. _____ muscles are ______ shaped _____ fibres that lie parallel to each other.

**3.** Muscles generate _____ as they_____; the ____ produce _____ triphosphate, giving the muscles the energy to contract.
**4.** The _____ is composed of dense _____ tissue that surrounds the _____ of the muscle.
**5.** _____ muscles have a rich _____ supply as they have large energy requirements, thus demanding a continual _____ supply.
**6.** Each _____ is surrounded by a layer of _____ tissue called the perimysium.

## Conditions

The following table contains a list of conditions. Take some time and write notes about each of the conditions. You may make the notes taken from text books or other resources; for example, people you work with in a clinical area or you may make the notes as a result of people you have cared for. If you are making notes about people you have cared for you must ensure that you adhere to the rules of confidentiality.

| Condition | Your notes |
|---|---|
| Myasthenia gravis | |
| Fibromyalgia | |
| Guillian–Barré syndrome | |
| Muscular trauma | |
| Ptosis | |



## Glossary

**Aerobic:** with oxygen.

**Anaerobic:** without oxygen.

**Anterior:** to the front side of the body.

**Aponeurosis:** a broad sheet of skeletal muscle fibres.

**Glycosomes:** granules of stored glycogen that produce glucose during muscle cell activity.

**Myofibril:** rod-like structures that are composed of sarcomeres. The contractile element of the skeletal muscle.

**Myoglobin:** red pigment that stores oxygen.

**Posterior:** to the back side of the body.

**Sarcolemma:** the plasma membrane of a muscle fibre.

**Sarcoplasm:** the cytoplasm of the muscle fibre.

**Tendon:** tissue that connects muscle to bone.

## References

Chamley, C., Carson, P., Randall, D., Sandwell, W. (2005) *Developmental Anatomy and Physiology of Children*, Churchill Livingstone.

Crawford, D., Hickson, W. (2002) *An Introduction into Neonatal Nursing Care*, Nelson Thornes, Cheltenham.
England, M. (1996) *Life Before Birth*, 2nd edn, Mosby Wolfe, London.
Goldsmith, K. (2003) *Assisted Ventilation of the Neonate*, 4th edn, Saunders, Philadelphia, PA.
Jarman, C. (2003) *The Complete Book of Muscles*, Lotus, Chichester.
Kingston, B. (2005) *Understanding Muscles. A Practical Guide to Muscle Function*, Nelson Thornes, Cheltenham.
Marieb, E.N., Hoehn, K. (2010) *Human Anatomy & Physiology*, 8th edn, Pearson, San Francisco, CA.
Martini, F.H., Nath J.L., Bartholomew, E.F. (2012) *Fundamentals of Anatomy and Physiology*, 9th edn, Pearson, San Francisco, CA.
Neu, C.M., Rauch, F., Rittwienger, F. *et al.* (2001) Influence of puberty on muscle development of the forearm. *American Journal of Physiology – Endocrinology and Metabolism*, **283**, E103–E107.
Peate, I., Nair, M. (2011) *Fundamentals of Anatomy and Physiology for Student Nurses*, 2nd edn, Wiley–Blackwell, Oxford.
Rauch, F., Bailey, D., Baxter-Jones, A. *et al.* (2004) The 'muscle–bone unit' during the pubertal growth spurt. *Bone*, **34** (5), 771–775.
Sanat, H.B. (2004) Neuromuscular disorder. In: Behrman, R.E., Kliegman, R.M., Jenson, H.B. (eds), *Nelson Textbook of Pediatrics*, 17th edn, Saunders, Philadelphia, PA.
Tortora, G.J., Derrickson, B.H. (2010) *Essential of Anatomy and Physiology*, 8th edn, John Wiley & Sons Ltd, Chichester.

# Chapter 17

# The skeletal system

Debbie Martin

*School of Health and Social Work, University of Hertfordshire, Hatfield, UK*

## Aim

The aim of this chapter is to provide the reader with an understanding of the anatomy and physiology of the skeletal system.

## Learning outcomes

On completion of this chapter the reader will be able to:

- Identify the bones in the skeleton.
- Describe the function of the skeleton.
- Identify factors that determine growth.
- Describe the structure of bone.
- Describe the healing process of bone.
- Identify different types of joints.

## Test your prior knowledge

- How many bones are there in the adult skeleton?
- There are six different classifications of bone types, can you name them?
- What functions does the skeletal system have?
- Where is the epiphysis found and what is its function?
- What substance provides some elasticity in bones?
- Why do babies have more bones than adults?
- What do you understand by a joint?
- Describe the function of osteocytes.
- At what age is bone growth complete?
- Name two essential minerals found in bone.

*Fundamentals of Children's Anatomy and Physiology: A Textbook for Nursing and Healthcare Students*, First Edition. Edited by Ian Peate and Elizabeth Gormley-Fleming.

Companion website: www.wileyfundamentalseries.com/childrensA&P

# Body map

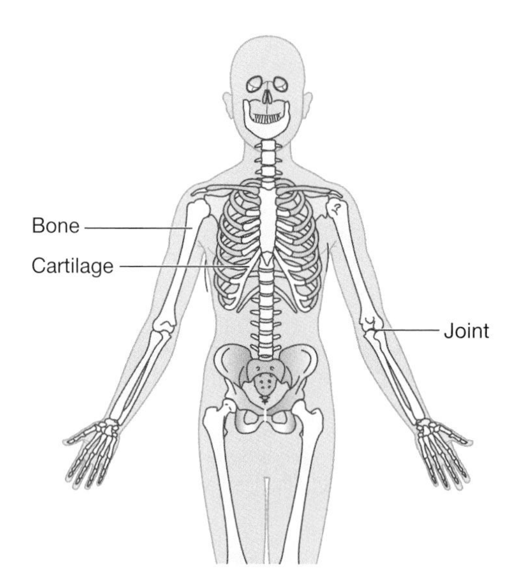




# Introduction

The skeleton is a remarkable part of the anatomy and has many functions. It is a tough flexible structure that provides support and shape for the body; it works as a lever system for the muscles, permitting movement, and acts as protection for the vital organs and structures within the body. Bone is living tissue that is constantly being renewed throughout life; it produces blood cells and stores minerals such as calcium and phosphorus. The skeleton works together with other structures in the body, such as muscles, ligaments, tendons and the nervous system; this permits a range of movements that are demanded by the body in order to survive and to carry out the activities of living.

Bones develop from cartilage, and this process begins *in utero* as early as 6 weeks of embryonic life. Some of this cartilage hardens (known as ossification) before birth, but some remains and it continues to develop and ossify after birth. This is why the bones of babies are softer and more pliable than those of adults. There are 206 bones in the adult skeleton of varying shapes and sizes and around 300 in the newborn baby. As the child grows, ossification of the bones takes place and some fuse together, mainly in the cranium, to form one bone, ending with fewer bones as an adult (Waugh and Grant, 2010).

The skeleton is divided into two parts:

- **The axial skeleton.** The axial skeleton forms the central bones: skull, ribs, vertebral column and sternum. There are 80 bones in the axial skeleton (Table 17.1).
- **The appendicular skeleton.** The appendicular skeleton forms the upper and lower limbs, along with the scapula, clavicle and pelvis; 126 bones in total (Table 17.2).

See Figure 17.1 for the axial and appendicular skeleton.

**Table 17.1** The bones of the axial skeleton. *Source:* Peate and Nair (2011), Table 8.1, p. 227. Reproduced with permission of John Wiley and Sons, Ltd.

| Structure | No. of bones |
|---|---|
| *Skull* | |
| Cranium | 8 |
| Face | 14 |
| Total | 22 |
| Hyoid | 1 |
| Auditory ossicles | 6 |
| Vertebral column | 26 |
| *Thorax* | |
| Sternum | 1 |
| Ribs | 24 |
| Total | 25 |
| Total number of bones in the axial skeleton | 80 |



# The function of the skeleton

## Support

The skeleton provides a strong flexible framework that provides and maintains shape; without it the body would not be able to stand or function. The skeletal framework also provides anchorage for the many ligaments, tendons and muscles essential for movement.

## Movement

Movement is achieved in partnership with the muscular system. Muscles attach to the ends of bone and stretch across joints to attach on another bone. These muscles then contract using the skeletal system for leverage; and as one bone moves the other remains stable, which causes controlled movement. Movements can be large, gross movement, such as standing or running, or they can be small, fine movement, allowing intricate tasks such as sewing or writing.

## Storage

The bones in the skeleton act as a storehouse for essential minerals such as calcium, phosphorus, magnesium, potassium, zinc and other trace elements. In order for bones to properly absorb these minerals, vitamin D must be present. Stored minerals are released from the bone when demanded by the body.

## Protection

The skeleton protects vital organs and soft tissues within the body, minimizing the risk of injury to them. For example, cranial bones protect the brain, vertebrae protect the spinal cord, the

**Table 17.2** The bones of the appendicular skeleton. *Source:* Peate and Nair (2011), Table 8.2, p. 227. Reproduced with permission of John Wiley and Sons, Ltd.

| Structure | No. of bones |
|---|---|
| *Pectoral girdle* | |
| Clavicle | 2 |
| Scapula | 2 |
| Total | 4 |
| *Upper limbs* | |
| Humerus | 2 |
| Ulna | 2 |
| Radius | 2 |
| Carpals | 16 |
| Metacarpals | 10 |
| Phalanges | 28 |
| Total | 60 |
| *Pelvic girdle* | |
| Pelvic bone | 2 |
| Lower limbs | 2 |
| Femur | 2 |
| Patella | 2 |
| Fibula | 2 |
| Tibia | 14 |
| Tarsals | 10 |
| Metatarsals | 28 |
| Phalanges | 62 |
| Total | |
| Total number of bones in the appendicular skeleton | 126 |
| Total number of bones in the adult human skeleton | 206 |

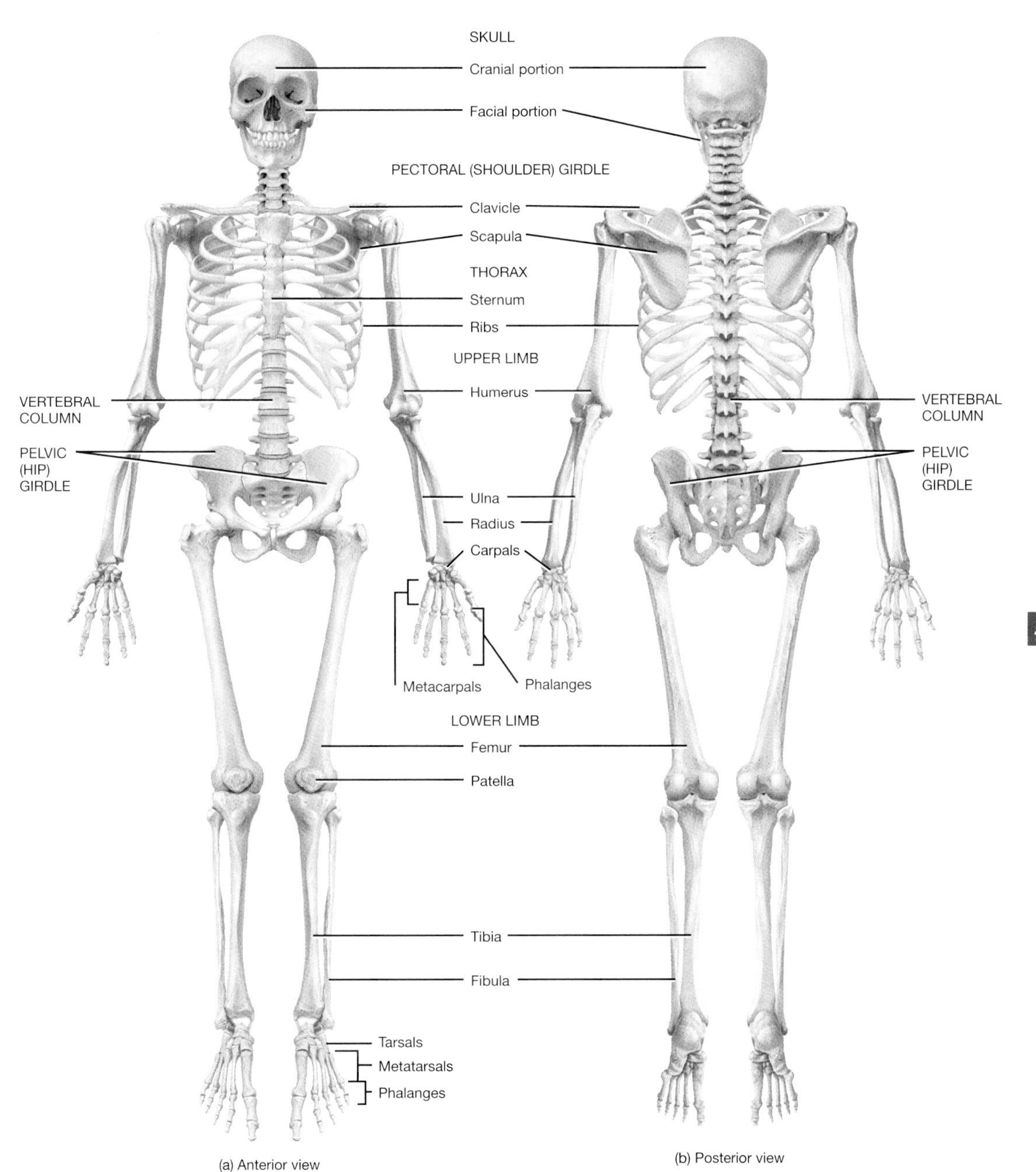


**Figure 17.1** The axial and appendicular skeleton.
*Source:* Tortora and Derrickson (2009), Figure 7.1, p. 200. Reproduced with permission of John Wiley and Sons, Inc.

ribcage protects the heart and lungs, and the pelvis protects the abdominal and reproductive organs.

### Production

Bones are essential in the production of red blood cells (erythrocytes). Red blood cells carry oxygen and waste products from cellular metabolism. They are produced in the bone marrow, which runs through the centre of some larger bones such as the femur. White blood cells (leucocytes) and platelets are also produced in the bone marrow. This process is known as haematopoiesis.

## Clinical application

Vitamin D is essential for the growth and health of bone as it helps the body to absorb calcium and phosphorus from the diet. Vitamin D is produced via the skin after exposure to sunlight; unlike other vitamins, very little is gained from diet. During the darker winter months the body uses its stores and dietary sources to maintain vitamin D levels (Scientific Advisory Committee on Nutrition, 2007). Rickets (osteomalacia) is a condition that affects bone development in children which leads to softening and weakening of the bones, which can lead to deformities such as bowed legs, curvature of the spine and thickening of the ankles, wrists and knees. In recent years rickets is being seen more often in the UK (Pearce and Cheetham, 2010). Although vitamin D deficiency can occur at any age, it is more common in infants and young children during periods of rapid growth, typically 6–24 months of age. There are various reasons that the body is unable to produce vitamin D: limited sun exposure all year round; people with darker skin tones, such as African or Caribbean; keeping skin covered when outside; high sun protection factor (SPF) in sun cream and some medications such as phenytoin and carbamazine can interfere with the way the body metabolizes vitamin D. As a children's nurse it is important to have an understanding of the causes and effects of vitamin D deficiency to ensure that the appropriate advice and care are provided for children, their families and carers.



## Bone structure and growth

Bone is made from specialized cells that form a matrix mainly of protein fibres, water and minerals. Bone consists of both living and nonliving tissue. The living part consists of blood vessels, nerves, collagen and cells called:

- osteoblasts – cells that build bone which later develop into osteocytes;
- osteocytes – mature bone cells that monitor and maintain bone tissue;
- osteoclasts – cells that clear away old bone and reabsorb bone to maintain its shape (Figure 17.2).

A balance of activity from osteoblasts and osteoclasts is essential to maintain normal bone function and structure. Osteoblasts are modified from fibroblasts, which have collagen fibres around them; calcium salts then collect here, which increases the bone size. Osteoclasts develop from bone marrow stem cells that continually remove excess material, which shapes the bone (Waugh and Grant, 2010). The nonliving elements are minerals and salts that are equally important for bone formation.

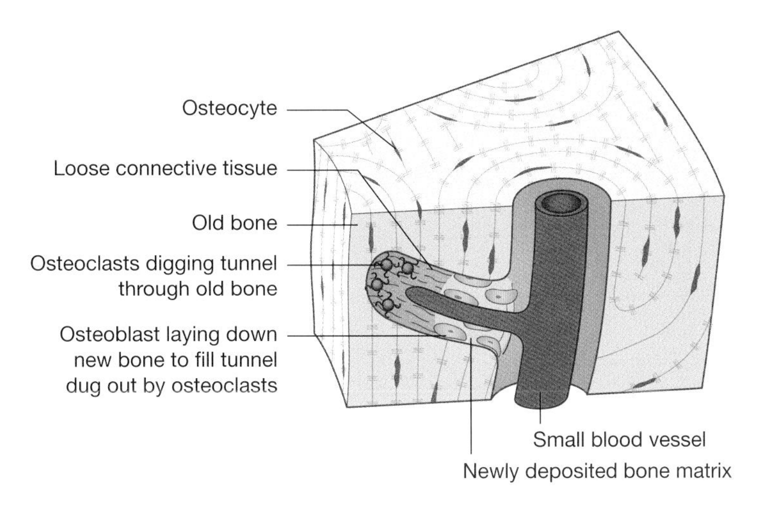


**Figure 17.2** Bone formation.

# Embryonic growth



Skeletal growth occurs early in fetal life from the mesoderm. This is connective embryonic tissue known as mesenchyme, which originates from mesodermal cells. The mesenchyme develops into different types of cells, such as chondroblasts (cartilage formation), osteoblasts (bone formation) and fibroblasts (connective tissue formation). By the end of the fourth week of gestation connective tissue begins to form and early cells begin to lay down a cartilage matrix. At 6 weeks early vertebrae are forming, and by 8 weeks the limbs, hands, fingers, feet and toes have a definite shape (MacGregor, 2008). At this stage all 206 bones are laid down, but the process of osetogenesis (bone development) has not progressed to ossification. There are two types of ossification that arise from mesenchyme:

- Intramembranous ossification (Figure 17.3), which is where osteoblasts form bones directly; for example, cranium and clavicle. After a period of growth a matrix is formed with collagen fibres, calcium and organic salts that are deposited by osteoblasts; this structure then calcifies (Oldfield, 2004).
- Endochondral ossification, which begins from cartilage. The cartilage is formed first from chondroblasts, which further develops into bone. An example of this is the long bones (Oldfield, 2004).

Ossification of many of the bones occurs in the second month of fetal life, and the starting point of this occurs in the primary centre of bone, which varies in different bones (Figure 17.4).

# Childhood

During childhood and into puberty the skeletal system continues to grow and develop. This is influenced by factors such as genetics, hormones, vitamins and exercise (MacGregor, 2008). Secondary centres of ossification develop (Figure 17.4) initially at the epiphysis, which continues throughout childhood until the epiphyseal plates become fused with the diaphysis, which occurs at the end of puberty and is completed around 16–18 in females and 18–21 in males (Oldfield, 2004). Long bones add width to the bone by adding layers on the outside to those that already exist (in a process known as subperiosteal apposition), whilst breaking down bone

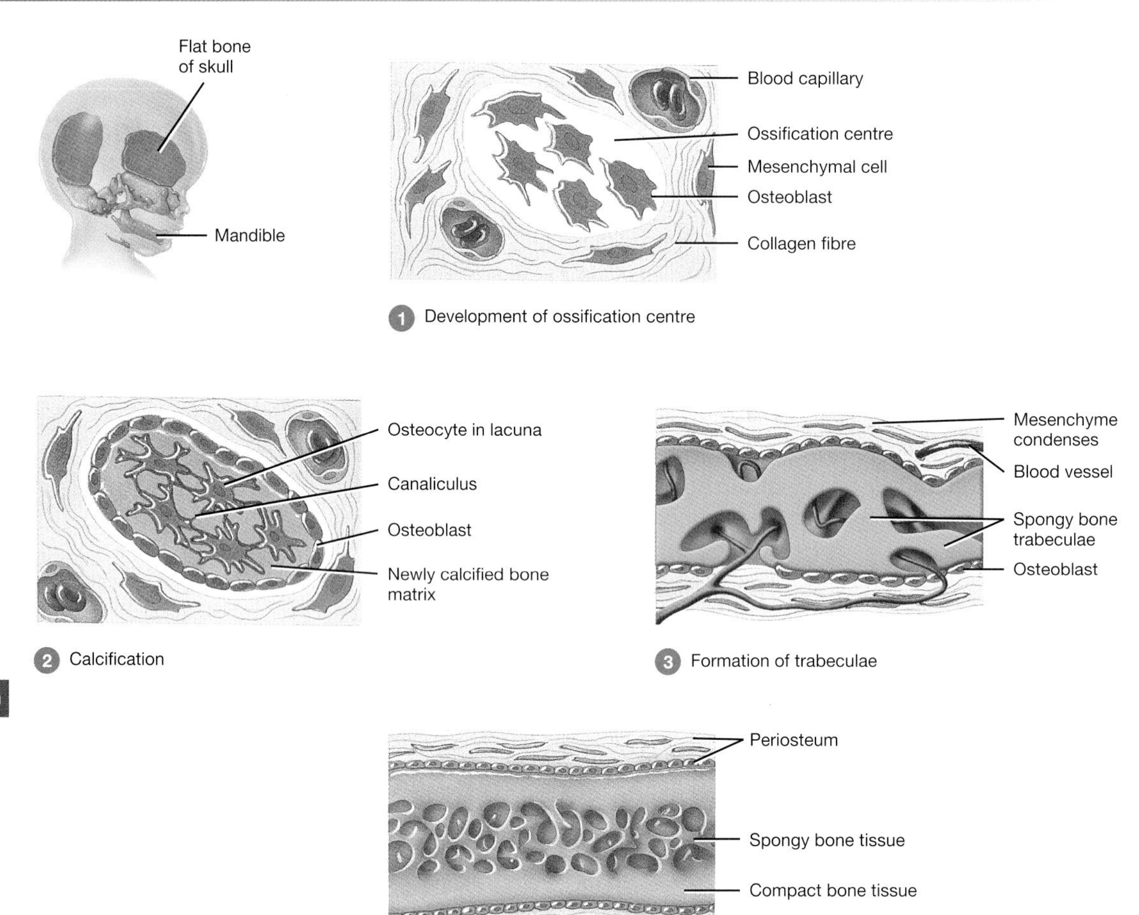




**Figure 17.3** Intramembranous ossification.
*Source:* Tortora and Derrickson (2009), Figure 6.5, p. 183. Reproduced with permission of John Wiley and Sons, Inc.

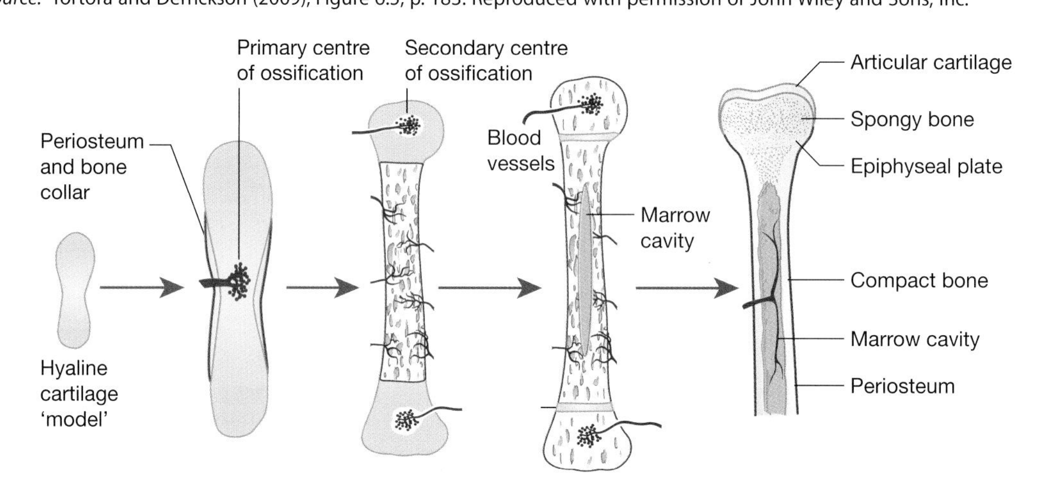


**Figure 17.4** Ossification of bone.

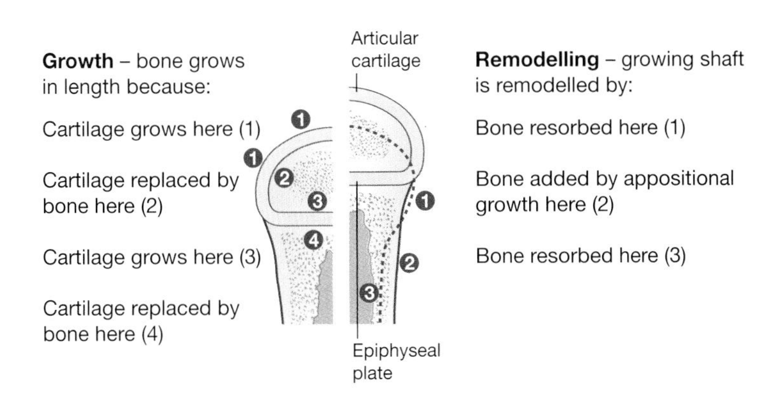


**Figure 17.5** Bone growth and remodelling.

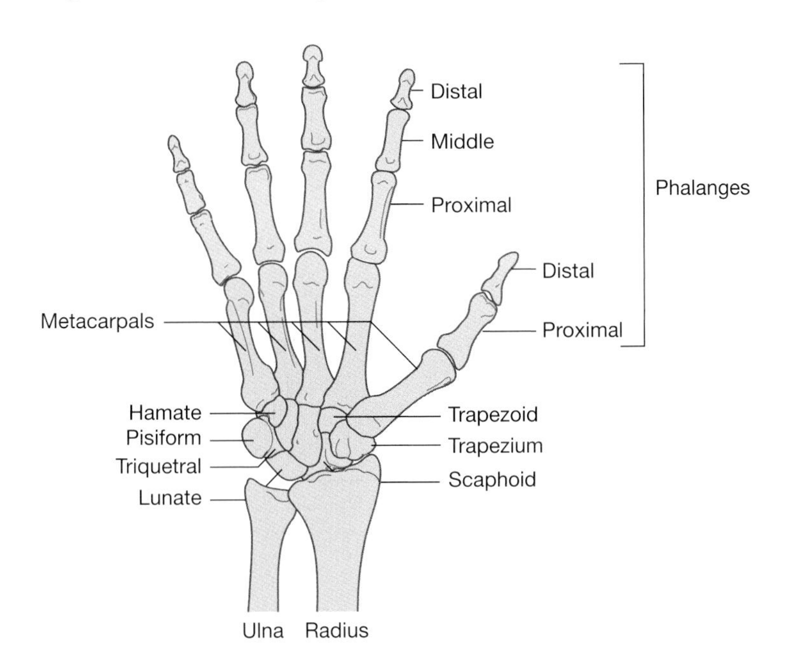


**Figure 17.6** Ossification of the wrist.
Adapted from Butler *et al.*, (2012).



and reabsorbing material from the inside (in a process known as endosteal resorption). Alongside this process the long bones add length by adding to the epiphyseal plate (Figure 17.5). As the bones lengthen they are constantly remodelling, therefore changing their outer shape. This continues throughout life, demonstrating that bone is a living, metabolizing organ (Peate, 2011).

This bone development is predictable, and ossification of the carpal bones begins with the capitate and ends with the pisiform. At birth there is no calcification in the carpals; therefore, this can be used as a measurement of bone age against chronological age. Skeletal age assessment is important in children that have growth and endocrine disorders. Skeletal age is measured from the left wrist and hand via an X-ray looking at the carpals, metacarpals, phalanges, radius and ulna and comparing them with a standard score to assess bone age (Figure 17.6 and Table 17.3).

**Table 17.3** Ossification. Adapted from Knipe *et al.*, (http://radiopaedia.org/articles/ossification-centres-of-the-wrist) as cited in Butler *et al.*, (2012).

| Ossification site | Bone | Time of ossification |
|---|---|---|
| Carpal bones | Capitate | 1–3 months |
| | Hamate | 2–4 months |
| | Triquetral | 2–3 years |
| | Lunate | 2–4 years |
| | Scaphoid | 4–6 years |
| | Trapezium | 4–6 years |
| | Trapezoid | 4–6 years |
| | Pisiform | 8–12 years |
| Centres of the distal radius and ulna | Distal radius | 1 year |
| | Distal ulna | 5–6 years |



## Clinical application

Osteogenesis imperfecta (OI), also known as brittle bone disease, is a genetic condition that is characterized by fragile bones that are prone to fracture with little or no trauma. OI varies in severity from person to person and can be mild to severe. There are four types of OI, which are identified mostly by the amount of fractures, their severity and other identified features: Type I, mild forms; Type II, extremely severe; Type III, severe; Type IV, undefined. Other health-associated problems with OI can include muscle weakness, hearing loss, fatigue, joint laxity, curved bones, scoliosis, blue sclera (white of the eye), dentinogenesis imperfecta (brittle teeth), short stature and skeletal malformations (http://www.oif.org/site/PageServer?pagename=AOI_Facts). OI occurs in about 1 in 20 000 people; although not that common, it is important for children's nurses to have an awareness of the condition. Children with OI require careful handling to minimize injury, and care will differ from child to chil,d so assessment of individual needs using a family-centred approach is essential. In addition, a multidisciplinary team approach is required to ensure all the child and family's needs are supported. The OI team should include a consultant paediatric neurologist, a consultant orthopaedic surgeon, a paediatric dental surgeon, clinical specialist occupational therapist and clinical specialist physiotherapists. OI is an important differential diagnosis for the children's nurse to consider when treating children with fractures where the history does not seem to fit with the injury presented.

## Blood supply

The blood supply in bones is provided in several ways. Figure 17.7 shows the Haversian canals (also known as central canals), which are small tubes that run longitudinally throughout the bone and carry blood vessels and nerves. The canals are surrounded by a network of expanding

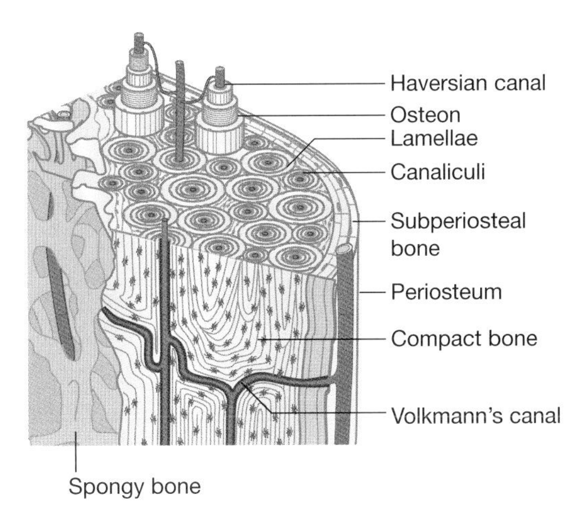


**Figure 17.7** Haversian canal.



rings that are made up of hard calcified matrix called lamellae. Alongside the lamellae are the lacunae, which contain osteocytes; these communicate with each other to allow circulation of interstitial fluid throughout the bone. Volkmann canals are microscopic structures found in compact bone. They run at right angles to the Haversian canals, and connect to each other and the periosteum. The periosteum allows the nerves, blood vessels and lymphatic vessels to enter the Volkmann canals and therefore penetrate the compact bone (Peate, 2011).

## Hormones and other factors required for bone growth

There are many other factors that are needed for the growth and structure of bone; any deficit will impact on the bones development, strength and growth. Table 17.4 lists some key elements.

# Bone healing

When a bone fractures, the ends of the bone and surrounding tissue bleed and a clot forms (haematoma). This triggers an inflammatory healing process to commence which releases macrophages that remove the dead tissue and haematoma. The initial strands from the haematoma change into osteoid tissue, forming a callus around the fracture site; any necrotic bone is cleared by osteoclasts and new bone is laid down by osteoblasts. It takes 6–12 weeks for the callus to harden and for the bone to begin to regain strength. The callus continues to mature, and by 1–2 years bone remodelling has occurred and normal bone shape has been restored. In children this whole process is much faster as their bones are already in a growing phase or osteogenic environment; because of this, children can restore a bony deformity up to a 30° malalignment through bone remodelling, leaving the bone with no signs of ever being broken (Lindaman, 2001). There are many types of fractures, and these are classified according to the nature of the injury and force that caused the fracture. Common types of fractures include (Duckworth and Blundell, 2010):

**Table 17.4** Some factors impacting on bone structure. *Source:* Peate and Nair (2011), Table 8.4, p. 236. Reproduced with permission of John Wiley and Sons, Ltd.

| Factor | Comment |
|---|---|
| Vitamins | A, C, D, K and $B_{12}$. There are a number of sources of these vitamins and their roles and functions vary. Too high amounts can be toxic and too low amounts can, for example, stunt growth. |
| Hormones | Human growth hormone (hGH), insulin-like growth factors (IGFs), oestrogens, androgens and thyroid hormones (for example, parathyroid hormone). Some of these hormones are produced in the pituitary gland; hence, any condition affecting this gland could result in problems related to bone formation and structure. IGFs are produced by the liver in response to hGH. If oestrogen (the sex hormone produced in the ovaries) and androgens (the sex hormone produced in the testes) are not released, there are potential complications associated with a number of body functions, including bone. |
| Minerals | Calcium, phosphorus, magnesium and fluoride. Calcium is a mineral that becomes available to the rest of the body when remodelling occurs. The amount of calcium in the bloodstream must be within precise limits otherwise serious problems can occur (the heart may stop if there is too much circulating calcium). |
| Exercise | All exercise is beneficial to the body, but weight-bearing exercise is particularly important for the stimulation of bone growth, particularly those exercises that place stress on the bones. When placed under stress, bone tissue becomes stronger as remodelling occurs; without this stress the bone will weaken and demineralization occurs. |



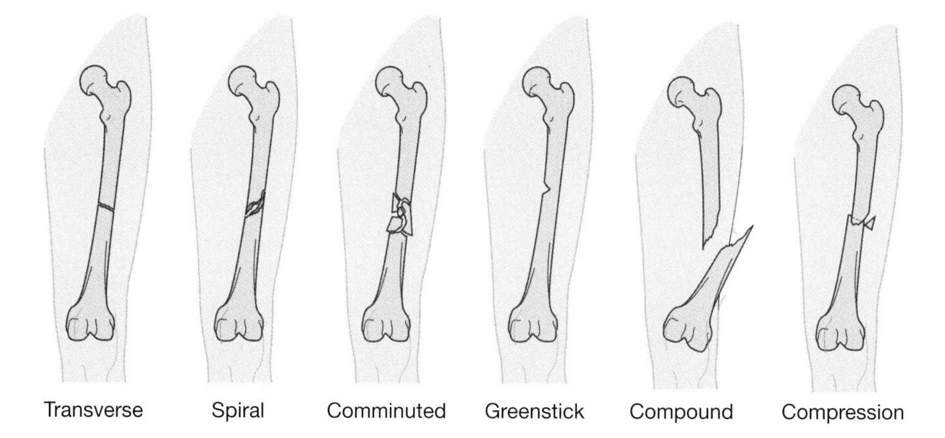


**Figure 17.8** Common types of fractures.

- Stable fracture – the bone is broken but the ends of the bone line up and are barely out of place.
- Open, compound fracture – the skin may be pierced by the bone or by a blow that breaks the skin at the time of the fracture. The bone may or may not be visible in the wound.
- Transverse fracture – a horizontal fracture line.
- Spiral fracture – this type of fracture spirals round the bone.
- Comminuted fracture – the bone shatters into three or more pieces.
- Compression fracture – where bones are compressed and a fracture appears, common in crush injuries.
- Greenstick fracture – fracture is incomplete and common in children.

## Clinical application

Fractures are common in children, and most are straightforward; however, there are concerns if a fracture extends through the epiphyseal growth plate as this can result in interference or complete cessation of growth, possibly causing length discrepancy in the limb. Epiphyseal injuries account for 15–30% of all skeletal injuries in children (Clark *et al.*, 2005). These fractures are known as epiphyseal fractures, or growth plate fractures, and have been classified by Salter and Harris (1963) (as cited in Clark *et al.*, 2005); the Salter–Harris classification system is the most recognized and widely used method of grading epiphyseal fractures. It is based on the involvement of the structures surrounding the fracture, such as the epiphysis and the joint; these are graded into five types of epiphyseal fractures (Figure 17.9).

It is crucial that epiphyseal fractures are correctly identified and managed to prevent growth impairment and long-term bone deformities that can persist throughout the treatment of the injury.

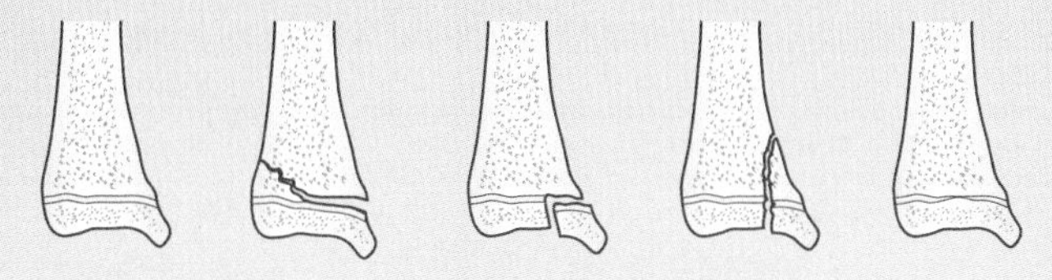

**Figure 17.9** Salter–Harris classification of epiphyseal fractures.



# Bone classification

There are five different bone types, and their shapes reflect their functions.

- Long bones, which are the femur, humerus, radius, ulna, clavicle, tibia, fibula, metacarpals, metatarsals and phalanges. Long bones act as levers, to raise and lower, and are classified by having a body that is longer than it is wide. They have a diaphysis, or shaft, which has growth plates, known as epiphyses, at each end of the bone and has a hard outer surface of compact bone which is covered by periosteum. Inside is spongy or cancellous bone, which contains the marrow, known as the medullary cavity. The ends of the bone are covered by hyaline cartilage, which acts as a shock absorber and protects the bone (Figure 17.10).
- Short bones are useful bridges, such as the talus bone in the foot. They are as wide as they are long; their main function is to provide support and stabilize, but with minimal movement. Examples of short bones are the carpals in the wrist and the tarsals in the foot. Their structure is mainly cancellous bone on the inside with large amounts of bone marrow covered by a thin layer of compact bone (Figure 17.11).
- Flat bones are strong flat plates of bone that form a protective shell; in addition, they provide a relatively large surface for muscle attachment. Examples of flat bones are the scapula, sternum, cranium, pelvis and ribs. These bones are formed from compact bone, providing strength for protection, and the centre consists of cancellous bone (Figure 17.12).
- Irregular bones are bones that do not fall into any other category owing to their irregular shapes. They mainly consist of cancellous bone with a thin outer covering of compact bone. Examples of irregular bones are the vertebrae, coccyx and mandible (Figure 17.13).

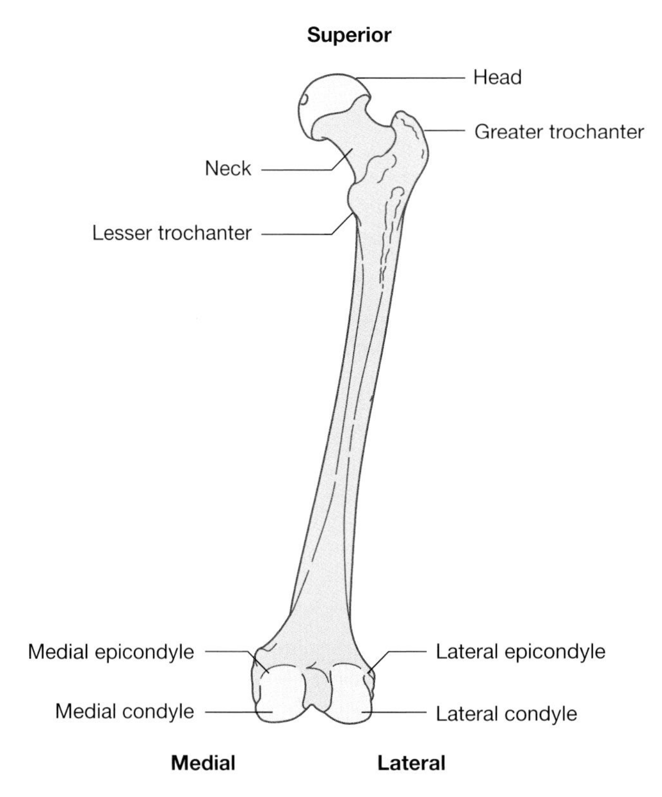




**Figure 17.10** A long bone – femur.

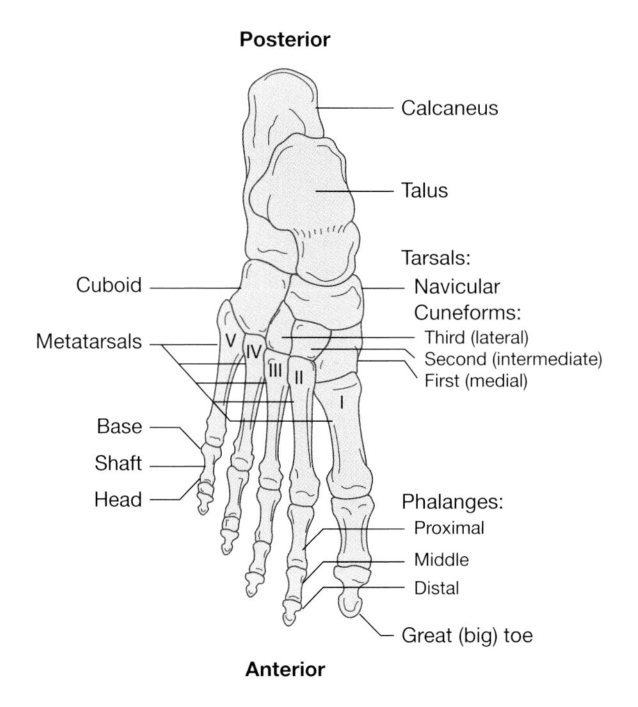


**Figure 17.11** The tarsal bones – short bone.

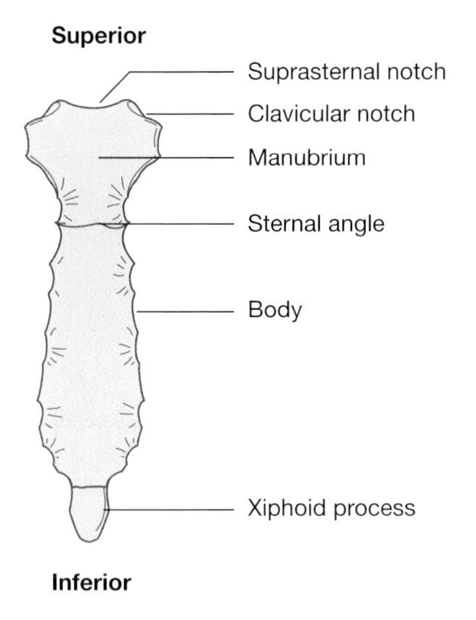


**Figure 17.12** A flat bone – the sternum.



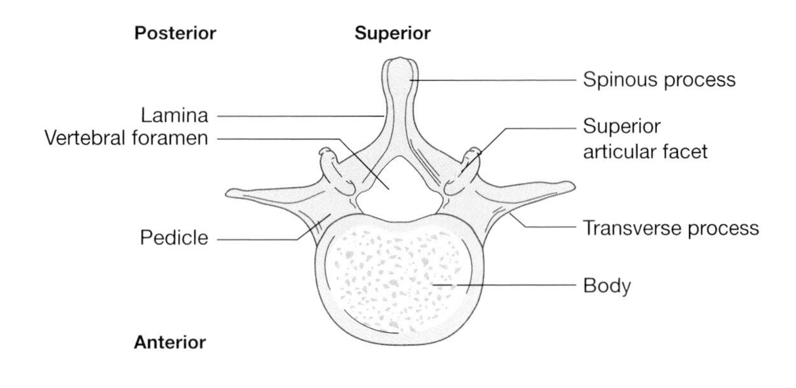


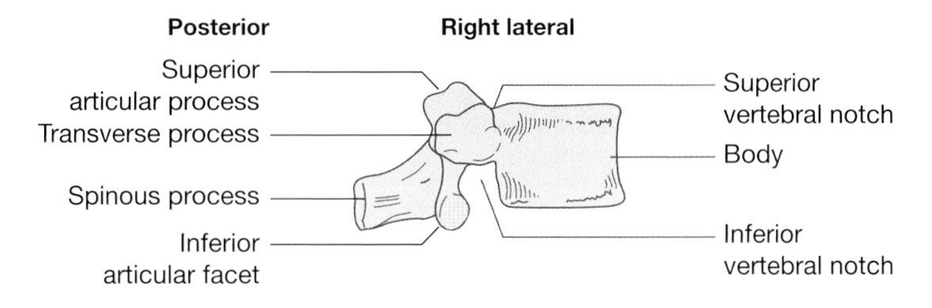


**Figure 17.13** The vertebrae – an irregular bone.

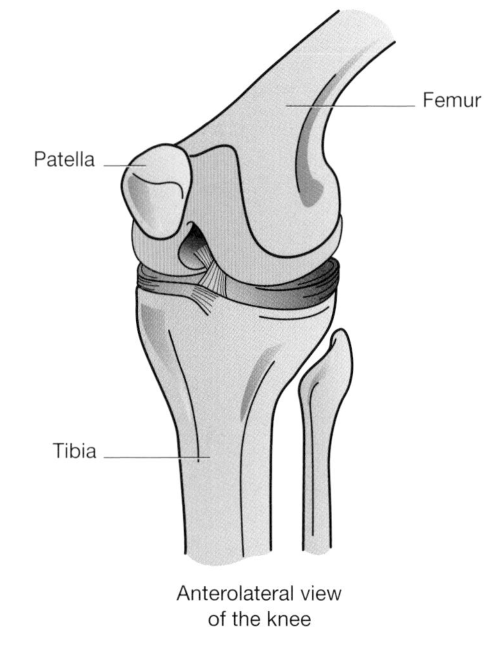




**Figure 17.14** The patella – a sesamoid bone.
*Source:* Peate and Nair (2011), Figure 8.10, p. 243. Reproduced with permission of John Wiley and Sons, Ltd.

- Sesamoid bones are small round shapes classified by being imbedded in a tendon; the most obvious is the patella. Sesamoid bones are usually present in a tendon that passes over a joint, which adds strength and protects the tendon (Figure 17.14).

# Joints

Joints are points where two or more bones meet and articulate or join together. There are three main types of joints: fibrous (immoveable), cartilaginous (partially moveable) and synovial (freely moveable).

## Fibrous joints

Fibrous joints, known as synarthrodial joints, are held together by one ligament, which is a tough fibrous material that prevents movement. Examples of these are the joints between the skull (sutures), teeth held in their bone sockets, and both the radioulnar and tibiofibular joints.

## Cartilaginous joints

Cartilaginous joints, or synchondroses, and symphyses occur where the connection between articulating bones is made up of cartilage; for example, between the vertebrae and the spine.

In children there are temporary joints, synchondroses, typically in the epiphyseal growth plates that are present until the end of puberty, when they ossify and growth is complete. Some of these joints allow some movement, such as between the vertebrae, where a fibrous disc separates, and the symphysis pubis, which is softened in pregnancy by hormones to allow childbirth.

## Synovial joints

Synovial joints, or diarthroses, are the most common joint in the human body and are identified by the presence of a joint space or synovial capsule, which is a collagenous structure surrounding the joint. They also contain synovial membrane, the inner layer of the capsule, which secretes synovial fluid, a thick sticky consistency providing nutrients for the structure and lubricating the joint and hyaline cartilage which lines the end of the articulating surfaces. There are six types of synovial joints, which are classified by their movement and shape. Table 17.5 (page 420) provides more detail.



# Conclusion

The chapter has provided an overview of the skeletal system. The skeletal system comprises all of the bones in the body; this is closely related to tissues such as tendons, ligaments and cartilage that connect them. The teeth are also deemed part of the skeletal system; they are not, however, counted as bones.

The key function of the skeleton is to provide support for the body. Without this support from the skeleton the body would be unable to remain upright and would collapse into a heap. Despite the fact that the skeleton is strong, it is also light. The skeleton also offers protection to the internal organs and fragile tissues of the body. The brain, eyes, heart, lungs and spinal cord are all afforded protection by the skeleton. It is the cranium that protects the brain and the eyes, the ribs protect the heart and the lungs, and the spine protects the spinal cord. Bones also provides the body with the ability to move.

Children are not young adults, and children differ significantly from adults with regard to skeletal anatomy and physiology. The differences in bone growth and modelling, and also remodelling, impact on the way in which conditions involving the skeleton can be viewed and managed.

**Table 17.5** Six types of joint. *Source:* Peate and Nair (2011), Table 8.5, pp. 244–246. Reproduced with permission of John Wiley and Sons, Ltd.

| Type of joint | Movement of joint | Examples | Structure |
|---|---|---|---|
| Hinge | A convex portion of one bone fits into a concave portion of another bone. The movement reflects the hinge-and-bracket movement of a household hinge and bracket: movement is limited to flexion and extension. The joint produces an open and closing motion. These joints are uniaxial. | Elbow<br>Knee 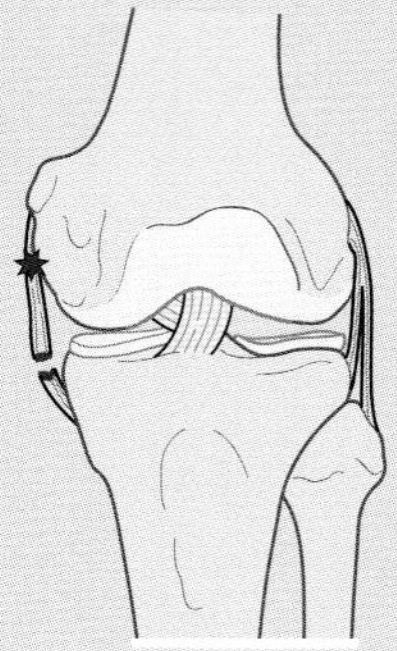 | 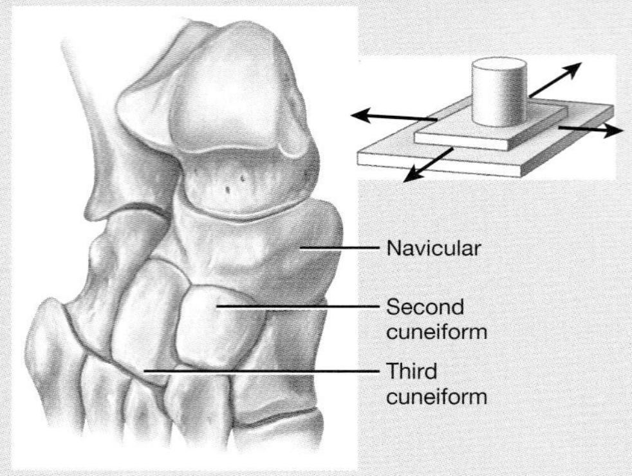 Navicular<br>Second cuneiform<br>Third cuneiform |
| Pivot | A rounded part of one bone fits into the groove of another bone. These joints will only permit movement of one bone around another – uniaxial movement | Radius and ulna<br>The atlas and the axis 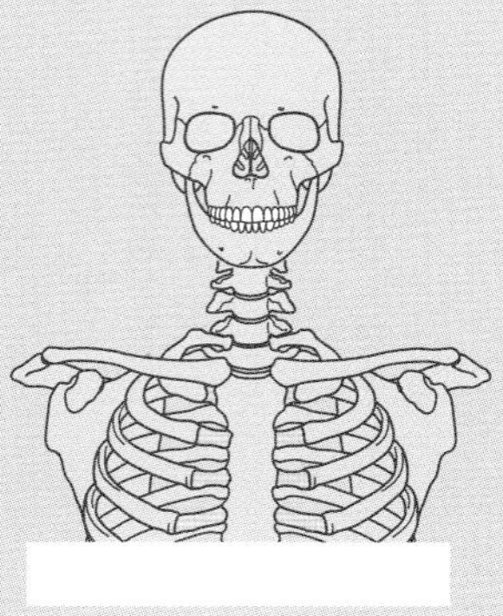 | 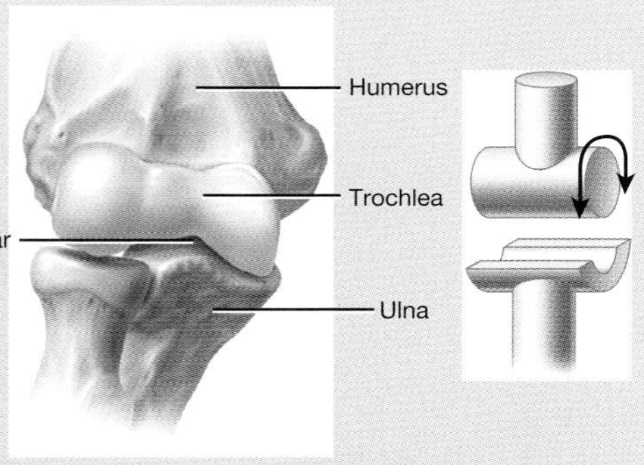 Humerus<br>Trochlea<br>Trochlear notch<br>Ulna |

| | | | |
|---|---|---|---|
| Ball and socket | The spherical end of one bone fits into a concave socket of another bone, hence ball and socket. Movement occurs through flexion, extension and adduction. This is a triaxial joint. | Hip<br>Shoulder | |
| Saddle | Similar to condyloid joints, but these joints permit greater movement. Allow flexion, extension and adduction. This is a triaxial joint. | The carpometacarpal joints of the thumb | |

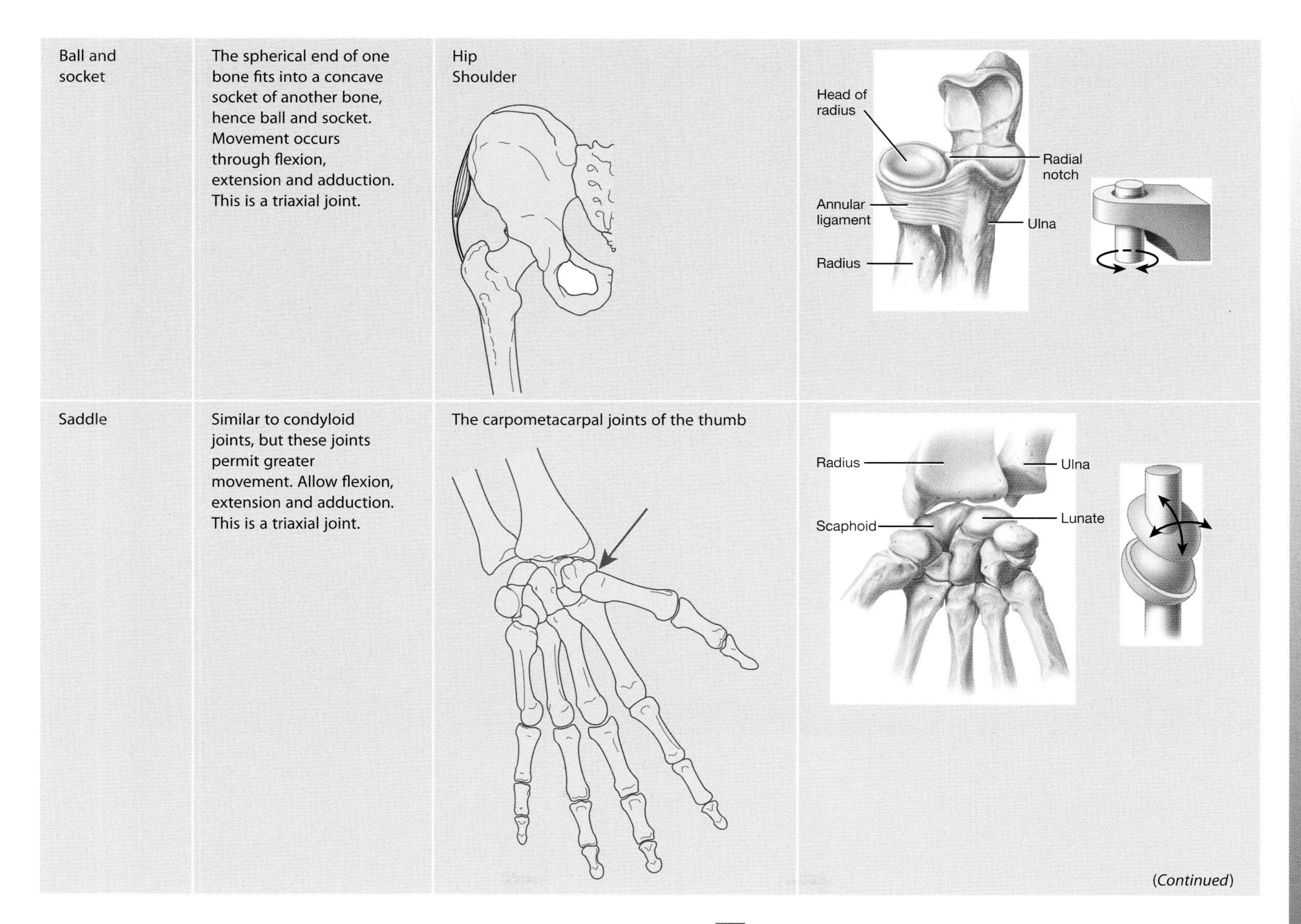


(*Continued*)

**Table 17.5** (*Continued*)

| Type of joint | Movement of joint | Examples | Structure |
|---|---|---|---|
| Condyloid | Where an oval surface of one bone fits into a concavity of another bone is where condyloid joints are found. Allows flexion, extension and adduction. This is a biaxial joint. | The radiocarpal and metacarpophalangeal joints of the hand 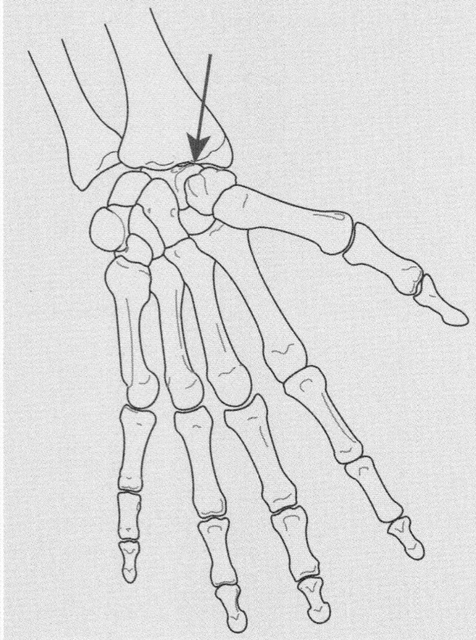 | 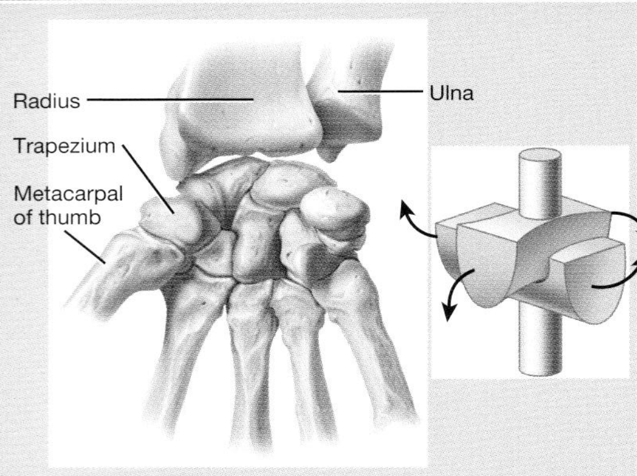 Radius<br>Ulna<br>Trapezium<br>Metacarpal of thumb |

| | | | |
|---|---|---|---|
| Gliding | These joints have a flat or slightly curved surface permitting gliding movement. The joints are bound by ligaments and movement in all directions is restricted. The joint moves back and forth and side to side. | Intertarsal and intercarpal joints of the hands and feet | 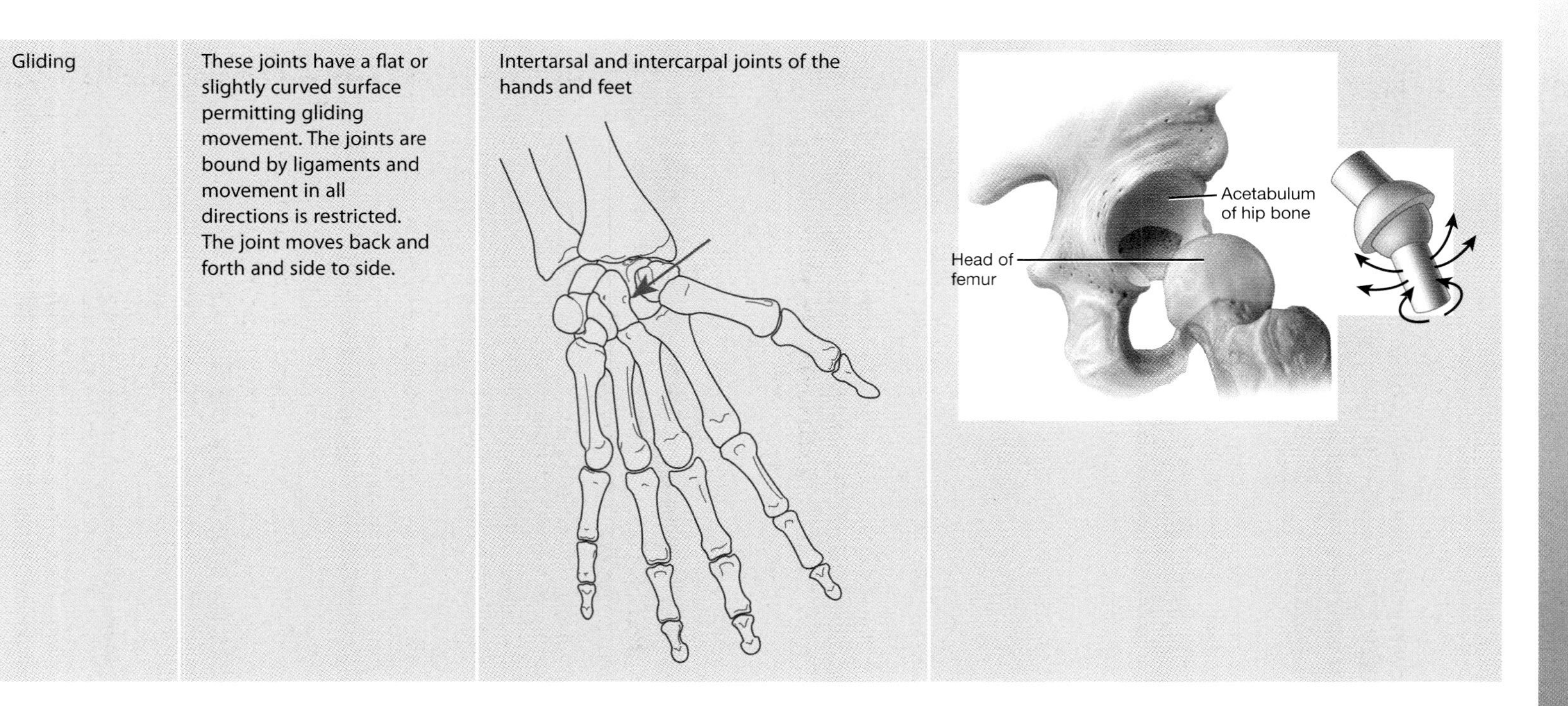 |

# Activities

**Now review your learning by completing the learning activities in this chapter. The answers to these appear at the end of the book. Further self-test activities can be found at www.wileyfundamentalseries.com/childrensA&P.**

## Wordsearch

There are several words linked to this chapter hidden in the square on page 424. Can you find them? A tip – the words can go from up to down, down to up, left to right, right to left, or diagonally.

| | | | | | | | | | | | | | | | |
|---|---|---|---|---|---|---|---|---|---|---|---|---|---|---|---|
| W | O | N | H | B | D | I | R | D | M | S | F | Y | R | O | A |
| P | S | N | E | A | H | R | S | U | G | D | H | E | S | H | N |
| Y | T | B | A | F | D | F | I | N | M | T | O | S | M | A | I |
| E | E | O | N | G | H | C | S | R | C | L | I | S | A | M | E |
| W | O | C | T | S | L | I | Y | O | W | C | A | K | N | O | I |
| E | B | N | H | A | O | N | H | I | L | G | R | N | W | T | G |
| T | L | P | C | E | K | U | P | E | A | T | A | S | E | A | E |
| H | A | V | E | R | S | I | A | N | C | A | N | A | L | M | P |
| S | S | L | G | R | E | P | I | C | O | N | D | Y | L | E | I |
| E | T | E | A | T | I | A | D | O | N | V | L | A | T | A | P |
| S | C | E | L | I | N | O | I | T | C | U | D | B | A | H | H |
| A | O | T | I | H | V | E | S | E | C | U | B | E | O | O | Y |
| M | I | I | T | U | I | O | E | T | M | H | L | E | E | E | S |
| O | D | I | R | N | A | L | N | O | E | I | R | R | D | E | I |
| I | O | K | A | L | A | C | I | Y | E | U | R | E | E | A | S |
| D | K | R | C | N | H | P | B | O | S | T | M | U | E | W | E |



## Match the bones to the body part

(a) metacarpal
(b) tarsal
(c) zygomatic
(d) humerus
(e) incus
(f) patella
(g) iliac crest

(A) knee
(B) face
(C) hand
(D) pelvis
(E) ankle
(F) upper arm
(G) ear

## True or false?

1. There are three bones in each finger of the hand.
2. The total number of bones in the axial skeleton is 80.
3. Cartilage that is replaced by bone is called epiphyseal ossification.
4. The radius and ulna are the same length.
5. Ossification of the bones is complete by the age of 3 years.
6. Rickets is a condition that occurs due to lack of vitamin D.
7. The acetabulum is part of the shoulder joint.
8. Osteocytes proliferate when a bone breaks.
9. The only two sesamoid bones in the human body are the patella and coccyx.
10. There are 26 bones in the vertebral column.

## Conditions

The following table contains a list of conditions. Take some time and write notes about each of the conditions. You may make the notes taken from text books or other resources; for example, people you work with in a clinical area or you may make the notes as a result of people you have cared for. If you are making notes about people you have cared for you must ensure that you adhere to the rules of confidentiality.

| Condition | Assessments | Conditions/cause |
|---|---|---|
| Osteomyelitis | | |
| Perthes' disease | | |
| Paget's disease | | |
| Osteoarthritis | | |
| Rickets | | |



# Glossary

**Adduction:** movement towards the body's midline.

**Anatomy:** the study of body structures and their relation to other structures in the body.

**Ball-and-socket joint:** a synovial joint in which the rounded surface of one bone fits within the cup-shaped depression of the socket of the other bone.

**Calcification:** deposition of mineral salts in a framework formed by collagen fibres in which tissue hardens.

**Cancellous:** a type of structure as seen in spongy bone tissue; resembles a latticework structure.

**Cartilage:** strong, tough material on the bone ends that helps to distribute the load within the joint; the slippery surface allows smooth movement between the bones; a type of connective tissue.

**Cartilaginous joint:** a joint where the bones are held together tightly by cartilage; little movement occurs in this joint. This joint does not have a synovial cavity.

**Collagen:** a protein that makes up most of the connective tissue.

**Condyloid joint:** a synovial joint that allows one oval-shaped bone to fit into an elliptical cavity of another.

**Diaphysis:** the shaft of a long bone.

**Epiphysis:** the ends of long bone.

**Fibrous joint:** a type of joint that allows little or no movement.

**Flexion:** movement where there is a decrease in the angle between two bones.

**Fracture:** a break in a bone.

**Gliding joint:** a synovial joint whose articulating surfaces are usually flat, allowing only one side-to-side or back-and-forth movement.

**Haematopoiesis:** the formation and development of blood cells in the bone marrow after birth.

**Hormone:** the secretion of endocrine cells that have the ability to alter the physiological activity of target cells in the body.

**Insulin-like growth factor (IGF):** produced by the liver and other tissues, this is a small protein that is produced in response to human growth hormone.

***In utero*:** within the uterus.



**Lacuna:** a small, hollow space found in the bones where osteocytes lie.

**Lamellae:** rings of hard, calcified matrix found in compact bones.

**Ligaments:** tough fibrous bands of connective tissue that hold two bones together in a joint.

**Macrophages:** cells that engulf and digest cellular debris and pathogens.

**Marrow:** a sponge-like material found in the cavities of some bones.

**Mesenchyme:** embryonic connective tissue from which nearly all other connective tissue arises.

**Metaphysis:** the aspect of long bone that lies between the diaphysis and the epiphysis.

**Ossification:** the formation of bone, sometimes called osteogenesis.

**Ossicles:** small bones of the middle ear – the malleus, stapes and incus.

**Osteoblasts:** cells that arise from osteogenic cells; these cells participate in bone formation.

**Osteoclasts:** large cells that are associated with absorption and removal of bone.

**Osteocytes:** mature bone cells.

**Periosteum:** membrane covering bone consisting of osteogenic cells, connective tissue and osteoblasts. This is vital for bone growth, repair and nutrition.

**Pivot joint:** a joint where a rounded or conical-shaped surface of a bone articulates with a ring formed partly by another bone or ligament.

**Remodelling:** replacement of old bone by new.

**Resorption:** absorption of what has been excreted.

**Saddle joint:** a synovial joint articulates the surface of a bone that is saddle shaped on the other bone that is said to be shaped like the legs of the rider.

**Spongy (cancellous) bone:** bone tissue comprised of an irregular latticework of thin plates of bone known as trabeculae. Some bones are filled with red bone marrow and these are found in short, flat and irregular bones as well as the epiphyses of long bones.

**Synovial cavity:** the space between the articulating bones of a synovial joint, filled with synovial fluid.

**Synovial fluid:** the sections of the synovial membranes that lubricate the joints and nourish the articular cartilage.

**Trabeculae:** a network of irregular latticework of thin plates of spongy bones.

# References

Butler, P., Mitchell, A., Healy, J.C. (2012) Ossification centres in the wrist, *Applied Radiological Anatomy*, Cambridge University Press.

Clark, T.J., Eberman, L.E., Cleary, M.A. (2005) Physeal growth plate fractures: implications for the pediatric athlete. In: *COERC 2005, Proceedings of the Fourth Annual College of Education Research Conference*, pp. 2–8.

Duckworth, T., Blundell, C.M. (2010) *Orthopaedics and Fractures (Lecture Notes)*, 4th edn, Wiley–Blackwell.

Lindaman, L.M. (2001) Bone healing in children. *Clinics in Podiatric Medicine and Surgery*, **18** (1), 97–108.

MacGregor, J. (2008) *Introduction to the Anatomy and Physiology of Children. A Guide for Students of Nursing, Child Care and Health*, 2nd edn, Routledge.

Oldfield, S. (2004) The development of movement. In: Neill, S., Knowles, H. (eds), *The Biology of Child Health. A Reader in Development and Assessment*, Palgrave Macmillan, pp. 87–121.

Pearce, S.H.S., Cheetham, T.D. (2010) Diagnosis and management of vitamin D deficiency. *BMJ*, **340**, b5664.

Peate, I. (2011) The skeletal system. In: Peate, I., Nair, M. (eds), *Fundamentals of Anatomy and Physiology for Student Nurses*, Wiley–Blackwell, pp. 224–257.

Salter, R., Harris, W. (1963) Injuries involving the epiphyseal plate. *Journal of Bone Joint Surgery*, **45**, 587–622.

Scientific Advisory Committee on Nutrition (2007) Update on vitamin D. Position statement by the Scientific Advisory Committee on Nutrition, The Stationery Office, London.

Tortora, G.J., Derrickson, B. (2009) *Principles of Anatomy and Physiology*, 12th edn, John Wiley & Sons, Inc., Hoboken, NJ.

Waugh, A., Grant, A. (2010) *Ross and Wilson Anatomy and Physiology in Health and Illness*, Churchill Livingston.

# Chapter 18

# The senses

Joanne Outteridge

*Department of Family and Community Studies, Anglia Ruskin University, Cambridge, UK*

## Aim

The aim of this chapter is to discuss the senses of smell, taste, hearing and sight, exploring how the structures involved receive environmental signals and process these into nerve impulses to be interpreted by the brain.

## Learning outcomes

By the end of this chapter the reader will be able to:

- State the anatomical, physiological and neurological requirements for the chemical senses of taste and smell.
- Outline how the anatomy of the ear enables it to perform the functions of both hearing and equilibrium (balance).
- Discuss how hearing and balance are sensed and translated into neurological signals to be processed by the brain.
- Recognize and label the anatomical structures of the eye.
- Explain how the retina processes visual images, and the role of the visual cortex of the brain in receiving and processing those images.
- Critically analyse how normal growth, development and family functioning are affected when a child has a hearing or visual disorder.

*Fundamentals of Children's Anatomy and Physiology: A Textbook for Nursing and Healthcare Students*, First Edition. Edited by Ian Peate and Elizabeth Gormley-Fleming.

Companion website: www.wileyfundamentalseries.com/childrensA&P

## Test your prior knowledge

- To what do the terms 'olfaction' and 'olfactory' refer?
- Where are the taste buds situated?
- What are the four main tastes that we can process?
- What is the Eustachian tube, and where is it?
- What is the organ of Corti responsible for?
- What are the semicircular canals responsible for?
- What is the coloured part of the front of the eye called, and what is its function?
- What are the rods and cones of the retina?
- What is your 'blind spot'?
- At what age do all children have routine visual screening, and why?

# Introduction

The way in which children interact with their environment relies on their senses receiving information, their brains processing that information and then providing either a physical or a culturally appropriate social response to that information. The aim of this chapter is to introduce the senses of smell, taste, hearing and sight, which along with the tactile senses of pain, pressure and temperature (Chapter 19) provide the child with the information that they need in order to make sense of their environment. Each of the senses will be considered separately, with relevant embryology, anatomy, physiology and application to clinical practice.



# The sense of smell (olfaction)

The nasal cavity contains folds of mucous membranes: the nasal conchae. These are responsible for warming and filtering inspired air (see Chapter 10). The mucus secreted allows chemicals in the environment to dissolve on those mucous membranes. The molecules in these chemicals are then sensed by the olfactory cells in the epithelium and translated into electrical impulses which are signalled via cranial nerve I (olfactory nerve) to the brainstem for processing. Olfactory interpretation in the temporal lobe of the brain is linked to the limbic area, meaning that certain odours may evoke memories and emotions.

## Embryology

In the embryo, the development of the face occurs between weeks 5 and 12, with the nose growing downwards between weeks 6 and 9. There are initially two nasal processes that will form the nostrils, which are then joined by the downward-growing pillar of tissue which forms the nasal septum. This occurs at the same time as the formation of the hard and soft palate in the mouth. If the tissue fails to meet in the midline of the face, the child may be born with cleft lip and/or palate.

An infant's vision is limited at birth (see the 'Sight' section). Therefore, other senses become more important in their interpretation of their environment, and their sense of smell is particularly acute. They use the smell of milk to locate the nipple for feeding, and quickly recognize their mother by her individual smell (Porter and Winberg, 1999). Equally, fathers and other significant carers/siblings can be encouraged to have skin-to-skin contact with newborns in order that they are recognized early. It has been suggested that perfumed bath products and fragrances can interfere with this infant recognition, and should be avoided in the first 4 weeks until facial recognition begins.

# The sense of taste (gustation)

Taste buds are physical areas of primarily the tongue, but also the epiglottis, pharynx and palate (Figure 18.1). Each taste bud has a group of receptor cells, and taste buds are grouped together in 'papillae'. Like the sense of smell, chemical compounds from the environment dissolve in the saliva, activate the receptor and produce electrical signals. These electrical impulses then transmit to the brainstem via cranial nerves VII (facial), IX (glossopharyngeal nerve) and X (vagus nerve). These signals pass through the medulla, then the thalamus and then onto the parietal lobes where interpretation takes place.

Much of the sense of taste is dependent upon the sense of smell working. We all know that when we have a cold our food does not taste the same; children may hold their nose when swallowing medicines that are unpalatable. The traditional four main tastes that a human can process are sweet, sour, bitter and salty. Metallic and umami (meaty) have more recently been added to this list (Thibodeau and Patton, 2012).



## Clinical application

As taste relies on chemicals dissolving on the tongue, some drugs can interfere with a person's sense of taste. The most common culprit is smoking; it is also known that some antibiotics and cancer chemotherapeutic agents will cause alteration in taste sensation.

## Embryology

The sense of taste develops *in utero* (Chamley *et al.*, 2005). The fetus swallows amniotic fluid continuously, and this is flavoured with what the mother has been eating and drinking. Strong flavours in particular, such as garlic and curry, are transmitted. It is suggested that we all need to be exposed to a taste a minimum of 14 times in order for us to 'like' a particular taste. Thus, the infant's sense of taste preference is thought to be correlated with what the mother has eaten during pregnancy (Bakalar, 2012). However, it has been shown that infants do naturally show a preference for sweet tastes; breast milk is naturally very sweet. When faced with a choice of foods, all humans would naturally choose the foods for which they have a preferred taste (given the absence of societal influences). This needs to be considered in conjunction with Chapter 12 ('Fetal development and infant nutrition' section) in the discussion on weaning diet. If a child is repeatedly exposed to only foods that they enjoy, they will not develop a liking for other foods. In addition, if they are not allowed to become hungry due to constant grazing, they will be unwilling to try foods that are not their favourite tastes. This is in conjunction with the social development issues as discussed in Chapter 12 ('Fetal development and infant nutrition' section) where food can be a mechanism of control over a very adult-led environment. However, the sense of taste is also genetic (Breen *et al.*, 2006). In particular, a genotype has been

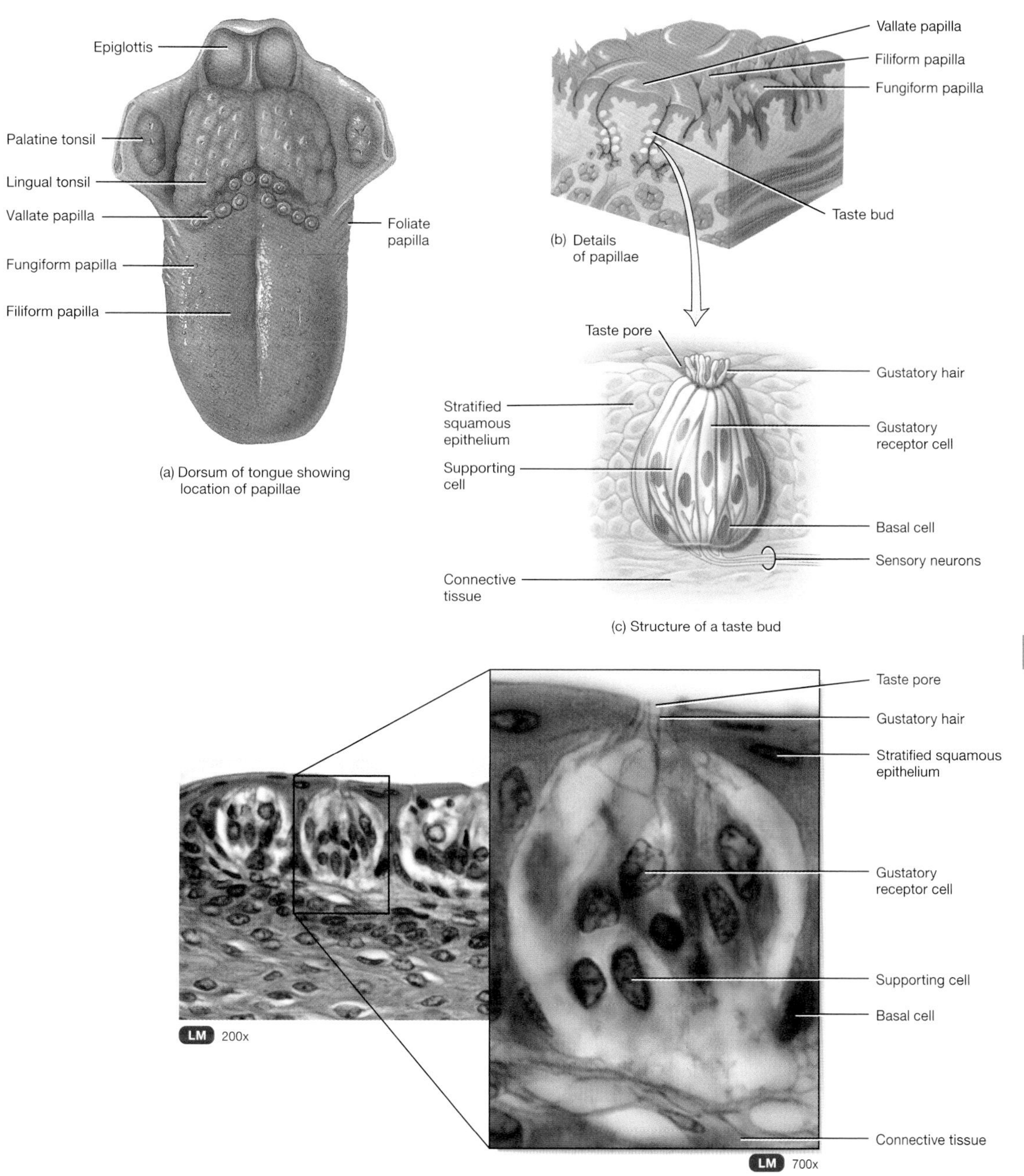


**Figure 18.1** Location and structures of tongue papillae and taste buds.
*Source:* Tortora and Derrickson (2009), Figure 17.3(a–c), p. 603. Reproduced with permission of John Wiley and Sons, Inc.

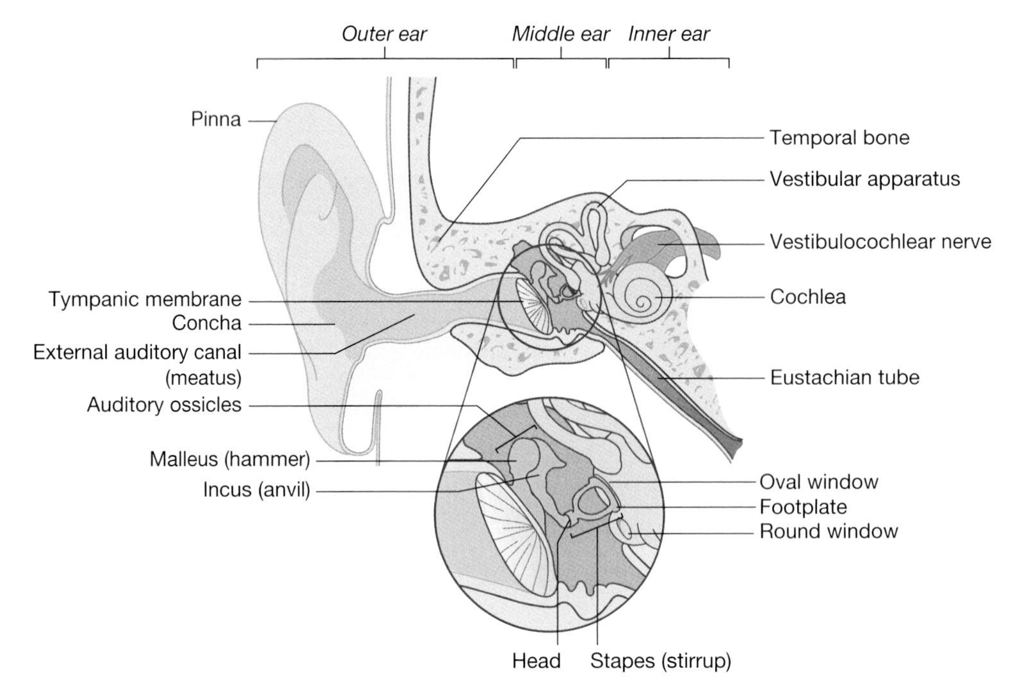




Figure 18.2 The anatomy of the ear.

identified that makes cruciferous vegetables (cauliflower, broccoli, sprouts) taste particularly unpleasant.

# The ear

The ear is responsible for the sense of hearing and equilibrium (balance). Each is controlled separately: hearing by the organ of Corti in the cochlea, and balance by the semicircular canals and vestibule (Figure 18.2).

## Hearing

### *Embryology*

By the fourth week of gestation, otic pits are present that will become the inner ear, and by the sixth week the external ear canal and pinna are formed. Ears are originally low set and move upwards until 24 weeks, and by 22–24 weeks' gestation the fetus will respond to noise as the sounds from the environment are transmitted through the mother's abdominal wall and through the amniotic fluid. As gestation progresses, the fetus begins to recognize regular voices, and at birth will be able to recognize the mother and other members of the same household by their voices. It is thought that this leads to development of the language centre of the brain and results in infants being born with a preference for their mother's language. Infants show a greater response to words spoken in their own language than a foreign language, although the vocal ability to speak any language exists up until 7 months of age. After this, if there has been

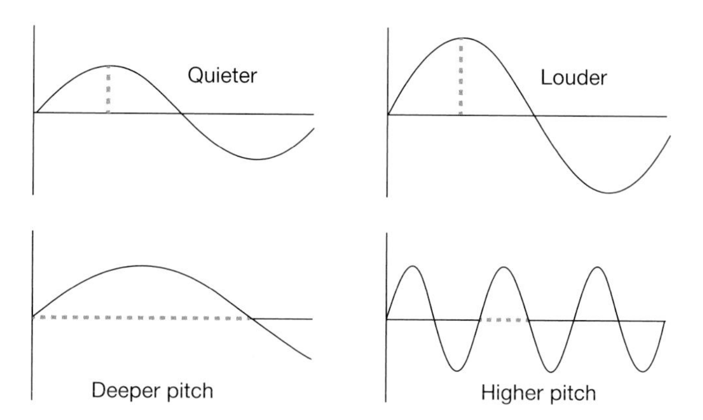


**Figure 18.3** Properties of sound waves with different pitch and volume.

no exposure to other languages with sounds not found in the infant's native language, the language centre of the brain diminishes its ability to reproduce these sounds (Kisilevsky *et al.*, 2009; Kuhl, 2010), and the majority of people will not be able to replicate the language exactly. (For example, some sounds in Chinese do not exist in Western vocalization. If the infant has not heard these sounds before 7 months of age, it is probable that they will never be able to exactly replicate these sounds.)



Sound travels as waves through the air. Sound waves are measured by frequency and amplitude. The frequency, or pitch (how high or how low the sound), depends on how many waves per second and is measured in hertz (Hz); the more waves, the higher the sound (Figure 18.3). The amplitude is how loud the sound is, and is dependent upon the size of the wave (big or small), and is measured in decibels (dB).

## External ear

This sound travels down the external auditory canal, and 'hits' the tympanic membrane. The frequency (pitch – high or low) determines how fast the membrane vibrates. The amplitude (volume or loudness) determines the extent of the deviation of the membrane. Think of the tympanic membrane like a trampoline: the frequency (pitch – high or low) is how fast you are jumping; the amplitude (volume or loudness) is determined by how hard you are landing on the trampoline and its deviation from its neutral position. This has now translated a sound wave into a mechanical vibration.

## Middle ear

This vibration is now transmitted from the tympanic membrane to the small bones of the middle ear: the hammer (malleus), anvil (incus) and stirrup (stapes). These vibrate against each other and transmit the signal to the oval window (Figure 18.2). This middle-ear cavity contains air, allowing the fast transmission of the sound signals. However, one problem with this is the way in which air expands or contracts in response to pressure. In order to allow the body to equalize the pressure in the middle ear when faced with increasing pressure (for example, diving to the bottom of a swimming pool), or with decreasing pressure (for example, when ascending in an aeroplane), there is a small tube, the Eustachian tube, joining the middle ear to the oropharynx. This allows air to enter and exit the middle ear.

## Clinical application

Owing to the shape of the infant's head and face, the Eustachian tube is fairly horizontal. As the child's face elongates the Eustachian tube becomes more vertical. However, during early childhood, the child is particularly prone to upper respiratory tract infections due to developing immunity (see Chapter 7). The excess mucus produced can travel up the Eustachian tube and into the middle ear, and the inflammation and swelling in the oropharynyx can prevent its drainage. A similar effect happens due to inflammatory responses triggered by environmental cigarette smoke in the home. This can lead to 'glue ear'. This prevents the sound from the tympanic membrane being transmitted through the malleus, incus and stapes to the oval window, leading to conductive hearing loss. If this is persistent in early childhood, it can interfere with speech development. Therefore, the mucus needs to be drained, usually by insertion of 'grommets'; a small plastic tube placed in the tympanic membrane.

### *Inner ear*

The oval window then leads directly to the cochlea (Figure 18.4). This is spiral shaped, with a bony outer labyrinth filled with a fluid called 'perilymph'. Within this is a membranous labyrinth, filled with fluid called 'endolymph', and containing the organ of Corti.



The frequency and amplitude of mechanical vibration of the oval window is now translated into fluid waves (frequency – how many waves per second; amplitude – how high are those waves). These travel from the oval window, to the outer perilymph, to the inner endolymph and move the small hairs on the organ of Corti. The extent to which these hairs move in the fluid

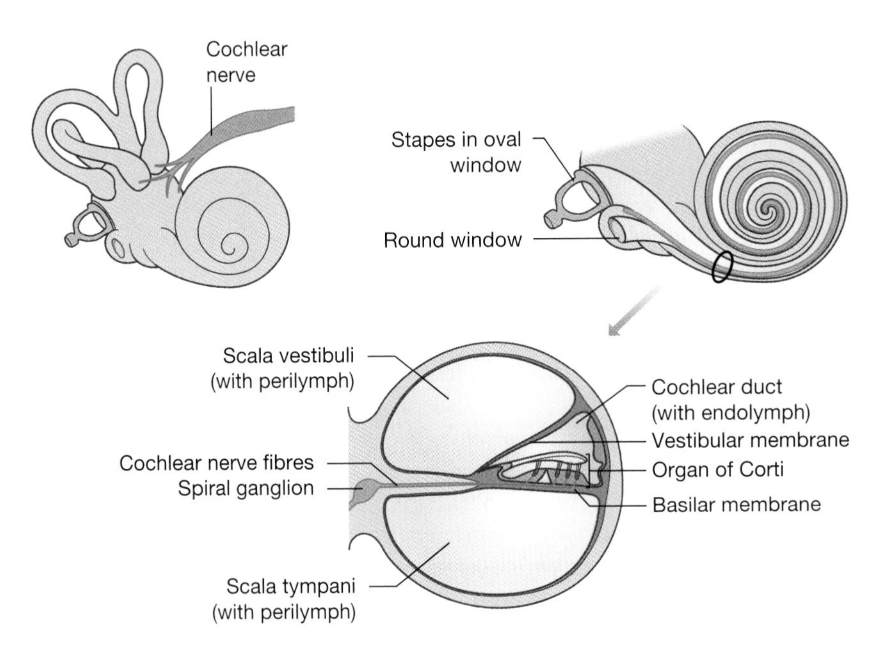


**Figure 18.4** Cochlear anatomy.

waves (think seaweed moving in sea currents) causes an electrical impulse. These impulses are then sent by cranial nerve VIII (auditory or vestibulocochlear) to the brainstem and are passed to the auditory centre on the temporal lobes of the brain for processing. Speech development in childhood relies on a child hearing a sound and repeatedly attempting to replicate it until they think it sounds the same. The social response from others will confirm or disprove their assumptions. Thus, language development relies upon repeated exposure to sound (through talking to young children), and encouragement of practice of sounds even when unintelligible. Criticism or constant correction can lead to children not trying sounds; development of intelligible speech relies on practice.

## Clinical application

Children diagnosed with autistic spectrum disorder have a cluster of neuro-typical behaviours which lead to this diagnosis; one of these is auditory hypersensitivity (Lucker, 2013). They become distressed at some sounds, which are individual for each child but typically include very sudden loud sounds and echoing sounds (for example, this is seen in toddlers when they scream every time there is a knock at the door or cannot tolerate going to parent and toddler groups in large halls).



## Equilibrium

The infant's sense of balance begins to develop as they learn head control, roll, then sit up, then crawl and then walk. Thus, it develops in line with the musculoskeletal system (Chapters 16 and 17). Some sense of balance remains under conscious control (for example, walking a tightrope), but other balance is subconscious (for example, not falling sideways off a chair).

The sense of equilibrium is our ability to sense one's personal space in relation to the physical surrounding environment. It allows us to know which way is up and which way is down, and to maintain body position in relation to this.

Equilibrium is controlled by the semicircular canals and the vestibule. The semicircular canals are three fluid-filled rings perpendicular to each other, providing a three-dimensional awareness of space. They have the outer space filled with perilymph and the inner space filled with endolymph, similar to the cochlea. Like the cochlea, there are hair-like projections into the endolymph, detecting movement of the fluid as your head moves. At the base of each half circle is a dilated area, the ampulla, with specialized receptor cells called a crista ampullaris that also help to detect this fluid movement (Figure 18.5).

The semicircular canals provide a rapid response to dynamic changes, and we then consciously alter our body posture to compensate for this movement (in conjunction with the musculoskeletal system – Chapters 16 and 17).

The vestibule also contains endolymph and a denser otolithic membrane. If the head moves, the fluid moves more quickly than the membrane, triggering the receptors. This is a slower response than the semicircular canals and informs our brain of our position relative to gravity, as signals from the left and right ears should be equal in a steady position. If the messages sent from the vestibule and the eyes do not match, this can cause nausea or disrupt balance (for example, motion sickness).

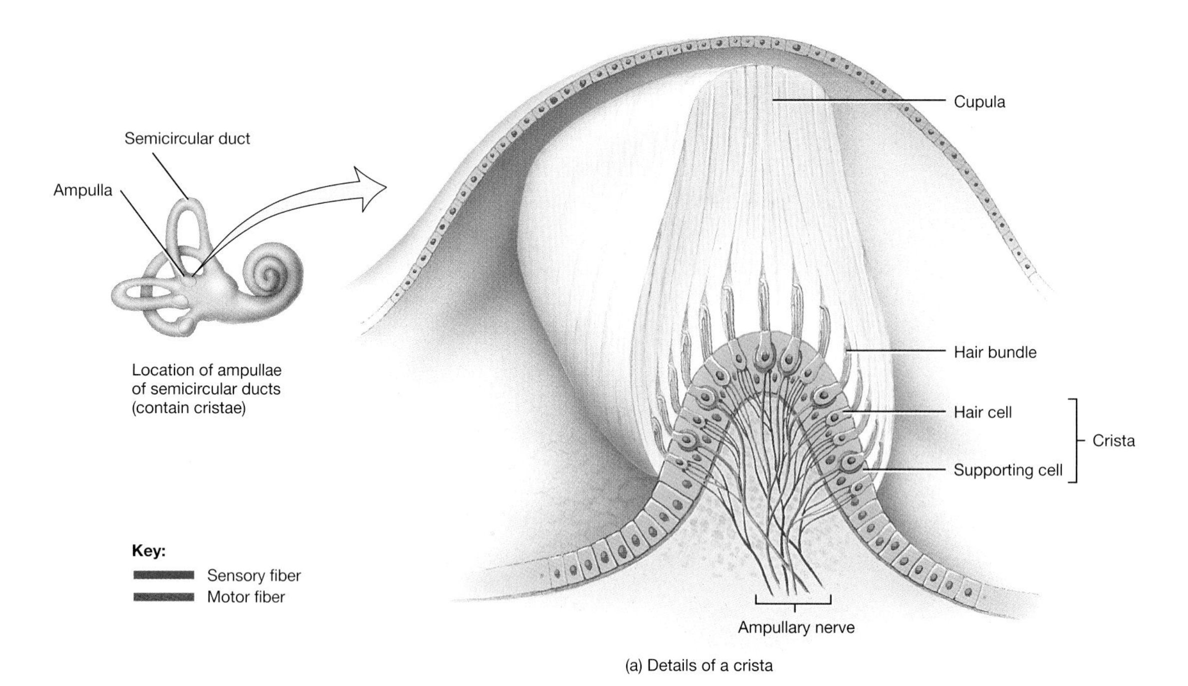




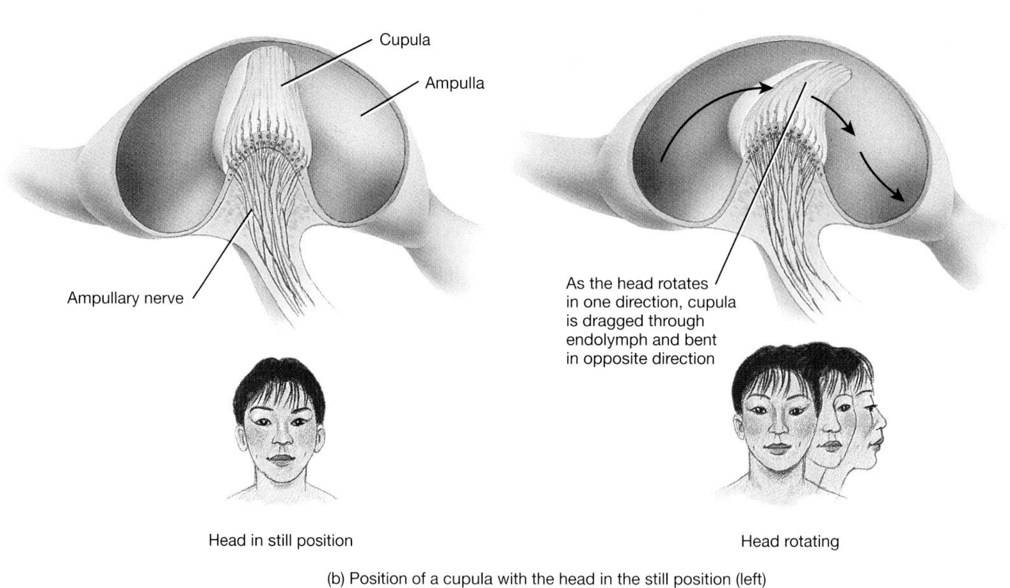


**Figure 18.5** Structure of a crista ampullaris.
*Source:* Tortora and Derrickson (2009), Figure 17.25, p. 630. Reproduced with permission of John Wiley and Sons, Inc.

## Activity idea

SAFETY ALERT – PLEASE ASSESS WHETHER THIS ACTIVITY IS SUITABLE FOR YOU. WE CANNOT BE LIABLE IF YOU CHOOSE TO DO THIS AND INJURE YOURSELF.

Stand in the middle of a room not touching anything, shut your eyes and stand on one leg. See how long it takes before you fall sideways.

The majority of people will fall within 30 s; this shows how important sight is in relation to maintaining balance. Those individuals who exceed this time are usually those who participate in sports activities requiring good balance skills; they are therefore able to use their perception of pressure from their foot on the ground to adjust their balance. This demonstrates that balance skills can continue to be learnt throughout life. Thus, the sense of balance is interpretation of data from the eyes, the vestibules of the ear, and the joints and muscles.

## Clinical application

Some programmes for older people work to improve sense of balance to try to reduce the incidence of falls. As more mothers are forced back to work due to financial pressures, grandparents are often the main carers during working hours. The safety of a child needs to be very carefully considered if a grandparent has had a previous fall.



# Sight

The sense of sight involves a visual image being sensed by the retina and then travelling via the optic nerve to the occipital lobe of the brain, where interpretation takes place. Thus, development of sight relies on both eye and brain development occurring concurrently.

## Embryology

At 4 weeks' gestation, the optic vesicles are present, which will then become the eyes. Eyelids grow towards each other and fuse by week 8, then reopen at around 24 weeks' gestation. Eyes are initially at the side of the developing head, but move towards the midline by week 14. The visual cortex of the brain develops rapidly during weeks 28–32, and myelination of the visual pathway starts shortly before birth and continues until 10 weeks postnatally. In the womb, fetuses respond to bright light from around 24 weeks onwards.

## Clinical application

The capillaries in the retina are extremely delicate and, therefore, liable to damage easily. This is especially relevant in the pre-term infant when administering oxygen, which can cause free-radical damage, leading to retinopathy of prematurity and subsequent reduced or absent vision. To try to prevent this, oxygen is used very cautiously in neonatal units, with infants' oxygen saturations not exceeding 94% when receiving supplemental oxygen.

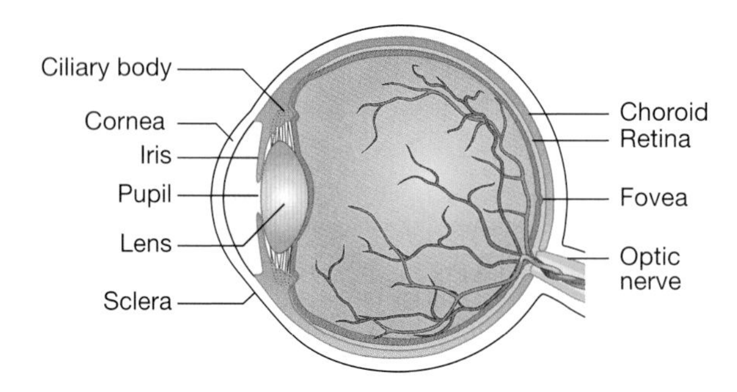


**Figure 18.6** Anatomy of the eye.

## The eye

The eye has three layers: the outer sclera, the choroid and the inner retina (Figure 18.6).

### The sclera

The outer part of the eye is the sclera: the 'white' of the eye. This is tough and fibrous and maintains the shape of the eyeball, providing protection and support. The front portion of the sclera is transparent – the cornea. The sclera is covered by a mucous membrane that is continuous with the inner aspect of the eyelids – the conjunctiva. This provides protection from environmental substances. The conjunctiva is kept moist by secretions – tears – from the lacrimal glands. The sclera is attached to the orbit of the skull by six muscles that control eye movement.



### The choroid

This is the middle layer of the eyeball, and it contains a dark pigment to prevent light from being reflected within the eyeball. The choroid is the vascular layer, providing the sclera with oxygen and nutrients. At the front of the choroid is the pigmented iris, which dilates or constricts to control the amount of light entering the eye. The ciliary muscle sits just behind the iris and attaches to the lens via suspensory ligaments. Constriction of the ciliary muscle causes the lens to bulge and focus on near objects; relaxation of the ciliary muscle causes the lens to curve less and focus on far objects.

### The retina

The retina is the innermost layer of the eye and contains the photoreceptors responsible for vision – the rods and cones. The innermost layer of the retina also contains blood vessels. The cones are concentrated around the fovea centralis, which is surrounded by the macula lutea, or yellow spot. This is where daytime vision is sharpest, as cones only work well in bright light. There are three types of cones, each responsive to either red, green or blue. The rods are responsible for vision in dim light, and they can only distinguish between black and white. The rods and cones contain chemical pigments that break down in the presence of these different colours, translating the visual image into a chemical signal.

In order to hold its shape, the eye is filled with fluid. Between the cornea and the lens this fluid is the aqueous humour, which is watery and constantly produced and reabsorbed. This is because the cornea and iris do not have a blood supply, so need the aqueous humour to deliver nutrients and remove waste products. Then to the rear of the lens in the main eyeball is the vitreous humour, which is more jelly-like and, with the tough outer sclera, helps to maintain the shape of the eyeball.

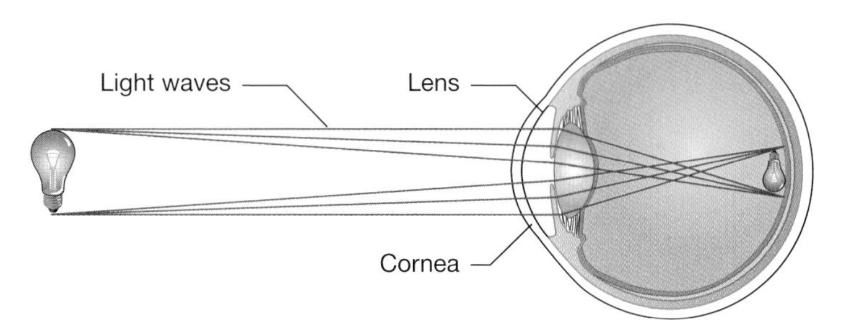


Figure 18.7 Refraction of light in the eye.

The eye develops throughout childhood. At birth, infants can only distinguish between different light intensities and still and moving objects. However, as myelination of the nerve pathways within the brain happen within 10 weeks of birth, sight rapidly improves. Newborns have difficulty focusing on far objects within the first few weeks of life, seeing best at 15–20 cm. All infants are born long sighted, and this resolves as the eyeball grows and lengthens, but children are still significantly long-sighted at age 2. This development of their sight takes place alongside their physical development, particularly of the musculoskeletal system, and often results in 'clumsiness' in young children. The ability to focus on near objects as the eyes continue to develop also coincides with the movement from large 'toddler' toys to more intricate fine motor skills involving greater hand–eye coordination; for example, threading beads, fastening own buttons and zips. Therefore, any assessment of delay in developing these skills should include a visual assessment.



Light enters the eye and is refracted (or bent) to form an image on the retina. Refraction happens as light passes through the cornea, aqueous humour, lens and vitreous humour (Figure 18.7).

## Neurological processing of visual signals

Once the retina has received the information about an image and transferred this to electrical impulses, the optic nerve fibres travel from the individual rods and cones to join together at the main optic nerve. In this pathway, some optical processing has already taken place. The nerve then exits on the posterior surface of the eyeball, sending visual signals along cranial nerve II (optic nerve) to the occipital lobe of the brain. As there are no rods or cones at the point where nerve fibres join to form the main optic nerve, images cannot be focused on this part of the retina; thus, it is called the 'blind spot' or optic disc (Figure 18.8). However, the eyes will work together to compensate for this in most situations.

As the image is sent along the optic nerve from each eye, these optic nerves then cross at the optic chiasma at the base of the brain, where the optic nerves divide their signals to detect any differences in the image from each eye. It is this that allows perception of distance and depth of vision. The left field of each eye is combined and sent to the left of the brain; the right field of each eye is combined and sent to the right side of the brain (Figure 18.9). Therefore, poor vision in one eye will lead to lack of three-dimensional perception, which can affect balance as well as vision (see 'The ear' section).

Therefore, as well as the eye functioning correctly, the optic nerve must also remain healthy, and the brain must be able to process the signals. As the eye develops, so does the visual cortex of the brain. The information from each eye is processed separately; therefore, the areas of the

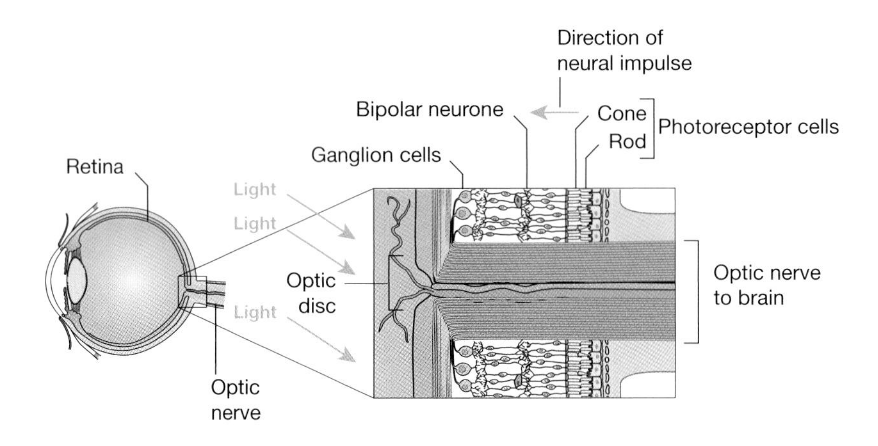


Figure 18.8 Anatomy of the optic nerve.



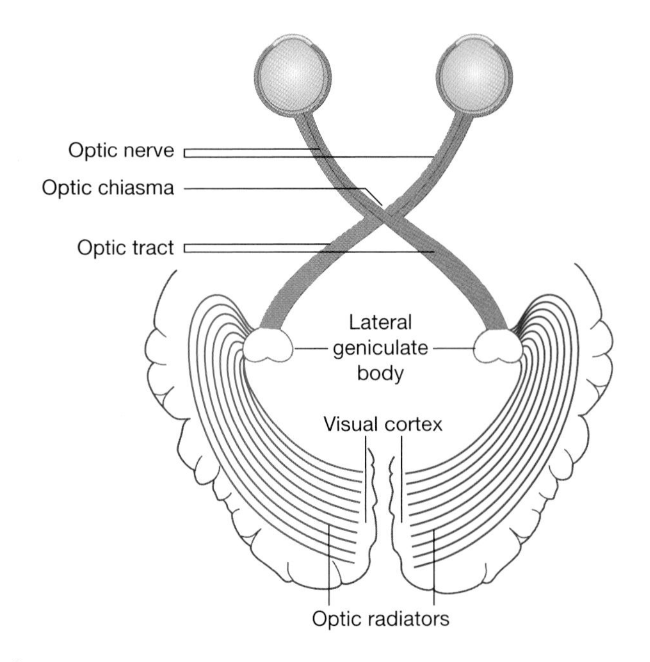


Figure 18.9 Optic chiasma.

brain receiving the sharpest picture develop the quickest. This means that if one or both eyes have difficulty in focusing on objects, the part of the visual cortex receiving that information will develop more slowly. The visual cortex stops developing around age 8 years of age. After this age, using spectacles to better focus an image on the retina of the eye will not necessarily lead to improvement in perceived vision, as it is too late for the cortex to develop the ability to process this information. Therefore, it is essential that all children receive screening at age 4, to allow for correction of vision through the use of spectacles (UK NSC, 2006). Glasses prescribed at this age normally need to be worn for the majority of a child's waking day to encourage development of the visual cortex. This requires education of the family and also involvement of the school; most importantly, the child needs to understand why they need to wear glasses and to be able to choose glasses that are comfortable and that they like. All glasses are free on the NHS up to the age of 16, and for those under 19 in full-time education.

## Clinical application

On some occasions, one eye develops considerably faster than the other. This leads to slow development of parts of the visual cortex. Therefore, the better eye will be 'patched' with an adhesive eye patch for a proportion of the day. This makes the weaker eye work harder, and subsequently develops the part of the visual cortex responsible for processing that image. This needs to be done before age 8 whilst the visual cortex is still developing.

# Conclusion

This chapter has explored the special senses. The senses of olfaction and gustation are detected by chemoreceptors; the senses of hearing and balance are detected by mechanoreceptors; and the sense of sight is detected by photoreceptors. Each of these specialized organs sends signals via cranial nerves to the cortex of the brain for processing. Therefore, development of both the organ itself and the area of the brain needed for interpretation are of concern during fetal and childhood development. Throughout this chapter there has been application to child health, with discussion of the role of the family in promoting healthy development.



# Activities

**Now review your learning by completing the learning activities in this chapter. The answers to these appear at the end of the book. Further self-test activities can be found at www.wileyfundamentalseries.com/childrensA&P.**

## Exercises

1. Why can certain smells produce an emotional response?
2. How does the tympanic membrane transmit sound from the outer to the middle ear?
3. How do the malleus, incus and stapes help to prevent damage to the organ of Corti?
4. Describe how a human maintains equilibrium.
5. Explain what is meant by the 'blind spot'.

## Complete the sentence

1. The four primary tastes that humans perceive are ________, ________, ________ and ________.
2. The cones detect the colours ________, ________ and ________.
3. The three specialized receptor types for the special senses are ________receptors (smell and taste), ________receptors (hearing and balance) and ________receptors (vision).

## True or false?

Rods work best in bright light. True or false?

## Conditions

The following table contains a list of conditions. Take some time and write notes about each of the conditions. You may make the notes taken from text books or other resources; for example, people you work with in a clinical area or you may make the notes as a result of people you have cared for. If you are making notes about people you have cared for you must ensure that you adhere to the rules of confidentiality.

| Condition | Your notes |
| --- | --- |
| Choanal atresia | |
| Otitis media | |
| Conjunctivitis | |
| Laryngitis | |
| Epistaxis | |
| Sinusitis | |

## Glossary

**Aqueous humour:** the fluid in front of the lens.

**Amplitude:** the height of the sound waves, determining the volume of the sound.

**Choroid:** the middle layer of the eyeball.

**Cochlea:** the structure responsible for detecting sound waves.

**Conjunctiva:** the mucous membrane covering the eye.

**Cones:** the specialized receptor cells within the eye, containing pigments that can detect either red, blue or green.

**Cornea:** the clear part of the sclera in the front of the eye.

**Cortex:** the outer area of the brain responsible for interpreting signals relating to the special senses.

**Cranial nerves:** nerves that go from the specialized sense organs directly into the brainstem.

**Crista ampullaris:** the structure at the base of the semicircular canals that contributes to the detection of head movement.

**Embryology:** the study of how the initial group of cells post fertilization develops into the recognizable human form with differentiated organs.

**Equilibrium:** the sense of balance.

**Eustachian tube:** the tube leading from the middle ear to the pharynx.

**Fetus:** the developing human before birth.

**Frequency (pitch):** the number of waves per second (measured in hertz), leading to the pitch (or tune).

**Grommets:** small plastic tubes inserted through the tympanic membrane, joining the middle ear to the outer ear and allowing for drainage of secretions.

**Gustation:** the sense of taste.

**Infant:** a child up to 1 year of age.

**Iris:** the coloured part of the eye.

**Newborns/neonates:** a child from birth to age 28 days.

**Olfaction:** the sense of smell.

**Optic chiasma:** the crossover of the optic nerves matching the left visual field from both eyes, and the right visual field from both eyes.

**Optic disc (blind spot):** the place where the optic nerve fibres join together to exit the eyeball.

**Organ of Corti:** the specialized organ that translates sound waves into chemical signals. It sits within the cochlea.

**Pupil:** the 'hole' in the centre of the iris through which light passes into the back of the eye.

**Retina:** the innermost layer of the eye, containing the rods and cones that detect light waves.

**Rods:** the specialized photoreceptors within the eye that differentiate between black and white in dim light.



**Sclera:** the outermost white layer of the eye.

**Semicircular canals:** three half-circle tubes at right angles to each other in the inner ear, responsible for the sense of equilibrium.

**Vestibule:** at the base of the cochlea and semicircular canals, responsible for equilibrium in terms of gravity.

**Vitreous humour:** the jelly-like fluid behind the lens.

## References


Bakalar, N. (2012) Sensory science: partners in flavour. *Nature*, **486** (7403), S4–S5.

Breen, F.M., Plomin, R., Wardle, J. (2006) Heritability of food preferences in young children. *Physiology and Behaviour*, **88**, 443–447.

Chamley, C.A., Carson, P., Randall, D., Sandwell, M. (2005) *Developmental Anatomy and Physiology of Children: A Practical Approach*, Elsevier, Edinburgh.

Kisilevsky, B.S., Hains, S.M.J., Brown, C.A. *et al*. (2009) Fetal sensitivity to properties of maternal speech and language. *Infant Behaviour and Development*, **32** (1), 59–71.

Kuhl, P.K. (2010) Brain mechanisms in early language acquisition. *Neuron*, **76** (5), 713–727.

Lucker, J.R. (2013) Auditory Hypersensitivity in children with autistic spectrum disorders. *Focus on Autism and other Developmental Disabilities*, **28** (3), 184–191.

Porter, R.H., Winberg, J. (1999) Unique salience of maternal breast odours for newborn infants. *Neuroscience and Biobehavioural Reviews*, **23** (3), 439–449.

Thibodeau, G.A., Patton, K.T. (2012) *Structure and Function of the Body*, 14th edn, Elsevier, St Louis, MO.

Tortora, G.J., Derrickson, B.H. (2009) *Principles of Anatomy and Physiology*, 12th edn. Wiley, Hoboken, NJ.

UK NSC (2006) *The UK NSC Policy on Vision Defects Screening in Children*, UK National Screening Committee, http://www.screening.nhs.uk/vision-child (accessed 23 October 2013).

# Chapter 19

# The skin

## Elizabeth Gormley-Fleming

*Learning and Teaching Institute, University of Hertfordshire, Hatfield, UK*
*Children's Nursing, School of Health and Social Work, University of Hertfordshire, Hatfield, UK*

## Aim

The aim of this chapter is to introduce you to the anatomy and physiology of the skin, which is also referred to as the integumentary system. There are important differences between the skin of a newborn and an adult. As a nurse it is important that you have knowledge of the skin as it changes. The condition of the skin may indicate underlying disease, and it is often the first indicator of an underlying health issue.

## Learning outcomes

On completion of this chapter the reader will be able to:

- Describe the anatomy and physiology of the skin from birth to maturity.
- List the functions of the skin.
- Describe the function of the accessory structures.
- Describe the thermoregulatory function of the skin.
- Understand how the skin matures from birth to adulthood.

## Test your prior knowledge

- Name the layers of the skin.
- Identify three of the layers in the epidermis.
- List the functions of the skin.
- What is the role of the skin in thermoregulation?
- Identify the skin appendages.
- Explain how vitamin D is synthesized by the skin?
- Name the two main networks of blood vessels that supply the skin with blood.
- Identify the glands in the dermis and their function.
- List five differences between an infant's skin and that of mature skin.
- Describe the function of melanin.

*Fundamentals of Children's Anatomy and Physiology: A Textbook for Nursing and Healthcare Students*, First Edition. Edited by Ian Peate and Elizabeth Gormley-Fleming.

Companion website: www.wileyfundamentalseries.com/childrensA&P

# Introduction

The skin, also known as the integumentary system, is a complex organ that is essential for human survival owing to it physiological functions. It undergoes significant changes from birth to adulthood, such as thickening of the dermis and increased activity of the sebaceous glands. The most dynamic changes occur with the first 3 months of life (Hoeger and Enzmann, 2002). However, at birth the skin of a term baby is developed to cope with extra-uterine life.

As a system it has contributions from basic germ layers: the ectoderm and the mesoderm. The ectoderm forms the surface epidermis and the associated glands, while the mesoderm forms the underlying connective tissue of the dermis and the subcutaneous layer (Chamley *et al.*, 2005). It is also populated with melanocytes and sensory nerve endings; thus, these different tissues perform many specific functions: thermoregulation, synthesis of vitamin D, excretion and immunity.

Frequently referred to as the largest organ in the body, the skin covers all of the body's external surfaces and is approximately 10% of the body mass. By adulthood the skin will be almost 2 $m^2$. The ratio of skin surface to body weight is highest at birth, and this will decline progressively during infancy. At birth the surface area is nearly three times greater than that of an older child, whereas at 37 weeks' gestation or less it is proportionally five times greater than that of a term baby (Wong, 1999).

The skin is the first line of defence against the environment. At birth the infant's skin is sterile, but immediately it becomes colonized with bacteria and fungi from the birth canal and anal region of its mother. Following birth, exposure to the environment and other humans quickly becomes an essential source in the long-term colonization of the newborn infant's skin.

The skin provides a tactile perception and is instrumental in the initial bonding between a mother and her infant during skin-to-skin contact.



The condition and appearance of the skin is a useful indicator when assessing health and well-being in the child and can assist in the diagnosis of medical conditions. Many factors affect the skin, such as nutrition, hygiene, immune status, congenital diseases and genetic traits. The skin may also be altered in the presence of acute illness and trauma. As an external organ, the skin may be damaged from environmental factors, such as exposure to ultraviolet (UV) light, chemicals, irritants and microbes.

The appearance of the skin may have an impact of the overall well-being of the child and its family, particularly in the presence of abnormalities. The long-term impact of this can have lifelong consequences.

# Clinical application

Assessing the child is a fundamental aspect of care. While elements of this involved collecting objective data, collecting subjective data is also important (Gormley-Fleming, 2010). A lot of information can be collected from the infant and child at a glance. However, this is a skilled activity, and one that will develop as your clinical experience develops.

Understanding the anatomy and physiology of the skin and its appendages will help you understand the data that you will gather from observing the child. Think about the first observation you make when you approach a child and its family. You will likely note the child's colour, facial expression, presence of skin rashes, lesions, child's behaviour/interactions and muscle tone before you even touch the child.

All of this forms an important aspect of the initial and ongoing assessment.

# The structure of skin

A versatile organ, the skin has two main structural layers: the epidermis and the dermis. Skin development is mostly complete by the fourth month of gestation. By the fifth and sixth months of gestation the fetus is covered in lanugo. Lanugo is a downy coat of colourless hair that is shed by the seventh month of gestation and the vellus hair appears (Marieb and Hoehn, 2010).

## The epidermis

The epidermis is the outermost superficial layer of the skin and serves as the physical and chemical barrier to the external environment and the interior body. It has no direct blood supply, receiving its nutritional supply and oxygen from the vascular network in the dermis by diffusion.

The development of the structures within the epidermis is directly proportional to the gestational age of the infant. Functionally immature in the premature infant, post-natal maturation is rapid in the first 2 weeks of life and the skin undergoes a period of adaptation by increasing epidermal cellularity (Hoeger and Enzmann, 2002).

In the newborn infant the epidermis will be 40–50 mm thick at birth compared with 20–25 mm thick in a premature infant (White and Denyer, 2006). This continues to increase in depth as the child grows. The thickness of the epidermis is similar to that of an adult, but it not as an effective barrier to transcutaneous water loss.

The epidermis varies in thickness in different parts of the body, with the soles of the feet having the thickest skin and the eyelids having the thinnest. At birth, the palmar and plantar surfaces are thickened in preparation for the wear and tear of walking. These are extra protective surfaces and result from epidermal cell differentiation.



The epidermis comprises keratinized stratified squamous epithelium that consists of four cell types (Figure 19.1):

- **Keratinocytes.** This typical epithelial cell forms the lining of all the internal and external body surfaces; for example, the mucosal tissue. Approximately 95% of the cells are keratinocytes. Keratinocytes originate from the division of stem cells in the stratum basale. Keratinocytes undergo continual mitosis, and the cells are constantly pushed upwards towards the surface due to the production of new cells beneath them (Marieb and Hoehn, 2010). By the time migration to the surface of the epidermis is achieved they are dead cells. These dead

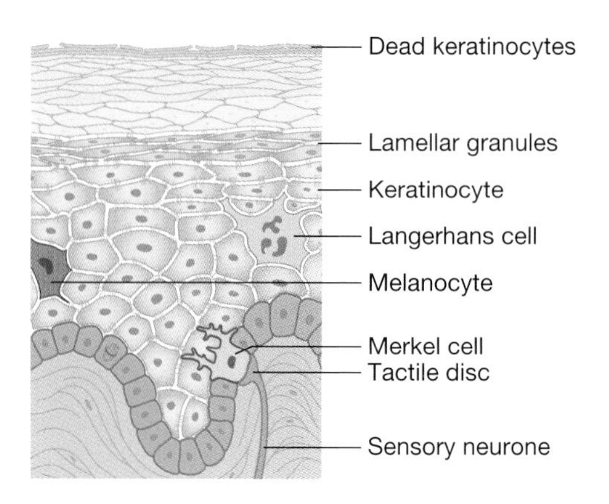


Figure 19.1 The cells of the epidermis.

cells are lost to the environment by a process referred to as desquamation, and every 25–40 days the epidermis is replaced in totality. The primary function of keratinocytes is to produce keratin, which is a fibrous protein that gives the epidermis its protective properties. These include protection from chemicals and micro-organisms.

- **Langerhan cells or dendritic cells.** These star-shaped cells arise from the bone marrow leucocytes and migrate to the basal region of the epidermis (White and Denyer, 2006). They form part of the immune system, particularly the immune reactions of the skin to environmental antigens. These dendritic cells are referred to as antigen-presenting cells (Romani *et al.*, 2003). As foreign proteins and micro-organisms make contact with the T-lymphocytes, the Langerhan cells regulate the production of antibodies by the B lymphocytes and this enables the activation of the macrophages. UV light is known to damage these cells.
- **Merkel cells.** These cells have a mechanoreceptor function and are present at the epidermal–dermal junction (Marieb and Hoehn, 2010). The Merkel cell is in contact with the sensory neurone, and this part of the cell is referred to as the tactile disc. It is this that allows the sensation of touch.
- **Melanocytes.** These cells are located in the basal region of the epidermis. They synthesize melanin pigment granules, and these are then transferred to the keratinocytes. They are present in greater number in the face, areola, nipples, penis and limbs. These melanocytes provide the hair, skin and eye with colour. Irrespective of skin colour, melanocytes are present in the same quantities in black skin as they are in white skin. It is the amount of the pigment melanin that is produced and how this is distributed that determines skin tone (Peate and Nair, 2013). They are present in the fetus from approximately the seventh week of gestation (England, 1996).

  At birth, a newborn of dark-skinned ethnicity will be only slightly darker than an infant with white skin. It is the response to light that increases the melanin production that causes the skin to darken. Consequently, newborns are more susceptible to the harmful effects of the sun. Melanocytes also protect the human body from the effects of UV radiation. The synthesis of melanin is controlled by both hormones and receptors. Exposure to sunlight will trigger melanin synthesis, as does the presence of oestrogens and progesterone.



It is the differences in these layers that distinguish the child's skin from that of an adult.

There are five distinct layers in the epidermis (Figure 19.2), all present at birth:

- stratum basale
- stratum spinosum
- stratum granulosum
- stratum lucidium
- stratum corneum.

As the skin varies in thickness, the thickest areas have five layers, while the thinner areas have four layers (Table 19.1). In the child the strata are all present but thinner.

### Stratum basale

The deepest layer of the epidermis, the stratum basale, is a wave-like border that rests on the basement membrane and consists of a single layer of keratinocytes. These cells rapidly divide, pushing one cell, a daughter cell, into the layer above, where it begins its journey to becoming a mature keratinocyte, and the other daughter cell remains in the layer below to continue the process of producing new keratinocytes. The basement membrane is made predominantly of collagen, which provides a secure foundation to support the epidermis and acts as a filter that regulates the passage of cells and nutrients from the dermis to the epidermis (White and

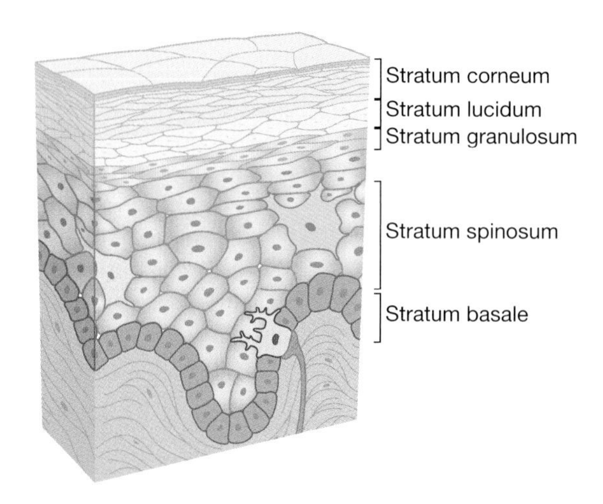


Figure 19.2 The layers of the epidermis.

Table 19.1 The layers of the epidermis.

| Epidermal Layer | Cell types | Number of layers |
|---|---|---|
| Stratum basale | Cuboidal or columnar keratinocytes | 1 |
| Stratum spinosum | Thorn like keratinocytes that fit closely together | 8–10 |
| Stratum granulosum | Flattened keratinocytes | 3–5 |
| Stratum lucidum | Dead, flat, clear keratinocytes | 2–3 (located only in finger tips, palms of hands and soles of feet) |
| Stratum corneum | Flattened dead keratinocytes | 25–30 |



Denyer, 2006). It is not yet firmly attached to the dermis; this dermo-epidermal junction develops with age. In a premature infant this is flat, but it is developed as a wave-like border by full gestational age. The implication of this incomplete attachment can be significant and lead to life-limiting conditions.

## Stratum spinosum

Several layers thick, the stratum spinosum lies above the stratum basale. The keratinocytes in this layer develop spines known as dermosomes as they shrink and may be called prickle cells. These cells are in abundance in this layer and are tightly packed together, which provides the skin with its strength, integrity and flexibility.

## Stratum granulosum

In this layer the keratinocytes lose their nuclei and begin to flatten and die. As a result of this, keratinization commences. There are three to five layers of these flattened cells. As disintegration occurs, the granules contained in these cells form both a water-resistant lipid called lamellar granules and keratohyaline granules. The lamellar granules move into the extracellular space. Their main function is to slow down water loss across the epidermis and prevent entry of microbes. The lipids released by these cells along with their thickened plasma make them more resistant to destruction. Along with the keratohyaline granules, which help to form keratin in the upper layers, both contribute to making the skin stronger and tougher.

### Stratum lucidium

Consisting of two or three rows of dead, flat, clear keratinocytes, with indistinct boundaries, the stratum lucidium appears as a thin band lying just above the stratum granulosum. The cells in the stratum lucidium are closely packed together and prevent fluid loss. It is only found in areas that are exposed to wear and tear, such as the soles of the feet and palms of the hands, as its function is to offer extra protection.

### Stratum corneum

This is the tough, waterproof outer layer of the epidermis. At birth it is considerably thinner than that of mature skin. It is composed of two subtly distinct strata: the lower stratum compactum and the exterior stratum dysjunctum (Bowser and White, 1985). The stratum corneum consists of 25–30 layers of dead cells. The stratum compactum contains cells that are fibrous in nature and contains keratin, which protects the skin from abrasions and penetrative injury. Glycolipid is present between the cells. It is the existence of this that gives the skin it near-waterproof properties. This is absent in an extremely premature infant. The stratum dysjunctum is the interface with the environment, and here the cells can reabsorb water, thus leading to maceration and damage. It is this layer that is constantly shed, and these cells are referred to as cornified or horny cells.

## The dermis

The primary function of the dermis is to provide the epidermis with nutrients and support. It accounts for approximately 15–20% of the total weight of the human body and varies in thickness from 1 mm in the eyelids to 5 mm in the back (White and Denyer, 2006).



A relatively acellular layer, the dermis is located under the basement membrane. Developed from the mesenchyme, it is predominantly made up of collagen, fibrous protein and elastin, which form a dense connective tissue that is evident from the 11th week of fetal gestation (England, 1996). In the newborn, the dermis will contain small collagen bundles and the elastin fibres are immature. Fibroblasts, mast cells and macrophages are also found in the dermis. The infant has a greater number of fibroblasts than an adult. At birth the dermis is very thin, oedematous and is loosely bound to the epidermis. The dermis of a newborn is 60% as thick as adult skin and takes 6 months to mature, making the infant skin more prone to damage. The rete pegs, which anchor it to the epidermis and dermis, are not yet formed. Within this structure are the vascular and neurological components.

Within the dermis (Figure 19.3) are:

- blood vessels
- hair follicles
- nerves
- lymph vessels
- sebaceous glands
- smooth muscle
- sweat glands.

Regarded as two layers with subtle variations, the dermis can be divided into the

- papillary dermis
- reticular dermis.

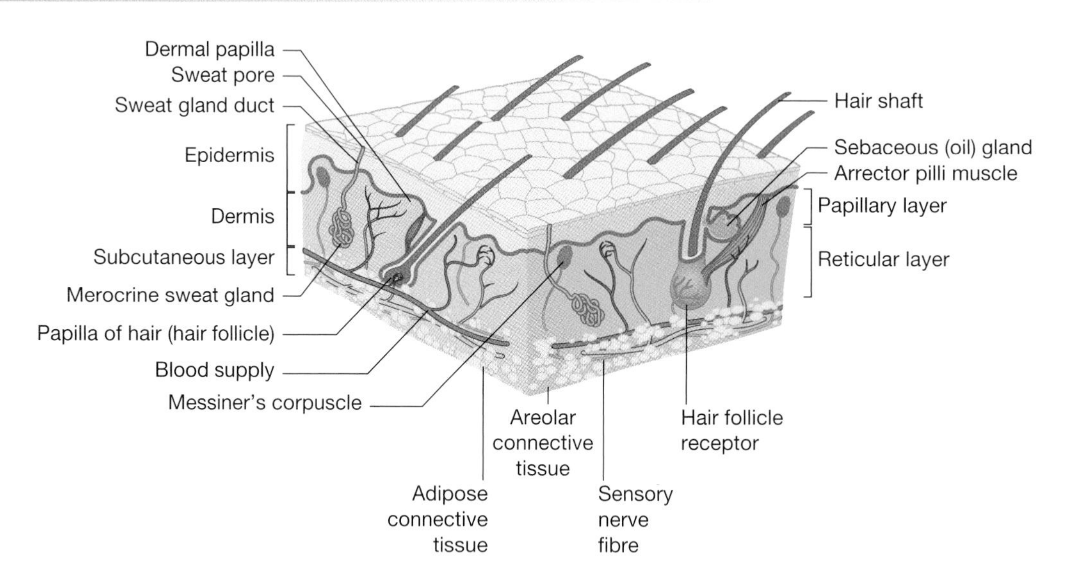


Figure 19.3 The structures of the dermis.



The papillary dermis is located immediately below the basement membrane and is the upper layer of the dermis. It is clearly demarcated from the epidermis by an undulated wave-like border. This thinner superficial layer is composed of areolar connective tissue. Within this layer there is fine interlacing collagen and elastic fibres that form a loose mat-like structure. This is then interlaced with small blood vessels. Phagocytes may be found circulating in this region, providing a defence mechanism against microbes. The upper surface of the dermis has protrusions that are called the dermal papillae, and these contain capillary loops.

In the palms of the hands and the soles of the feet these papillae are called dermal ridges, which in turn form epidermal ridges in the epidermis lying above. The function of these ridges, referred to as frictional ridges, is to enable the gripping ability of the hands and toes. These frictional ridges, which will become the unique identifier of the individual (the fingerprint), will have been present in the fetus from approximately 24 weeks' gestation. These epidermal ridge patterns may be use diagnostically to detect abnormal chromosome complements. In the presence of Down's syndrome there is an abnormal epidermal ridge pattern and there is one palmar crease instead of two.

Other areas of the skin contain organized sensory nerve endings such as the Meissner corpuscles. These are closely packed together at birth, but within a few months of life these become more spread out as the skin grows, particularly on the dorsal surfaces (MacGregor, 2012). Hence, a newborn baby has acute skin sensation.

The thicker reticular layer is mostly composed of thick bundles of collagen. This accounts for approximately 80% of the dermis. The collagen in this layer is constantly being broken down and being synthesized into dermal fibroblasts. These will eventually become mature collagen cells. In the newborn infant, type III collagen is more prevalent than in the human adult skin, which will only have 15% type III collagen and 85% type I collagen. It is this collagen that gives the skin its tensile strength, thus offering protection from direct injury. Collagen and elastin are produced more rapidly in children; as a result of this, granulation tissue forms more quickly. This impact on wound healing and scars may not lengthen at the same rate that the child's growth occurs.

Within this layer, elastin fibres are present throughout, and their role is to give the skin its elastic recoil. The number of these fibres degenerates with age and from exposure to UV radiation.

The dermis modifies itself to produce not only dermal ridges but also cleavage lines and flexure lines. Cleavage lines are collagen fibres that separate, are invisible and run longitudinally to the skin. Flexure lines are dermal folds that occur near the joints and palms. Here, the dermis is tightly secured to the layers below to assist with movement.

This layer is nourished by blood vessels. The cutaneous plexus lies between the dermis and the subcutaneous layer. This blood supply is established by the first trimester of fetal gestation.

The spaces between the collagen and elastin fibres contain adipose tissue, hair follicles, sebaceous glands and the sudoriferous glands.

## The appendages

### Clinical application

Consider the young child who has a significant loss of skin tissue due to burns. What principles of wound care should you consider in managing their wounds?



The skin appendages include the nails, sweat glands, sebaceous glands, hair follicles and hair. Essential in the production of any of the skin appendages is the formation of an epithelial bud.

### *The nails*

The function of the nail is to provide protection from trauma to the finger tips, as a mean of scratching and to assist with picking up very small objects.

The nails are formed from ectoderm that covers the dorsal tip of the digit which then thickens to become the nail field. Growth is slow, and eventually the surrounding epidermis covers the proximal and lateral part of the nail field by forming nail folds. These cells grow over the nail field, keratinization occurs and a nail plate forms.

A thin layer of epidermis (the stratum corneum) covers the developing nail. This degenerates and what remains is referred to as the cuticle or the eponychium. It is here that nail production occurs.

The visible area of the nail is referred to as the nail body and the area beneath this is the nail bed (Figure 19.4). The nail body is deeply recessed into the lateral surrounding epithelium on both sides by the lateral nail grooves and the lateral nail depressions. The free edge of the nail extends over the hyponychium – the distal end of the nail.

The pink colour of the nail is as a result of the underlying blood vessels that lie beneath the nail (Pringle and Penzer, 2002). If these vessels are obscured near the area of the nail root a pale crescent shape may be noticed; this is called the lunula.

The nail reaches the fingertip by 32 weeks' gestation and the top of the toe by 36 weeks' gestation, so nails not reaching the finger or toe tips indicate prematurity.

Nail growth is more rapid in the infant and decreases with age. A newborn baby may be born with scratches on their face from their own nails. An infant born after 42 weeks will have keratinized nails, which may also be long and stained green due to the passage of meconium *in utero*.

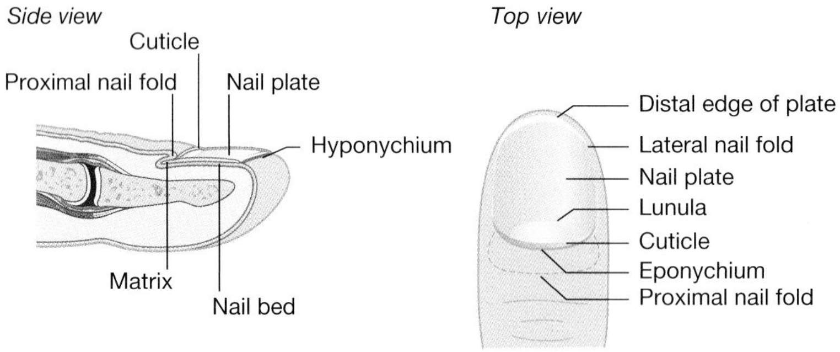

Side view
Cuticle
Proximal nail fold
Nail plate
Hyponychium
Matrix
Nail bed
Top view
Distal edge of plate
Lateral nail fold
Nail plate
Lunula
Cuticle
Eponychium
Proximal nail fold

Figure 19.4 The nail.

4 mm
Sebaceous gland
Arrector pilli muscle
Bulge
Hair bulb
Follicular papilla



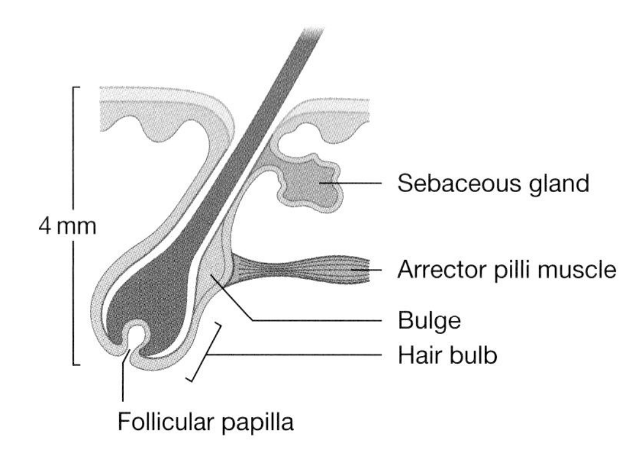

Figure 19.5 The hair follicle.

## Hair

Very fine hairs appear at 20 weeks' gestation, first on the eyebrow, chin and upper lip, and then followed by appearance on the forehead and scalp. This fine downy unpigmented hair is referred to as lanugo, and by 26 weeks' gestation this is replaced by the vellus hair (secondary hair) which covers the entire body. The lanugo is shed both before and after birth, with the scalp, eyebrows and eyelashes shed last (England, 1996). At birth the newborn's hair lies flat and has a silky feel. Individual strands are identifiable.

Terminal hair is the hair on the head, eyebrows and eyelashes; it remains present throughout the lifespan and is more deeply pigmented than vellus hair. The vellus hair that is present in the axillae, pubic area and limbs remains there until puberty, but then the circulating sex hormones stimulate the hair follicles to produce terminal hair, which is often curly.

Each hair is composed of columns of dead keratinized epidermal cells that are connected together with extracellular protein. The hair follicle consists of three distinct areas (Figure 19.5): the lowest section extends from the base of the hair follicle (the bulb) to the insertion of the arrector pili muscle; the mid-section extends from the muscle to where the sebaceous gland inserts; and finally, from the sebaceous gland to where it protrudes into the skin's surface and is referred to as the shaft.

The hair growth cycle determines the growth and shedding of hair. A hair in the scalp grows for 2–5 years and growth occurs at the rate of 0.33 mm per day. When the hair is growing the cells of the hair root absorbs the nutrients from the body and these are then incorporated in the hair thus maybe analysed for diagnostic purposes. DNA fingerprinting may also be done on hair.

The colour of the hair is determined by the amount of melanin present. Hair is a distinguishing feature of personal appearance, and in the young person may be of prime importance. Loss of hair colour can happen in young people but is generally part of the aging process. This is thought to be due the presence of air bubbles in the melanin and the lack of pigment. Hair growth is determined by genetic factors and by hormones (Tortora and Derrickson, 2009).

Hair has a limited protective function, but its main function is to protect against heat loss. The human body is covered in hair. Every hair follicle appears onto the surface of the skin through pores. Heat leaves the body through the skin by convection and it immediately becomes trapped in the hair. In the presence of cold air, the arrector pili muscle contracts causing the hair shaft to become erect, thus causing goose bumps to appear. These also appear in the presence of fear and strong emotions.

The eyelashes and eyebrows protect the eyes from foreign bodies entering. The nasal hairs and the hairs in the external auditory canal also prevent the entry of foreign matter from entering these cavities.

## The glands in the dermis

A range of glands with secretory properties are located in the dermis (Table 19.2).

There are three types of sweat glands in the dermis: the eccrine, apocrine and apoeccrine. The function of these glands is to release sweat into the hair follicles and onto the skin surface.



Table 19.2 The location and function of the glands of the dermis.

| | Eccrine | Apocrine | Apoeccrine | Sebaceous |
|---|---|---|---|---|
| Distribution | Throughout the skin with the exception of nail beds, labia, glans penis and ear canal | Axilla, groin, areolae, ear canal, breast and eye lid | Axilla | Throughout the skin with the exception of palms of the hands and the soles of the feet |
| Location | Lies in the lower third of the dermis and may extend down to the subcutaneous fat layer; resurfaces at the surface of the epidermis | Mostly in the subcutaneous layer and the duct ends at the hair follicle | Not found in the axillae until puberty and develop from eccrine and apocrine glands | Associated with hair follicles generally, but located in areolae, nipples, labia and inner prepuce as free structures. |
| Secretion | Sweat – water, sodium chloride, urea, uric acid, ammonia, glucose, enzymes and lactic acid | Sweat, but also a lipid substance | Sweat | A lipid-rich secretion called sebum |
| Function | Thermoregulation | Scent gland | Sweat | Moisturize hair follicle and skin |
| Onset of function | Shortly after birth | Puberty | Puberty | Well developed at birth and active for a few months, after which they atrophy and then become active again at puberty |

In total there are approximate 3–4 million sweat glands in the skin (Tortora and Derrickson, 2009). Functionally, newborn babies have the ability to sweat, but both thermal and emotional sweating is reduced compared with the older child (Harper *et al.*, 2000).

### Eccrine glands

The primary purpose of eccrine glands is to produce sweat; hence their role in the regulation of body temperature (Sato *et al.*, 1989). They consist of a simple coiled tube and lie in the lower third of the dermis but may extend down to the subcutaneous fat layer. The eccrine glands are found in abundance on the forehead, palms of the hands and the soles of the feet. They are capable of producing up to 1.8 L of sweat per hour. Sweat is composed of water (99%), sodium chloride, potassium, vitamin C, antibodies and the waste products of metabolism. Drugs may be secreted via sweat. The diet consumed will also add to its constituents, as will hereditary factors.

### Apocrine gland

Located in the axillae and the anogenital region, a modified version of the apocrine gland is found in the eyelid, ear canal and the breast. These glands are larger than the eccrine gland and lie deeper in the dermis. They secrete a similar substance to the eccrine gland but also a fatty substance that is odourless until it comes into contact with the bacteria of the skin (Marieb and Hoehn, 2010). The apocrine gland or scent gland is present at birth and becomes active at puberty.

### Apoeccrine gland

This third gland is not found in the axilla until puberty. At the age of 8 years the eccrine glands increase in size, and by puberty the number of apoeccrine glands accounts for 45% of the total number of glands in the axilla. It is thought that they originate from the eccrine gland (Sato *et al.*, 1989).



### Sebaceous glands

Located throughout the body with the exception of the palms of the hands and the soles of the feet, the sebaceous glands are primarily found near the hair follicles as they form as a bud from a hair follicle root sheath, except those in the glands of the penis and the labia, which form from the epidermis (England, 1996). They are mainly located in the face, back and in the neck (Pringle and Penzer, 2002). These glands secrete sebum, which is a lipid-rich fluid that contains triglycerides, cholesterol, protein and salts.

Active at birth due to the presence of the maternal hormones, they become dormant within a few months and atrophy until puberty commences. Stimulated by the sex hormones at puberty, the glands increase in size and produce sebum. There may be overactivity of the sebaceous glands at puberty, which leads to the overproduction of sebum. The arrector pili muscles contract and this forces the sebum to be ejected onto the surface of the skin (Marieb and Hoehn, 2010). Sebum covers the surface of the hair shaft, and this prevents them from becoming brittle and dry. Because of the oily nature of sebum, this acts as an inhibitor of water evaporation from the skin. This helps maintains the softness and suppleness of the skin. *In utero* it is this oily sebum that forms the part of the vernix caseosa.

In the newborn, milia are formed if these glands become blocked.

### Vernix caseosa

At birth, the newborn infant is covered in vernix caseosa. This is produced by the sebaceous glands during the last trimester of pregnancy. Vernix is composed of water (80%), protein and lipid (Hoeger *et al.*, 2002). Composed of two structures, wax produced by the sebaceous glands and barrier lipids derived from keratinocytes, the function of vernix is to protect the skin of the fetus *in utero* from maceration of the developing epidermis from the amniotic fluid.

### *Ceruminous glands*

These are modified apocrine glands and are found in the lining of the external ear. They produce cerumen (ear wax), which is a stick yellow substance (Tortora and Derrickson, 2009). The function of this substance is protective, as it acts as a sticky barrier to prevent foreign bodies entering the ear canal, along with some antimicrobial properties due to its waterproofing ability.

## Blood vessels

The blood supply in the dermis is divided into two main networks: the superficial plexus and the deep plexus. The superficial plexus is composed of interconnecting arterioles and venules that wrap themselves around the structures of the dermis. They supply oxygen and nutrients to the cells, which metabolize very rapidly in this area, and then they branch off to carry blood to the epidermis.

The deep plexus is located at the border between the subcutaneous fat layer and the dermis. It supply's these layers with blood along with smaller tributaries that supply the hair follicles and the other structures within the dermis.

## Subcutaneous layer

Immediately below the dermis is the subcutaneous layer, also referred to as the hypodermis. While this may not be strictly considered to be part of the integumentary system, it must be considered in terms of the skin function of thermoregulation. Adipose tissue is a loose connective tissue that consists of adipocytes that store triglycerides. A specialized adipose tissue referred to as brown fat is found in the nape of the neck, posterior to the sternum and perineal area. This forms during the 17th–20th weeks of gestation. These cells have a high number of mitochondria. As the fatty acids break down, energy is released in the form of heat. This is essential for thermoregulation in the newborn.



# Functions of the skin

As an organ the skin has several important functions. A vulnerable organ owing to its constant visibility, and therefore exposure to harm from pathogens, chemicals, the environment and trauma, the skin is an easily observed organ that can indicate the general state of health at a glance but also abnormalities that are taking place within the body. The functions of the skin are:

- protection
- thermoregulation
- sensation
- synthesis of vitamin D
- excretion and absorption
- non-verbal communication.

## Protection

The skin serves as the main protective barrier. This barrier function can be subdivided into three distinct barriers:

- physical
- chemical
- biological.

### Physical barrier

It is the continuity of the skin that forms the physical barrier between the environment and the internal organs of the human body. As the infant grows, the skin increases in strength due to the hardening of the keratinized cells. The transepidermal lipid barrier is effective at birth in the mature infant, thus offering immediate protection from the environment (Hoeger and Enzmann, 2002).

There are many substances that penetrate the skin, such as fat-soluble vitamins, steroids, carbon dioxide, oxygen, solvents, resins from plants, salts from heavy metals (lead), selected drugs and drug penetration enhancers. Some of these are helpful and necessary but may also be toxic to the human body in undesired quantities.

The water-resistant nature of the skin, due to the presence of glycolipids in the epidermis, blocks most of the diffusion of water and water-soluble substances between the cells (Tortora and Derrickson, 2009). This prevents the drying out of the tissues and internal organs and the gain of water to and loss from the body through the skin. In the premature infant, transdermal water loss is significant in terms of overall wellbeing. A small of amount of water will be lost through the epidermis, and long-term submersion in water (not salt water) will cause the body to swell; immersion in salt water will cause the skin to dehydrate as water is drawn out of the body, so the skin is not entirely waterproof.

### Chemical barrier

The chemical barriers are the skin's ability to secrete an acid mantle (the pH) and the presence of melanin. At birth the skin is sterile, but it quickly colonizes with flora – bacteria and fungi from the birth canal, environment and human contact. This initial colonization may not be reflective of the longer term pattern of colonization and representative of the normal skin flora for that individual.



The normal pH of the skin is approximate 5.5 (Mims *et al.*, 1995). The normal flora of the skin metabolize fatty acids and sebum. This produces an acid pH of 5.5 and this inhibits the growth of pathogens. However, in the first 3 months of life the infant has a skin pH of around 6.6, so this reduces the protective function provided by the skin.

The sebaceous glands do not function until puberty also. This, along with the smaller body surface area of infants, less developed ecological environment and the need to be handled by their carers, is thought to increase the ability of pathogens to spread (Carr and Kloos, 1977).

By puberty there is an increase in the propionibacterium acne which acts on sebum, thus producing an inflammatory response resulting in acne (Mims *et al.*, 1995). This usually subsides by early adulthood.

Melanin acts as a chemical barrier against UV light, and this prevents damage to the skin cells.

### Biological barrier

The biological barrier includes the dendritic cells of the epidermis and the macrophages in the dermis along with the DNA. The dendritic cells and the macrophages play an important role in the immune system. The macrophages constitute the second line of defence to bacteria and viruses that have successfully penetrated the epidermis. The dendritic cells play the role of an antigen to foreign substances.

The function of the DNA is to act as a natural protector against the UV radiation. This includes UVA and UVB rays. It is the UVB rays that penetrate the epidermis and cause sunburn, and in the longer term this may led to skin cancer developing. UVA rays lead to skin aging. Melanin acts as a pigmentation barrier, and a protein barrier in the stratum corneum enables the skin to have this protective function against UV light. The DNA absorbs the harmful radiation, where it converts harmful radiation into harmless heat and circumvents the potential genetic mutation

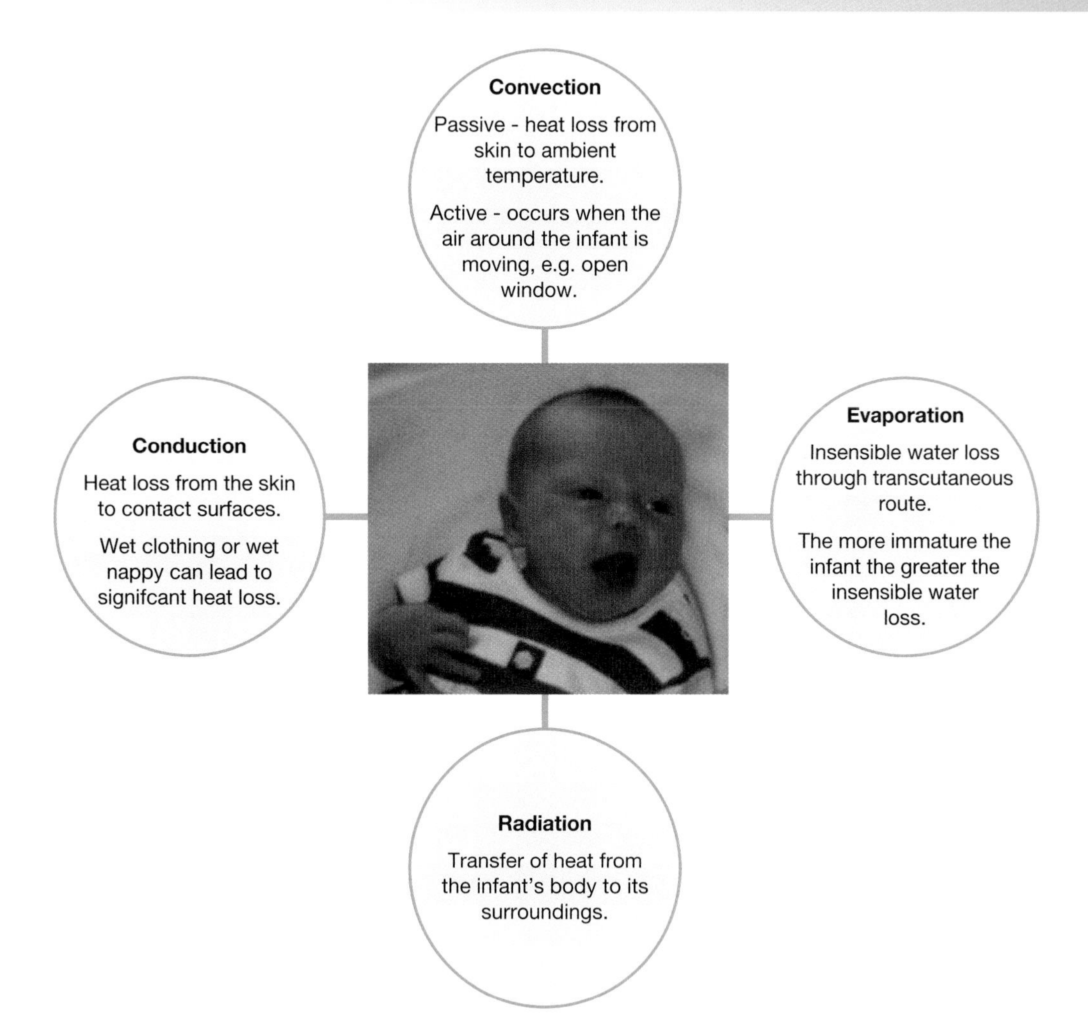


Figure 19.6 Heat loss from a newborn baby. Adapted from Crawford and Hickson (2002).



that would manifest as skin cancer, which is rare in children but nonetheless does occur. Hence, in the immature, skin protection is required.

## *Thermoregulation*

The skin has a direct role in thermoregulation for the infant and child. Excess heat is lost through the skin by convection, conduction, radiation and evaporation (Figure 19.6). Damage to this skin function in severe conditions can lead to collapse in the child (e.g. scalded skin syndrome).

The human being maintains their core temperature within narrow parameters, and the metabolic rate of the human body decreases with age. The temperature of the fetus is higher than that of its mother due it having a higher metabolic rate, and it passes the heat energy to its mother through the placental interface and the amniotic fluid (Power, 1989). As soon as the baby is born it is essential that it can generate heat to deal with the hostile environment it has been born into where the ambient temperature is lower that the *in-utero* temperature. Heat

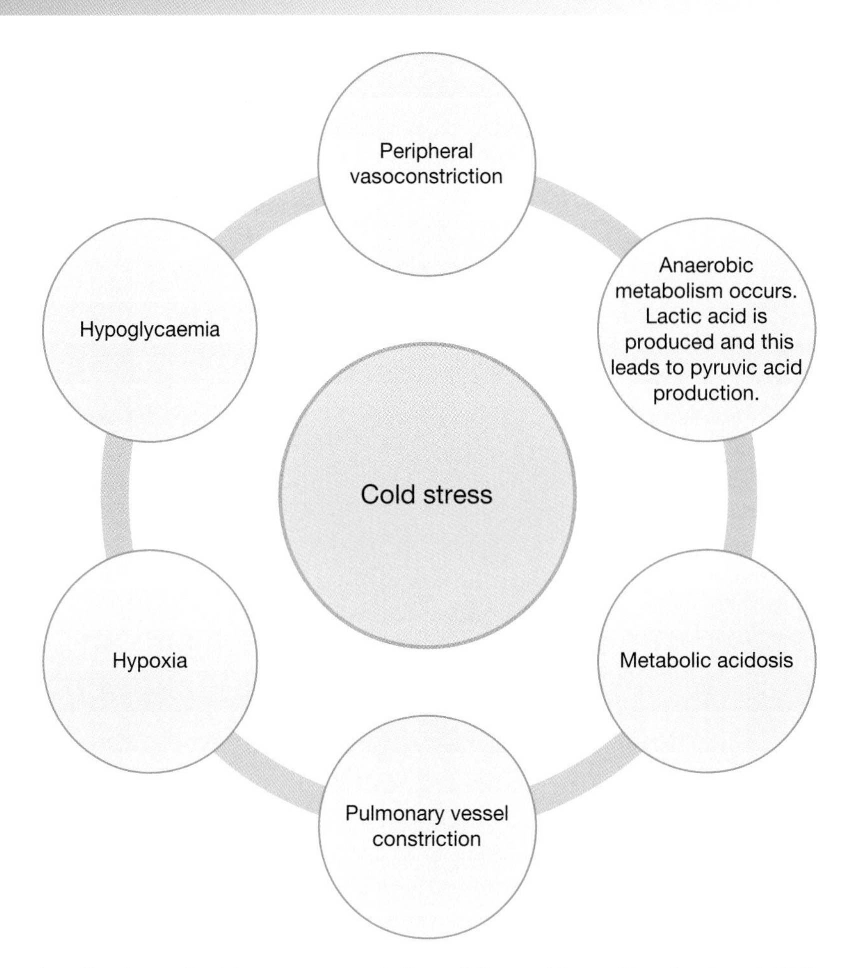


Figure 19.7 Implication of cold stress on an infant. Adapted from Crawford and Hickson (2002).



generation is essential for survival as the newborn needs to compensate for this loss of energy. Cold stress is the term used to describe the infant's response to cold, and this has serious physiological implications if left untreated (Figure 19.7).

Heat generation in the newborn infant occurs through shivering and non-shivering thermogenesis. However, the shivering response of the newborn is immature owing to immaturity of the musculature system, so this is unable to achieve the desired response. Therefore, it is important to dry the baby at birth and maintain a thermoneutral environment.

It is the presence of brown fat in non-shivering thermogenesis that enables heat to be generated. Brown fat has a plentiful supply of mitochondria, blood and a well-developed sympathetic nerve system (Asakura, 2004). There is an increase in noradrenaline in the presence of cold, and this acts on brown fat, causing lipolysis. This breaks down the brown fat into free fatty acids, producing heat in the process and thus raising the core temperature due to rich blood supply of brown fat. The heat that has been generated is rapidly transported around the body.

The newborn has significantly more sweat glands than the adult does, but it does not have the same ability to sweat and so is unable to maintain its core temperature by this mechanism.

The eccrine glands are immature. The ability to sweat does not occur until the infant is 2–3 weeks old, so they cannot lose heat through this mechanism.

All newborns have the ability to lose water through the transcutaneous route. The more immature the infant is, the greater the insensible water loss is through their skin, and thus the greater potential for temperature loss as the evaporation of water requires energy. Infants are unable to constrict their surface blood vessels, unlike older children, so cannot conserve heat in this manner.

As the infant matures, the neural network in the skin provides an accurate measure of the ambient temperature. This is done by the activation of the heat receptor in the dermis, which then passes a signal along the sensory nerve pathway to the hypothalamus. This will then either initiate the shivering reflex or deactivate the sweating mechanism.

The thermoregulatory function in children is immature, and because children also have a large body surface area they are at an increased risk of hypothermia. This risk decreases with age.

### Sensation

There is a rich supply of receptors in the skin that are part of the nervous system. These enable a range of sensations from the environment. This is known as the cutaneous sensation. Meissner's corpuscles and tactile discs enable the sensation of touch: the feeling of clothing and objects as they come into contact with the skin. This is important as the primary method of bonding between the infant and its mother should be through skin-to-skin contact. It is thought there may be some analgesic effect through skin-to-skin contact in the young infant (Gray *et al.*, 2000).

Pacinian corpuscles, which are located deep in the dermis, alert us to increased pressure from direct contact. Hair follicle receptors are also present that alerts us to hair being pulled or air currents.



The number of receptors in the lips, genitals and finger tips are greater than found in the rest of the body. There are over 1 million never fibres in the skin.

### Vitamin D synthesis

Vitamin $D_3$ (cholecalciferol) is an essential component in calcium regulation in the body and impacts on the serum calcium level and the bone deposition. Direct sunlight, UVB, on the skin will modify the cholesterol molecules (7-dehydrocholesterol) in the circulating blood vessels to a vitamin D precursor (White and Butcher, 2006). This is transported through the body via the blood. Enzymes present in the liver and kidneys then alter the molecules to produce the active form of vitamin D. This hormone, calcitriol, is essential for the absorption of calcium into the body.

### Excretion and absorption

There are limited amounts of waste substances excreted through the skin, such as water, sodium, urea, uric acid and ammonia. There is high transepidermal water loss in the preterm baby, and this may be up to 10 times greater than in a term infant.

The skin has the ability to absorb substances both directly from the environment, such as carbon dioxide and heavy metals, and from the direct application to the skin in the form of topical medication.

### Non-verbal communication

The skin does a very effective job at communicating many important pathologies that are occurring both locally and deeper within the body. Observations of the skin will involve the colour, presence of rashes or lesions, temperature, texture, and breaks in its continuity.

The skin is also good at conveying our mood and emotions. It is the primary method of identification as it is highly visible and thus has both aesthetic and cultural significance. It has an important role in the psychological well-being of the person.

# Conclusion

In this chapter the anatomy and physiology of the skin has been examined and its functions identified. A good understanding of this is a prerequisite to the provision of safe and effective nursing care of the infant and child.

# Activities

**Now review your learning by completing the learning activities in this chapter. The answers to these appear at the end of the book. Further self-test activities can be found at www.wileyfundamentalseries.com/childrensA&P.**



## Complete the sentence

1. The integumentary system consists of __________ __________ or ______, which includes the __________, _________ and the _________ __________.
2. Beneath the _______________ lies the _________ _________.
3. The functions of the skin are ____________, _____________, _________ __ __________ __, __________ __________, __________ ____ __________, and __________.
4. The innermost layer of the epidermis is the _________ ___________ and the outermost layer is the _________ ___________.
5. The middle layer of the epidermis is the ________ _________.
6. ___________, which are located in the stratum basal, manufacture ____________.
7. Located in the dermis, _____ is composed of ____________ dead cells.
8. The _______ _____ muscle contracts when ___________, forcing the hair to _______ _____.
9. The ____________ glands are holocrine glands that discharge a ______ secretion into hair follicles.
10. There are three types of sweat glands: ___________ sweat glands, ____________ and __________ sweat glands.

## Case study

Atopic eczema is a chronic, relapsing inflammatory condition of the skin. It is associated with epidermal dysfunction. It manifests from approximately 3 months of age but may occur at any age. It is pruritic and causes great distress. Currently, there is no cure for atopic eczema, but it may often disappear by adolescence. In order to be classified as atopic eczema the child must have an itchy skin condition and any three of the following (British Association of Dermatologists, 2009):

- history of itch in the skin creases/folds or cheeks of the face;
- history of asthma or hay fever;
- dry skin in the last year;
- visible flexural eczema or on the cheeks, forehead and outer limbs in children under 4 years;
- Onset in the last two years of life.

## Pathophysiology

There is a genetic predisposition to developing atopic eczema. However, the aetiology is multifactorial, with the environment, immunological and physical factors all possibilities in its development. Corneocytes separate in the epidermis and reduce the epidermal lipids. The epidermis already has an inability to retain water, so fluid leaks from the cells. This is what makes the skin dry and the skin barrier is now less effective. Bacteria can penetrate, and this now results in an inflammatory response. As a result of the inflammatory response, erythema and oedema occur in the epidermis. There is increased blood flow, and this in turn causes the white blood cells to leak into the dermis. It is this that causes vesicles and blistering of the epidermis. When this is scratched and the skin breaks, weeping occurs.

1. How will a child with eczema present to the health-care professional?
2. What do you think are the important factors in managing a child with atopic eczema to ensure the quality of their life is not affected by their eczema?
3. What practical help might a family require if they have a child with atopic eczema?



## Answers

1. The infant or child will present with pruritus, dry, scaly skin on the cheeks, flexures of the elbows and behind the knees. Other areas of skin may also be affected. The skin may be broken and oozing vesicles present. There will be a history of scratching and possibly crying.
2. The aim of the treatment of a child with atopic eczema is to hydrate the skin, reduce inflammation, and promote comfort to maintain a normal quality of life.
   This will include daily skin care regimes-bathing, use of emollients, topical steroids during flare ups and antihistamines to provide relief from pruritus. Antibiotic may be required when infection is present.
   A full nutritional assessment should be undertaken on the initial consultation, including weighting and height of the child, to identify any possible food triggers.
   RAST may be required to plan on going management if it thought specific food and pollen may be the cause
3. Caring for a child with eczema may be challenging for the family. Treatments are time consuming and repetitive, yet it is essential that good compliance is achieved so motivation is required. Explanation of this to the parents is important. There may be many restriction on the family in terms of managing laundry, preparing food, restrictions on having a pet, sleep deprivation, behavioural problems due to the child being eased or bullied, and having to deal with the psychological impact of an altered body image combined with parental guilt and anguish.
   The practical advice the family need are to avoid extremes of temperature, minimise skin damage from scratching by keeping finger nails short, avoid biological detergents and fabric softener. Wear cotton clothing and use cotton bedding. Gloves may be worn to bed. Advice on house dust mite is essential.
   Ensure parents know where and how to use the treatments and where to get help.

## Crossword

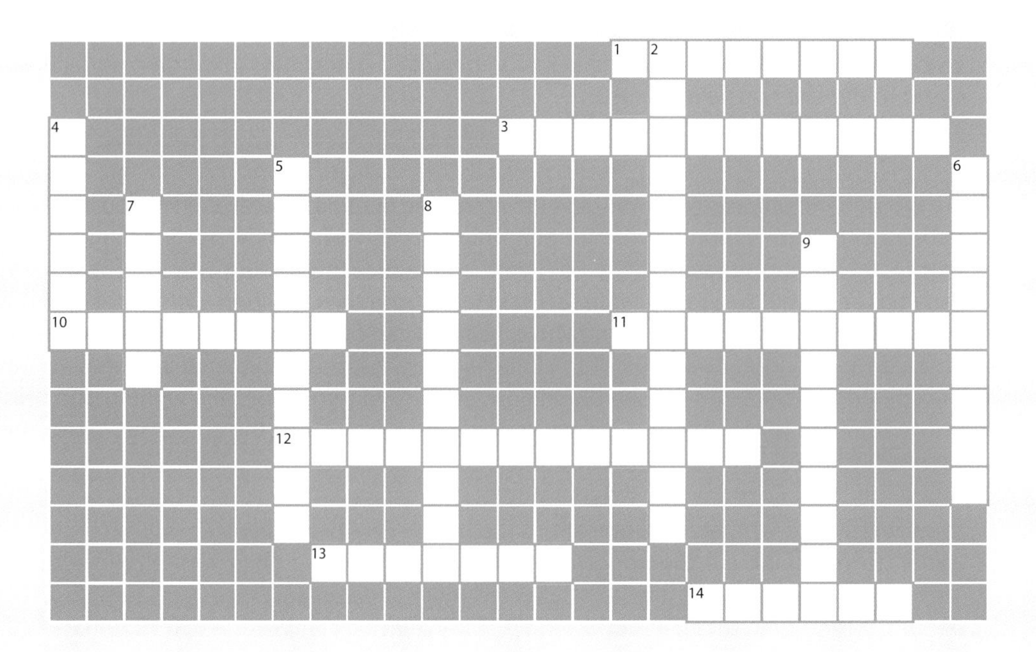




**Across**

1. Synthesis of this vitamin begins in the skin
3. Cell found in the epidermal layer
10. A flattened cell
11. Pigment producing cell in epidermal layer
12. Another name for the sweat gland
13. Pigment produced by melanocytes
14. Waxy covering substance on the new-born skin

**Down**

2. Term for the skin and its appendages
4. Lower layer of skin
5. Secondary hair
6. Uppermost layer of the skin
7. A substance produced by sebaceous glands
8. Cell that allows the sensation of touch
9. Function of the skin

## Conditions

The following table contains a list of conditions. Take some time and write notes about each of the conditions. You may make the notes taken from text books or other resources; for example, people you work with in a clinical area or you may make the notes as a result of people you have cared for. If you are making notes about people you have cared for you must ensure that you adhere to the rules of confidentiality.

| Condition | Your notes |
| --- | --- |
| Burns | |
| Psoriasis | |
| Head lice | |
| Dermatitis | |
| Acne | |

## Glossary

**Absorption:** intake of fluid or a substance into the tissues.

**Apocrine:** a gland found in the dermis associated with sweating.

**Apoeccrine:** a gland found in the axillae during puberty.

**Arrector pilli muscle:** bundle of smooth muscle associated with the hair follicle that inserts into the hair follicle via the dermal shaft.

**Calcitonin:** a hormone that assists in the metabolism of calcium.

**Cerumen:** secreted by the ceruminous glands and known as ear wax.

**Collagen:** a protein that is found in connective tissue.

**Dermis:** the middle layer of the skin, composed mainly of connective tissue and consists of two layers – the papillary and the reticular layers.

**Epidermis:** the superficial layer of the skin that covers the entire body and is composed of stratified keratinized squamous epithelium.

**Excretion:** elimination of waste products from the body

**Hair:** consists of vellus and terminal hair that is produced by the hair follicle.

**Keratin:** an insoluble protein that is found in the hair and nails.

**Keratinocyte:** a cell that is responsible for forming protein called keratin.

**Lanugo:** down-like hair that covers the fetus until shortly after birth.

**Lanula:** the moon-shaped area at the base of the nail.

**Melanin:** pigment produced by melanocytes that provides protection from UV light.

**Melanocytes:** pigmented cells that produce melanin, which gives the skin and hair their colour.

**Merkel cells:** cells that are located in touch-sensitive areas of the epidermis.

**Meissner's corpuscles:** found in abundance in the finger tips and associated with touch.

**Nail:** a compacted plate of keratin.

**Papillae:** projection of cells in the dermis into the epidermis.

**Pore:** the opening of the gland duct onto the surface of the skin.

**Stratum:** a layer.

**Stratum basale:** deepest layer of the epidermis.

**Stratum corneum:** the most superficial layer of the epidermis and often referred to as the horny layer.

**Stratum lucidium:** consists of five layers of flat dead cells.

**Stratum granulsosum:** three to five layers of flattened keratinocytes.

**Stratum spinosum:** a layer that has tightly packed keratinocytes that have spine-like projections.

**Sebaceous gland:** consists of secretory epithelial cells derived from the same tissues as hair follicles.

**Sebum:** an oily substance produced by the sebaceous glands.

**Thermoreceptor:** a sensory receptor that can detect changes in heat.

**Thermoregulation:** the ability of the skin to regulate heat.

**Vernix:** a wax-like substance that covers the fetus until birth in order to protect its skin.

## References

Asakura, H. (2004) Fetal and neonatal thermoregulation. *Journal of Nippon Medical School*, **71**, 360–370.

Bowser, P.A., White, R.J. (1985) Isolation, barrier properties, and lipid analysis in the stratum corneum, a discrete region of the stratum corneum. *British Journal of Dermatology*, **112**, 1–4.

British Association of Dermatologists (2009) *British Association of Dermatologists' Management Guidelines*, http://www.bad.org.uk/healthcare-professionals/clinical-standards/clinical-guidelines (accessed 21 July 2014).



Carr, D.L., Kloos, W.E. (1977) Temporal study of the staphylococci and micrococci of normal infant skin. *Applied and Environmental Microbiology*, **34**, 673–680.

Chamley, C., Carson, P., Randall, D., Sandwell, W. (2005) *Developmental Anatomy and Physiology of Children*, Churchill Livingstone.

Crawford, D., Hickson, W. (2002) *An Introduction to Neonatal Care*, Nelson Thornes.

England, M.A. (1996) *Life before Birth*, 2nd edn, Mosby-Wolfe, London.

Gormley-Fleming, E. (2010) Assessing and vital signs – a comprehensive review. In Glasper E.A, Aylott, M., Batterick C. (eds), *Developing Skills for Children's and Young People's Nursing*, Elsevier.

Gray, L., Watt, L., Blass, E.M. (2000) Skin to skin contact is analgesia in healthy newborns. *Pediatrics*, **105** (1), e14.

Harper, J., Oranje, A., Prose, N. (2000) *Textbook of Pediatric Dermatology*, Blackwell Science, Oxford.

Hoeger, P.H., Enzmann, C.C. (2002) Skin physiology of the neonate and young infant: a prospective study of functional skin parameters during early infancy. *Pediatric Dermatology*, **19** (3), 256–262.

Hoeger, P.H, Schreiner, V., Klaasen, I.A. (2002) Epidermal barrier lipids in human vernix caseosa. Corresponding ceramide patterns in vernix and fetal epidermis. *British Journal of Dermatology*, **146**, 194–201.

Marieb, E., Hoehn, K. (2010) *Human Anatomy & Physiology*, 8th edn, Pearson Benjamin Cummings, San Francisco, CA.

MacGregor, J. (2012) *Introduction to the Anatomy and Physiology of Children*, 2nd edn, Routledge, London.

Mims, C., Dimmock, N., Nash, A., Stephens, J. (1995) *Mims' Pathogenesis of Infectious Diseases*, 4th edn, Academic Press, London.

Peate, I., Nair, M. (2013) *Fundamentals of Anatomy and Physiology for Student Nurses*, 2nd edn, Wiley–Blackwell, Oxford.

Power, G.G. (1989) Biology of temperature: the mammalian fetus. *Journal of Developmental Physiology.*, **12**, 259–304.

Pringle, F., Penzer, R. (2002) Normal skin: its function and care. In: Penzer, R. (ed.), *Nursing Care of the Skin*, Butterworth Heinemann, Oxford.

Romani, N., Holzman, S., Tripp, C.H. (2003) Langerhan cells – dendritic cells of the epidermis. *Acta Pathologica et Immunologica Scandinavica*, **111** (7–8), 725–740.

Sato, K., Kang, W.H., Saga, K.T., Sato, K.T. (1989) Biology of sweat glands and their disorders. I. Normal sweat gland function. *Journal of American Academy of Dermatology*, **20** (4), 537–563.
Tortora, G.J., Derrickson, B. (2009) *Principles of Anatomy and Physiology*, 12th edn, John Wiley & Son., Inc., Hoboken, NJ.
White, R., Denyer, J. (eds) (2006) *Paediatric Skin and Wound Care*, Wounds UK.
White, R., Butcher, M. (2006) The structure and function of the skin: paediatric variations. In: White, R., Denyer, J. (eds), *Paediatric Skin and Wound Care*, Wounds UK.
Wong, D. (1999) *Whaley & Wong's Nursing Care of Children*, 6th edn, Mosby, London.

# Self-assessment answers

## Chapter 2

### Wordsearch

1. Afferent, arteriole, creatinine, diuretic, efferent, haematuria, interstitial, ion, medulla, cortex, solute, urea, urinalysis, urine, vasopressin.

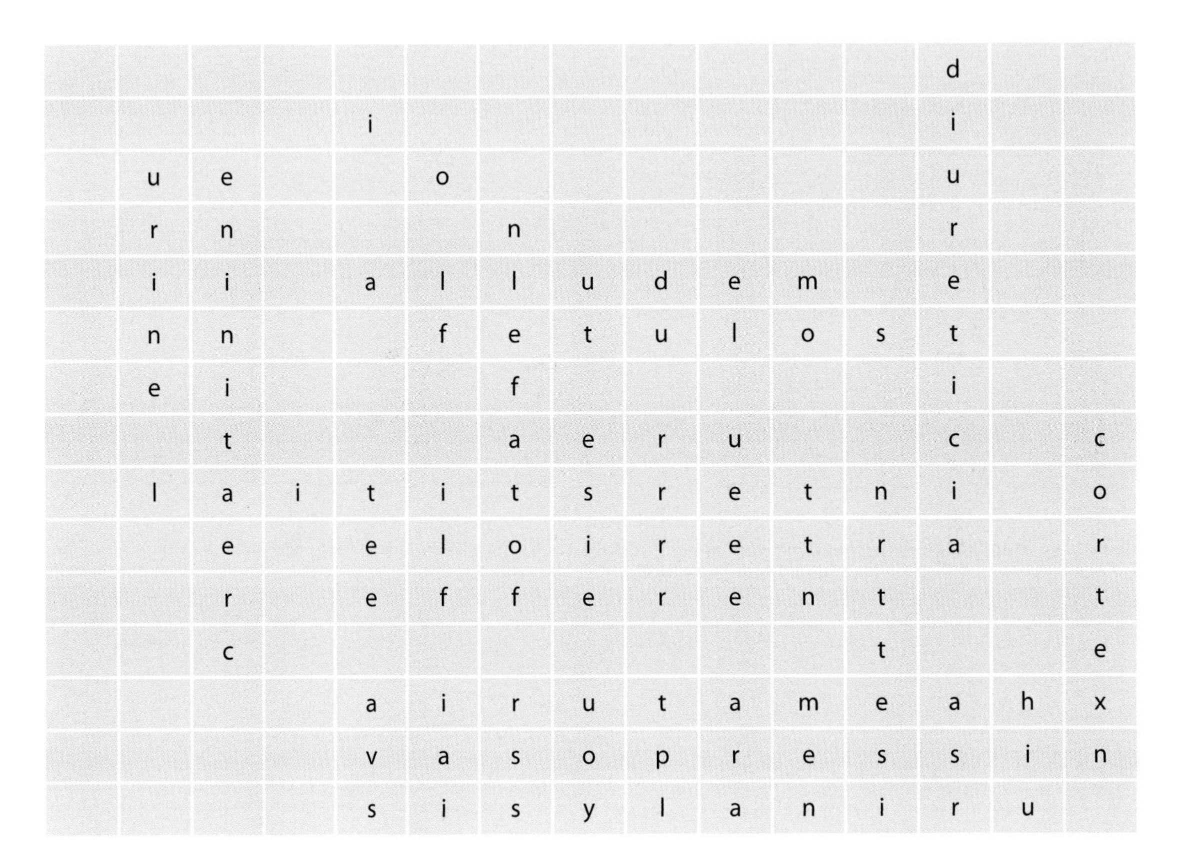

| | | | | | | | | | | | | | | |
|---|---|---|---|---|---|---|---|---|---|---|---|---|---|---|
| | | | | | | | | | | | | d | | |
| | | | | i | | | | | | | | i | | |
| | u | e | | | o | | | | | | | u | | |
| | r | n | | | | n | | | | | | r | | |
| | i | i | | a | l | l | u | d | e | m | | e | | |
| | n | n | | | f | e | t | u | l | o | s | t | | |
| | e | i | | | | f | | | | | | i | | |
| | | t | | | | a | e | r | u | | | c | | c |
| | l | a | i | t | i | t | s | r | e | t | n | i | | o |
| | | e | | e | l | o | i | r | e | t | r | a | | r |
| | | r | | e | f | f | e | r | e | n | t | | | t |
| | | c | | | | | | | | | t | | | e |
| | | | | a | i | r | u | t | a | m | e | a | h | x |
| | | | | v | a | s | o | p | r | e | s | s | i | n |
| | | | | s | i | s | y | l | a | n | i | r | u | |

*Fundamentals of Children's Anatomy and Physiology: A Textbook for Nursing and Healthcare Students*, First Edition. Edited by Ian Peate and Elizabeth Gormley-Fleming.

Companion website: www.wileyfundamentalseries.com/childrensA&P

2. Aldosterone, angiotensin, calyx, collecting duct, cortex, glomerulus, kidney, loop of Henle, medulla, nephron, parenchyma, pelvis, pyramid, reabsorption, renin, secretion, ureter, urethra.

| | l | | | e | | | | n | c | a | l | y | x | |
|---|---|---|---|---|---|---|---|---|---|---|---|---|---|---|
| | o | | | n | | | o | | | | | | | |
| | o | | | o | | i | | | | | | | | |
| | p | | r | r | t | | | | | | | | | |
| c | o | l | l | e | c | t | i | n | g | d | u | c | t | |
| | f | | r | t | a | m | y | h | c | n | e | r | a | p |
| | h | c | | s | | b | m | e | d | u | l | l | a | |
| x | e | t | r | o | c | | s | p | y | r | a | m | i | d |
| s | n | k | i | d | n | e | y | o | r | n | i | n | e | r |
| | l | | | l | s | u | l | u | r | e | m | o | l | g |
| | e | | | a | | | | | | p | t | | | |
| a | n | g | i | o | t | e | n | s | i | n | t | e | | |
| | i | | | | | | | | | | | i | r | |
| u | r | e | t | h | r | a | p | e | l | v | i | s | o | u |
| | | | n | o | r | h | p | e | n | | | | | n |

## Crossword

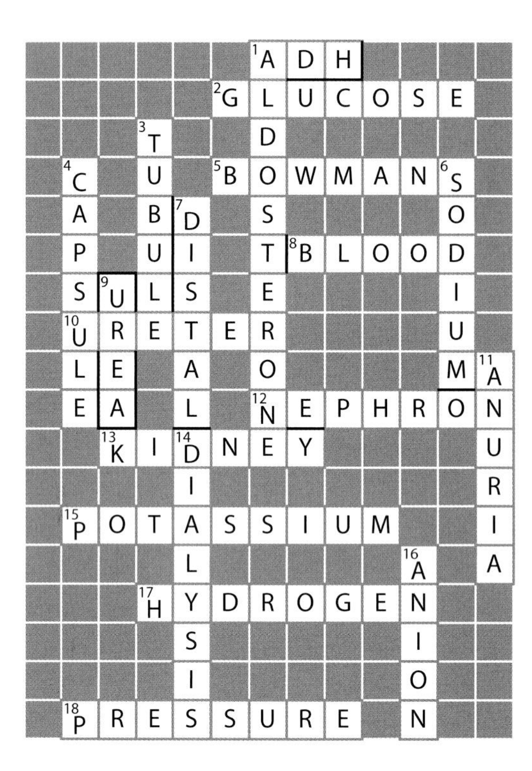



# Chapter 3

## Crossword

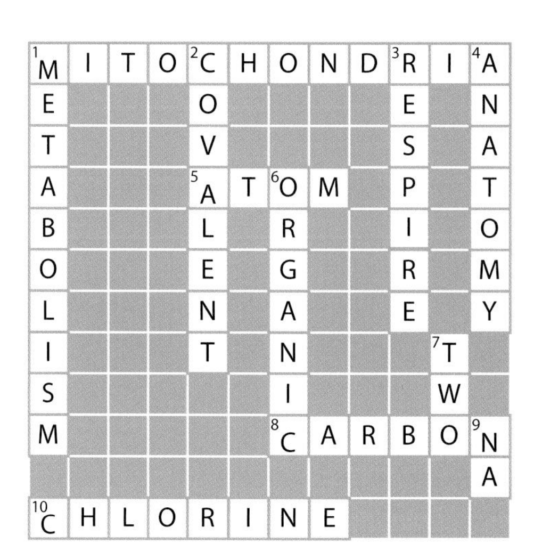

## Wordsearch

Atom, chemical bond, chlorine, covalency, digestion, electron, elements, energy, enzyme, growth, heat, ion, molecule, mitochondria, organic, oxygen, protein.

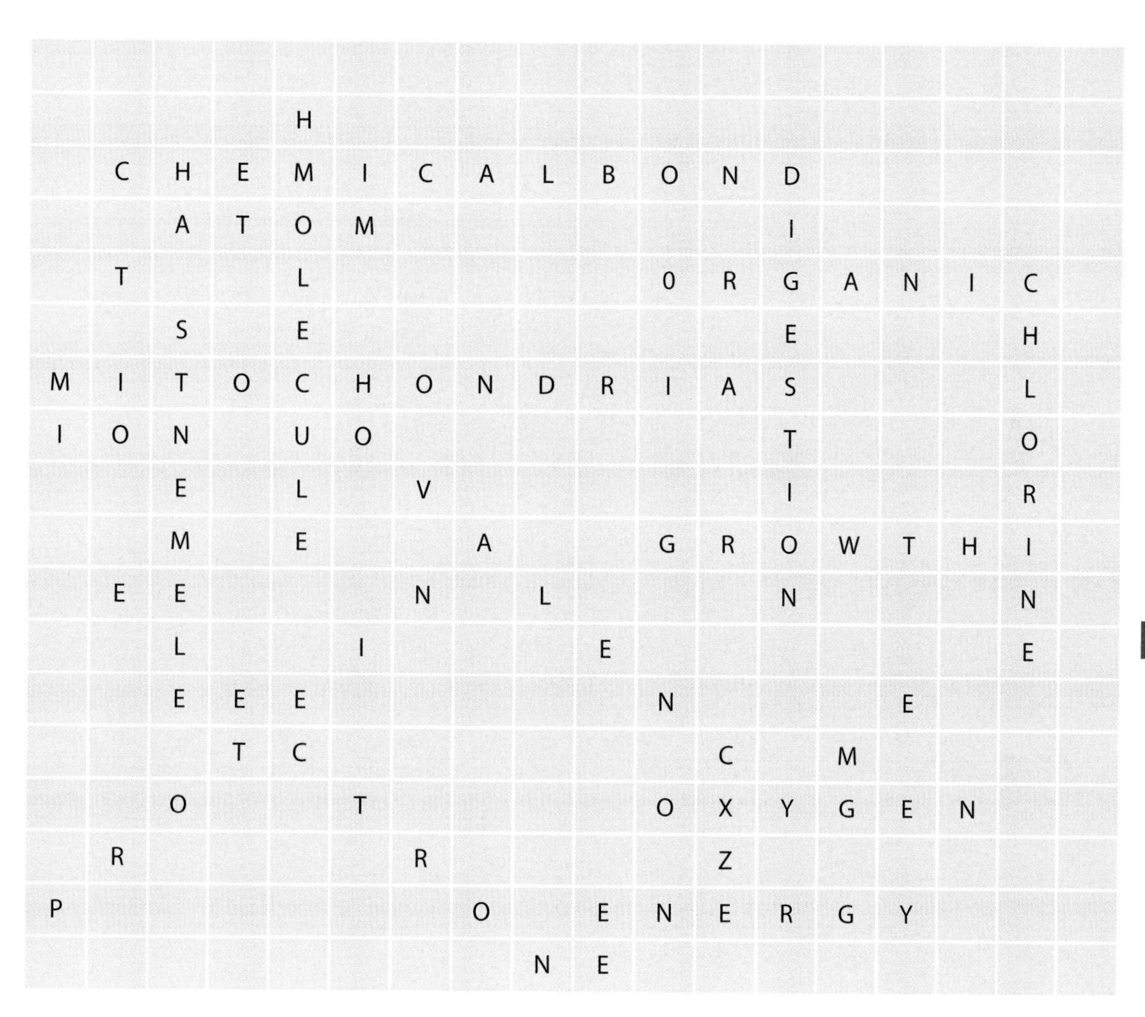



## Which is the odd one out?

1. (b) water – all the others are examples of organic substances
2. (b) anion – all the others are parts of an atom
3. (c) lipids – all the others are examples of inorganic substances
4. (b) equatorial – all the others are methods of bonding of atoms/molecules

## Exercise

1. $SO_3 + H_2O \rightarrow H_2SO_4$ (=sulphuric acid)
2. $H_2O$ is the chemical formula for water

# Chapter 4

## Crossword

| 1 E | R | | | | 2 A | T | 3 P | | | | |
|---|---|---|---|---|---|---|---|---|---|---|---|
| N | | | | 4 N | | | R | | | | |
| 5 D | I | F | F | U | S | I | O | N | | | |
| O | | | | C | | | T | | 6 S | | 7 M |
| C | | | | 8 L | Y | S | O | S | O | M | E |
| Y | | 9 C | | E | | | P | | L | | M |
| T | | A | | O | | | L | | V | | B |
| O | | L | | P | | | A | | E | | R |
| S | | 10 C | E | L | L | | S | | N | | A |
| I | | I | | A | | | M | | T | | N |
| S | | U | | S | | | | | | | E |
| | | M | | 11 M | A | T | R | I | X | | |

## Fill in the gaps



1. In order for the body to function properly, it must be able to maintain electrolyte levels within very narrow limits. Controlled by signals from hormones, these electrolyte levels are maintained by the movement of electrolytes into, and out of, cells, as required.
2. Although facilitated diffusion is the commonest form of protein-mediated transport across the cell membrane, it tends to be overshadowed by active transport. Rather than solutes moving down their concentration gradients to reach equilibrium, in active transport they are actively 'pumped' up a gradient using energy from another source – adenosine triphosphate (ATP).

## Wordsearch

1. ADP, ATP, cell, cell membrane, cilia, cycle, cytoplasm, electrolytes, exocytosis, flagellae, golgi, hydrophobic, lipid, matrix, mitochondria, osmosis, pH, pinocytosis, ribosomes, solution, solvent, synthesis.

| E | | | | | H | Y | D | R | O | P | | | | | | | |
|---|---|---|---|---|---|---|---|---|---|---|---|---|---|---|---|---|---|
| | X | | | | | | | | P | H | | | | | M | | |
| | | O | S | M | O | S | I | S | | O | | E | | | I | | S |
| | | | C | E | L | L | M | E | M | B | R | A | N | E | T | | I |
| | | | Y | Y | | | | | | I | | L | | L | O | | S |
| | | | T | | T | | | | | C | | L | | E | C | | O |
| | | | O | | G | O | L | G | I | | | E | | C | H | | T |
| | A | D | P | | | | S | | L | | | G | | T | O | | Y |
| | C | I | L | I | A | | | I | | S | | A | | R | N | | C |
| | E | | A | | | | P | | S | Y | | L | | O | D | | O |
| | L | | S | | | I | | N | | N | | F | | L | R | | N |
| | L | | M | | D | | O | | | T | | | | Y | I | | I |
| C | | | | A | | I | | | | H | | | | T | A | T | P |
| | Y | | | | T | | | | | E | | | | E | | | |
| | | C | | U | | R | I | B | O | S | O | M | E | | | | |
| | | | L | | | | I | | | I | | | | | | | |
| | | O | | E | | | | X | | S | O | L | V | E | N | T | |
| | S | | | | | | | | | | | | | | | | |

2. Calcium pump, cytoskeleton, glycoprotein, nucleoplasm, organelle, passive transport, prokaryote, protein.

| T | N | U | C | L | E | O | P | L | O |
|---|---|---|---|---|---|---|---|---|---|
| R | Y | C | O | P | R | O | T | A | R |
| O | L |  | P | R | O | K | E | S | G |
| P | G | P | P | R | O | A | I | M | A |
| S |  | M |  | N | T | R | N | C | N |
| N |  | U |  | I | E | Y |  | Y | E |
| A |  | P |  | E | T | O | C | T | L |
| R |  | M | U | I | C | L | A | O | L |
| T | N | O | T | E | L | E | K | S | E |
| E | V | I | S | S | A | P |  |  |  |

# Chapter 5



## Fill in the gaps

Genes that occupy corresponding loci and code for the same characteristic are called alleles, which are found at the same place in each of the two corresponding chromatids, and each one determines an alternative form of the same characteristic.

## Crossword

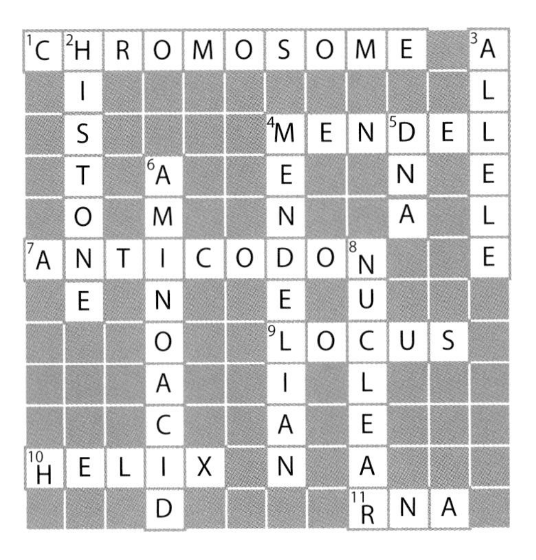

## Wordsearch

Allele, amino, autosomal, centromere, codon, crossover, diploid, DNA, gene, guanine, haploid, helix, histone, interphase, mendelian genetics, metaphase, mutation, nuclear spindle, spontaneous, synthesis, RNA

| H | I | S | T | O | N | E | | | | | | | | | | | |
|---|---|---|---|---|---|---|---|---|---|---|---|---|---|---|---|---|---|
| | N | | C | | | | | | M | U | T | A | T | I | O | N | |
| | T | G | E | N | E | C | | | E | | | | N | | | U | |
| | E | | N | | | R | | | N | | | | | R | | C | |
| | R | | T | | | O | | | D | N | A | | | | | L | |
| | P | | R | | | S | | | E | I | | | X | I | L | E | H |
| | H | | O | | | S | | | L | | P | | | | | A | |
| | A | A | M | I | N | O | P | | I | | | L | C | | | R | |
| | S | | E | | | V | | O | A | | | | O | | | S | |
| H | E | | R | | | E | | U | N | | | | D | I | | P | S |
| | A | | E | | | R | T | | G | T | | | O | | D | I | Y |
| | | P | | | | O | | | E | | A | | N | | | N | N |
| | | | L | | S | | A | | N | | | N | | | | D | T |
| | | | | O | | | L | | E | | | | E | | | L | H |
| | | | M | | I | | L | | T | | | | | O | | E | E |
| | | E | | | | D | E | N | I | N | A | U | G | | U | | S |
| | | | | | | | L | | C | | | | | | | S | I |
| E | S | A | H | P | A | T | E | M | S | | | | | | | | S |



## Exercise

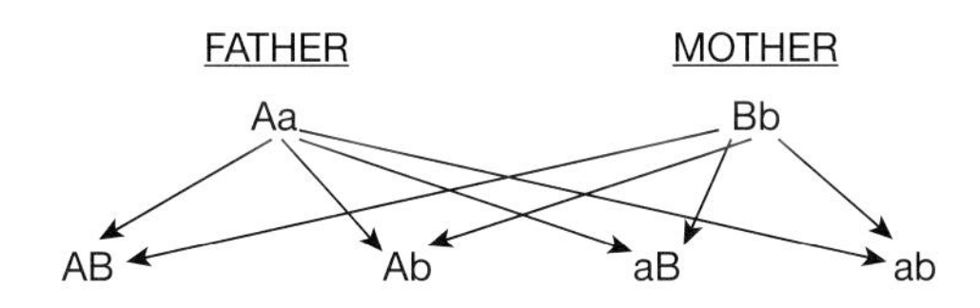

# Chapter 6

## Complete the sentence

1. The main type of cell that is found in nervous tissue is the neurone.
2. Nervous tissue does not normally undergo mitosis to replace damaged neurone
3. In open or large wounds, the process of granulation occurs using granulation tissue which is perfused, fibrous connective tissue, which replaces the initial fibrin clot.
4. In early childhood there is a rapid increase in body size for the first 2 years.
5. An adult has a total of 206 bones which are joined to ligaments and tendons, whilst babies, at birth, have 270 bones.
6. Cartilage is found in only a few places in the body; for example, hyaline cartilage – which supports the structures of the larynx.
7. Bone is the most rigid of the connective tissues and is composed of bone cells surrounded by a very hard matrix containing calcium and large numbers of collagen fibres.
8. Plasma cells produce antibodies in response to invading substances, prior to the body's immune system destroying them.
9. Connective tissue is not present on the body surfaces.
10. The most common function of connective tissue is to act as the framework on which the epithelial cells gather to form the organs of the body.

## Wordsearch

Glycoprotein, connective, epithelial, parenchyma, fibroblast, exocytosis, diffusion, cartilage, avascular, synapse, mitosis, tissue, neuron, stroma, gland, bone

| f | n | | | | | | | | | | c | | t | | e | | | n | |
|---|---|---|---|---|---|---|---|---|---|---|---|---|---|---|---|---|---|---|---|
| i | | o | | | | | | | m | | a | | | i | | n | e | | |
| b | e | p | i | t | h | e | l | i | a | l | r | | | | s | u | o | | |
| r | x | | | s | | | t | | | | t | | | | r | s | | b | a |
| o | o | | | | u | o | | | | | i | | | o | | | u | v | |
| b | c | | | | s | f | | | | | l | | n | | | | a | e | |
| l | y | | | i | | | f | e | s | p | a | n | y | s | | s | | | |
| a | t | | s | | | | | i | | | g | | | | c | | | | |
| s | o | | | | | | | | d | | e | | | u | e | | | | |
| t | s | | | | | | | | | | | p | l | | v | | | | |
| | i | | | s | | | | | | | a | a | | | i | | | | |
| | s | | | | t | | | | | r | r | | | | t | | | | |
| g | l | y | c | o | p | r | o | t | e | i | n | | | | c | | | | |
| g | l | a | n | d | | | o | n | | | | | | | e | | | | |
| | | | | | | | c | m | | | | | | | n | | | | |
| | | | | | | h | | | a | | | | | | n | | | | |
| | | | | | y | | | | | | | | | | o | | | | |
| | | | | m | | | | | | | | | | | c | | | | |
| | | | a | | | | | | | | | | | | | | | | |
| | | | | | | | | | | | | | | | | | | | |

## Complete the table

| Connective tissue type | Primary blast cell | Connective tissue cell |
|---|---|---|
| Connective tissue proper | Fibroblast | Fibrocyte |
| Cartilage | Chondroblast | Chondrocyte |
| Bone | Osteoclast | Osteocyte |

# Chapter 7

## Crossword

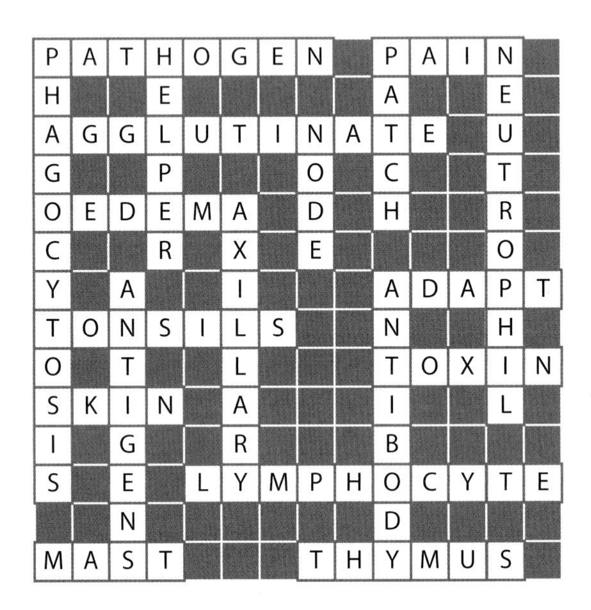




# Chapter 8

## Which is the odd one out?

1. (b) HCO system.
2. (d) Protection against injury
3. (a) Blood cell

## Complete the sentence

1. Vasoconstriction occurs as a result of vascular spasm which causes the smooth muscle of the blood vessel wall to contract, which in turn constricts the small blood vessels. This process is a result of the sympathetic nervous system restricting blood flow.

2. The aorta is the largest artery in the body and oxygenated blood leaves the heart through it.
3. Blood pressure is maintained by means of baroreceptors which are found in the arch of the aorta and the carotid sinus. When blood pressure increases, this sends signals to the cardioregulatory centre, which increases parasympathetic activity to the heart, reducing heart rate and inhibiting sympathetic activity to the blood vessels.

## Wordsearch

Albumin, aorta, blood, BP, cell, clotting, Hb, hormone, leucocyte, pH, plasma, platelet, type, vein.

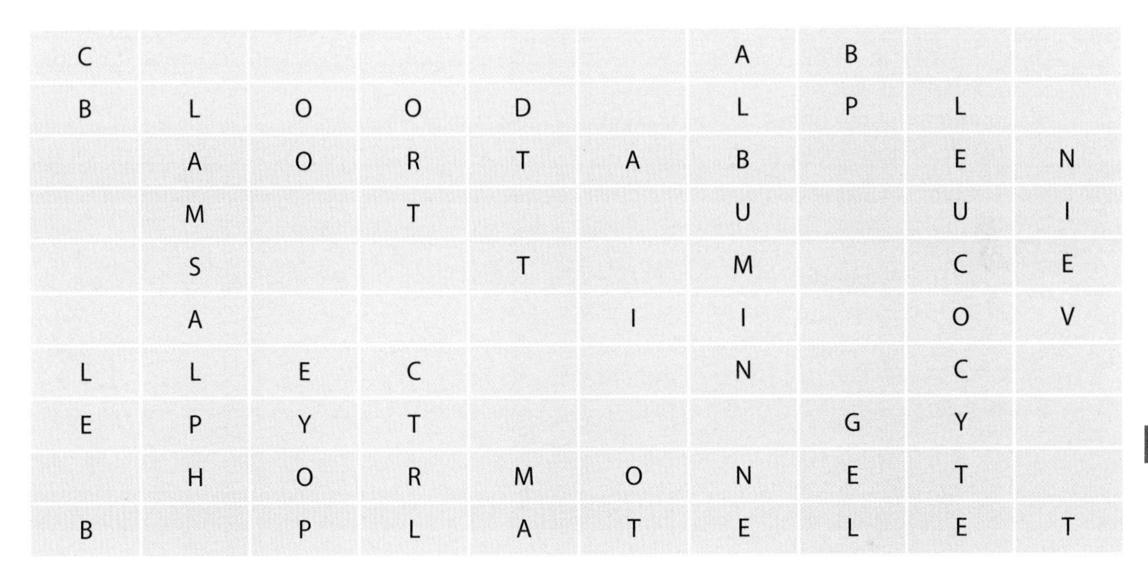

| C | | | | | | A | B | | |
|---|---|---|---|---|---|---|---|---|---|
| B | L | O | O | D | | L | P | L | |
| | A | O | R | T | A | B | | E | N |
| | M | | T | | | U | | U | I |
| | S | | | T | | M | | C | E |
| | A | | | | I | I | | O | V |
| L | L | E | C | | | N | | C | |
| E | P | Y | T | | | | G | Y | |
| | H | O | R | M | O | N | E | T | |
| B | | P | L | A | T | E | L | E | T |



## Crossword

| H | | | | | | | | | | | |
|---|---|---|---|---|---|---|---|---|---|---|---|
| A | R | T | E | R | Y | | | A | | C | |
| E | | | | | | | | R | | O | |
| M | | C | | V | | A | N | T | I | A | |
| O | X | Y | G | E | N | | | E | | G | |
| G | | T | | I | | | G | R | O | U | P |
| L | | E | | N | | | | I | | L | |
| O | | | C | | | R | | O | | A | |
| B | L | O | O | D | C | E | L | L | | T | |
| I | | | 2 | | | D | | E | | I | |
| N | | | | | | | | | | O | |
| | | | | | | | I | R | O | N | |

# Chapter 9

## Complete the sentence

1. Blood returning from the body enters the heart via the inferior and superior vena cavae, into the right atrium across the tricuspid valve to the right ventricle and then leaves to go to the lungs via the pulmonary artery.
2. Many of the structures within the heart are known by more than one name; complete the following:
   right atrioventricular valve = tricuspid valve
   left atrioventricular valve = bicuspid valve
   atrioventricular bundle = bundle of His.
3. The structure within the heart which connects the pulmonary artery to the aorta is the ductus arteriosus.
4. Connecting the right atrium to the left atrium is the foramen ovale.
5. The work of the liver in the unborn baby is carried out by the placenta; therefore, blood bypasses the liver via the ductus venosus.

## True or false?

1. True. (1) The pulmonary circulation, from the heart to the lungs and back and (2) the systemic circulation from the heart to the body and back.
2. False. The fetal adaptations (ductus arteriosus, foramen ovale and ductus venosus) are essential to the circulation system of the unborn infant.
3. False. The sympathetic nervous system increases the heart rate.

## Wordsearch

Artery, atrium, bicuspid, bundle of His, cardiac, cyanotic, diastole, hypoxia, lubdub, myocardium, output, placenta, preload, pulmonary, semilunar, sinoatrial node, stroke volume, tricuspid, vein, ventricle.

| | | | | | | | | | | | | | | |
|---|---|---|---|---|---|---|---|---|---|---|---|---|---|---|
| | | | | | P | L | A | C | E | N | T | A | | S |
| A | R | T | E | R | Y | | A | I | X | O | P | Y | H | T |
| T | U | P | T | U | O | | | | E | | | R | | R |
| L | U | B | D | U | B | | | L | | V | A | | B | O |
| M | U | I | D | R | A | C | O | Y | M | N | E | U | | K |
| | | Y | | | | T | | | U | | N | I | D | E |
| | C | | R | | S | | | L | | D | | A | N | V |
| S | I | N | O | A | T | R | I | A | L | N | O | D | E | O |
| C | T | | I | | N | M | | E | | L | | N | A | L |
| A | O | D | | | E | O | O | | E | | T | T | | U |
| R | N | | | S | | F | M | R | | R | R | | | M |
| D | A | | | | H | | P | L | I | I | | | | E |
| I | Y | | | I | T | R | I | C | U | S | P | I | D | |
| A | C | | S | | | | L | M | | P | | | | |
| C | | | | | | E | D | I | P | S | U | C | I | B |

# Chapter 10

| Complaint | Assessments | Possible causes |
| --- | --- | --- |
| Croup | • Assessed through a full respiratory assessment, listening for 'a seal pup bark', inspiratory stridor and overall air entry<br>• Also pyrexia and possible reduction in SaO2 | Viral infections |
| Bronchiolitis | • Assessing respiratory sounds: dry cough, possible wheeze, clear mucus visible<br>• Fluid assessment may indicate dehydration<br>• Mood may be irritable | Viral infections, often RSV |
| Pneumonia | • Airway assessment may indicate increased rate<br>• Auscultation may indicate a localized area of reduced air entry<br>• There may be a productive cough, also localized chest pain<br>• A CXR may be used as part of the diagnosis | Usually caused by a bacterial infection, occasionally viral or fungal pathogens |
| Tetralogy of Fallot | • Assessment is generally cardiovascularly led- SaO2, echocardiogram, auscultation listening for murmurs, signs of cardiac failure in the older undiagnosed infant | Maternally based-diabetes, rubella, drug usage (prescribed or recreational), alcohol ingestion<br>Other causes are genetic or spontaneous |
| Asthma | • Airway assessment, increased rate, decreased air entry, wheeze, possible reduction in SaO2 | Inflammation of small airway caused by triggers; house dust mite, cigarette smoke, animal fur, pollen, exercise or viral infections |



# Chapter 11

## True or false?

1. False
2. True
3. False
4. True

## Match the hormone with its function

1. i(h), ii(e), iii(f), iv(a), v(b), vi(d), vii(c), viii(g)
2. i(d), ii(h), iii(a), iv(g), v(b), vi(e), vii(f), viii(c)

# Chapter 12

## Exercises

1. Stratified epithelium has several layers, providing protection in areas where there may be hard foods. Simple columnar epithelium is only one cell thick, and therefore ideal for absorption.

2. (a) Chief cells – secrete pepsinogen and prorennin. (b) Parietal cells – secrete hydrochloric acid. (c) Gastric glands - secrete mucus, water and mineral salts.
3. Vitamins A, D, E and K are only soluble in fat, so require ingested fat and the ability to absorb this ingested fat in order to be absorbed from the gastrointestinal tract. All other vitamins are water soluble.
4. Refer to Figure 12.2.
5. (a) In the liver. (b) In the gallbladder. (c) Into the duodenum.
6. Endocrine glands secrete hormones directly into the circulation. Exocrine glands have a duct through which the secretions are expelled.
7. Ileo-caecal valve, caecum with appendix, ascending colon, transverse colon, descending colon, sigmoid colon, rectum, anus.
8. See Tables 12.2 and 12.3.

## Complete the sentence

The pH of the gastrointestinal lumen changes from acidic in the stomach to alkaline in the duodenum.

# Chapter 13

## True or false?

1. True.
2. False. Renin is an enzyme.
3. True.
4. False. It is micturition.
5. True.
6. True.
7. False. It increases the permeability.
8. True.
9. False. It is the volume of filtrate formed by both kidneys each minute.

## Wordsearch

Aldosterone, bladder, bowmans, capsule, detrusor, filtration, glomerulus, maximum, medulla, muscle, nephron, pelvis, potassium, renin, sodium, transport, trigone, ureters, urine.

| P | F |  |  |  |  |  |  |  | S | S |  |  |  | E |
|---|---|---|---|---|---|---|---|---|---|---|---|---|---|---|
| U | O | I |  |  |  |  |  | U |  | O |  |  | L |  |
| C | R | T | L |  |  |  | L |  |  | D |  | C |  |  |
|  | A | E | A | T |  | U |  |  |  | I | S |  |  |  |
| P |  | P | T | S | R | D | E | T | R | U | S | O | R | E |
|  | E |  | S | E | S | A |  |  | M | M |  |  | N |  |
|  |  | L | M | U | R | I | T |  |  |  |  | O |  | S |
|  |  | O | V | R | L | S | U | I |  |  | R |  | N | R |
|  | L |  |  | I |  | E |  | M | O | E | E | A | E | M |
| G | T | R | A | N | S | P | O | R | T | N | M | D | P | A |
|  |  |  |  | E |  |  |  | S | O | W | D |  | H | X |
| M | E | D | U | L | L | A | O | G | O | A |  |  | R | I |
| R | E | N | I | N |  | D | I | B | L |  |  |  | O | M |
|  |  |  |  |  | L | R |  | B |  |  |  |  | N | U |
|  |  |  |  | A | T |  |  |  |  |  |  |  |  | M |

# Chapter 14

| Method | Advantages | Disadvantages |
|---|---|---|
| *Surgical*<br>Vasectomy/<br>tubal ligation | Relatively permanent solution | Surgical intervention with all of the risks of surgery<br>Difficult to reverse if required later |
| *Hormonal*<br>Birth control pills | Easy method of birth control with a 99% rate of success when taken correctly | May have side effects such as headache, dizziness, nausea, breakthrough bleeding, blood clots, mood swings<br>These often go away with continued use |
| Birth control injection | Similar advantages to the birth control pill | May also have similar side effects<br>Regular visits to health practitioner |
| Birth control patch | Easy to use and apply | Delivers 60% more oestrogen than a low-dose pill, so an increased risk for blood clots |
| Birth control ring | Delivers oestrogen and progestin similarly to the pill | Women who smoke, have blood clots or certain cancers should not use this method as it increases the risks<br>May be difficult to position correctly for some |
| *Barrier*<br>Male condom | Provides an effective barrier when properly applied and helps to prevent sexually transmitted diseases | May decrease sensation and may not be used if either person has an allergy to the materials used |
| Diaphragm (with spermicide) | Provides an effective barrier when used with a spermicide | Needs to initially be fitted by a health professional<br>Some women may have difficulty inserting correctly<br>Fluctuating weight gain or loss may alter the fit |
| Contraceptive sponge (with spermicide) | Just used prior to intercourse<br>The spermicide soaked sponge prevents sperm entering the uterus by covering the cervix<br>90% effective when used correctly | Needs inserting prior to intercourse<br>Increases the risk of yeast infections and urinary tract infections when used regularly |
| *Other*<br>Spermicide | Usually used alongside other methods | May cause irritation |
| Fertility awareness | This refers to the method of being aware when the woman ovulates and avoiding intercourse during those times<br>Some women experience pain on ovulation otherwise may be determined by temperature as it becomes slightly raised during ovulation | Abstinence and withdrawal play a part in this method and are not always correctly used, so resultant pregnancies are more likely |



# Chapter 15

## True or false?

1. True.
2. False.
3. False.
4. True.
5. False.

## Wordsearch

Axon, brainstem, cation, cerebrum, dendrite, diencephalon, effector, ganglia, hemispheres, hippocampus, lobe, microglia, midbrain, myelin sheath, nerve, neuroglia, nuclei, pons, synapse, thalamus

| | | | | | | | | | | | | | | | | | | | | | | | | |
|---|---|---|---|---|---|---|---|---|---|---|---|---|---|---|---|---|---|---|---|---|---|---|---|---|
| | | B | R | A | I | N | S | T | E | M | | | | | | | R | | | | M | | | |
| H | | | | | | | | | | | | | | | | | O | | | | I | | | |
| E | | | | | | | N | E | U | R | O | G | L | I | A | | T | | | | C | | | |
| M | | P | | | | | | | | | | | | | | | C | | | | R | | | G |
| I | | O | | | | | A | | | | | | | | | | E | | | | O | | | A |
| S | | N | | | | | | X | | | | | | | | | F | | | | G | | | N |
| P | | S | | | | | | | O | | | | | | | | F | | | | L | | | G |
| H | | | | | | | N | | | N | | | | | | | E | | | | I | | | L |
| E | | | | | | U | | | | | | | | | | | | | | | A | | | I |
| R | | | | | C | | | | | | M | I | D | B | R | A | I | N | | | | | | A |
| E | | | | L | | | | | | | | | | | | | E | S | P | A | N | Y | S | |
| S | | | E | | | | M | | | | | | | | | | | | | | | U | | |
| | C | I | | | | | | Y | | | | | | | | | L | | | | M | | | |
| | A | E | | | | | | | E | | | | | | | O | | | | A | | | | |
| | T | | R | | | | | | | L | | | | | B | | | | L | | | | | |
| | I | | | E | | | | | | | I | | | E | | | | A | | | | | | |
| | O | | | | B | | | | | | | N | | | | | H | | | | | | | |
| | N | | | | | R | | | | | | | S | | | T | D | E | N | D | R | I | T | E |
| | | | | | | | U | | | | | | | H | | | | | | | | | | V |
| | | | | | | | | M | | | | | | | E | | | | | | | | | R |
| | | | | | | | | | | | | | | | | A | | | | | | | | E |
| D | I | E | N | C | E | P | H | A | L | O | N | | | | | | T | | | | | | | N |
| | | | | | | | | S | U | P | M | A | C | O | P | P | I | H | | | | | | |
| | | | | | | | | | | | | | | | | | | | | | | | | |

# Chapter 16

## True or false?

1. True.
2. False – it is divided into four areas.
3. False – it is one of the facial muscles.
4. True.
5. True.
6. False – it is under voluntary control.
7. True.
8. False – it does regenerate, but slowly.
9. False – its function is to hold a reserve supply of oxygen in the muscle.
10. True.

## Wordsearch

Triceps brachii, gastrocnemius, smooth muscle, myofilament, myofibrils, insertion, abduction, diaphragm, purkinje, extensor, adductor, skeletal, fusiform, deltoid, muscle, origin, flexor, fascia, fibre, ramus.

| t | s | m | o | o | t | h | m | u | s | c | l | e | r | n |
|---|---|---|---|---|---|---|---|---|---|---|---|---|---|---|
| | r | | s | e | j | n | i | k | r | u | p | | o | o |
| | | i | | l | d | e | l | t | o | i | d | | s | i |
| | | | c | | i | r | a | m | u | s | | u | n | t |
| | | | | e | | r | | | | | i | r | e | c |
| | | | | | p | | b | | | m | | o | t | u |
| n | o | i | t | r | e | s | n | i | e | | | x | x | d |
| | | | | | | | b | n | f | | | e | e | b |
| m | u | s | c | l | e | | c | r | | o | | l | | a |
| e | r | b | i | f | | o | | | a | | y | f | f | n |
| d | i | a | p | h | r | a | g | m | | c | | m | a | i |
| | | r | o | t | c | u | d | d | a | | h | | s | g |
| | | | | | | | | | | | | i | c | i |
| | | a | m | r | o | f | i | s | u | f | | | i | r |
| | g | m | y | o | f | i | l | a | m | e | n | t | a | o |

## Complete the sentence

11. Smooth muscle contains small, thin spindle-shaped cells of variable size that have one centrally located nucleus and is arranged in parallel lines.
12. Skeletal muscles are cylindically shaped striated fibres that lie parallel to each other.
13. Muscles generate heat as they contract; the cells produce adenosine triphosphate, giving the muscles the energy to contract.
14. The epimysium is composed of dense irregular tissue that surrounds the whole of the muscle.
15. Skeltal muscles have a rich blood supply as they have large energy requirements, thus demanding a continual oxygen supply.
16. Each fascicle is surrounded by a layer of connective tissue called the perimysium.

# Chapter 17

## Wordsearch

Haversian canal, epicondyle, periosteum, osteoblast, epiphysis, diaphysis, cartilage, abduction, haematoma, sesamoid, synovial, ossicle, calcium.

| | O | | | | | | | | M | | | | | O | |
|---|---|---|---|---|---|---|---|---|---|---|---|---|---|---|---|
| | S | | | | | | S | U | | | | | S | | |
| | T | | | | | | I | | | | | S | | A | |
| | E | | | | | C | S | | | | I | | | M | |
| | O | | | | L | | Y | | | C | | | | O | |
| | B | | | A | | | H | | L | | | | | T | |
| | L | P | C | | | | P | E | | | | | | A | E |
| H | A | V | E | R | S | I | A | N | C | A | N | A | L | M | P |
| S | S | L | G | R | E | P | I | C | O | N | D | Y | L | E | I |
| E | T | | A | | I | | D | | | | | | | A | P |
| S | | | L | I | N | O | I | T | C | U | D | B | A | H | H |
| A | | | I | | V | | S | | | | | | | | Y |
| M | | | T | | | O | | T | | | | | | | S |
| O | | | R | | | | N | | E | | | | | | I |
| I | | | A | | | | | Y | | U | | | | | S |
| D | | | C | | | | | | S | | M | | | | |

## Match the bones to the body part

(a) (C); (b) (E); (c) (B); (d) (F); (e) (G); (f) (A); (g) (D).

## True or false?

1. False. The thumb only has two bones.
2. True.
3. False. It is endochondral ossification.
4. False. The ulna is longer.
5. False. It is complete by 2years of age.
6. True.
7. False. It is part of the hip joint.
8. True. A fracture in a bone stimulates production.
9. False. It is the patella and hyoid.
10. True.

# Chapter 18

## Exercises

1. The sense of smell is linked to the limbic functions on the temporal lobe.
2. The sound waves hit the membrane causing it to vibrate. The pitch or frequency is how fast it vibrates, and the amplitude or volume is the height deviation from neutral. This causes the bones of the middle ear to vibrate, transmitting the sound across the middle ear from the tympanic membrane to the oval window.
3. These bones only just touch each other, and are held in place by muscles. If the deviation of the tympanic membrane is too great (a sound is too loud), the muscles move these bones apart slightly, reducing the amplitude of vibration that hits the oval window. This takes several minutes to work, meaning that sound in a loud room can be reduced, but not protecting against sudden loud explosive sounds.
4. The vestibule and semicircular canals contain a sac of endolymph, suspended in an outer labyrinth of perilymph. As the head moves, so does this fluid. The fluid movement causes deviation in hair-like projections which transmit a signal to the brain about our position relative to gravity. The brain integrates this with perception from the eyes, and joints and muscles, about any objects and our position within our environment.
5. The point where the optic nerve exits the eye has no rods or cones, leading to a 'blind spot'. The physical area is called the optic disc.



## Complete the sentence

6. The four primary tastes that humans perceive are sweet, sour, bitter and salty.
7. The cones detect the colours red, blue and green.
8. The three specialized receptor types for the special senses are *chemo*receptors (smell and taste), *mechano*receptors (hearing and balance) and *photo*receptors (vision).

## True or false?

False. They work best in dim light.

# Chapter 19

## Complete the sentence

1. The integumentary system consists of cutaneous membrane or skin which includes the epidermis, dermis and the accessory structures.
2. Beneath the dermis lies the subcutaneous layer.
3. The functions of the skin are excretion, thermoregulation, synthesis of vitamin D, sensory detection, excretion and absorption, and communication.
4. The innermost layer of the epidermis is the stratum basal and the outermost layer is the stratum corneum.
5. The middle layer of the epidermis is the stratum granulosum.
6. Melanocytes, which are located in the stratum basal, manufacture melanin.
7. Located in the dermis, hair is composed of keratinized dead cells.
8. The arrector pili muscle contracts when stimulated, forcing the hair to stand erect.
9. The sebaceous glands are holocrine glands that discharge a lipid secretion into hair follicles.
10. There are three types of sweat glands: apocrine, apoeccrine and eccrine sweat glands.

## Crossword



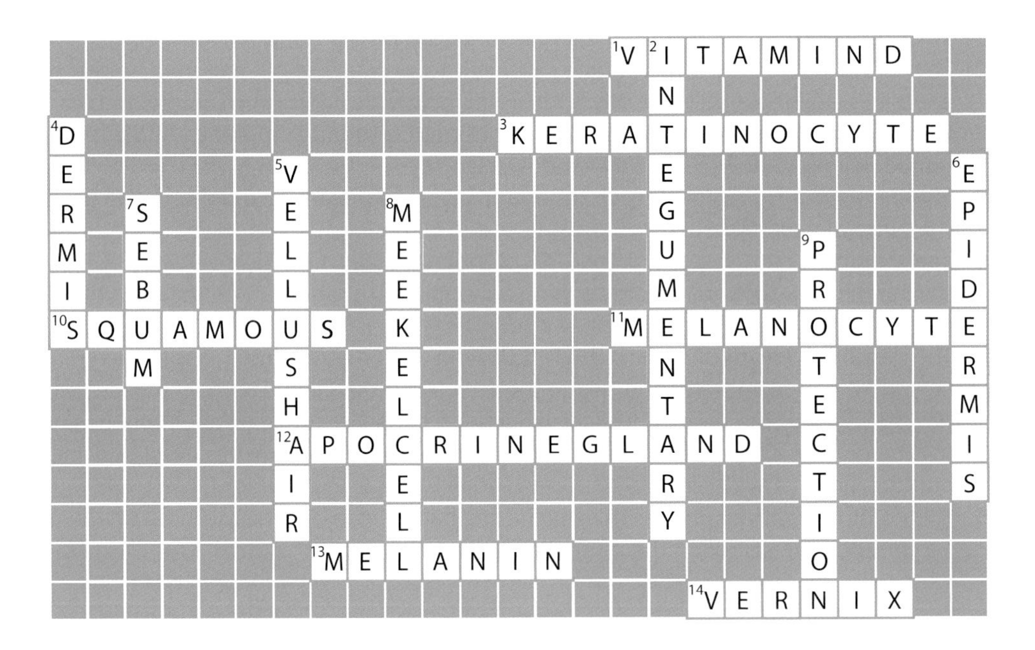

# Index

Page numbers in *italics* denote figures, those in **bold** denote tables.




*Fundamentals of Children's Anatomy and Physiology: A Textbook for Nursing and Healthcare Students*, First Edition. Edited by Ian Peate and Elizabeth Gormley-Fleming.



Companion website: www.wileyfundamentalseries.com/childrensA&P